W9-AFH-964

TECHNICAL COLLEGE OF THE LOWCOUNTRY
3 2992 00010 6302

TRANSFORMATION AND TRADITION IN THE SCIENCES

TRANSFORMATION AND TRADITION IN THE SCIENCES

ESSAYS IN HONOR OF I. BERNARD COHEN

Edited by
EVERETT MENDELSOHN

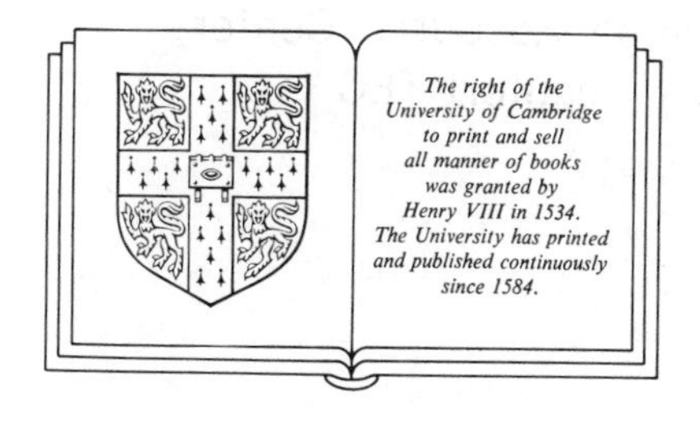

CAMBRIDGE UNIVERSITY PRESS
CAMBRIDGE
LONDON NEW YORK NEW ROCHELLE
MELBOURNE SYDNEY

Published by the Press Syndicate of the University of Cambridge
The Pitt Building, Trumpington Street, Cambridge CB2 1RP
32 East 57th Street, New York, NY 10022, USA
296 Beaconsfield Parade, Middle Park, Melbourne 3206, Australia

First published 1984

Printed in the United States of America

Library of Congress Cataloging in Publication Data
Main entry under title:
Transformation and tradition in the sciences.
Includes index.
1. Science – History – Addresses, essays, lectures.
2. Science – Philosophy – Addresses, essays, lectures.
3. Mathematics – History – Addresses, essays, lectures.
4. Mathematics – Philosophy – Addresses, essays, lectures.
5. Science – United States – History – Addresses, essays, lectures. 6. Cohen, I. Bernard, 1914– .
I. Cohen, I. Bernard, 1914– . II. Mendelsohn, Everett.
Q126.8.T73 1984 509 84-7832
ISBN 0 521 26724 2

This book is dedicated to
Frances Davis Cohen (1908–1982)
whose spirit was so important
to the career of her husband
I. Bernard Cohen

Contents

Contributors

George Basalla
Department of History
University of Delaware
Newark, Delaware 19711

Peter S. Buck
Program on Science, Technology, and Society
Massachusetts Institute of Technology
Cambridge, Massachusetts 02139

Harold L. Burstyn
U.S. Geological Survey
Reston, Virginia 22092

Joseph T. Clark, S. J.
Canisius College
Buffalo, New York 14208

William Coleman
Department of History of Science
University of Wisconsin
Madison, Wisconsin 53706

Joseph W. Dauben
Department of History
Herbert H. Lehman College
City University of New York
Bronx, New York 10468

Allen G. Debus
The Morris Fishbein Center for the History of Science and Medicine
University of Chicago
Chicago, Illinois 60637

Yehuda Elkana
Institute for the History and Philosophy of Science
Tel Aviv University
Tel Aviv, Israel

Martin Fichman
Division of Natural Science
Glendon College
York University
Toronto, Ontario, Canada M4N 3M6

Judith Grabiner
School of Social and Behavioral Sciences
Department of History
California State University, Dominguez Hills
Carson, California 90246

Erwin Hiebert
Department of the History of Science
Harvard University
Cambridge, Massachusetts 02138

Victor Hilts
Department of History of Science
University of Wisconsin
Madison, Wisconsin 53706

Frederic L. Holmes
Department of History of Medicine
Yale University
New Haven, Connecticut 06510

Gerald Holton
Physics Department
Harvard University
Cambridge, Massachusetts 02138

Everett Mendelsohn
Department of the History of Science
Harvard University
Cambridge, Massachusetts 02138

Uta C. Merzbach
Division of Physical Sciences
Museum of History and Technology
Smithsonian Institution
Washington, D.C. 20560

John E. Murdoch
Department of the History of Science
Harvard University
Cambridge, Massachusetts 02138

Nakayama Shigeru
College of General Education
Tokyo University
Komaba Meguro-Ku
Tokyo, Japan

Richard Olson
Department of Humanities and Social Science
Harvey Mudd College
Claremont, California 91711

Stanley Joel Reiser
Health Science Center
University of Texas
Houston, Texas 77225

Shirley A. Roe
The Wellcome Institute for the History of Medicine
London, NW1 2BP, England

Duane Roller
Department of History of Science
University of Oklahoma
Norman, Oklahoma 73069

Barbara Gutmann Rosenkrantz
Department of the History of Science
Harvard University
Cambridge, Massachusetts 02138

A. I. Sabra
Department of the History of Science
Harvard University
Cambridge, Massachusetts 02138

Robert Schofield
Department of History
Iowa State University
Ames, Iowa 50011

Nathan Sivin
Department of Oriental Studies
University of Pennsylvania
Philadelphia, Pennsylvania 19104

Edith Sylla
Department of History
North Carolina State University
Raleigh, North Carolina 27650

Arnold Thackray
Department of History and Sociology of Science
The University of Pennsylvania
Philadelphia, Pennsylvania 19104

Preface

These essays have been written and published to honor the career of I. Bernard Cohen. Although his activities in the history of science have touched the work of scholars in many parts of the world, this festschrift represents the efforts of those of us who have been Bernard Cohen's students and teaching colleagues in the Department of the History of Science at Harvard University.

The practice of presenting a festschrift to a senior colleague at an important career or life turning point seems to us a particularly appropriate way to express personal thanks for shared experiences through the activities that scholars most value, that is, research and production of the written word. A festschrift allows us not only to honor the intended individual but also to share more broadly the fruits of scholarly labors with the learned community. It reflects in a fundamental way the very role of the teacher and researcher who is being celebrated. The appearance of this volume should be a source of pride to Bernard Cohen ("I. B." as he is known to so many in the field) because it shows the degree to which his example has been successfully emulated by those whom he has trained and influenced.

The diversity of interests displayed in these essays aptly reflects the breadth of Bernard Cohen's interests; it also demonstrates the variety of directions and approaches currently taken in the history of science. Indeed, these productions can serve as an anthology of recent scholarship in the field.

The decision to create this volume was taken by the Department of the History of Science at Harvard, and invitations to submit contributions were sent to former students of Bernard Cohen and to past and present teaching colleagues. Many current members of the department read and commented on the papers as they were received and revised, and their help is gratefully acknowledged. The final responsibility, however, rests with the editor. As often happens in an enterprise involving so many individuals there have been false starts and difficult delays. All of these essays were written expressly for this festschrift, and the patience and understanding of the authors whose papers faced delayed publication deserve explicit recognition. Several contributions were published elsewhere before this volume actually appeared, and we appreciatively acknowledge permission to include them here.

Alongside the intellectual production involved in publishing a book there are the critically important technical tasks. The administrative and secretarial staff of the department willingly shared the numerous jobs of corresponding with authors, retyping manuscripts, and otherwise helping to prepare the volume for publication. Betsy A. C. Smith, whose tenure as the department's administrative assistant spans the full history of the department in its modern form, deserves special appreciation for her continued interest and encouragement, not only of the editor but, of equal importance, of many of the authors as well. Ruth Bartholomew, who has worked closely in the department with the editor on many tasks over the years, was particularly important in overseeing many of the technical details and keeping up the flow of correspondence. Valerie Lester, who was able to stay with the department only one year before leaving for Washington, was very helpful in the final stages of manuscript preparation.

It seems completely appropriate that Cambridge University Press, which has been so deeply involved in publishing much of Bernard Cohen's important Newtonian scholarship, should publish this festschrift. Richard Ziemacki, science editor for the Press in New York, immediately perceived the project's value and gave full encouragement through all its stages. The manuscript editor, Mary Byers, deserves our full gratitude for the careful and thoughtful manner in which she attended to the many details of manuscript preparation.

But finally it is to Bernard Cohen that we owe our fullest gratitude. His role in the department and in the discipline as a whole has made this volume possible. And as editor, I have special reason to thank him as my teacher, colleague, and friend for these many years.

Everett Mendelsohn

Berlin, 1984

Introduction

Transformation and tradition are the dialectic of scientific activity. The tension that exists between these two necessary attributes has been the focus of study for the modern history of science. If change is valued, as it so obviously is in science, and rewarded, as the annual Nobel Prize pageant attests, there is also a secret admiration of stability, a reluctance to part with comfortable and often useful current understandings and practices. It is the examination of the fine structure of the processes of change that has shaped I. Bernard Cohen's work in the history of science.

Although the work of very few scholars shows an undeviating commitment from the earliest to the most recent productions, central lines of development are often apparent. This certainly is true for Bernard Cohen. The history of the physical sciences in the seventeenth and eighteenth centuries is the locus, and the approach has included both careful textual scrutiny and the analysis of concepts and practices. No attempt will be made here to analyze in detail the full corpus; rather we identify important high points in Bernard Cohen's work, especially those that have influenced his students and the field as a whole.

Benjamin Franklin's electrical experiments were the basis of Cohen's first significant publication. He prepared a new edition of Franklin's *Experiments and Observations on Electricity* and added to it a substantial historical introduction setting this electrical experimentation and theory construction in the context of the development of experimental traditions in eighteenth-century science.[1]

This publication set the major theme for Cohen's sustained examination of Franklin and also of the links of the experimental approach of the eighteenth century to the strong theories propounded by Isaac Newton a century earlier. It would be nice to periodize neatly Bernard Cohen's work and thereby identify a progression backward in time from Franklin to Newton. The careful bibliographer will note, however, at least several early if brief articles on Isaac Newton that rapidly followed the publication of the Franklin edition.[2]

Several other diversions from the "main path" occurred during, and immediately after, the Second World War. They are worth mentioning because they reflect a secondary theme of Bernard Cohen's interests, namely, the relations between science and society, particularly in the context of United States history. In a set of four brief articles, clearly influenced by the war and the role of science in it, Cohen explored the relations between science and war in the United States from the Revolution to 1942.[3] Another book dealt with other aspects of American science: instruments and collections, as exemplified by those held in Harvard's museums, attics, and basements.[4] This concern with science in the American context remained an interest, if subdued, in Bernard Cohen's scholarly and teaching career over the years and periodically brought forth suggestive articles, including an important series that explored the nineteenth-century conflict between the tradition of the basic and applied sciences in the United States.[5]

But the real commitment of Cohen's scholarly life was made clear by the full and detailed book he published in 1956, *Franklin and Newton, An Inquiry into Speculative Newtonian Experimental Science and Franklin's Work in Electricity as an Example Thereof.*[6] The eighteenth-century experimental tradition is explored at length and is related directly to theoretical sources in the seventeenth century. Several themes emerge in the book that remain commanding ones in Cohen's subsequent work: his interest in "the scientific personality" of key figures in science (Franklin and Newton in this case),[7] and an explicit concern for the conditions of the emergence of novelty, treated in the book in an appendix.[8] If it was Benjamin Franklin who captured Bernard Cohen's imagination in the first phase of his career,[9] it became clear with this book that Isaac Newton would dominate the next phase. Newton obviously fascinated and challenged Cohen. In his preface he alluded to the Newton projects then underway (W. H. Turnbull was undertaking publication of Newton's correspondence) and closed by indicating that he had agreed to participate in the preparation of a critical and variorum edition of the *Principia* in collaboration with Alexandre Koyré.[10] Just one year earlier Koyré had identified the need for such an edition of Newton's works and an important collaboration was begun.[11] It brought together two scholars with some obviously shared interests but also with different backgrounds and scholarly approaches. Koyré's influence as a historian of philosophy and ideas focusing on conceptual analysis in the sciences was just beginning to be felt in the Anglo-American community of historians of science and the commitment to a joint project with Bernard Cohen brought his work even greater attention.[12]

For Bernard Cohen the Newtonian phase of his career involved not only the joint preparation of the critical variorum in two volumes, but also the construction of a 380-page introduction to the text of the *Principia.* These were joined by a steady stream of papers examining aspects of Newton's scientific development and on occasion involving a form of historical detective work as in the joint paper with Koyré, "The Case of the Missing *tanquam:* Leibniz, Newton, and Clarke."[13]

The text of the *Principia* and the *Introduction* were published in 1972 and 1971, respectively, and mark one of the high points of what became a veritable "Newton industry" among historians of science.[14] Unfortunately, Koyré died in 1964 and Cohen saw the jointly edited *Principia* through the publication process and prepared the *Introduction* on his own. It is fair to say that although he has been joined by many other talented scholars Bernard Cohen remains one of the recognized doyens of Newtonian studies.

If attention to detail was the necessary mark of the work on the *Principia* critical edition and the *Introduction,* Cohen was given an opportunity to step back from the editor's role and take a broader look at the meaning of Isaac Newton's work when in 1966 he was invited to give the Wiles Lectures at Belfast University. These lectures as originally delivered and then revised over some dozen years, prior to publication in 1980, intertwined two important themes, namely, the development and meaning of Newtonian science and the concept of scientific change. In a supplement to the main text of his book, *The Newtonian Revolution,* Cohen added a personal account of his efforts to come to grips with what he referred to at the time as "transformations" of scientific ideas.[15] In fact, in the bibliography of the *Introduction* he refers to the forthcoming publication of the Wiles Lectures, giving them the provisional title, *Transformations of Scientific Ideas: Variations on Newtonian Themes in the History of Science.*[16] But Cohen also tells us that in the Autumn of 1966, he had privately circulated texts of the lectures, including one dealing with the "doctrine of transformation," under the title, *Isaac Newton: The Creative Scientific Mind at Work.*[17] This locates for us Cohen's meaning, as he links the creative scientific personality directly to the processes of scientific change and transformation.

By the time of publication of the Wiles Lectures in 1980, now in much altered form, Cohen had recast and refined his ideas of change and moved from the word "transformation" to the word "revolution" to denote the Newtonian impact. But he has gone further and has added many comments on the concept of revolution as it refers to the sciences in general, and promises that in a forthcoming book, *Revolution in Science: History, Analysis, and Significance of a Name and a Concept,* he will deal at length with ideas of scientific change as they have developed in history.[18] Several sections of the *Newtonian Revolution,* his "supplement," and several articles give a foretaste of his foray into the unscrambling of the concept of scientific revolution.[19] He is quite precise in stating his view, a focus on the "fine structure of the scientific revolution" produced by a work like the *Principia,* not on a macroscopic scale but rather on the microscopic scale of the history of science.

> My concern is with the role of the individual in scientific change, even in scientific revolution, as a means of understanding how science may undergo radical alterations of its systems of concepts, laws, and explanations. The analysis of revolutions into a series of transformations shows the continuity within the change, but does not thereby diminish the magnitude of the net change itself.[20]

That his concepts of revolution and transformation will be challenged is certain as this aspect of the history of science has been rife with dispute. But by locating his claims firmly in the microanalysis of the work of Isaac Newton, he has ensured that the role of the creative scientific personality cannot easily be cast aside.

To the scholarly world at large Bernard Cohen is now best known for his significant contributions to Newtonian studies. To historians of science, however, there are other important contributions that have brought him recognition. *Isis,* the official journal of the History of Science Society founded and originally edited by George Sarton, Bernard Cohen's own teacher at Harvard, plays a central role in publications in the discipline in the United States. First as managing editor (1947–52) and then as editor (1953–8), Cohen took the journal through transition to its modern form. He has been active as well in the affairs of the History of Science Society in the United States as a long-term member of the executive council, as vice-president, and ultimately as president. He has similarly served the discipline at an international level through terms as vice-president and president of the International Union of the History and Philosophy of Science.

For several generations of students at Harvard University, Bernard Cohen is best known as an outstanding lecturer and dedicated apostle of the history of science. He taught large undergraduate courses, first in the general education program, where he used the history of science to introduce nonscience students to the nature of the physical sciences, and later in the Core program, where he focused on the Scientific Revolution. Physical demonstrations, audiovisual materials, and a high degree of lecturing drama made his courses memorable. But in addition to "teaching out" to nonspecialist students he also played a major role in departmental courses at both undergraduate and graduate levels. The courses and seminars he taught were wide ranging, from the traditional surveys, (Aristotle to Einstein) through science in the United States, to such specialized interests as seventeeth-century physics, the history of computing, and relations between the natural and social sciences. He regularly impressed students in his courses, especially in seminars, with the depth and breadth of his knowledge and with his vast acquaintance with the literature in the field. There were often "surprises": Alchemy was not one of his favorite fields, but he became knowledgeable about it and ultimately even produced a significant article about one of its American practitioners, Ethan Allen Hitchcock.[21]

Although he has not published extensively in the field of science policy, Bernard Cohen has been a regular teaching participant, in Harvard's Program on Science, Technology and Public Policy, since its foundation in 1960. Working together with the program's directors, Dean Don K. Price of the Kennedy School of Government and Professor Harvey Brooks, Cohen has added the historical dimension to their project. This activity picked up an early interest that Cohen had demonstrated in a book written for the public on the role that scientific discovery has played in modern societies.[22]

Bernard Cohen's whole academic career has been linked to Harvard University. He arrived in Cambridge as a freshman in 1933 and received his undergraduate degree in 1937. He stayed on as a graduate student with the Committee on Higher Degrees in the History of Science on Learning; he remained in Cambridge during the Second World War teaching physics and mathematics in the program arranged for the navy; and in 1947 he received his doctorate and took up his post in history of science and general education. His professorship came in 1959 and in 1977 he was given additional recognition by apointment as Victor S. Thomas Professor of the History of Science, becoming emeritus in June 1984. His second intellectual home was England, initially as a Special University Lecturer at University College, London (1957), and subsequently at Cambridge University (the source of so many Newton manuscripts), where he has returned on numerous occasions with longer stays as Visiting Fellow at Clare Hall (1965) and as Overseas Fellow of Churchill College (1968).

Although it is customary in accounts of academic careers to give pleasant passing reference to the devoted spouse, in Bernard Cohen's case it is not possible to fully understand his career without understanding the role of his wife, Frances Davis Cohen. She was, as expected, a vigorous supporter of the history of science and of the Harvard department. But she was also a tough-minded individual with pronounced views of her own who could and did at one and the same time provide a lovely dinner and a rich and pointed discussion. She was herself an author and almost certainly a sharp and constructive critic of Bernard Cohen's written words.[23]

The history of science itself has come through important transformations in the years since the Second World War. As a subject of teaching and research it has grown from an exotic interest of a few scholars to a strong discipline represented in many centers of learning. Its publications, both journal articles and books, now enjoy a readership that crosses the boundaries between the "two cultures" and finds itself firmly fixed as an important element in the history of cultures. Intellectually and conceptually the field has developed new approaches and methods that incorporate philosophical analysis and sociological scrutiny. Bernard Cohen's own career has been a part of this transformation and his own teaching and scholarship have made him a vigorous participant in it. The essays that follow honor his contributions and demonstrate the intellectual vitality and the emerging perspective in the history of science.

NOTES

1 I. Bernard Cohen (ed.), *Benjamin Franklin's Experiments. A New Edition of Franklin's "Experiments and Observations on Electricity,"* Edited with a critical and historical introduction (Cambridge, Mass.: Harvard University Press, 1941).

2 I. Bernard Cohen, "Newton and the Modern World," *American Scholar* II (1942): 328–38; "Isaac Newton (1643–1727)," *Sky and Telescope* 2 (1943): 3–5; "Authenticity of Scientific Anecdotes," *Nature* 157 (1946): 196–7. The last deals with the story of Newton and the falling apple.

3 I. Bernard Cohen, "American Physicists at War: From the Revolution to the World Wars," *American Journal of Physics* 13 (1945): 223–35; "American Physicists at War: From the First World War to 1942," *American Journal of Physics* 13 (1945): 333–46; "Science and the Revolution; the Vital Interplay of Engineering and Science with Government had its Beginning in War Necessities," *Technology Review,* 47 (1945): 367–8, 374–8; "Science and the Civil War, First Large-scale Organizations of Technical and Scientific Resources of Manpower during the Civil War marks that Conflict as the turning Point in the Technology of Warfare," *Technology Review* 48 (1946): 167–70, 192–3.

4 I. Bernard Cohen, *Some Early Tools of American Science, An Account of the Early Scientific Instruments and Mineralogical and Biological Collections in Harvard University* (Cambridge, Mass.: Harvard University Press, 1950).

5 I. Bernard Cohen, "The New World as a Source of Science for Europe," *Actes de IXe Congres International d'Histoire des Sciences* (Barcelona, 1959), 95–130. I. Bernard Cohen, *Science and American Society in the First Century of the Republic* (Columbus: Ohio State University Press, 1961)

6 I. Bernard Cohen, *Franklin and Newton, an Inquiry into Speculative Newtonian Experimental Science and Franklin's Work in Electricity as an Example Thereof* (Philadelphia: American Philosophical Society, 1956).

7 See Ibid. Chapter 3, "The Scientific Personality of Franklin and Newton," pp. 42–88.

8 Ibid. Appendix 2, "Originality in Scientific Discovery, with Special Reference to Franklin's Experiments and his Concepts of the Electric Fluid," pp. 590–600.

9 He prepared a book-length biography in 1953, *Benjamin Franklin: His Contribution to the American Tradition.* Makers of the American Tradition Series. (New York: Bobbs-Merrill, 1953). It has been followed by other biographical volumes and, most recently, the Franklin entry in the *Dictionary of Scientific Biography.*

10 I. Bernard Cohen, *Franklin and Newton,* Preface, p. xii. The book is jointly dedicated to Alexandre Koyré (who represented the new interest in Newton) and Perry Miller, the Harvard literary and intellectual historian, whose works on the development of the American mind and culture can be linked to Cohen's interests in Franklin and American science.

11 Alexandre Koyré, "Pour une édition critique des oeuvres de Newton," *Revue d'Histoire des Sciences* 8 (1955): 19–37.

12 See especially Alexandre Koyré, *Etudes Galiléenes* (1939; reprint Paris: Hermann, 1966) and also his, "The Significance of the Newtonian Synthesis," *Archives Internationales d'Histoire des Sciences* 3 (1950): 291–311. His most influential text for American students was, *From the Closed World to the Infinite Universe* (Baltimore: Johns Hopkins University Press, 1957).

13 Alexandre Koyré and I. Bernard Cohen, "The Case of the Missing *tanquam:* Leibniz, Newton, and Clarke," *Isis* 52 (1961): 555–66.

14 Alexandre Koyré and I. Bernard Cohen (eds.), *Isaac Newton's Philosphiae Naturalis Principia Mathematica,* 2 vols. The third edition (1726) with variant readings assembled by Alexandre Koyré, I. Bernard Cohen, and Anne Whitman. (Cambridge: Cambridge University Press, and Cambridge, Mass.: Harvard University Press, 1972); I. Bernard Cohen, *Introduction to Newton's "Principia"* (Cambridge: Cambridge University Press, and Cambridge, Mass.: Harvard University Press, 1971).

15 I. Bernard Cohen, *The Newtonian Revolution with Illustrations of the Transformation of Scientific Ideas* (Cambridge: Cambridge University Press, 1980); "Supplement: History of the Concept of Transformation: a Personal Account," pp. 280–9.

16 I. Bernard Cohen, *Introduction,* p. 358. The Preface to the *Introduction* also gives a useful account of Cohen's developing Newtonian scholarship.

17 I. Bernard Cohen, "Supplement," p. 280.

18 I. Bernard Cohen, *Newtonian Revolution,* p. xiv; see also his discussions of "transformation" in Part II, esp. pp. 194–221.

19 See, for example, I. Bernard Cohen, "The Eighteenth-Century Origins of the

Concept of Scientific Revolution," *Journal of the History of Ideas* 37 (1976): 257–88; and "The Copernican Revolution from an Eighteenth-Century Perspective," in Yasukatsu Maeyama and W. Saltzer (eds.), ΠΡΙΣΜΑΤΑ: *Festschrift für Willy Hartner* (Wiesbaden: Steiner, 1977), 43–54.

20 I. Bernard Cohen, *Newtonian Revolution,* p. 219.

21 I. Bernard Cohen, "Ethan Allen Hitchcock, Soldier, Humanitarian Scholar, Discoverer of the 'True Subject' of the Hermetic Art," *Proceedings of the American Antiquarian Society,* 61 (1951): 29–139, published separately by the society in 1952.

22 I. Bernard Cohen, *Science, Servant of Man: A Layman's Primer for the Age of Science,* (Boston: Little, Brown, 1948).

23 Frances Davis, *My Shadow in the Sun* (New York: Carrick and Evars, 1940); Frances P. Davis, *A Fearful Innocence* (Kent, Ohio: Kent State University Press, 1981).

PART I

The history and philosophy of the exact sciences and mathematics

1

Compounding ratios

Bradwardine, Oresme, and the first edition of Newton's Principia

EDITH SYLLA
North Carolina State University

I propose in this paper to look at one link between fourteenth-century science and Newton – namely, a concept of compounding ratios common to Thomas Bradwardine's *De proportionibus velocitatum in motibus,* to Nicole Oresme's *De proportionibus proportionum,* and to the first edition of Isaac Newton's *Principia.* My purpose in doing this is not to give Bradwardine or Oresme credit for any of the achievement represented by Newton's *Principia.* In fact, the link between Bradwardine, Oresme, and Newton that I want to demonstrate concerns an area in which Newton was old-fashioned and in which he was rapidly superseded, he himself conforming, at least superficially, to newer ways almost immediately after 1687. I do want to show, however, that the "old-fashioned" concept of the compounding of ratios common to Bradwardine, Oresme, and Newton was not false, misguided, or inconsistent, but rather was a mathematically viable alternative to the concept of the compounding of ratios that prevailed after 1700. Although I cannot even broach the subject here, this history of changes in the dominant concept of compounding ratios should eventually be incorporated into the history of the transition from the ancient and medieval custom of representing physical relationships in terms of proportionalities to the modern use of equations and functions for the same purpose.

In what follows, I will first look at the concept of ratio and of compounding ratios used in Newton's *Principia* and at changes Newton made in the *Principia*'s terminology of ratios as the result of an early complaint about the book. In the second main section of this essay I will examine two traditions within the Greek and medieval treatment of ratios, one associated with theoretical mathematics, with music, and with physics, particularly as found in Bradwardine's *De proportionibus,* and the second associated with practical calculations using ratios and with astronomy. In the last section I will make a preliminary survey of the conflicts between these two traditions that developed in the seventeenth century with the resulting eclipse of the first tradition concerning compounding and the emergence of the modern concept.

GILBERT CLERKE AND THE REVISION OF THE *PRINCIPIA*

Newton's *Principia* is a work belonging to the transitional period between the medieval and modern treatment of ratios. This can be seen very easily thanks to the work of I. Bernard Cohen and Alexandre Koyré in producing their edition of the *Principia* with variants.[1] I hope that this essay will pay a small tribute to that edition and its editors in showing that there are terminological and notational changes in the *Principia* that would be very hard to find without the help of the new edition but which in fact echo noisy struggles within seventeenth-century mathematics, in the process of which much of the distinctively medieval science of ratios was, along with some inferior seventeenth-century conceptions of ratio, left behind.

Soon after the publication of the first edition of the *Principia* in 1687, Newton received a letter from one Gilbert Clerke, mathematician and Presbyterian minister, Fellow of Sydney Sussex College, Cambridge, 1648–55, and author of *Oughtredus explicatus* (1682), a commentary on William Oughtred's *Clavis mathematicae* (1631).[2] Clerke is generally overlooked by historians of mathematics as a very minor figure, but he may claim our attention as possibly the first critic to propose changes in the *Principia* after its publication.[3]

On the whole, the modern reader, aided by Clerke's own expressed modesty, may be inclined to smile indulgently at his questions concerning the text of the *Principia.* Thus the passage about which Clerke first wrote concerned a continued equation to which Newton had added a complicated expression.[4] Clerke added this expression to the quantities on either side of the last equals sign, found that the algebra did not work out as Newton stated, and so suspected a printer's error.[5] But, just as the modern reader would expect and as Newton explained in his reply to Clerke, Newton had meant that the expression should be added to the original quantity before the continued equation, as well as to the last quantity.[6]

But although Clerke was a mathematician of modest ability, he did lead Newton to make systematic changes throughout the *Principia* in the terminology concerning the ratios. In the first edition of the *Principia,* Newton had, for the most part, used the same medieval terms to refer to powers and roots of ratios as he used to refer to multiples of integers or fractions. Thus *dupla* or double was the square of a ratio in modern terms, but it referred to two times an integer or fraction. *Dimidiata,* or half, was the square root of a ratio, but an integer or fraction divided by two. *Sesquialtera* was a ratio to the 3/2 power, in modern terms, but it was a number multiplied by 3/2.[7] Newton did have a different set of terms, namely *duplicata, subduplicata, sesquiplicata,* and so forth, which he might have used to distinguish taking the powers or roots of ratios from taking multiples of numbers, but he did not do this consistently.[8]

That Newton did not distinguish between what appears to us quite different operations is not as surprising as it may seem at first sight. For Newton, ratios are relations and, as such, are different from numbers, just as, for instance, lines are different from numbers. The "same operation" is quite naturally expected to have different applications to different sorts of entity. Thus, in practice, "adding" lines is different from "adding" numbers, and producing a rectangle from two lines is different from forming the product of two numbers by multiplication. Although in theory adding integers is the same as adding fractions, in practice different procedures must be used – fractions but not integers must be reduced to a common denominator, and so forth. So too, then, "doubling" a ratio might be different from "doubling" a number, with the difference being ascribed not to a difference of operation in theory, but to a practical difference resulting from the sort of entity being operated upon. Thus the terminology for "doubling" in these two cases would be identical because the operation was conceived of as theoretically identical.

If ratios and integers or fractions are identified, however, then the same operation cannot be imagined to have two different results corresponding to multiplying and taking powers when applied to the same sorts of things. This is, implicitly, Gilbert Clerke's point of view in criticizing Newton's terminology. Clerke was initially confused when Newton used the unfamiliar term *sesquiplicata* to refer to the 3/2 power of a ratio.[9] In his reply to Clerke Newton explained, "By *sesquiplicata* I mean *sesquialtera*, a ratio and an half or ye root of ye *ratio triplicata.*"[10] Clerke was at first still confused as to what this meant. Knowing that the medieval term *sesquialtera* meant the ratio of 3 to 2, he reasoned that the *sesquialtera* or *sesquiplicata* of a ratio A would be $3/2A$:

You say you meant $\sqrt{Ac}$ & that indeed will doe, But first whereas you say by *sesquip;* you meane *sesquialtera,* a ratio and a halfe, yt is not $\sqrt{Ac}$. for if A be 4, ye *sesquialtera* is .6. but 8 = $\sqrt{Ac}$. I thought you might meane so & tried $A/1 + A/2$ but yt would not doe: I also multiplied ye whole by the halfe viz: $A \times A/2 = A/2$; that would not doe either. now I perceive I should have multiplied ye whole not be ye halfe, but by ye *ratio dimidiata* viz $A \times \sqrt{A} = \sqrt{Ac}$.[11]

Thus in the end, Clerke apparently figured out Newton's meaning correctly. But in a postscript, he came to the point that led Newton to make changes in the second edition of the *Principia:*

I confesse I did not very well approve of your calling ye root by ye *dimidiata ratio,* for *dimidiare* is properly to divide by .2. So Oughtr. c. 15, 11: *si dimidianum sit* $\sqrt{c32}$, *vel dividendum per 2.* so Clavius lib. 5. towards ye beginneing *de proportione – unitatem quae est pars dimidiata numeri binarij:* therefore I thought that for most mens understanding, this ratio had been better called *sub-duplicata;* but if use, amongst you virtuoso's hath authorized ye other way, because there is as well 4 × 4 as 4 + 4, I am content. . . .

I think yt wch you mean by *sesquipl:* should have a name for it & yt *sesquipl:* is a good name; but it cannot be alloweable to call it *sesquialtera,* there being a knowne ratio by addition as 3/2 wch is another thing, so called: by all meanes let divers things have diverse names![12]

Newton's reply on November 2 to this second letter from Clerke is lost, but from Clerke's quoting of it in his third letter, it is easy to gather much of its content. Newton replied, in fact, by distinguishing between ratios and integers or fractions and by claiming that "ratios are summed up by multiplication," so that, for instance, "ye quantity 4/1 doubled is 8/1, but ye ratio 4/1 *doubled* is 4/1 × 4/1 or 16."[13] Clerke's first response was that:

When particular quantities are named & set downe, it is another thing, than if a man being asked in general what is halfe ye duplicate ratio? should answer, that, that ratio is compounded of two equal ratio's, wch being twice taken, one of them may in yt regard, be called halfe of it.[14]

He also repeated in several ways that different terms should be used for different things. To Newton's remark that "ye ratio $4/1$ *doubled* is $4/1 \times 4/1$ or 16," Clerke replied that "certainly mathematicians doe not use to call that *doubleing* but duplicateing in contradistinction to doubleing."[15] To Newton's remark that, "ye ratio *dimidiata* of $16/1$ is ye root thereof namely $4/1$," Clerke replied, "I think *ratio subduplicata* is a more proper, more usual & better known word."[16] Clerke insisted that the *ratio sesquialtera* of $16/1$ is 24, whereas Newton had stated it to be 64 [i.e., $= 16^{3/2}$], and Clerke claimed that 64 should only be called the *sesquiplicate* of 16. Clerke concluded:

For never any man called $2A$ but Aq ye duplicate of A & I am persuaded that not one Mathematician in ten but would say being asked, yt .6. is ye *sesquialtera* of $4/1$, without ye least thought of distinguishing between quantities and ratios.[17]

Clearly, Newton and Clerke disagreed whether ratios and quantities were to be distinguished from each other. Newton thought they were different sorts of entities; Clerke made no distinction between them. Furthermore, despite Clerke's claims to the contrary, use among virtuosos *had* authorized Newton's use of the same terms for powers of ratios as for multiples of numbers, as will be seen later. Nevertheless, since there were already established words that could be used to distinguish between the "doubling" or "halving" of ratios and the "doubling" or "halving" of integers or fractions, Newton and Clerke could agree on this distinction of vocabulary in order to avoid confusion.

In his revisions to the first edition of the *Principia*, therefore, in the fourth corollary to Lemma XI of Book I, which is the passage that had first confused Clerke about the meaning of *sesquiplicatam*, Newton added the explanation, "I call the ratio *sesquiplicate* which is the subduplicate of the triplicate, and which is compounded from the simple [ratio] and its subduplicate, and which elsewhere they call sesquialterate."[18] In the third edition, he dropped the last clause as if by then he thought the need to admit a now-extinct usage no longer remained.

Sesquiplicate was a neologism. In the other common cases, however, the alternate words to be applied to powers and roots of ratios were more familiar. Thus Newton could simply refrain from using the terminology *in ratione dupla, tripla, subdupla, subtripla,* and so on, and instead use *in ratione duplicata, triplicata, subduplicata,* and *subtriplicata.* This he did in his revision to the *Principia* writing at one point, as instructions to his editor, "p. 42, 55, 57, 58, and elsewhere for *dimidiata ratione* write *subduplicata ratione.*"[19] He also frequently changed *dupla* to *duplicata.*

Even the third edition of the *Principia* was not completely consistent in this matter, however. In some places, presumably, Newton and his editors simply missed a case where a change should have been made. In other cases it happened, for instance, that *duplicata* had been used in the first edition to refer simply to the doubling of a quantity, and the inverse change to *dupla,* which should have been made for consistency, was not made. On the very first page of the *Principia,* Newton changed "Air doubly (*duplo*) dense in a double (*duplo*) space is quadruple" to "Air with density duplicated (*duplicata*) in a space also duplicated (*duplicata*) becomes quadruple,"[20] even though he clearly meant that the density of the air and the space should be doubled and not squared. Thus *duplicata* remained for Newton a synonym of *dupla,* perhaps to be preferred where an actual physical doubling and not simply a multiplication by 2 was at issue, and it was not limited to the meaning "squared."[21]

On the basis of this evidence and of other evidence to be found in Newton's mathematical works, I believe that although Newton changed the terminology of the *Principia* to answer Clerke's complaints, he never did give up the view that ratios are different from numbers. Despite the rapid development of analytic geometry in his day and despite his own signficant contributions to it, Newton was not a mathematician who easily identified the entities and operations of the different branches of mathematics. Rather, he attempted to preserve or restore the purity or rigor of each branch of mathematics and had plans to write a work of geometry using only strictly geometrical methods.[22] Thus the changes in the *Principia* introduced at Clerke's instigation were cosmetic and superficial rather than representing a fundamental change in point of view.

This can be seen more clearly by looking at one further change Clerke led Newton to make. This change involved notation rather than terminology. As we saw earlier, Newton wrote to Clerke that "ratios are summed up by multiplication." In saying this, Newton was thinking about the operation on ratios that in the Middle Ages was most often called *compounding.* As long as mathematics was written only in words, it might be left ambiguous whether compounding ratios was to be identified with any of the standard arithmetic operations. When notation was first introduced to represent the arithmetic operations, some authors chose to represent the compounding of ratios by the notation for addition of numbers, that is, by the plus sign. This notation was used by Isaac Barrow[23] and it is found in the first edition of the *Principia,* where, as might be expected, it confused Gilbert Clerke.

In Book I, Proposition 11, in proving the important result that when a body revolved in an elliptical orbit, this is caused by an inverse square centripetal force toward the focus of the ellipse, Newton wrote:

> . . .erit $L \times QR$ ad $L \times Pv$ ut QR ad Pv, id est, ut PE seu AC ad PC; & $L \times Pv$ ad GvP ut L ad Gv; & GvP ad Qv *quad.* ut PC *quad.* ad CD *quad.* & (per corol. 2. lem. VII.) Qv *quad.* ad Qx *quad.* punctis Q & P coeuntibus est ratio aequalitatis; & Qx *quad.* seu Qv *quad.* est ad QT *quad.* ut EP *quad.* ad PF *quad.* id est, ut CA *quad.* ad PF *quad.* sive (per lem. XII) ut CD *quad* ad CB *quad.* Et conjunctis his omnibus rationibus, $L \times QR$ fit and QT *quad.* ut AC ad $PC + L$ ad $Gv + CPq$ ad $CDq + CDq$. ad CBq. id est ut $AC \times L$ (seu $2CBq$.) $\times$ CPq. ad $PC \times Gv \times CBq$., sive ut $2PC$ ad Gv.[24]

Put into somewhat more modern notation, and arranged more conveniently on the page, the first part of what Newton has said here is that:[25]

$$
\begin{array}{llllllll}
L \times QR & : & L \times Pv & :: & AC & : & PC \\
L \times Pv & : & GvP & :: & L & : & Gv \\
GvP & : & Qv^2 & :: & PC^2 & : & CD^2 \\
Qv^2 & : & Qx^2 & :: & 1 & : & 1 \\
Qx^2 & : & QT^2 & :: & CD^2 & : & CB^2
\end{array}
$$

It is not necessary to know here the justifications of these proportionalities, since the issue concerns only what Newton does next, which is to compound all the ratios on the left-hand sides of these proportionalities and say that the result is proportional to the compound of all the ratios on the right-hand sides of the proportionalities. The ratios on the left-hand sides of the original proportionalities are continuous (i.e., the consequent of the first ratio is the same as the antecedent of the second ratio, etc.), and Newton takes the compound of the ratios on the left-hand side to be simply the ratio of the first antecedent to the last consequent, namely, $L \times QR : QT^2$.[26] Compounding the ratios on the right-hand sides, Newton first simply writes out the ratios with plus signs in between: $AC : PC + L : Gv + Cp^2 : CD^2 + CD^2 : CB^2$. He then takes the ratio of the antecedents of all these ratios multiplied together to the product of all the consequents (omitting CD^2, which would appear in both expressions): $AC \times L \times CP^2 : PC \times Gv \times CB^2$. Substituting $2CB^2 = AC \times L$, and cancelling further, this becomes $2PC : Gv$.

The problem of notation here, involving Newton's writing the compound of a series of ratios using plus signs, should be clear enough without going into the mathematics further. Gilbert Clerke was confused because he tended to identify compounding with the sort of multiplication Newton did on the right-hand sides of his proportionalities and because he was most familiar with the work of William Oughtred, who used a multiplication sign to represent compounding. For him the use of plus signs for compounding was unexpected. He wrote:

I was much troubled, at your *conjunctis rationibus,* by wch I perceive you meane multipli: viz: *p:* 51. 1. 6. how to find $PC + L$. a 4th proportional. at last I perceived it was but a method of writeing & no analogism. so contracting ye ratios by changeing ye Quad's, & getting ym in rank & file one under another, I easily saw how I was to multiply & divide

LQR.LPU :: AC.PC
LPu.GuP :: L.GU, &c.

and indeed you should have sett ym so, for your booke is hard enough, make it as easie as you can.[27]

Here Clerke is using a single dot rather than a colon to separate the terms of his ratios, he has represented Newton's multiplications simply by writing the letters together, and he has formed his letters slightly differently, but in the end he has read Newton correctly. In the beginning, however, he had started by trying to add $PC + L$, which was far from Newton's intention. To deal with this possible misunderstanding, Newton simply omitted the confusing passage containing plus signs from the later editions of the *Principia.* Elsewhere this problem was avoided because ampersands were used rather than any mathematical notation for the compounding of ratios.[28]

Here again, then, Newton revised the *Principia* to meet Clerke's complaint, at least in part. I do not believe, however, that his view of the operations involved was changed. Part of his view, as he wrote to Clerke, was that "ratios are summd up by multiplying." In line with this understanding, one could both indicate the summing of ratios by connecting them with plus signs and then go on to multiply the antecedents and consequents to form a new ratio as Newton in fact did with the ratios on the right-hand sides of his proportions.

But Newton also compounded continuous ratios simply by taking the ratio of the first antecedent to the last consequent without any multiplication whatsoever. Newton's proof of Proposition 11 of Book I is first derived by carefully finding a continuous series of ratios on the left-hand sides of the proportions and then simply taking the ratio of the first antecedent to the last consequent. As will be seen, this was theoretically the normal mode of compounding whereas compounding by multiplication was a theoretically less justified algorism. Newton could easily have proved the same theorem without using continuous ratios and he did so in a letter to John Locke in 1689/90.[29] I suggest that Newton would have considered the compounding of a continuous series of ratios simply by taking the ratio of the extremes as more obvious and as theoretically more justified than compounding by multiplication. Clerke and modern commentators on the text miss this when they assume that in compounding or conjoining ratios here, Newton multiplies and cancels on *both sides* of his proportions or equations.[30]

In these exchanges between Newton and Clerke concerning the terminology and notation for ratios and the compounding of ratios, Newton was obviously by far the better mathematician, yet Clerke led Newton to make changes in the *Principia* reflecting Clerke's views. This must have been partly because the view of ratio reflected in the terminology of the first edition of the *Principia* was rapidly becoming obsolete, making way for what was to become the modern view of ratio. What then were these views and what history lay behind them?

The period immediately before Newton saw a great proliferation of theories of ratio, only some of which I will be able to mention in the third section here. But first, in the second section, I want to sketch briefly and characterize two ancient and medieval traditions concerning ratios and the compounding of ratios. These two traditions may not encompass all ancient and medieval concepts of ratio. Neither were these traditions always separate – in fact, they were often strangely mingled. Nevertheless, they represent two poles of the ways in which ratios and the operations on ratios could be treated.

TWO ANCIENT AND MEDIEVAL THEORIES OF COMPOUNDING RATIOS

I plan to publish elsewhere a paper tracing the traditions concerning the compounding of ratios from Euclid to Nicole Oresme. Here I can only give a summary of its contents to show that in using terminology and notation that linked the compounding of ratios with addition and that considered the compound of a ratio with itself as double the original ratio, Newton was following a theoretical and Euclidean tradition concerning the compounding of ratios. Gilbert Clerke, on the other hand, in not distinguishing ratios from integers or fractions, in considering compounding as multiplication, and in calling the result of a ratio compounded with itself the original ratio *duplicata,* was the heir of a second practical tradition.

The first ancient and medieval tradition concerning compounding

Euclid's *Elements,* at least in some of its editions, contains roots of both traditions. The roots of the first tradition are in Book V of the *Elements,* which presents the general theory of ratio and proportion applicable both to ratios of numbers and to ratios of continuous quantities of any sort, or, alternately, applicable both to ratios that can be expressed as occurring between two integers and to those that cannot.

None of the standard arithmetic operations are applied to ratios in Book V, though these operations are applied to the terms of ratios. The closest approaches to standard arithmetic operations on ratios in Book V are in definitions 9–10 and in definitions 17–18. In definitions 9 and 10, Euclid says:

> **9.** When three magnitudes are proportional, the first is said to have to the third the *duplicate ratio* of that which it has to the second.
> **10.** When four magnitudes are continuously proportional, the first is said to have to the fourth the *triplicate ratio* of that which it has to the second, and so on continually, whatever be the proportion.[31]

And in definitions 17 and 18, he says:

> **17.** A ratio ex *aequali* arises when, there being several magnitudes and another set equal to them in multitude which taken two and two are in the same ratio, as the first is to the last among the first magnitudes, so is the first to the last among the second magnitudes.
> **18.** A *perturbed proportion* arises when, there being three magnitudes and another set equal to them in multitude, as antecedent is to consequent among the first magnitudes, so is antecedent to consequent among the second magnitudes, while, as the consequent is to a third among the first magnitudes, so is a third to the antecedent among the second magnitudes.[32]

In definitions 9 and 10 proportional magnitudes are set out and the ratio between nonadjacent terms said to be *duplicata* the ratio between adjacent terms if one term is skipped over, *triplicata* if two terms are skipped, and so forth. If one writes the terms of a geometric proportion in a line, for instance:

$$2 \quad 4 \quad 8 \quad 16 \quad 32$$

then the ratio of 8 to 2 is said to be equal to the ratio of 4 to 2 *duplicata.* The ratio of 16 to 2 is the ratio of 4 to 2 *triplicata,* and so on. Looking at this way of representing ratios and proportions, one sees that the interval between 8 and 2 contains the interval between 8 and 4 and the interval between 4 and 2. Since the ratios of 8 to 4 and of 4 to 2 are equal, one can say that the interval or ratio of 8 to 2 contains the interval or ratio of 4 to 2 twice, or that it equals the ratio of 4 to 2 *duplicata.*

In definition 17 a set of arbitrary terms is given, *A, B, C, D, E,* and a second related set *P, Q, R, S, T,* such that the ratio of *A* to *B* equals the ratio of *P* to *Q;* the ratio of *B* to *C* equals the ratio of *Q* to *R;* the ratio of *C* to *D* equals the ratio of *R* to *S;* and the ratio of *D* to *E* equals the ratio of *S* to *T.* Then a ratio *ex aequali* arises when the ratio *A* to *E* equals the ratio of *P* to *T.* In this definition, one also considers the ratios between adjacent terms and the ratios between the extreme terms of a series, but the ratios between adjacent terms within a single series need not be equal. The terminology *ex aequali* seems to refer to the distance or interval between the terms being considered.[33] Definition 18 is a slight variant of 17.

Nowhere in Book V does Euclid give a name to the ratio of the extreme terms of a series when the ratios between adjacent intervening terms are not equal. One might suppose that the name *ex aequali* would do, but the wording of the definition suggests that it is a proportion rather than a ratio that is being defined.[34] In Book VI, however, Euclid remarks in the course of proving Proposition 23 that:

> the ratio of *K* to *M* is compounded of the ratio of *K* to L and of that of *L* to *K.*[35]

Reading definition 17 in the light of this, one may well agree to call the ratio of the extreme terms in any series the "compound" of the ratios of the adjacent intervening terms. This was the term most commonly used in the later tradition, although it did not take the status of a technical term, and it should not be confused with "composition of a ratio," which *was* a technical term defined by Euclid, and referred to taking the sum of the antecedent and consequent of a given ratio in relation to the consequent.

This understanding of compounding appears quite often in the later tradition associated with the representation of the terms of a proportion in a line and with harmonic proportions, in which the ratios between succeeding terms were not the same. In his *De institutione arithmetica,* II, 3, for instance, Boethius shows that a double ratio is compounded of a sesquialter (3/2) ratio and a sesquitertiate (4/3) ratio by an example using a row of numbers with curved lines connecting them:

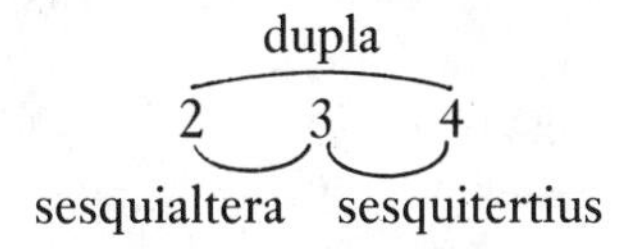

Four, he says, is sesquitertiate to 3, but it is double 2, whereas 3 is sesquialterate to 2. In this way, therefore, a sesquialter and a sesquitertius compound a double.[36] Since an octave in music is associated with a double ratio, whereas a musical fifth is associated with a sesquialter and a musical fourth with a sesquitertius, this same display of numbers shows that an octave is compounded of a fourth and a fifth.

Boethius sometimes uses the word compound (*componere*) in connection with examples like that given here, but he also uses a wide range of other terms with connotations of joining (*coniungere, adglomere*) or of birth or creation (*procreare, creare, nasci, exoriri*).[37] In his *De institutione musica* he uses the term "add" (*addere*).[38] There is no algorism for adding or compounding continuous ratios on this understanding; one simply determines by inspection of a row of terms what is compounded of what.

This first tradition concerning the compounding of ratios has its best medieval expositions in Thomas Bradwardine's *De proportionibus velocitatum in motibus* and in Nicole Oresme's *De proportionibus proportionum* – in fact, up to now, historians have observed the tradition mainly in these two works. Bradwardine's work has received attention from historians of medieval science as containing a new theory of the relations of forces, resistances, and velocities in motions. When it was first discovered by historians, this theory was misunderstood. Later, approximating more nearly Bradwardine's intent, it was expressed as the logarithmic relationship:

$$V = k \log F/R$$

or as the exponential relationship:

$$\frac{F_2}{R_2} = \left(\frac{F_1}{R_1}\right)^{\frac{V_2}{V_1}}$$

where F represents force; R, resistance; and V, velocity.[39]

But both of these modes of expressing Bradwardine's ideas are anachronistic, and recently the attempt has been made to express Bradwardine in his own terms, something that involves a preliminary exposition of his theory of ratio.[40]

In fact, Bradwardine follows the method of compounding based on Book V of the *Elements* and on the sort of thing Boethius did, but he takes seriously the notion that compounding ratios is akin to or identical to addition. He states, for instance:

> If there is a ratio of greater inequality of a first term to a second and of a second term to a third, the ratio of the first term to the third will be precisely double (*dupla*) the ratio of the first to the second and of the second to the third.[41]

This is simply Euclid's definition 9 of Book V, but it should be noted that, like the first edition of the *Principia*, Bradwardine calls this ratio double and not duplicated (*duplicata*), as did Euclid and as did Newton in the second edition of the *Principia* at Clerke's instigation. In his proof of this conclusion, Bradwardine refers to the definition of a double ratio he has given earlier, and it is clear that he means double in the usual sense according to which something that contains a smaller thing twice is called its double.

This is extended in Bradwardine's seventh conclusion and its proof:

> No ratio is greater or less than a ratio of equality.[42]

As part of his second proof of this, Bradwardine says:

> Similarly, it can be proved that a ratio of equality is not less than a ratio of greater inequality, because then if it were taken some number of times, it would equal or exceed

the ratio of greater inequality. But this consequent is false because no matter how many times equals are taken [in a series] the ratio of the first to the last is not greater than the ratio of the first to the second, but all remain entirely in equal ratios of equality.[43]

The basis of this argument seems to be something like definition 4 of Book V of the *Elements*.[44] If a ratio of equality were to have a ratio to some other ratio, say one of greater inequality, then if the ratio of equality were multiplied often enough the result should equal or exceed the given ratio of greater inequality. But no matter how far a series of terms equal to each other is extended the ratios between any terms are still ratios of equality. Thus Bradwardine regards the extension of terms in a geometric series as akin to multiplying, and he regards the finding of geometric means between these terms as akin to dividing. This fits nicely with the notion that compounding ratios is akin to addition. In a series of terms in proportion A, B, C, D, E . . . , the ratio of A to C can be formed by "adding" the equal ratios of A to B and of B to C, or it can be formed by "multiplying" the ratio of A to B by 2, accomplished by finding the third proportional C.

With this view of the operations on ratios established, Bradwardine could then propose his solution concerning the relationships between force, resistance, and velocity in motion by saying simply that when the ratio of force to resistance was multiplied by any number, so too the velocity was multiplied by the same number: if F to R was doubled or tripled or halved, V would be doubled or tripled or halved. Translated into modern conceptions this amounts to something like a logarithmic or exponential relationship, but it has no such connotations in Bradwardine.

Nicole Oresme's *De proportionibus proportionum* is a work that accepts Bradwardine's version of the first tradition concerning compounding ratios and carries it further, developing a concept of "the ratios of ratios."[45] This concept makes more explicit the notion of "multiplying" ratios found in Bradwardine – or, in modern terms, it involves operations on roots or powers of ratios. The concept of the "ratios of ratios" confused some later mathematicians familiar only with the second tradition concerning compounding, who treated ratios simply as integers or fractions, in which case there is nothing special about the ratios of ratios.

The question of the ratios of ratios brings in a whole new area, of great interest in itself, but too large to be dealt with here. If it could be treated more fully, however, it would help to underline that in a fundamental sense Oresme was an heir of the first tradition concerning ratio, that is, compounding ratios by taking the extremes of a continuous series of terms and considering this operation as addition.

An obvious sign of Oresme's adherence to the first tradition is his use of the word *addere* for compounding and of *subtrahere* for its opposite. This has often been remarked upon, and Oresme has been suggested as the originator of this terminology, but this has not been fully understood. We saw earlier that, in fact, Boethius on occasion used *addere* for compounding ratios, and scattered, unsystematic uses of *addere* for compounding might be found elsewhere. Perhaps

Oresme was the first to use the terminology systematically, but we should not suppose that he uses it because in multiplying ratios exponents are added, even if Oresme was important in the development of fractional exponents (in fact, according to Oreseme's rules for compounding ratios with exponents, no special attention is paid to the case when the bases are the same).[46] *Addere* is used because compounding by taking a continuous series of terms was viewed as addition, and Oresme's terms "adding" and "subtracting" come straight from the first tradition concerning ratio.

So far, then, I have sketched a first tradition concerning the compounding of ratios. It is this tradition that I claim Newton had in mind in the last passage from the *Principia* quoted earlier. In finding a continuous series of ratios on the left-hand sides of his proportionalities and in calling the ratio of the first antecedent to the last consequent on the left-hand side of the proportionalities the compound of that series of ratios, he has done simply that, with no multiplication involved whatsoever.

Newton goes beyond this sort of compounding, however, when he starts multiplying together the antecedents and consequents of the ratios on the right-hand sides of his proportionalities. How is such multiplication justified? All that I have explained about the theory of ratios and of compounding ratios up to this point is what I call the first ancient and medieval tradition concerning ratios and it is a tradition that Bradwardine, Oresme, and Newton share, as well as being founded in Euclid. There is, however, a second tradition concerning the compounding of ratios that can also be found in some editions of Euclid, though the best modern editors now agree in excluding the relevant passage.

The second ancient and medieval tradition concerning compounding

In some editions of Euclid, Book VI has a fifth definition:

> A ratio is said to be compounded of ratios when the sizes (πηλικότητες) of the ratios multiplied together make some [?ratio].[47]

This definition appears only in the margin of the best manuscript of Euclid,[48] and it does not appear in Campanus's translation from the Arabic. It is never referred to by Euclid or by any of the other Greek geometers, who seem to compound ratios as in the first tradition.[49] It is, however, referred to by Theon of Alexandria in his commentary on Ptolemy's *Almagest,* I, 13.[50] On the basis of this connection with Theon, Robert Simson excluded the definition as a Theonine interpolation, distorting the truly Euclidean sense of compounding.[51] This was confirmed when a manuscript reflecting the pre-Theonine tradition was discovered.

But however theoretically objectionable this definition might be, a second ancient and medieval tradition concerning the compounding of ratios was based on it, introducing both the notion of the "sizes of ratios" and the operation of multiplying together these sizes as the method of compounding two ratios.

In this second tradition every ratio was associated with its "size," or in modern terms, simply with the number expressing the ratio in lowest terms. To compound ratios – which even in this second tradition were initially conceived of as relations – one could multiply together their sizes. The resulting product would then be the size of the desired compounded ratio. In Latin, these associated sizes (Heath's translation of πηλικότητες) were called by various authors *denominationes, denominators,* or *exponentes,* as well as other names.[52] In what follows, I will call them "denominations" after their most common medieval name.

The second tradition concerning the compounding of ratios almost always appears in the Middle Ages in close or remote connection with astronomy. The roots of this tradition go back to the transversal theorem of Menelaus, particularly as this appeared in Ptolemy's *Almagest,* I, 13, along with Theon's commentary on that chapter. The Arabic works on the subject most influential in the West were the *Epistola de proportione et proportionalitate* of Ametus filius Josephi (Aḥmed ibn Yūsuf)[53] and Thābit ibn Qurra's *De figura sectore,*[54] both translated into Latin by Gerard of Cremona.

In the medieval Latin West similar treatises *De proportionibus* are ascribed to Jordanus Nemorarius and to Campanus.[55] Roger Bacon's *Communia Mathematica*[56] and Richard of Wallingford's *Quadripartitum*[57] each contains a section of a similar nature and there are other works of the same genre. Here I take only Jordanus as an example.

The treatises are all related to a case of the compounding of ratios found in the *Almagest* in which Ptolemy compounds one ratio out of two others with no term in common (thus it was not the sort of case to which the first tradition immediately applied). In the medieval treatises on ratio related to this passage, the purpose is to show all the valid transformations of such a relationship, often considered quite independently of its origin. Thus, given six magnitudes such that the ratio of the first to the second is compounded of the ratios of the third to the fourth and of the fifth to the sixth, what other relationships will hold true between the ratio of any two of these magnitudes and the compound of two ratios formed from the other four magnitudes?

Superficially, in these cases, one has to do with the compounding of ratios that are not continuous, so it might seem that the understanding of compounding found in the first tradition would be impossible to apply. Nevertheless, everything could be accomplished either geometrically or using only the first tradition. In the treatise of Ametus filius Josephi, all of the valid transformations of the original relationship are derived geometrically, while in Thābit's *De figura sectore* they are derived as in the first tradition of compounding ratios. The treatise by Jordanus and Campanus follow Thābit in that their proofs are based on the first tradition for compounding ratios.

Both treatises, however, preface their descriptions of the valid modes with definitions and propositions that belong to the second tradition. The work ascribed to Jordanus begins:

1. A ratio is a relationship of one thing to another determined with respect to quantity.
2. The determination is manifested by the denomination.
3. The denomination of a ratio of this to that is what arises from the division of this by that.
4. For a ratio to be produced or compounded from ratios is for the denomination of the ratio to be produced from the denominations of the ratios, one drawn (*ductus*) into the other.

[Conclusions:]

1. Any ratio that is produced from two [ratios] is shown to be produced from one of them and from what produces the other. For let the denomination of a ratio A be produced from two denominations C and B, and again [let] B [be produced] from D and E. Since, therefore, E into D produces B and B into C produces A, it is clear that E, D, C are three sides (*latera*) of that which is A. Therefore it is produced from these.
2. If between any two extremes there is posited a mean which has a ratio to either extreme, the ratio of the first to the third is compounded from the ratio of the first to the second and of the second to the third.[58]

The proof of this second conclusion is important:

For let E be a mean between D and F. I say that the ratio of D to F arises from the ratio of D to E and the ratio of E to F. Demonstration: divide D by E and H arises, which is the denomination of the ratio of D to E. Also divide E by F and K arises, which is the denomination of the ratio of E to F. Divide D by F and G is produced which is the denomination of the ratio of D to F. Draw (*ducatur*) H into K and T is produced. I say that T is equal to G, for K times H and F produces T and E. Therefore the ratio of H to F is the same as the ratio of T to E. But the ratio of H to F is the same as the ratio of G to E because H times E produces D (*ex ductu H in E fit D*) and F times G produces the same. It is clear, therefore, that T is equal to G and that from the product of H and K arises the denomination of the ratio of D to F.[59]

This proof makes extensive use of denominations, which tends to link it with the second tradition, but more importantly, these denominations are used and multiplied together. When Jordanus argues that $H : F :: G : E$ because $H \times E = F \times G$, it is difficult to see how this should follow except on the grounds that when four terms are proportional, the product of the extremes equals the product of the means and vice versa. But this is not a proposition proved in the general Euclidean theory of ratio found in the first tradition. And, in general, from the point of view of the first tradition one might argue against the propriety of dividing the antecedent by the consequent to find the denomination and against taking the ratios of H to F and G to E, because the terms of these ratios may well not be of the same sort.

After a third preliminary conclusion based on the second, Jordanus begins his proofs of the validity of the modes of transforming the compounding of one ratio from two others. His second conclusion and its proof are good examples of the rest of his treatise:

If, of six quantities, the ratio of the first to the second is produced from the ratio of the third to the fourth and of the fifth to the sixth, then the same, that is [the ratio] of the first to the second is produced from the ratio of the third to the sixth and of the fifth to the fourth. Let there be six quantities, A the first, B the second, C the third, D the fourth, E

the fifth, F the sixth. I say as set out before that the ratio of A to B is compounded of that which is of C to F and that which is of E to D. For let that which is G ratio be [compounded] from C to F and E to D. But that which is C to D and that which is D to F compounded that which is C to F by the second [preliminary conclusion]. Therefore these same [ratios] together with that which is E to D compound that which is G, by the first [preliminary conclusion]. But those two which are D to F and E to D compound that which is E to F again by the second [preliminary conclusion]. Therefore, those two which are C to D and E to F compound G. But these are those ratios which compound that which is A to B by hypothesis.[60]

This proof lacks a last step. As it stands it proves only that what is known follows from what is unknown. It can, however, easily be completed. Jordanus's basic technique is simply to divide one ratio into two by inserting as a mean the consequent of the second ratio (here he divides the ratio of C to F by inserting the mean D, given that for any D having a ratio to C and F, the ratio of C to F will be compounded of the ratios of C to D and of D to F). He then compounds the one of these two ratios having the inserted term as antecedent with the original second ratio having the same term as consequent (in this case he compounds the ratios of D to F and of E to D to obtain the ratio of E to F). But then this compound and the ratio remaining from the first ratio are the ones he wants. To complete the proof in the order he has given it here he ought to have added that since both G and A to B are compounded of C to D and E to F, therefore G is the same as A to B. And since G is compounded of C to F and E to D, then A to B will also be compounded of C to F and E to D.

This is all done only by compounding continuous ratios and could fall within the first tradition on ratios, but Jordanus clearly thought that the proof rested on the proof of his second preliminary conclusion. The same can be said for all his later conclusions. In sum, Jordanus's treatise on ratios has an introductory section of definitions and preliminary conclusions associating ratios with denominations and arithmetic operations offered as a foundation for the rest of the treatise, which, however, within the first tradition on compounding, might have stood on its own. Bradwardine, who had Jordanus's treatise before him when he wrote his own, recognized this when he omitted Jordanus's fourth definition, which described producing or compounding in terms of the denomination of the ratio being the product of the denominations of the ratios compounding it, and when he included Jordanus's second and third preliminary conclusions in his own work, not as conclusions, but as suppositions – indeed, in the first tradition, the compounding of continuous ratios is not something to be proved.

In line with Jordanus's arrangement, however, multiplication often came to be understood as the fundamental method of compounding, albeit without taking full advantage of this and without, for instance, saying that the second conclusion, just quoted, is obvious on the grounds that multiplication is commutative. But not all medieval mathematicians followed Jordanus in making multiplication more fundamental, and Nicole Oresme, along with Bradwardine, made the first tradition the fundamental one.

Bradwardine made the first tradition fundamental by pretending that the second tradition did not exist. Oresme, on the other hand, recognized both traditions while giving priority to the first. Besides his theoretical work on ratios, Oresme also wrote an *Algorismus proportionum,* using the terms of the first tradition, but giving practical rules for compounding ratios based on the multiplication of the second tradition.[61] He understood that *theoretically* ratios are compounded as in the first tradition, but said that *practically* – as an algorism is understood as a practical technique – one can often operate as in the second tradition. In addition to his choice of the title *Algorismus proportionum* for this second-tradition material, this is shown by the fact that in the *De proportionibus proportionum,* when Oresme inserts material from the second tradition, he prefaces it by saying that he will show how to do *per artem* what he has just discussed theoretically.[62]

Within the second tradition Oresme also made significant contributions, showing that the techniques of the second tradition could be extended to cover not only rational ratios, but also all irrational ratios that could be expressed as a "ratio of ratios," that is, in modern terms, as some rational ratio raised to a power that is also a rational ratio, or $(A/B)^{m/n}$, where A, B, m, and n are all integers.

Thus Oresme made significant contributions to both traditions concerning the compounding of ratios, but it is interesting to note that in his theoretical work he was somewhat conservative. Some earlier mathematicians, including Thābit, Campanus, and, in the fourteenth century, Richard of Wallingford, although following techniques of the first tradition, nevertheless allowed the compounding of ratios with arbitrary means, with the result that a ratio compounded or other ratios might be smaller than the ratios generating it. Oresme rejected this possibility and insisted that only ratios of greater inequality could be compounded with each other in the usual way, guaranteeing that the "sum" would always be greater than the parts.[63] Perhaps it was this conservatism that allowed him to conceive of compounding so clearly as addition and to derive from this identification the far-reaching implications that he did.

In the next section it will be clear that in the seventeenth century few of the leading mathematicians scrupled about such details, most either adopting the more general form of the first tradition allowing the compounding of any ratios or else abandoning the first tradition entirely in favor of the second. Newton was drawing on the second tradition when he multiplied together the antecedents and consequents on the right-hand sides of his proportionalities in the proof of Proposition 11, Book 1, quoted earlier, though it might be noted that he multiplies together not the ratios but only their antecedents and consequents.[64] And despite its dubious theoretical basis as far as the Euclidean conception of ratio is concerned, this second tradition is closely related to the modern concept of ratio, represented by Clerke in his letters to Newton, in which ratios are not only paired with numbers "measuring their sizes," they are simply identified with these numbers.

THE HISTORY OF COMPOUNDING RATIOS FROM ORESME TO NEWTON

As far as I know very little of the later history of theories of compounding ratios has been written, particularly with regard to distinguishing what I have been calling the first and second traditions.[65] What follows is only a rough sketch of part of the terrain.

Since about the beginning of the eighteenth century, most mathematicians have belonged to the second tradition. Certainly, at least, they have compounded ratios by multiplying and not by forming a continuous series of terms and taking the ratio of the first to the last. When the first tradition has been revived or carried only, it has largely been by theoretically inclined mathematicians who, especially in connection with a study of Euclid's *Elements,* have concluded that Euclid's general theory of ratio does not justify the compounding of ratios by multiplication. One of the earliest of many modern revivers of the first tradition was Robert Simson, whose edition of the *Elements* in 1756 (not long after the general demise of the first tradition and its replacement by the second), recalled and emphasized the first tradition by rejecting definition 5 of Book VI and all that went with it. A sustained source of support for the first tradition has come from those theorists who emphasize that a general theory of ratio should cover incommensurables, a view that may be accompanied by the inference that if all ratios cannot be expressed in numbers then the compounding of ratios by multiplication is unjustified. Another view is that ratios of quantities other than numbers or lines must be covered, and another inference is that the taking of cross products – or any multiplication of terms by each other – is unjustified.[66]

Most mathematicians today consider ratios as rational numbers and, in effect, identify ratios with what medieval mathematicians called the denominations of ratios. A strong force promoting the identification of ratios with fractions was the introduction of mathematical notation. On the one hand, there was the question of how to represent compounding notationally, which we have seen in the interchanges between Newton and Gilbert Clerke. Isaac Barrow as well as Newton represented it by a plus sign as the first tradition might imply. Ignoring the first tradition, Florian Cajori speaks of this use of the plus sign for compounding ratios as a "strange misapplication" and a "curious notation," giving examples of its use not only from Barrow, but also from J. F. Lorenz's translation of Euclid's *Elements* from the Greek, published in 1824.[67] Quite appropriately, it appears in Lorenz in a first-tradition substitution for definition 5 of Book VI.

It became much more common, however, to represent compounding by whatever sign was used for multiplication. Thus William Oughtred, who introduced the × sign for multiplication, used it also for compounding.[68]

Even more basic, however, was the question of how to represent ratios and proportions themselves. In line with the first tradition, ratios should not have been represented in the same way as fractions, and proportions should not have been represented as equations, but in line with the second tradition they might be. G. W. Leibniz, then, was fully representing the second tradition when he advocated that a proportion be written: $a : b = c : d$, since, in his notation, the colon was also the sign for division.[69] This notation was adopted by the *Acta eruditorum* in 1708.[70]

In England, however, a separate notation was often used for ratios. For a while William Oughtred's symbolism introduced in his *Clavis mathematicae* in 1631 was the most common one: $a \cdot b :: c \cdot d$.[71] In adopting a special notation for ratios, Oughtred consciously distinguished between ratios on the one hand and indicated divisions or fractions on the other.[72] After 1650 Oughtred's notation was rivaled by the use of the colon: $a : b :: c : d$ introduced by the astronomer Vincent Wing in 1651.[73] At the same time the British tended to use ÷ and not a colon to represent division. In general, special symbols for ratios and proportions were retained in England and the United States much longer then elsewhere.[74]

But, of course, notation, especially in the hands of its creators, reflects concepts more than creating or influencing them. In using the equals sign in writing proportions, Leibniz did so because he thought that ratios were quantities that were in fact equated in a proportion:

> I have always disapproved of the fact that special signs are used in ratio and proportion, on the ground that for ratio the sign of division suffices and likewise for proportion the sign of equality suffices. Accordingly, I write the ratio a to b thus: $a : b$ or a/b just as is done in dividing a by b. I designate proportion, or the equality of two ratios by the equality of the two divisions or fractions. Thus when I express that the ratio a to b is the same as that of c to d, it is sufficient to write $a : b = c : d$ or $a/b = c/d$.[75]

In the Leibniz-Clarke correspondence, in arguing that situation and order can have quantity even though they are relative, Leibniz used ratios as an example of a relative thing with quantity:

> Relative things have their quantity, as well as absolute ones. For instance, ratios or proportions in mathematics, have their quantity, and are measured by logarithms, and yet they are relations.[76]

As quantities, ratios could be equal or unequal.

In his reply, Samuel Clarke argued that ratios are not quantities:

> If they were quantities, they would be the quantities of quantities; which is absurd. Also, if they were quantities, they would (like all other quantities) increase always by addition: but the addition of the proportion[77] of 1 to 1, to the proportion of 1 to 1, makes still no more than the proportion of 1 to 1; and the addition of the proportion of half to 1, to the proportion of 1 to 1, does not make the proportion of 1 and a half to 1, but the proportion only of half to 1. That which mathematicians sometimes inaccurately call the quantity of proportion is (accurately and strictly speaking), only the quantity of the relative or comparative magnitude of one thing with regard to another: and proportion is not the comparative magnitude itself, but the comparison or relation of the magnitude to another. . . . The (§ 54) logarithmic expression of a proportion, is not (as this learned author styles it) a measure, but only an artificial index or sign of proportion; 'tis not the expressing a

quantity of proportion, but barely a denoting the number of times that any proportion is repeated or complicated . . . Duplicate or triplicate proportion, does not denote a double or triple quantity of proportion, but the number of times that the proportion is repeated.[78]

In these exchanges in October 1716, we see some of the ways in which the two traditions concerning compounding of ratios has evolved. Leibniz is an heir of the second tradition, but he goes beyond it in simply identifying ratios with fractions. Clarke is an heir of the first tradition, considering ratios as relations, and considering compounding as addition, but, perhaps in reaction to Leibniz, he has a conservative version of it, denying that ratios are quantities of any sort, whereas Oresme would have admitted that they were a sort of quantity. Clarke's conservatism is also indicated by his use of the word "proportion" to mean ratio, thus adopting the medieval terminology, which, by the seventeenth century, had largely been supplanted by the use of "ratio" derived from the new translation of Euclid from the Greek.

How did the change from the first tradition concerning ratio to an even more extreme version of the second tradition come about? A reasonable guess might be that it was connected with the development of analytic geometry or more generally of analysis, and of algebra or *arithmetica speciosa,* but not enough work has been done on this subject yet to give a definitive answer. If this essay has succeeded in showing that the first tradition concerning ratio has a long and not undistinguished history, perhaps it will justify further research into its fate in the postmedieval period.

Here I can only mention a few of the mathematicians closest to Isaac Newton. A brief inspection of Descartes's *Geometrie* and of Franciscus Schooten's *Principia Matheseos Universales seu Introductio ad Geometrie Methodum Renati Des Cartes,* Schooten's edition of Descartes being among the mathematical works Newton read in 1664,[79] indicates that these works have a very different approach, seemingly much nearer to modern conceptions, than the works we have been examining – and there are no apparent traces of a struggle to attain that new approach.

The works of the British mathematicians closest to Newton have, however, much more relevant material than might have been expected given the near oblivion into which the first tradition sank. Most striking in this regard is the work of John Wallis. In his *Mathesis Universalis seu Opus Arithmeticum,* first published in 1657, Wallis says many things concerning ratios that put him squarely in the second tradition. Geometric ratios, he says, between any sorts of quantities, are all homogeneous, because the units divide out.[80] For this reason, he goes on, from the identity of ratios there arises not only equality, which is an affectation of quantity, but also similitude, which belongs more to quality than quantity.[81] The word πηλικότητα (of Book VI, def. 5 and also Book V, def. 3 of the *Elements*), he argues, means *quotient.*[82] In subsequent chapters he identifies these quotients with the denominations of the ratios, which he calls the *denominatores.*[83] These fractions are the indices (*indicia*) of the ratios.[84] And ratios, no less than numbers, are true quantities.[85] Book V of the *Elements* and the whole doctrine of ratios, he says, are more arithmetical than geometrical.[86]

Thus Wallis has a conception of ratios that emphasizes their quantitative rather than relational aspects. He distinguishes ratios from their denominations, but then he says that really there is very nearly an identity between ratios and their denominations:

But here also is clear either the identity or the greatest affinity of ratios and fractions. Since it is all the same whether we say, for example, that ¾ indicates three-quarters of one unit . . . or one quarter of three units, or the quotient of the number 3 divided by 4 . . . or, finally, the ratio of 3 to 4 or that portion of a unit which has that ratio to the whole . . . it is clear that by the same act (*opera*) both fractions, and quotients of divisions, and ratios, or at least the denominations of ratios are designated.

And whatever things should be said about the addition, subtraction, multiplication, division or even ratio, etc. of fractions, I wish exactly the same to be understood about ratios or rather about the denominations of ratios. In fact, fractions (whether proper, or improper, or even irrational) are nothing else but the denominations of ratios.[87]

Despite his near identification of ratios with fractions, Wallis recognized that the ancients had preferred to represent ratios by lines rather than by numbers. This, he said, was partly because they hardly recognized any numbers but integers (whereas in his day the boundaries of arithmetic had been expanded to include, not only integers, but also fractions and surds or irrationals), so that the ancients thought that not all ratios could be expressed by numbers. But their preference for lines was mostly, he thought, to be explained by the fact that they had not yet achieved a symbolic method or even the use of Indian numerals. Thus they had difficulty operating with numbers.[88]

On the basis of his own view of ratios, Wallis went on to complain about Gregory of Saint Vincent's *Opus geometricum* (1647):

If one paid close enough attention to Gregory of Saint Vincent, it would not be clear why he offers for sale all of this material on the ratios of ratios or on proportionalities as if it were new, nor why he seems to exhibit the comparison of unequal ratios as if it were something unknown and unheard of.[89]

Indeed, examination of the Gregory's work reveals that he has spent eighty-eight folio pages elaborating many complex cases of the ratio of ratios, when in fact he has done nothing that is unlike combinations of the usual arithmetic operations on fractions.[90] Thus Wallis is right that the whole business is superfluous, since the operations are well-known from arithmetic and there is nothing new. It is no wonder that he protests. I cannot see any explanation for what Gregory has done except that in the *first* tradition concerning ratios as elaborated by Oresme, ratios of ratios do involve something beyond simple arithmetic, and so there would be a purpose in devoting special attention to them. But since Gregory is operating in the second tradition, his whole treatment belabors the obvious.

If this were all Wallis had to say about the relationship of ratios and fractions, then the development of the second tradition concerning the ratios in his work might seem to be a smooth and natural one. But in fact there is much more, stimulated it would appear by the *De proportionibus Dialogus* of Marcus Meibom, published in 1655, a work that takes the first tradition concerning ratio to a logical extreme.

Marcus Meibom seems to have come at the question of compounding ratios from music. In 1652 he has published an edition and Latin translation of seven classical works on music,[91] and in his *De proportionibus Dialogus* he says that the misunderstanding of ratio going back to Theon results in part from a failure to study the use of ratios in music.[92] At one point Meibom suggests that ratios might be compared by reducing them to a common measure, and he suggests for this measure the ratio of 81 to 80, which, he says, is the difference between a major tone, corresponding to the ratio of 9 to 8, and a minor tone, corresponding to the ratio of 10 to 9. The difference between these two ratios will perhaps be the smallest interval or ratio of musical interest. Larger intervals or ratios can be measured by how many times they contain this ratio. He calculates and prints in a table the powers of the ratio of 81 to 80 from 1 to 112, providing a sufficient basis for comparison of any ratio up to 4 to 1, the ratio corresponding to two octaves.[93]

Thus Meibom is clearly within the first tradition for compounding ratios. When he says that the differences between 9:8 and 10:9 is 81:80, "difference" (*differentia*) must be construed as it is in the first tradition rather than in an arithmetic sense. Like Bradwardine and Oresme, Meibom treats what is in modern terms the square of a ratio as if it were double the ratio, and he argues that the word *duplam* should be used for this rather than *duplicatam*.[94] But he also goes further. He likens a ratio of equality to zero and claims that the ratio of 1 to 2 is equal to the ratio of 2 to 1 since in both cases there is the same interval between the antecedent and the consequent.[95] This makes sense because he has musical intervals in mind: If one considers the strings corresponding to notes an octave apart, one might say indifferently that they have a ratio of 1 to 2 or of 2 to 1. Either ratio corresponds to the same interval. Furthermore, Meibom claims that the same relationship of greater and less hold for ratios of lesser inequality as hold for the corresponding ratios of greater inequality, so that he contends that the ratio of 2 to 7 is greater than the ratio of 2 to 5 just as the ratio of 7 to 2 is greater than the ratio of 5 to 2.[96] When Oresme applied his theory to ratios of lesser inequality, he fell into error trying to preserve the usual relations of greater and less for corresponding fractions.[97] A better extension of the first tradition might indeed include Meibom's claims.

From the perspective of his view of ratio, Meibom claims that Gregory of Saint Vincent had 117 false conclusions – indeed, Gregory's conclusions are false if ratios of ratios are understood in Oresme's and Meibom's sense. As perpetrators of false views concerning ratio Meibom also lists Theon of Alexandria, Eutocius, Nichomachus, Heron, Cardan, Rodulpho Volumnio, Christopher Clavius, and "following Clavius the whole cohort of mathematicians up to our day."[98] This is a not unreasonable list of followers of the second tradition concerning ratio. Cardan's *De proportionibus* does advocate an extreme version of the second tradition, even admitting to some extent ratios of unlike quantities, and in his edition of Euclid Clavius does argue against the Bradwardinian view that what is in modern terms the square of a ratio should be called its double.[99] In support of the second tradition, Clavius refers to Theon, Eutocius, and

Witelo, as well as to Rodulpho Volumnio, *Disputatio de proportione proportionum.*[100] Much more reasonable than Wallis, Clavius points out that propositions using compounding of ratios will be valid even if the multiplication of the denominations of the ratios together does not produce the denomination of the compound.[101] This seems to be a concession to the validity of the first tradition at least in certain respects.

Thus Meibom suggests an extended version of the first tradition concerning ratios. One might criticize him on the grounds that he is too long-winded and overestimates his own originality, but one has the sense that he feels he is fighting determined opponents. A valuable aspect of Meibom's *Dialogus* is his printing of extended excerpts in parallel columns of Latin and Greek of a wide selection of relevant passages from Greek mathematical works including ancient scholia. He claims that his views are supported in part, but not entirely, by the ancient works. He does not reject definition 5 of Book VI of Euclid's *Elements* – the definition usually used as a basis for the interpretation of compounding as multiplication – but he tries to reinterpret it. He does reject definitions 8 and 10 of Book V of the *Elements* as false because they do not fit with his ideas about comparison of size for ratios in which the consequent is greater than the antecedent. Among the speakers in his dialogue are Euclid and Archimedes, who are portrayed as listening to and considering seriously his ideas on ratio.[102]

A point of view similar to that of Meibom was defended by Marin Mersenne in the general preface to his *Cogitata Physico-Mathematica.* At the end of Book IX of his edition of the *Elements,* Clavius had argued that the compounding of ratios was not like compounding a whole from parts, which is the business of addition, because otherwise a whole would be less than its parts when ratios of lesser inequality are compounded. To this Mersenne had replied that in the addition of negative quantities, too, the result is less than the parts and that an analogy must be made between ratios of greater inequality and positive numbers, between a ratio of equality and nothing, or zero, and between ratios of lesser inequality and negative numbers.[103]

Barrow felt that the style of Meibom's work was not a little immodest.[104] If Wallis had this same reaction, this may help to explain the startling acrimony of his response. In the same year as his *Mathesis Universalis* (1657), Wallis also published a work *Adversus M. Meibomii de Proportionibus Dialogum, Tractatus Elenchticus,* in which he castigates Meibom, going over his *Dialogus* phrase by phrase and using quite inflammatory language. Even at the time this must have seemed to be so, since Isaac Barrow refers to Wallis's work as a *diatriba.*[105] I have not observed anyone but Barrow referring to the body of Wallis's treatise, but Wallis himself saw fit to reprint it in his *Opera Mathematica* in 1695 even though at the same time he decided against reprinting his similarly violent criticism of the mathematics of Thomas Hobbes. In the *Opera* the treatise against Meibom covers sixty large pages. Perhaps tellingly, the first twenty-six pages consist of a dedication to William Brouncker in which Wallis develops some new methods for resolving cubic equations. Newton himself took notes on the dedication, so he obviously had the treatise in hand at some point.[106]

I have not found any treatment of this controversy between Wallis and Meibom in the histories of mathematics I have looked at, possibly both because the practical identification of ratios with numbers is now so complete that historians of mathematics have taken little interest in the first tradition concerning ratios, and because whatever Meibom's merits, the episode shows Wallis personally in a bad light. It is commonly conceded, however, that for all Wallis's good results in arithmetizing ratios (and geometry in general), his work lacked a solid theoretical foundation, since real numbers had not been defined.[107]

Thus Wallis was firmly within the second tradition concerning ratios, as too was his teacher William Oughtred. So far as I know Newton never gave a reason for agreeing to the changes in the *Principia* that Gilbert Clerke wanted. He might have agreed with Wallis that analysis could be pushed into new areas if ratios and fractions were identified, but he might also have acceded to Clerke's complaints simply from having observed the acrimony of Wallis's treatise against Meibom.

In contradistinction to Oughtred and Wallis, Newton's own teacher, Isaac Barrow, was more conservative. I have already indicated that it was from Barrow that Newton derived the first-tradition-linked use of the plus sign for compounding ratios. But like Newton, too, Barrow did not only compound ratios by taking the extremes of a continuous series of terms. He might also do this on the left side of an equation and compound ratios by multiplying on the right.[108] In his edition of Euclid Barrow speaks of knowing the quantity of a ratio by dividing the antecedent by the consequent.[109] Sometimes he writes a proportion as an equation $A/B = C/D$.[110] And he includes definition 5 of Book VI defining compounding in terms of the multiplication of the ratios.[111]

Barrow's *Lectiones Mathematicae* of 1666 reveal a sense of fatigue brought on by the noise of recent controversies concerning ratios, which he punningly refers to in Greek as *logomachies.* Already in 1620 Henry Savile had stated that there were two warts or blemishes on the body of geometry over which the industry of both ancient and recent authors had spent sleepless nights, the first being the parallel postulate and the second being the compounding of ratios.[112] And beyond all the authors I have referred to here, others such as Thomas Hobbes and Petrus Ramus had in the period before Barrow sought to criticize Euclid's definitions of ratio and of the equality of ratios, which often they did not understand.[113] Even Galileo had written a "fifth day" on ratios, which was published after his death.[114] In Barrow's view, both Meibom and Gregory of Saint Vincent had written enormous works on ratio while adding very little.[115]

The root of all the trouble, Barrow concluded, was that all of the various authors of different persuasions who sought to make innovations concerning ratio – he refers to Gregory of Saint Vincent, Taquetus, Borellus, Mersenne, Meibom, and others – treated ratios as quantities, whereas in fact they were not quantities, but relations, with terms implying quantity being used by ratios only by metonomy, applying to ratios attributes that are true only of their denominations.[116] On the basis of this conservative view, Barrow rejected the ideas, espoused by Wallis, that ratios are numbers and that ratios have more to do with arithmetic than geometry.[117]

He then drew several conclusions. To the question whether compounding should more correctly be considered addition or multiplication, he replied that in fact compounding corresponded to neither operation. As far as common usage was concerned, however, since in compounding one multiplied denominations, he concluded that it was better to speak of multiplying than adding.[118]

To the controversy connected with the claim of some, including Meibom and Mersenne, that the ratio of equality could be considered as nothing, while ratios of greater inequality rise above zero (*nihilum*) and ratios of lesser inequality are below zero, Barrow replied that since no ratios are quantities the whole controversy had no basis. If, on the other hand, one looked at the denominations of ratios, it was clear that they were all positive, a ratio of equality being represented by the number 1 and ratios of lesser inequality being represented by fractions between zero and 1.[119]

Barrow simply rejected Meibom's claim that ratios of greater and lesser inequality between the same terms are equal because in each case the same interval is involved, rejecting also Meibom's view of how the size of proportions of lesser inequality should be determined.[120]

To the whole "logomachy" concerning the use of the terms *dupla* or *duplicata*, and so on, for what is, in the modern sense, squaring, he replied that it is obvious that ratios that are *duplicata* are not in general double in their base,[121] thus agreeing with Clavius and Wallis and a host of modern authors against Bradwardine and Oresme.

Thus Barrow, like Newton later, acceded to most of the terminological preferences of Wallis and the advocates of the second tradition, despite the fact that he did not accept the view, from which Wallis was only infinitesimally distant, that ratios are, after all, really numbers.[122] Private views on the nature of ratio came to mean very little when the terminology and notation adopted implied arithmetization, although in the succeeding decades the first tradition concerning ratio now and again showed its head. Perhaps the tradition would have suffered a better fate if its advocates had not attempted to save it by saying that ratios were not quantities – like Oresme, they might have asserted that ratios were quantities but of a different sort than fractions.

Let me end with one last example of the first tradition concerning ratio, appropriately coming from Roger Cotes, and written at the same time he was seeing the second edition of the *Principia* through the press, appearing in 1714, the year after the publication of the latter volume. Cotes had sent the paper, which was entitled "Logometria," meaning the measurement of ratios, to Newton for his comments in 1712, and in submitting it to Edmund Halley as secretary of the Royal Society, he said he did so with Newton's encouragement.[123]

The paper is known for the solutions Cotes was able to find using his measures of ratios, but the point I want to make here involves only the introductory conceptual basis Cotes gave for logarithms, something in which he may not have been original, although it was an entirely different basis than that given by Napier in his introduction of logarithms a hundred years before.[124]

Cotes began his paper by defining measures of ratios as quantities of any genus whatsoever, of which the magnitudes were proportional to the magnitudes of the ratios, and which would be equal when the ratios were the same, double when the ratios were duplicated, subdoubled when the ratios were subduplicated, and so forth.[125] These measures, he claimed, were such that however a ratio was augmented or decreased by composition or resolution, the measure would be similarly augmented or decreased.[126] A ratio of equality, he said, had no magnitude because when it was added or subtracted it induced no change, whereas ratios of greater and lesser inequality had contrary magnitudes: If the former was positive, then the latter would be negative.[127]

Thus Cotes began from something quite similar to the ideas of Meibom and Mersenne, described previously, and something related to the ideas of Bradwardine and Oresme on ratios, but he went on to make this the basis for a system of logarithms. The essential steps in constructing this system were as follows. First, one took a series of numbers and considered them not as numbers but as ratios of the same number to unity. Second, one adopted a convenient *modulus* (a word Cotes introduced) for the system, and calculated what each of the original series of ratios equaled in terms of some power of this modulus. These powers would be the logarithms. Third, if one wanted to "add" the original ratios in the sense of the first tradition, one added their logarithms; to "subtract" the ratios one subtracted the logarithms; to "multiply" the ratios one multiplied the logarithms, and to "divide" one divided.[128]

Thus from the perspective of the first tradition concerning ratios, logarithms could be considered as alternate measures of ratios with the advantage over the time-worn denominations that when ratios were "added" or "subtracted" or "multiplied" or "divided" in the sense of the first tradition, these measures were also added and subtracted, and so forth. Even Oresme, who developed the first tradition of ratios a long way, when confronted with an algorism for ratios had to resort to the multiplication of their denominations. But logarithms provided an algorism for operations on ratios, which used in practice the same operation that theory said was involved. To anyone who worked within the terminology of the first tradition and who converted his ratios to powers of a single modulus and observed the behaviors of the exponents under the usual operations on ratios, the step to the theory of logarithms would have been a small one indeed.[129] Oresme did not convert all his ratios to a single modulus. But beyond that, the real achievement in producing logarithms lay not in conceiving their theoretical basis, but in finding practical ways to calculate, for a series of numbers, what the logarithms should be.

CONCLUDING REMARKS

The preceding discussion is only an introduction to the history of ratios and of compounding ratios from Euclid to Newton. Many authors whose ideas were important to this history have been mentioned only in passing while others have been mentioned not at all. I am not sure that I have mapped correctly even the major steps in this history, though I hope I have argued successfully for the existence of two distinct, though very often not separate, traditions. Islamic authors were clearly more important to this history than my treatment of them would indicate.

I have a sense that some historians of mathematics may consider this history of ratios to be not very important, first because it does not involve the discovery of new results, and, second, because the modern identification of ratios and fractions is so complete that questions about the nature of ratio may seem uninteresting. Some might support this second view on the grounds that mathematicians should not argue over definitions since they are, at root, arbitrary and conventional.

Third, the seventeenth-century struggle concerning the nature of ratio was an embarrassingly vituperative one, in the face of which historians may wish politely to avert their eyes.

Nevertheless, in this case, mathematicians did argue over definitions, and so historians who pass over their arguments in silence ignore part of the historical record. Furthermore, even if mathematical definitions are arbitrary and conventional, mathematical communities as groups do often adopt certain definitions and neglect others already proposed, so that the choice between mathematical definitions does affect the course the history of mathematics takes. One of Wallis's arguments against Meibom's claim that a long list of ancient and modern mathematicians had been in error about ratios was that, if it was an error, it was a common error, and in law a common error makes law (*communis error facit jus*), particularly concerning the meanings of terms and forms of speech.[130]

And so, finally, although this history may not involve the discovery of new results, it provides good evidence concerning the dynamics of scientific change in at least this period, helping to explain why some ideas come to dominate over others when there is not a simple matter of truth or falsity involved, but rather a matter of style, or of the varying weights to be attached to different criteria of mathematical merit such as theoretical soundness or practical use. The evidence I have provided here is too sketchy to provide a full understanding of why things took the turn they did, but the amount of available evidence that I have not been able to incorporate into this preliminary survey leads me to believe that further study of the question would be rewarded.

NOTES

1 *Isaac Newton's Philosophiae Naturalis Principia Mathematica: Third Edition (1726) with Variant Readings,* assembled and edited by Alexandre Koyré and I. Bernard Cohen with the assistance of Anne Whitman, 2 vols. (Cambridge, Mass.: Harvard University Press, 1972).

2 H. W. Turnbull (ed.), *The Correspondence of Isaac Newton,* vol. 2, 1676–1687 (Cambridge: Published for the Royal Society at the University Press, 1960), pp. 485–6.

3 I. Bernard Cohen, *Introduction to Newton's Principia* (Cambridge, Mass.: Harvard University Press, 1971), p. 158.

4 The passage occurs in Book I, prop. 17. Koyré and Cohen (eds.), *Newton's Principia,* vol. 1, p. 132. By continued equation I mean something of the form $A = B = C = D$, etc.

5 Cf. Turnbull, *Correspondence of Newton,* vol. 2, p. 488.

6 Ibid., p. 487

7 In the medieval theory of ratio, which used very little notation, principles were laid out for giving ratios names. These were the medieval names of some of the simpler ratios.

8 *Duplicata* and *triplicata* appear in Latin translations of Euclid's *Elements,* Book V, definitions 9 and 10, which, in modern terms, deal with the square and cube of a ratio. Advocates of a distinctive terminology for powers of ratios could appeal to Euclid in support of their view.

9 Turnbull, *Correspondence of Newton,* vol. 2, letter 312 (September 26, 1687), p. 485.

10 Ibid., letter 313 (September ?, 1687), p. 487.

11 Ibid., letter 314 (October 3, 1687), p. 489. In this passage A stands for a ratio, Aq stands for the square of the ratio, and Ac stands for the cube of the ratio. The square root sign is familiar, but here and elsewhere it must be understood as applying to the whole following expression. Thus $\sqrt{Ac}$ stands for $A^{3/2}$. A cube root would be written $\sqrt{cA}$. This is standard notation derived from William Oughtred and common to Clerke, Newton, and Wallis, although the preferable exponential notation of Descartes was sometimes adopted.

12 Ibid., p. 490.

13 Ibid., letter 315 (November 7, 1687), p. 491. Clerke quoting Newton. Newton also evidently explained the addition of ratios "in lines" (ibid.), which was commonly done in the medieval tradition represented by Boethius and Bradwardine and which made the notion of compounding as addition seem natural, as will be seen later.

14 Ibid.

15 Ibid.

16 Ibid.

17 Ibid.

18 Koyré and Cohen (eds.), *Newton's Principia,* vol. 1, p. 85. "Rationem vero sesquiplicatam voco triplicatae subduplicatam, quae nempe ex simplici & subduplicata componitur, quamque alias sesquialteram dicunt."

19 Koyré and Cohen (eds.), *Newton's Principia,* (Book I, sect. 2, prop. 4, 11, 17–19), vol. 1, p. 99: "p. 42, 55, 56, 57, 58 & alibi pro dimidiata ratione scribe subduplicata ratione." David Gregory noted that Newton had made this change in May 1694. See Cohen, *Introduction,* pp. xxii–iv, 192.

20 Koyré and Cohen (eds.), *Newton's Principia,* vol. 1, p. 39. "Aer duplo densior in duplo spatio quadruplus est" was changed to "Aer densitate duplicata, in spatio etiam duplicato, fit quadruplus."

21 Cf. Koyré and Cohen (eds.), *Newton's Principia,* vol. 1, p. xxiv, for interchangeable uses of *sesquialtera* and *sesquiplicata.* In revisions of the second edition Newton still used *dupla* and *duplicata* both to mean double or doubled. See Koyré and Cohen, *Newton's Principia,* vol. 2, pp. 792–3.

22 See D. T. Whiteside (ed.), *The Mathematical Papers of Isaac Newton,* vol 7, 1691–95 (Cambridge: Cambridge University Press, 1976), part 2: "Researches in Pure Geometry and Quadrature of Curves," esp. e.g., pp. 406–9 and 470–4. Newton often used the Greek "application of areas" where a modern mathematician would simply divide. See, e.g., Koyré and Cohen (eds.), *Newton's Principia,* vol. 1, Book I, sect. 2, prop. 4, and its corollaries and variants, pp. 96–101. Interestingly, despite using this obviously geometrical terminology, Newton called not it, but the language of compounding ratios "geometri-

cal," saying, for example, "*Corol.* 1. Igitur cum arcus illi sint ut velocitates corporum, vires centripetae sunt ut velocitatum quadrata applicata ad radios circulorum: hoc est, ut cum Geometris loquar, vires centripetae sunt in ratione composita ex duplicata ratione velocitatum directe, & ratione simplici radiorum inverse" (p. 98, variant from $E_1 iiE_2$). In this quotation and many other places Newton reveals a transitional position between expressing such relationships as proportionalities and expressing them as equations – he is moving toward expressing them as equations, but he constantly slips back into the terminology of proportionalities, speaking not of one velocity equal to the product or quotient of other quantities, but of the ratio of velocities following a ratio compounded of other ratios. Fascinating in this regard is the scholium he added to Book I, sect. 1, lemma 10. Ibid., p. 83.

23 Isaac Barrow, *Lectiones opticae et geometricae,* (London, 1674), p. a2 verso (facing p. 1): "Brevitatis gratia notae quaedam adhibentur, quarum hic subjungitur interpretatio. $A + B$. hoc est A & B simul acceptae; $A \times B$. A multiplicata, vel ducta in B. $A/B \cdot A$ divisa per B, vel applicata ad B. $A \cdot B :: C \cdot D$. A ad B eandem rationem habet, quam C ad D . . . $A \cdot B + C \cdot D \equiv M \cdot N$. Rationes A ad B & C ad D compositae adequant, excedunt, deficiunt a ratione M and N." See also Florian Cajori, *A History of Mathematical Notations,* vol. 1 (Chicago: Open Court Publishing, 1928), pp. 248–9.

24 Koyré and Cohen (eds.), *Newton's Principia,* vol. 1, p. 119.

25 Cf. Cohen, *Introduction,* p. 161.

26 That this is what he did will be argued further later.

27 Turnbull, *Correspondence of Newton,* vol. 2, letter 315 (November 7, 1687), p. 492.

28 Cf. Koyré and Cohen (eds.), *Newton's Principia* (Book I, sect. 2, prop. 10), vol. 1, p. 115.

29 Turnbull, *Correspondence of Newton,* vol. 3, letter 353 (March 1689–90), esp. p. 75.

30 Clerke as quoted earlier (Turnbull, *Correspondence of Newton,* vol. 2, p. 492). See also ibid., p. 495, and Cohen, *Introduction,* p. 161.

31 Thomas Heath (ed.), *The Thirteen Books of Euclid's Elements,* 2 ed., vol. 2 (Cambridge: Cambridge University Press, 1926; reprint, New York: Dover, 1956), p. 114.

32 Ibid., p. 115.

33 Ibid., p. 136.

34 Ibid.

35 Ibid., p. 248.

36 Godofredus Friedlein (ed.), *Anicii Manlii Torquati Severini Boetii, De Institutione Arithmetica, Libri Duo; De Institutione Musica, Libri quinque* (Leipzig: Teubner, 1867), p. 84. For some reason, Friedlein has II III III, which is an obvious mistake.

37 Ibid., pp. 84–5.

38 Ibid., p. 241.

39 See, e.g., Anneliese Maier, *Die Vorläufer Galileis im 14. Jahrhundert.* Studien zur Naturphilosophie der Spätscholastik, vol. 1, 2d ed. (Rome: Edizioni di Storia e Letteratura, 1966), pp. 89–95.

40 The ground-breaking papers on this subject are those of A. G. Molland, "The Geometrical Background to the 'Merton School,'" *British Journal for the History of Science, 4* (1968), pp. 115–21, and "An Examination of Bradwardine's Geometry," *Archive for History of Exact Sciences, 19* (1978), pp. 150–60. See also John Murdoch and Edith Sylla, "The Science of Motion," in David Lindberg (ed.), *Science in the Middle Ages* (Chicago: The University of Chicago Press, 1978), pp. 224–6.

41 H. Lamar Crosby, Jr. (ed.), *Thomas of Bradwardine. His Tractatus de Proportionibus. Its Significance for the Development of Mathematical Physics.* (Madison: University of Wisconsin Press, 1961), p. 78. I have made my own translations of Bradwardine to avoid Crosby's modernizations of terminology. In the Middle Ages the term *proportio* was used to mean ratio and the term *proportionalitas* was used to mean proportion in the modern sense. Normally, in discussing the medieval theories I prefer to use anglicized versions of the medieval terminology, and perhaps some insight into the fate of the various traditions concerning ratio might be gained by paying scrupulous attention to this terminology, but in this essay, which covers also the later period when *ratio* was used, it seemed too complicated to use this variant terminology in some places but not in others.

42 Ibid., p. 80.

43 Ibid., p. 82.

44 "Magnitudes are said to *have a ratio* to one another which are capable, when multiplied, of exceeding one another." Heath (ed.), *Euclid's Elements,* vol. 2, p. 114. It should be noted, however, that in some medieval translations of Euclid a different definition of continuous ratios was substituted for this one. See John Murdoch, "The Medieval Language of Proportions," in A. C. Crombie (ed.), *Scientific Change* (London: Heinemann, 1963), pp. 240–51.

45 I cannot here do justice to Oresme's very important work on ratios, but for fuller details see Edward Grant (ed.), *Nicole Oresme De proportionibus proportionum and Ad pauca respicientes* (Madison: University of Wisconsin Press, 1966).

46 See Edward Grant, "Part I of Nicole Oresme's Algorismus proportionum," *Isis, 56* (1965), pp. 335–41.

47 Heath (ed.), *Euclid's Elements,* vol. 2, pp. 189–90. Cf. pp. 132–3.

48 Ibid., pp. 189–90. The manuscript in question is P (= Vat. MS 190, 4to) labeled after Peyrard, who recognized it as representing a pre-Theonine text. The scribe of this manuscript had two earlier versions before him and consistently preferred the older, presumably pre-Theonine one. So the marginal presence of the definition only means that the later version had it.

49 Ibid. cf. Apollonius, *Conics,* I, props. 39, 40; III, props. 54, 66; Archimedes, *On the Sphere and Cylinder* II, prop. 4.

50 Heath (ed.), *Euclid's Elements,* p. 190.

51 Robert Simson, *Euclidis Elementorum Libri Priores sex item undecimus et duodecimus ex versione latina Federici Commandini* (Glasgow, 1756), pp. 372–7. Cf. Ioannes Guilelmus Camerer, *Euclidis Elementorum Libri Sex Priores* (Berlin, 1825), vol. 2, pp. 367–79.

52 Cf. Johannes Tropfke, *Geschichte der Elementar-Mathematik,* vol. 3, *Proportionen, Gleichungen,* 2d ed. (Berlin: Walter de Gruyter, 1922), p. 14.

53 Sister Walter Reginald Schrader, O.P., "The Epistola de proportione et proportionalitate of Ametus filius Josephi" (Ph.D. diss., University of Wisconsin, 1961).

54 A. Björnbo, H. Suter, and K. Kohl, "Thabits Werk über den Transversalensatz," *Abhandlungen zur Geschichte der Naturwissenschaften und der Medizin,* Book VII (Erlangen, 1924).

55 H. L. L. Busard, "Die Tractate *De proportionibus* von Jordanus Nemorarius und Campanus," *Centaurus, 15* (1970), pp. 193–227.

56 R. Steele (ed.), *Communia Mathematica Fratris Rogeri Partes Prima et Secunda, Opera hactenus inedita Rogeri Baconi,* fasc. 16 (Oxford: Clarendon Press, 1940).

57 John D. North, *Richard of Wallingford. An Edition of his Writings,* 3 vols. (Oxford: Oxford University Press [Clarendon Press], 1976), esp. vol. 1, pp. 59–94; vol. 2, pp. 53–65. Despite the fact that Richard incorporated Campanus's work into his own essentially verbatim, there are noteworthy additions, and North's commentary is very useful.

58 Busard, "Die Tractate De proportionibus," pp. 205–6. My translation.

59 Ibid., p. 206. As far as I can tell, such a close analogy was drawn between multiplying two numbers together and "drawing" one line into another to produce a rectangle that it is rarely necessary to distinguish *ducere* from *producere.* Nevertheless, I have indicated where the more obviously geometrical word is used.

60 Ibid., pp. 207–8.

61 Edward Grant, "Oresme's Algorismus proportionum," pp. 335–41. The Latin text is in Maximilian Curtze, *Der Algorismus Proportionum des Nicolaus Oresme; zum ersten Male nach der Lesart der Handschrift R. 4°. 2. der königlichen Gymnasialbibliothek zu Thorn* (Berlin, 1868). Edward Grant, "The Mathematical Theory of Proportionality of Nicole Oresme" (Ph.D. diss., University of Wisconsin, 1957), edits Part I from thirteen manuscripts, pp. 331–9.

62 Edward Grant, *Nicole Oresme De proportionibus proportionum,* pp. 142–3, 152–3.

63 Ibid., pp. 138–9, 150–3, 310–12, 317–22. He then went on to allow the compounding of ratios of lesser inequality with each other, but insisted erroneously that, given three proportionals *A, B, C,* in order of decreasing size, then the ratio of *B* to *A* would equal the ratio of *C* to *A duplicata.* Ibid., pp. 150–51.

64 It might be noted also that something similar holds for Euclid. See Heath (ed.), *Euclid's Elements,* vol. 2, p. 248, note to lines 1, 6, 19, 36, and p. 354, note to lines 1, 5, 29, 31.

65 North, *Richard of Wallingford,* vol. 2, p. 57.

66 See, for example, Rev. Baden-Powell, *On the Theory of Ratio and Proportion as Treated by Euclid, Including an Inquiry into the Nature of Quantity* (Oxford; The Ashmolean Society, 1836).

67 Cajori, *A History of Mathematical Notations,* vol. 1, pp. 248–50.

68 William Oughtred, *Clavis mathematicae,* 3d ed. (Oxford, 1952), p. 18: "7 · 9 :: x{7 · *A* : *A* · 9."

69 Cajori, *A History of Mathematical Notations,* vol. 1, pp. 271, 295–9.

70 Ibid., pp. 219, 295.

71 Ibid., pp. 275, 285–92.

72 Ibid., p. 277.

73 Ibid., p. 275.

74 Ibid., pp. 293–97.

75 G. W. Leibniz, *Mathesis Universalis pars prior, de Terminis incomplexis,* reprinted in *Gesammelte Werke,* 3. Folge, II^3, Band VII (Halle, 1863), p. 56, as quoted by Cajori, *A History of Mathematical Notations,* p. 295.

76 H. G. Alexander (ed.), *The Leibniz-Clarke Correspondence* (Manchester: Manchester University Press, 1956).

77 Throughout this whole passage "proportion" means ratio. Clarke is using the medieval terminology.

78 Alexander, *The Leibniz-Clarke Correspondence,* pp. 105–6.

79 I. B. Cohen, "Newton, Isaac," *Dictionary of Scientific Biography,* vol. 10 (New York: Scribner, 1974), p. 45.

80 John Wallis, *Opera Mathematica* (Oxford, 1695), vol. 1, p. 136.

81 Ibid., p. 137. From the point of view of the first tradition, similitude is assumed, but it is equality that might be questioned.

82 Ibid.

83 Ibid., p. 153.

84 Ibid., p. 154.

85 Ibid.

86 Ibid., p. 183.

87 Ibid., p. 212. "Sed & hinc etiam patet, Rationum et Fractionum, sive identitas, sive affinitas maxima . . . manifestum est eadem opera tum Fractionem, tum divisionis Quotientem, Denominatorem, designari.

Adeoque, quae de Fractionum Additione, Subductione, Multiplicatione, Divisione, aut etiam Proportione, &c. dicenda erunt: Eadem etiam de Rationibus, sive Rationum potius Denominatoribus, intelligenda vellim. Quippe nil aliud sunt fractiones (sive propriae sive impropriae, sive etiam irrationalis) quam Rationum Denominatores."

88 Ibid., p. 183.

89 Ibid., p. 212.

90 Gregory of Saint Vincent, *Opus geometricum quadraturae circuli et sectionum coni* (Antwerp, 1647).

91 Marcus Meibom, *Antiquae Musicae Auctores Septem.* (Amsterdam: Ludovicum Elzevirium, 1652).

92 Marcus Meibom, *De Proportionibus Dialogus* (Copenhagen: Melchior Martzanus, 1655), p. 101. "Sed, ut vides, nec Theo hoc intellexit, nec ex posterioribus mathematicis, ad nostra usque tempora, ullus. Illos autem errores, ut dicere solet Euthymius, ob neglectum Musices studium quae in rerum omnium rationibus ac proportionibus investigandis occupatur, mathematicis Deus immisit; . . . In naturalibus autem vix ulla evidentiora exempla reperiemus omnium quae in rationum doctrina disquiruntur, quam quae Musica suppeditat."

93 Ibid., pp. 70–1 (table) and 76–7. "Nunc alium tradam, quo rationum magnitudinem evidenter explorabimus. Scilicet minuta mensura adhibenda est, qua illas mensuremus. Haec mensura, rationibus adplicanda, nonnisi ratio esse potest. Uti enim longitudines mensuramus longitudinibus; superficies, superficiebus; corpora, corporibus; sic rationes, rationibus. Ratio quippe rationem superat, non numero, aut alia quacunque magnitudine, sed ratione. Mensuram autem, vulgarem usum secutus, adsumsi satis minutam, rationem $\frac{81}{80}$, qua rationem $\frac{9}{8}$, in qua tonus major spectatur, superare vidimus rationem $\frac{10}{9}$, in qua consideratur tonus minor, quam excessivam rationem $\frac{81}{80}$, antiquo vocabulo, comma vocant. Intervalla igitur Harmonica, seu rationes, in quibus illa intervalla spectantur, & universim omnes rationes, ipsa mensurali ratione $\frac{81}{80}$ non minores, hac metimur."

94 Ibid., pp. 172–3. Wallis, *Opera Mathematica,* vol. 1, pp. 194–5.

95 Ibid., p. 104, and Wallis, *Opera Mathematica,* vol. 1, p. 274.

96 Ibid., pp. 131–4, and compare Isaac Barrow, *Lectiones Mathematicae,* Lect. XVIII (1666), in William Whewell (ed.), *The Mathematical Works* (Cambridge: Cambridge University Press, 1860), p. 328.

97 Grant, *Nicole Oresme De proportionibus proportionum,* pp. 150–3, 320–4.

98 Meibom, *Dialogus,* pp. 165–7. Wallis, *Opera Mathematica,* vol. 1, p. 272.

99 Jerome Cardan, *Opus novum de proportionibus* (Basel, 1570), pp. 2–6. Christopher Clavius, *Opera Mathematica* (Mainz, 1611), vol. 1, pp. 216–18.

100 Clavius, *Opera Mathematica,* vol. 1, pp. 218, 243–5. For Witelo, see Sabetai Unguru, *Witelonis Perspectiviae Liber Primus. Book I of Witelo's Perspectiva. An English Translation with Introduction and Commentary and Latin edition of the Mathematical Book of Witelo's Perspectiva.* Studia Copernicana XV. (Wroclaw: Ossolineum, 1977), pp. 47, 54–5, 168–9, 173–4, 215, 221–2.

101 Clavius, *Opera Mathematica,* vol. 1, p. 245.

102 The participants in the dialogue are Euclid, Archimedes, Apollonius of Perga, Pappus, Eutocius, Theon, and Hermotimus, the first six of whom represent the classical mathematicians, while Hermotimus presents Meibom's views, which are ascribed to one Euthymius. Euclid says to Archimedes, p. 4, "Nova quaedam de Proportionibus inventa, quae ex Euthymio se didicisse ait, nobis exponet: illa autem, sola veritatis indagandae cupiditate inductum, nobis, ut acerimis censoribus examinanda prius proponere se voluisse, quam iis adsentiret." Concerning Book VI, def. 5 of the *Elements,* see pp. 78ff. Concerning Book V, defs. 8 and 10, see pp. 126ff.

103 Clavius, *Opera Mathematica,* vol. 1, pp. 216–18. Wallis, *Opera Mathematica,* vol. 1, pp. 289–90.

104 Barrow, *Lectiones Mathematicae,* p. 327, "stylo certe nimis quam inverecundo."

105 Ibid., p. 293.

106 D. T. Whiteside (ed.), *The Mathematical Papers of Isaac Newton,* vol. 1, 1664–66 (Cambridge: Cambridge University Press, 1967), pp. 119–21.

107 Cf. Carl B. Boyer, *A History of Mathematics* (New York: Wiley, 1968), p. 416.

108 Isaac Barrow, *Lectiones opticae et geometricae,* pp. 50–51. See also Cajori, *A History of Mathematical Notations,* vol. 1, p. 249, and n. 23.

109 *Euclidis Elementorum Libri XV breviter demonstrati, opera Is. Barrow* (London, 1678), p. 92.

110 Ibid., p. 93.

111 Ibid., p. 116.

112 Sir Henry Savile, *Praelectiones tresdecim in principium Elementorum Euclidis, Oxonii habitae MDCXX* (Oxford, 1621), lect. VII, p. 140. "In pulcherrimo Geometriae corpore duo sunt naevi, duae labes, nec quod sciem plures, in quibus eluendis et emaculandis cum veterum tum recentiorum ut postea ostendam vigilaverit industria. Prior est hoc postulatum [Post. 5, I], posterior pertinet ad compositionem rationum, scilicet *A* ad *B* componi ex ratione *A* ad quodcunque extrinsecus sumptum, verbi gratia *C,* et illius extrinsecus sumpti ad *B.*"

113 Petrus Ramus, *Scholae mathematicae* (Frankfurt, 1559). For reference to some of Hobbes's works, see Joseph F. Scott, *The Mathematical Work of John Wallis* (London: Taylor and Francis, 1938), ch. X, pp. 166–72.

114 In Vincenzo Viviani, *Quintro libro degli Elementi d'Euclide* (Florence, 1674).

115 Barrow, *Lectiones mathematicae,* p. 316. "Integrum quinquennium impendisse se profitetur M. Meibomius huic speculationi, neque praeter leviculos quosdam criticismos sani quicquam aut solidi videtur elephantinus iste partus in lucem protulisse. Diutius, opinior, et gravius eidem incubuit (an fere succubuit dicam?) maximus vir, et recentium Geometrarum nulli posthabendus Gregorius Vincentius, attamen ut rem meo judicio reliquerit, haud minus obscuram quam invenerit, fusissime licet et elaboratissime pertractam. Quid igitur a paucularum horarum studio, quid (ut caetera taceam) ab hac extemporanea pene scriptione circa materiam ejusmodi contumaciter perplexam merito possit expectari?"

116 Ibid., pp. 317–18.

117 Ibid., p. 322.

118 Ibid., p. 326.

119 Ibid., p. 327.

120 Ibid., pp. 327–8.

121 Ibid., p. 328.

122 Cf. ibid., pp. 292–3.

123 J. Edelston (ed.), *Correspondence of Sir Isaac Newton and Professor Cotes . . .* (London, 1850), p. 116; *Philosophical Transactions, XXIX* (1714), p. 5.

124 Napier had generated his logarithms by two motions, one with constant velocity and one in decreasing proportion. Ioanne Nepero, *Mirifici Logarithmorum Canonis descriptio* (Edinburgh, 1614). He had no concept of a base or modulus for logarithms. Boyer, *A History of Mathematics*, p. 343.

125 Roger Cotes, "Logometria," *Philosophical Transactions, 29* (1714), p. 5.

126 Ibid., pp. 5–6.

127 Ibid., p. 6.

128 Ibid., pp. 6–7. Cotes says, for example, "Datis enim duobus quibuscumque numeris in se multiplicandis, si quaeratur numerus ex multiplicatione productus; quoniam rationes numerorum datorum ad unitatem conficiunt simul *additae* rationem producti ad unitatem, & rationum componendarum mensurae simul additae conficiunt rationis compositae mensuram: Logarithmus producti aequabitur Logarithmis numerorum datorum simul sumptis" (my italics).

129 Before inventing logarithms Napier had been thinking of the sequences of the successive powers of a given number. Boyer, *A History of Mathematics*, p. 342.

130 Wallis, *Opera Mathematica*, vol. 1, p. 272.

2

Atomism and motion in the fourteenth century

JOHN E. MURDOCH
Harvard University

I

We are now aware that there was considerable opposition in the fourteenth century to any number of Aristotelian contentions in natural philosophy, even if, in face of that opposition, natural philosophy then remained basically Aristotelian as a whole. Still, many of these non-Aristotelian thrusts and parries were features characteristic of the fourteenth century and should therefore be examined in some depth if we are to gain a proper picture of late medieval thought.

Measured in terms of the amount of critical literature (together, of course, with that belonging to the Aristotelian counterattack), one of the most important areas of fourteenth-century controversy was undoubtedly that concerned with the problems of continua, especially of the correct view of their composition. Here renegades who disagreed with the Aristotelian orthodoxy and maintained the composition of continua out of indivisibles, out of extensionless atoms, were, it seems, more numerous than many others who were non-Aristotelian in some more incidental way. What is more, they were considered particularly worthy of attention, at least if one judges from the amount of criticism they received from keepers of the standard view of the issues and conceptions relevant to the analysis of continua.

It is not yet clear precisely why there arose this somewhat untidy band of indivisibilists at the beginning of the fourteenth century. There is, admittedly, some indication that one of the reasons may have been the attempt to provide a satisfactory account of angelic motion (since angels are nonextended immaterial substances who move from place to place). Alternatively, some may have been led to embrace indivisibilism as a consequence of new views of the infinite that derived, in one way or another, from the rather furious late thirteenth- and early fourteenth-century examination of the possibility of an eternal world. In any event, almost all of the fourteenth-century literature of an indivisibilist persuasion seems to have flourished with precious little attention paid to the motives for it. Indeed, in terms of present evidence, the question of such motives provides something like a field day for historical speculation.

To be sure, almost all fourteenth-century scholars who touched natural philosophy (whether they did so in works of that genre or in theological works, no matter) necessarily treated the problems of continuity. But a certain select group formed the core of the controversy over these problems. Here the "author and place map" of the controversy yields indivisibilists both English and Continental: Henry of Harclay and Walter Chatton being the most notable in England, with Gerard of Odo followed by Nicolas Bonetus and (an otherwise unknown) John Gedo, on the Continent. There is a similar geographic split among the critics of these indivisibilists: William of Alnwick first draws fire on Harclay, while William of Ockham does the same (but much less extensively) relative to Chatton. Next, Adam Wodeham critically examines both Harclay and Chatton, although his criticism of the former is largely parasitic on Alnwick's earlier treatment. Thomas Bradwardine's *Tractatus de continuo* is also critical of these two English indivisibilists, but in a quite different way, ignoring their specific arguments and instead basing his attack upon several isolated features that were absolutely essential to their positions. On the Continent, the most sustained criticism of Odo's views was carried out by John the Canon, although Walter Burley also expressed specific disagreement with some of his contentions. Slightly later, the indivisibilism of Crathorn and Nicholas Autrecourt does not seem to have drawn critical notice, at least not by name. John Wyclif did have later English critics as well as supporters, but the most notable fact about his indivisibilism was its condemnation at the Council of Constance early in the fifteenth century. Undoubtedly a case of guilt by association (how else explained the heresy of composing a line of points?), this seems to have been the only official condemnation medieval atomism every received.

Notice should be taken, however, not simply of the identity of these protagonists and their locales, but of the kinds of works that provided the sources for this fourteenth-century continuum controversy. Save for the case of John the Canon, these works were not commentaries on the *Physics,* an otherwise expected place considering that it was there that Aristotle had devoted almost all his attention to the problems at hand. Instead, we find ourselves faced with treatises on the *logica vetus* (by Bonetus and Wyclif, for example), where the treatment of continuous quantity in the *Praedicamenta* was standard. Alternatively, we have to do with tracts specifically devoted to indivisibilism and continua by Harclay, Bradwardine, Wodeham, and Gedo. Lastly, and most significantly, much of the relevant material is found in the theological context of *Sentence Commentaries:* Chatton, Odo, Crathorn, and (again) Wodeham. Finally, some note should be made of the rather considerable number of fourteenth-century Franciscans on both sides in this controversy. As far as I can see, this is circumstantial and without doctrional overtones.[1]

II

The many issues central to what might be called the medieval continuum problem are discussed, often to a rather immodest extent, by almost all the protagonists I have just named. Such issues, for example, as the existence of indivisi-

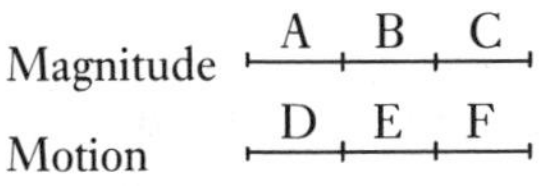

	A	B	C
Magnitude			
Motion	D	E	F

Fig. 1

bles, the relations that can obtain between them in continua, the nature of the divisibility of continua, the ascription of first and last elements within continua, and the bearing of mathematics upon one's view of the composition of continua. However, I shall limit myself to but a very small, but I think significant, part of this content and concentrate on a single problem in the relevant literature, a single problem that is closely tied to the analysis of a single argument – or, perhaps, a single species of argument – in this literature. The problem is one of motion; namely, if as an indivisibilist, one holds that both space and time are composed of indivisibles, how can one account for different speeds? The argument, or species thereof, is one that employed mobiles moving at different speeds to refute any indivisibilist conception of continua (i.e., of space, time, and motion, and latitudes of qualities as well if alterative motion be considered).

To the problem itself: first, something of its history. As one traces the atomist attempts from antiquity through the fourteenth century to account for the possibility of mobiles with different speeds, almost all of the "ways out" of the dilemma are found to have received serious attention. Not simply the expected assertion of discontinuous motion consisting of jerks, jumps, and slices (like a cinematic film), but also the postulation of such oddities as absolutely fastest and slowest motions, of a mobile moving while remaining in the same place or being in two places at the same time. All of these find appropriate indivisibilist expression.

Viewed abstractly, the challenge facing the atomist was to establish some kind of relation between his indivisibilist space and time that would allow of a variation in speed for mobiles moving over such a space in such a time. The most appropriate place to begin to see how such a challenge was met is to note an argument set forth by Aristotle in the first chapter of Book VI of the *Physics*[2] and then to examine the atomist reaction to which it gave rise. (This is not yet, it should be noted, the "model argument" whose medieval history will be documented shortly.)

The purpose of this particular argument of Aristotle was to prove that space, time, and motion cannot be composed of indivisibles. (Atomists, it might be noted, who had indivisible bodies traveling through truly continuous, nonindivisibilist, space and time would not be affected.) Consider (Fig. 1), this argument urges, a mobile moving over a magnitude *ABC* composed of indivisibles *A*, *B*, and *C*. Each corresponding part of its motion *DEF* will also be indivisible. However, the argument continues, "a thing that is in motion from one place to another cannot at the moment when it was in motion both be in motion and at the same time have completed its motion at the place to which it was in motion: for example, if a man is walking to Thebes, he cannot be walking to Thebes and at the same time to have completed his walk to Thebes." Consequently, if our mobile "actually passed through *A after* being in process of passing through, the motion must be divisible, . . . while if it is passing through and has completed its passage *at the same moment,* then that which is walking will at the moment when it is walking have completed its walk and will be in the place to which it is walking; that is to say, it will have completed its motion at the place to which it is in motion." In such a case, however, "the motion will consist not of motion but of starts and will take place by a thing's having completed a motion without being in motion" (οὐκ ἐκ κινήσεων ἀλλ' ἐκ κινημάτων καὶ τῷ κεκινῆσθαί τι μὴ κινούμενον'). Thus, one must conclude that motion, space, and time cannot be composed of indivisibles but are divisible into always further divisible parts.

The first atomist reaction to this argument of which we know is that of Epicurus, preserved for us by Themistius and Simplicius in their commentaries on the *Physics.*[3] Epicurus appealed to a remedy, Themistius tells us, that was less palatable than the sickness he was trying to cure, for he maintained that "the mobile moves over the whole segment *ABC;* but over each of the indivisible segments constituting the whole, it does not move, but rather *has* moved" (οὐ κινεῖται ἀλλὰ κεκίνηται). Motion is, then, composed of indivisibles, of "has moveds," of κεκίνηται.

Although not as a result of struggling with Aristotle's argument, a similar "has moveds only" conception of motion can be found as part of the atomism of early Islamic *kalām.* Thus, we are told that "the motion of a body consists in its mutation from one place to another; the movement is the first generation in the second place; motion exists in the second place, that is, the place toward which the mobile tends, and the first place where the motion begins is the rest."[4]

Indivisibilists in the fourteenth-century Latin West maintained much of the same, even though they may not have been quite as explicit in saying so. For they do espouse discontinuous motion, a motion that is, as often as not, made up of *mota* or *mutata esse,* clearly the Latin mates to κεκινῆσθαί, even though they have "detensed" their *mutata esse* and allow them to function as the indivisibles of motion in general without consistent and explicit reference to the completed motions of which they literally speak.

III

Let us now move beyond the indivisibilist problem of accounting for different speeds and focus on the specific argument, or species thereof, whose medieval history provides the *locus classicus* for fourteenth-century indivisibilist attempts to solve this problem. Again, the argument was devised by Aristotle: *Physics* VI, Chapter 2.[5] In terms more directly comprehensible than the Aristotelian text itself, the argument amounts to the following.

Given the fact that a mobile moving at a certain rate passes over a magnitude s_1 in time t_1, then a faster mobile can move over s_1 in some time t_2 less than t_1; hence, t_1 is divided. But in the time t_2 of the faster mobile, the slower mobile will have passed over magnitude s_2 less than s_1, thus dividing s_1. Furthermore, if we

m_s = slow mobiles	m_s traverses s_1	in t_1
m_f = fast mobiles	m_f traverses s_1	in $t_2 < t_1$
	m_s traverses $s_2 < s_1$	in t_2
	m_f traverses s_2	in $t_3 < t_2$
	m_s traverses $s_3 < s_2$	in t_3
	and so on.	

Fig. 2

ask what time it takes for the faster mobile to traverse s_2, the time will again be divided, giving t_3 less than t_2. Once more, in t_3 the slower mobile will traverse s_3 less than s_2, and so on; alternately taking faster and slower mobiles, the faster will continuously divide the time, the slower the magnitude. Therefore, time and magnitude must be continuous and infinitely divisible. The whole argument can be conveniently tabulated as in Figure 2.

The implication of Aristotle's argument was, of course, that the application of the fast and slow mobiles to any intervals of space or time held to be composed of indivisibles would divide such indivisibles since it demonstrated the infinite divisibility of all spaces and times. This was made explicit in the many medieval versions of the argument that "particularized" Aristotle's more generalized reasoning by phrasing the whole in terms of three- (or five-) indivisible space and time to which the two mobiles were applied.[6] There, the fast mobile would divide the middle instant of the three- (or five-) instant time, the slow mobile the middle indivisible of the three- (or five-) indivisible space. QED. Indivisibilism refuted.

Before we return to an examination of the assumptions involved in this argument and in Aristotle's more generalized version of it, as well as to a new medieval "two-mobile argument" that evolved from this Aristotelian base, it would be useful to say something of the fourteenth-century indivisibilists' reply to the more restricted "three-indivisible" version itself.

Overturning this restricted version was naturally particularly important to indivisibilists who like Odo, Bonetus, and Chatton, were finitists, that is, who maintained that only a *finite* number of indivisibles entered into the composition of any finite continuum.[7] Accordingly, their replies continually appeal to finitist, often rather physical, notions. Odo, for example, maintains (and Bonetus follows him idea by idea if not word for word) that the division of the middle indivisible implied by the argument is not at all a division *of* an indivisible, but rather *in* an indivisible. This sort of division *removes* the middle indivisible (thus leaving two equal halves of one indivisible each) and does not saddle one with the contradiction of a divided indivisible.[8]

Chatton, on the other hand, conflates the continuous with the discrete (a move that was rather congenial to his finitism) and claims that just as one cannot divide the number three into two equal numbers, so we should not expect to be able to divide a three-indivisible space or time into two equal halves. We can, however, perform such a division if an even number of indivisibles compose the continua in question. Further, faced with the commonly held belief that any continuous quantity (a line, for example) can indeed always be divided into equal halves, Chatton counters by saying that there the use of the term "half" might simply be based on the fact that we cannot *perceive* the excess of one part over another and hence infer that we do have equal halves.[9]

IV

Let us now direct our attention to the enabling assumptions of the three indivisible two-mobile argument and its more general Aristotelian original. Clearly both assume the infinite divisibility of space and time since they assume that one can always move faster over any given space and slower in any given time. Let us call this the "assumption of the possibility of velocity variation." Its presence did not mean, however, that these two-mobile arguments were not effective against indivisibilists or that they did not receive their serious consideration. For, by and large, indivisibilists did not wish to abandon the possibility of faster and slower motions.[10]

They also readily agreed to the other crucial assumption made by these arguments: Namely, that there is some kind of one-to-one correspondence between the parts or indivisibles of a space over which motion occurs and the parts or indivisibles of the time in which it occurs, even though their acceptance of this assumption was almost always a qualified one. The most frequently cited unqualified specific form of this (let us call it) "space-time correspondence assumption" derived from Aristotle and was almost always identified as such. In any instant of its motion, the assumption read, a mobile is in a space equal to itself and in diverse instants is in diverse spaces of such a sort.[11]

Now both of these assumptions are also the crucial ones in a variant two-mobile argument that became standard fare among fourteenth-century indivisibilists and their critics.[12] Put as concisely as possible, the new argument was the following:

1. Assuming the possibility of velocity variation,
2. Let a slow mobile move over space s in time $2t$ and a fast mobile over space $2s$ in time t.
3. Let n_t = the number of instants in time t
   n_{2t} = the number of instants in time $2t$
   n_{2s} = the number of points or "mobile-equal" spaces in $2s$
   n_s = the number of points or "mobile-equal" spaces in s.
4. There is a one-to-one correspondence between the points or "mobile-equal" spaces and instants of: (a) $2s$ and t, (b) s and $2t$.

Hence,

5. $n_t = n_{2s}$ and $n_{2t} = n_s$

But, since $2t > t$,

6. $n_{2t} > n_t$

Thus,

7. $n_s > n_{2s}$

But this means that

8. There are more points or "mobile-equal" spaces in s than in $2s$ and consequently for an indivisibilist (for whom a greater number of component indivisibles entails a greater size for that composed) $s > 2s$ and part > whole.[13]

This new argument was explicitly claimed to be derived from Aristotle's fast–slow mobile argument, and its absurd "part greater than its whole" conclusion was viewed as a refutatory alternative to the equally absurd division of an indivisible that served as conclusion in the Aristotelian original.[14] I know of no explicit statement explaining just why this new argument replaced that of Aristotle in combating indivisibilism, but I think it probable that one of the central reasons was the appearance of indivisibilists like Harclay who maintained that continua were composed of an *infinite* number of indivisibles. Before him, Robert Grosseteste had maintained that there were different infinites of indivisibles *in* (but not composing) continua of different sizes, and Harclay followed Grosseteste's idea, quoting him at length in support of his own view, even though he held, as Grosseteste did not, that these diverse infinites of indivisibles composed the continua in which they occurred.[15] Such developments were, then, likely behind the rise and eventual general acceptance of the new medieval two-mobile argument.

V

The issue to which we must now turn, however, is the way in which the major fourteenth-century indivisibilists resolved this new fast–slow mobile argument. Basically, these resolutions consisted in denying or, more often, qualifying or changing one or both of the two assumptions – that of velocity variation possibility and that of space–time correspondence – that we have noted as essential to the argument.

Thus, Gerard of Odo is quite explicit in noting the velocity-variation assumption,[16] but he then attacks the assumption if it be taken to mean that moving faster or more slowly in a given time entails that greater or less distance be traversed. Under such an interpretation, the assumption is false, because there are cases of local motion in which no change of place is involved. Gerard supports this rather mind-boggling claim by citing the motion of the indivisible centers of rotating wheels or the indivisible poles of revolving spheres. We might object that such indivisible centers or poles, being indivisible, do not move at all; after all, that is rather standard in most accounts of rotary motion. But we are countered by another feature of Gerard's indivisibilism: He maintains that extensionless indivisibles have different "sides" to them (called *differentie loci*), indivisible though they be. There is, for example, an *ante* and a *retro* to the indivisible center of a rotating wheel and these "differentiae" change position when the wheel rotates, even though this change involves no change of place. Further, this kind of local motion without change of place can occur at different speeds: The poles of the primum mobile and of the orb of the moon, for example, so move at different rates.[17] But if this is true, then the assumption of the possibility of velocity variation is false if it be taken to entail the traversal of different distances or spaces. And since it must be taken in this sense in the fast–slow mobile argument it serves, that argument is not effective. So Odo's indivisibilism stands.

Somewhat less awe inspiring is Crathorn's rejection of the velocity variation possibility assumption (even though he does not reject it as part of a refutation of our two-mobile argument). It is denied outright by one of his major conclusions, namely, that there can be no motion faster than a continuous motion, although a continuous motion can be faster than some discontinuous motion, and discontinuous motions can be faster or slower than one another.[18] This rather puzzling assertion lessens somewhat in its power of amazement when we realize that there is only one continuous motion for Crathorn. It is the fastest motion possible, one in which, and the only one in which, there is a one-to-one correspondence between spatial and temporal indivisibles (a claim that naturally implies the rejection of our argument's space–time correspondence assumption). Differences in speed, then, obtain only when discontinuous motions are at stake (a contention shared by almost all indivisibilists, but in a fashion as explicit as Crathorn's by Autrecourt and Wyclif in particular who also, incidentally, follow Crathorn's view of a fastest motion, which they specify is that of the heavens).[19] As might be expected, the discontinuity, and hence the relative variation in speed, results from the intervention of moments of rest.[20]

Moving more to the heart of the whole controversy surrounding indivisibilism and the possibility of faster and slower motions, it should be noted that the more extensive refutations of our argument at the hands of Chatton and Harclay do not have the assumption of velocity variation, but rather the space–time correspondence assumption, at their focus. Yet, before turning to this concluding part of the story, mention should be made of the fact that the nonindivisibilist William of Alnwick employed a slight variation of our argument for a quite different purpose. Not to refute indivisibilism (although Alnwick also does that with the argument), but to *establish* the overall one-to-one correspondence between the infinity of always further divisible parts in any part of a continuum with the infinity of such parts in the whole of that continuum. Thus, in place of having the slower mobile traverse half the distance traversed by the faster mobile in half the time, one simply specifies that it traverse that same half distance in the same time (thus, t replaces $2t$ in premise [2]). But this in turn means that the number of instants in time t is equal to the number of mobile-equal spaces in s (so premise [5] reads $n_t = n_s$), which then automatically leads to the desired equality between the infinity of parts in s and the infinity of parts in $2s$. (Conclusion 7 is replaced by $n_s = n_{2s}$.) QED.[21]

VI

To Walter Chatton's resolution, then, of our argument. Inasmuch as there are two separate *determinationes* – one before, the other after, Easter, he tells us – in his whole treatment of the problems of continua, there are two separate refutations to examine. Being extraordinarily diffuse, it is difficult enough to sift a measurable amount of sense out of one of them, let alone to attempt to do the same for both. But one should try.

An appropriate note on which to begin is perhaps Chatton's claim that our argument is not just against finitists like himself, but also against those who hold with an infinity of parts in continua. As such, if effective, it would establish the (Zenonian, he says) conclusion that "motion is nothing." Relative to indivisibles, however, it is simply not to the point.[22] Nevertheless, in proper scholastic fashion, he proceeds with its refutation.

Its core, indeed the central move behind his specification of just which space-time-motion correspondence relations he allows, is his appeal to his notion of *motio passiva.* Developed in response to Ockham's view that motion involves naught but the *res permanentes,* which are the mobile and (to speak only of local motion) the different places it successively occupies, Chatton's *motio passiva* is, as its name implies, a property, a *passio,* possessed by the mobile and distinct, or at least distinguishable, from the mobile itself and from the diverse locations it suffers.[23] Further, at every moment of local motion a mobile has a different *motio passiva;* they succeed one another, Chatton avers, continuously.

In the first resolution he gives of our two-mobile argument, Chatton applies his notion of *motiones passive* to the heavens; so applied, they provide him with one meaning of the term "instant." In this meaning, a *motio passiva celi is* an instant. Thus, given their connection with the uniform rotation of the heavens, the successive continuous series of these *motiones passive* serve as a kind of clock. This in hand, Chatton resolves our argument by pointing out that (1) while the mobile-equal spaces in s are equal to the *motiones passive celi* (functioning as instants) in $2t$, (2) the mobile-equal spaces in $2s$ are greater than the *motiones passive celi* in t. These two assertions specify just which type of correspondence relations are licit and the second of them replaces the equality asserted by premise (5) of our argument with an inequality, thus frustrating the inference of its absurd, indivisibilist-refuting, conclusion.[24] At the same time, Chatton also explains just how the inequality in (2) can obtain, namely, by reference to another sense of the term instant, a more familiar one identifying its referent with the occupation of a specific place (*mobile in puncto suo*) by a continuously moving mobile. Understood in this sense, a mobile – specifically a slower mobile – can "coexist" with more than one point in the same instant. That is to say, a moving body can, while moving, occupy diverse places at the same time; this occurs through its simultaneously touching these diverse places.[25]

Chatton's second resolution of our argument operates through his application of his notion of *motiones passive,* not to the heavens, but to the mobiles of the argument themselves. So applied, this gives one sense of motion, the only proper one in Chatton's eyes, since the opposing sense is the rejected Ockhamist view.[26] Given this much, Chatton then attempts to overturn our argument by claiming that, although it is false that (3) the mobile-equal spaces in s are equal in number to the instants in $2t$, it is true that (4) the *motiones passive mobilis* involved in traversing *s* *are* equal in number to the instants in 2*t*. But the falsity of (3) denies premise (5b) of our original argument, thus rendering it lame.

The raising and subsequent resolving of an objection fills in the implications of Chatton's contentions, especially with respect to just which correspondence relations are permissible when viewed from the standpoint of *motiones passive mobilis.* The objection urges the standard of space–time correspondence for locomotion, to wit, that, in any instant, a mobile is in one and another space equal to it. Chatton replies by noting that this is true only for the fastest motion possible (thereby limiting and criticizing, as Crathorn was later to do, the assumption of the possibility of velocity variation). It is false, he concludes, when the parts of space equal to the mobile are less numerous than the *motiones passive mobilis* it suffers in moving through a given distance (in which case, of course, the mobile moves more slowly). However, taken in conjunction with the equality of *motiones passive mobilis* and instants asserted by (4), these claims imply that the number of *motiones passive mobilis* always corresponds to the number of instants involved in the motion at hand, but that this only entails correspondence with the number of mobile-equal spaces during *motus velocissimus.* Finally, Chatton uses his *motiones passive mobilis* to explain how motion can occur when, as he has just said, the mobile-equal parts of space are fewer than the *motiones passive mobilis.* In such a case, he tells us, a mobile can undergo continuous motion (considered as successively occurring *motiones passive mobilis*) while occupying the same part of space *per tempus.* The implication is that this occurs when the mobile possesses different *motiones passive* throughout some interval of time, even though it is in a single part of space throughout that interval.[27]

Astonishing though Chatton's dual refutation of our two-mobile argument may be, in essentials its putative force derives from the fact that he has altered the space–time correspondence needed for the argument in such a way that they are no longer operative and, in the bargain, has explained how it can be that motion occurs with mobiles being in the same place for diverse instants and in diverse places at the same instant. As outlandish as such contentions are, they are nevertheless fitting with the discontinuous motion demanded by indivisibilist premises, especially those, like Chatton's, of the finitist variety.

Adam Wodeham's criticism of Chatton is less impressive than might be expected. He does not, for example, say much at all about the altered correspondence relations at the center of his position. Still, he does point out that some of the particulars of these relations require that the motion have jumps, and that the retention of a single place *per tempus* by a mobile under *motiones passive* really amounts to rest, not motion. In fact, Wodeham shows dissatisfaction with the non-Ockhamist idea of *motio passiva* in general.[28]

VII

Although we shall examine it last, Henry of Harclay's resolution of the two-mobile argument is the earliest, most extensive, and most complex. At the outset, we should note that, in the version of this argument Harclay must refute, all of the numbers in premise (3) are infinite, but different infinite numbers, since the existence of unequal infinites is something at the very center of Harclay's contention that continua of various sizes are all composed of an infinity of extensionless indivisibles, indivisibles that are immediately next to one another.[29]

Next, we should realize that the key move made by Harclay in altering the force of the relevant space–time correspondence relations in our argument is his distinction of two different kinds of time: *tempus continuum* and *tempus discretum.* Traditionally, continuous time was taken as time insofar as temporal intervals were infinitely divisible and, as such, reflecting the continuity of the motion with which any time is inextricably connected, were continuous *magnitudes* like straight lines. On the other hand, time was viewed as discrete insofar as it was the *number* of the motion of which it was the measure. This is, basically, the distinction to which Harclay appealed. Both kinds of time serve as measures of motion, but in different ways: discrete time does so through the repetition of some *prima mensura* (usually the diurnal rotation of the heavens serving as a kind of clock), whereas continuous time does so by being "applied" or "superposed" to the motion it measures. It is in this latter sense that time is viewed as having "extension," an extension directly relatable to that of the motions such a continuous time measures and to that of the spatial magnitudes over which this motion occurs. As so "extended," the time can be viewed as superposed to this motion and magnitude. Taken in this sense, then, it is clear that time satisfies the one-to-one correspondence assumption specified in premise (4) of our original argument.[30] Premise (4) intact allows premise (5) to stand.

However, there is another factor about continuous time that must be taken into consideration. It is that:

[A] when two time intervals considered as continuous measure two motions, the infinite number of instants contained in these intervals are *equal* when, and only when, the spaces over which the motions occur are equal.[31]

Thus, if one mobile takes 100 years to go from (say) Oxford to London, while another takes only two hours to make the same trip, it follows that 100 years and two hours, considered as continuous time intervals measuring these motions, will contain the same (infinite) number of instants. Alternatively, if the distance traversed by one mobile is greater than that traversed by another mobile, then the continuous time interval measuring the motion of the first mobile must contain a greater (infinite) number of instants than that measuring the motion of the second mobile. And this obtains even if the length of the first time interval is equal to or less than that of the second interval (which is to say, of course, that this obtains at no matter what relative rates the mobiles move).[32] Hence there is no way to measure the differing speeds of motions relative to continuous time, which in turn means that, even granting premises (4) and (5) one cannot infer (6), because the spaces involved in the time specified in (6) are unequal.

Turning to discrete time, the argument is rendered invalid in another way. For then, Harclay claims that:

[B] if one discrete time interval is greater than, equal to, or less than another time interval, then the infinite number of instants contained in the first of these intervals is, correspondingly, greater than, equal to, or less than that contained in the second.

As a corollary to this, it follows that, when different motions over the same space are measured by such discrete time intervals, the slower motion will always contain a greater infinite number of *mutata esse* than a faster one. But this means that the measure of the speeds of motions only occurs by means of discrete time, not continuous time.[33]

Under such a situation, premise (6) of the original argument will be true; but premises (4) and (5) are true only under continuous time, not discrete time, so once again the argument is inconclusive. Harclay had presumably therefore protected his indivisibilism from at least the two-mobile attack.

As an appendix, however, one should note another use to which Harclay put his distinction of continuous versus discrete time. If it is true that, as stipulated by (B), any time interval t_1 considered as discrete contains a greater infinite number of instants than any time interval t_2 less than t_1, then, given one mobile m_1 moving over a given space s in t_1 and another mobile m_2 moving over the same space s in t_2, it follows that m_2 will move two or more points of space s in a single instant of t_2. This will, of course, explain just how, taking time as discrete, the faster mobile in the original two-mobile argument manages to traverse the greater space in less time, something that fits well with Harclay's stipulation that fast and slow are only measured by discrete time.[34]

The criticism Harclay received from his two most "chapter-and-verse" citing opponents is surprisingly weak. William of Alnwick, for example, gives a faithful reproduction of the argument Harclay was answering, citing it as he does from Harclay himself (and Wodeham does the same taking it second-hand from Alnwick). But Alnwick's report of Harclay's refutation of the argument is extraordinarily brief and does little more than mention the crucial distinction of continuous versus discrete time and its relevance. Further, his reply to Harclay's refutation is really no reply at all. He merely steadfastly refuses to admit Harclay's central contention of unequal infinities for the parts and wholes of continua (possibly because he had argued against it at length elsewhere).[35]

Wodeham's critical remarks are even more disappointing. In effect, he opens by calling Harclay's view "irrational," continues with a few words about there being more indivisibles in larger continua than in smaller ones (which in no way is itself damaging to Harclay), and concludes by saying, in effect, that the refutation of the two-mobile argument can be left to those who have nothing better to do.[36]

Thomas Bradwardine's criticism of both Harclay and Chatton is much more satisfactory and of a quite different sort. He says absolutely nothing of their specific refutations of the fast–slow mobile argument or of their own arguments pro or con anything. Instead, he reveals the inconsistency between indivisibilism on the one hand and geometry and what it has to say about the composition of continua on the other. For Chatton the inconsistency is with his finitism; for

Harclay it is with his belief that his indivisibles are immediately next to one another.[37]

VIII

As is abundantly evident, from the history just traced of the fortunes of a single problem and a single argument, late medieval indivisibilism was not especially remarkable, as were other aspects of fourteenth-century philosophy, for its brilliance. Unlike the physical atomism of Democritus or the seventeenth century, which did carry some explanatory power, however problematic, relative to objects and events in nature, the central task of medieval Latin atomism lay not in such explaining, but simply in the formulation of a consistent and convincing account of the atomistic structure of continuous quantity in the abstract and especially in what such an account would have to be to render the Aristotelian opposition innocuous. The fact that it was a mathematical atomism of unextended indivisibles made this task an extraordinary difficult one and frequently resulted in philosophically and scientifically outrageous doctrines and conceptions of the sort of which samples have just been given.

The most notable impression one derives from reading the efforts of medieval indivisibilists is that they certainly had the courage of their convictions. Lemming-like, they often seemed willing to follow these convictions into whatever sea of troubles may have been awaiting them. As one observes them in such situations, treading water at best, one occasionally wishes that they had received the kind of scolding suffered by an earlier proponent of atomistic arguments about motion: Diodorus Cronos. Sextus tells us that Diodorus had dislocated his shoulder, and that, when he went to Herophilus for treatment, he was greeted with the following diagnosis: "Your shoulder has been put out either in the place where it was or where it was not; but it was put out neither where it was nor where it was not; therefore it has not been put out."

NOTES

1 I have treated aspects of the discussions of continuity, infinity, and indivisibilism in many of these authors in various earlier articles: "Superposition, Congruence and Continuity in the Middle Ages," *Melanges Alexandre Koyré,* vol. I, L'aventure de la science (Paris, 1964), pp. 416–41; "Two Questions on the Continuum: Walter Chatton, O.F.M. and Adam Wodeham, O.F.M.," *Franciscan Studies,* 26(1966), pp. 212–88 (with Edward A. Synan); "*Mathesis in philosophiam scholasticam introducta:* The Rise and Development of the Application of Mathematics in Fourteenth-Century Philosophy and Theology," in *Arts libéraux et Philosophie au Moyen Age* (Actes du Quatrième Congrès International de Philosophie Médiévale), Montreal/Paris, 1969, pp. 215–254; "Naissance et développement de l'atomisme au bas moyen âge latin," *Cahiers d'études médiévales. I La science de la nature:* Théories et pratiques (Montreal/Paris, 1974), pp. 1–30; "William of Ockham and

the Logic of Infinity and Continuity," in N. Kretzmann (ed.), *Infinity and Continuity in Ancient and Medieval Thought* (Ithaca, N.Y., 1982), pp. 165–206; "Infinity and Continuity," in N. Kretzmann, A. Kenny, and J. Pinborg (eds.), *The Cambridge History of Later Medieval Philosophy* (Cambridge, 1982), pp. 564–91; "Henry of Harclay and the Infinite," in A. Maieru and A. Paravicini-Bagliani (eds.), *Studi sul XIV secolo in memoria di Anneliese Maier* (Rome, 1982), pp. 219–61.

2 *Physics* VI, 1, 231b18–231a17

3 Themistius, *In Aristotelis physica paraphrasis,* ed. H. Schenkl, *Commentaria in Aristotelem graeca,* V, 2 (Berlin, 1900), pp. 185, 9–28; Simplicius, *In Aristotelis physicorum commentaria,* ed. H. Diels, *Commentaria in Aristotelem graeca,* X (Berlin, 1895), pp. 934, 18–30. The same texts are quoted from earlier editions in Hermann Usener, *Epicurea* (Leipzig, 1887), pp. 197–8.

4 Albert N. Nader, *Le système philosophique des Mu'tazila* (Beirut, 1956), pp. 170–1.

5 *Physics* VI, 2, 232b26–233a10.

6 For example, John Buridan (*Questiones super octo libros phisicorum* [ed. Paris, 1518], Lib. VI, Q. 1, fol. 94rb) uses a five-indivisible version of the argument: "ponamus gratia exempli quod *A* moveatur dupliciter velocius quam *B* et utrumque super spacium quinque punctorum et incipiunt simul movere; tunc ergo in quo tempore A pertransit spacium suum, quod est quinque punctorum, *B* solum pertransit dimidium neque plus neque minus, et non est dare medium quinque punctorum nisi duo puncta cum dimidio, et ita unum punctum habet duas medietates et est divisible." What else this argument can establish is indicated by Buridan's next words: "Multa autem inconvenientia sequerentur que esset longum et superfluum enumerare." In the next *questio* (*Questiones,* VI, Q. 2, fol. 95va), Buridan uses a three-indivisible version of the argument to establish parallel indivisibilism or nonindivisibilism for lines, motions, and times. Similarly, in his *Questiones physicorum,* Nicole Oresme uses a three-indivisible version to support his conclusion: "Quod nullus motus localis vel extensionis componitur ex mutatis esse" (MS Sevilla, Colomb. 7-6-30, fol. 66vb).

7 See the article on Chatton (and Wodeham) cited in n. 1 and V. P. Zoubov, "Walter Catton, Gerard d'Odon et Nicolas Bonet," *Physis,* 1 (1959), pp. 261–78.

8 Gerard Odo, *Comm. Sent.,* I, dist. 37 (MSS Naples, Bibl. Naz. VII.B.25, fol. 241v; Valencia Cated. 139, fol. 124r): "Defficit iterum ratio, quia dicit quod tempus compositum ex tribus instantibus sive indivisibilibus non potest dividi in partes equales. Impossibile enim est quod aliquod compositum ex tribus indivisibilibus dividatur nisi in partes equales, sicut apparet de linea. Quia, si intelligatur componi ex tribus punctis et dividatur, necessario dividetur in duobus punctis, quia divisio fiet in puncto continuante qui est continuatio duorum punctorum extremorum. Et ille punctus continuans trahitur per divisionem, non quod ille punctus dividatur, sed in illo puncto fit divisio. Et ita etiam si tempus compositum ex tribus instantibus intelligatur dividi, necessario intelligetur dividi in duo instantia, quia ymaginatio fundabitur super unum instans super quod fiet ymaginaria divisio et terminabitur in duo instantia extrema."

9 Walter Chatton, *Comm. Sent.,* II, dist. 2, quest. 3 (MSS Paris BN lat. 15887, fol. 95v; Florence, Bibl. Naz., conv. soppr. C.5.357, fol. 189v): "Item argumentum supra insolutum est, vel non factum etiam, et ita est de multis sequentibus. Accipiamus motum mensuratum tribus instantibus precise, et accipiamus motum in duplo velociorem; ergo, si tempus componeretur ex instantibus, indivisibile esset divisibile, quia ille motus in duplo velocius mensuraretur instanti et in dimidio. Istud non cogit, quia eque est contra negantes indivisibilia ex quo nihil raptim transiens ponunt extra animam in rebus. Valeret etiam eque contra ponentes indivisibilias set non componeretur quantum, quia quodlibet

indivisibile habet suam partem sibi correspondentem. Similiter, nisi haberet quantum datum secundum tantam multitudinem partium quod non maiorem, non posset illa multitudo excedi. Similiter, non cogit contra aliquem hominem, quia, sicud est dare quantum discretum – puta ternarium vel quinarium – cui repugnat dividi in duas medietates, ita dicam ego de quanto alico continuo; quia ex quo non habet medietates, non potest in medietates dividi. Et quod oppositum est in communi usu, dico quod communis usus vocat medietatem ubi in altera parte respectu alterius non percipitur excessus. Eodem modo, dico in proposito ad argumentum quod eo ipso quod ponis tempus esse componi ex tribus tantum instantibus, dico quod non habet medietates sicud nec ternarius habet."

10 It does not appear that fourteenth-century indivisibilists felt attracted to the Epicurean (*Letter to Herodotus,* 61) view that all indivisibles move with equal speed.

11 See the assertion of this assumption ("Accipio istam propositionem . . .") in the text quoted in the next note.

12 One can find versions of this argument in both indivisibilists (like Harclay, Chatton, Odo, Wyclif) – who naturally attempt to refute it – and their critics (Alnwick, Wodeham). Since Harclay's version is the earliest, I cite it here from his *questio:* Utrum mundus poterit durare in eternum a parte post (MSS Tortosa, Cated. 88, fol. 87v; Florence, Bib. Naz. fondo princ. II.II.281, fol. 97v): "Preterea, arguitur ex parte instantium indivisibilium in tempore continuo successivo: Eadem ratione in tempore prolixiori erunt plura instantia quam in tempore breviori, puta in duobus diebus quam in uno die. Probo tunc ex hoc istam conclusionem: Quod una quantitas pedalis sit maior quam alia quantitas bipedalis, et pars maior quam totum vel equalis toti. Probatio consequentie: Accipio unum mobile tardum quod movetur per spacium ⟨bi⟩ pedale in duobus diebus, mobile velox per spacium ⟨bi⟩ pedale in uno die. Accipio istam propositionem ex VI *Phisicorum:* Mobile dum movetur in quolibet instanti signato vel signabili est in spacio sibi equale, et in alio et alio instanti est in alio et alio spacio, quia aliter quiesceret et non moveretur continue. Quot igitur contingit assignare instantia in tempore illo in quo mobile movetur per aliquod spacium, tot contingit signare spacia equalia mobili in eadem magnitudine, et tot puncta terminantia illorum spaciorum. Cum igitur per te plura sunt instantia in duobus diebus quam in uno, plura erunt puncta in uno pede pertransito in duobus diebus quam in duobus pedibus pertransitis in uno die, quod est impossibile; igitur et cetera."

13 Note that the number of "mobile-equal spaces" is equivalent to the number of indivisibles since the end, or beginning, points of these spaces are indivisibles.

14 Adam Wodeham states Aristotle's argument and then, on the basis of it (*ex hiis*), gives his version of the new argument we are considering here (*Tractatus de indivisibilibus,* Q. 1; MSS Florence, Bibl. Naz. conv. soppr. A.3.508, fol. 136r; Florence, Bibl. Naz. conv. soppr. B.7.1249, fol. 134r): "Et procedit racio secundum veritatem et secundum Philosophum, accipiendo a velociori semper tardius moveri per minorem magnitudinem in equali tempore, et accipiendo a tardiori semper velocius moveri per equalem magnitudinem in minori tempore. Et ita per velocius dividetur tempus, et per tardius magnitudo, hoc est, ostendetur dividi, sicud deductum est. Et virtus rationis stat in hoc: Quod velocius in minori tempore pertransit equale spacium, et tardius in equali tempore transit minus de spacio; ergo, infert Philosophus, omne tempus et omnis magnitudo continua est divisibilis in infinitum. Nec dependet hec ratio, sicud ymaginatur Commentator, ex hoc quod omne mobile possit moveri velocius suo motu naturali – quamvis hoc apud theologum sit verum – sed ex hoc quod in eodem tempore moveatur mobile velocius et tardius, et econtra. Hoc autem sufficit ad intentum, sicud prius. Ex hiis adduco propriam rationem, scilicet quod, si tempus componatur ex instantibus, quod spacium pedale esset equalis spacio bipedali, ymo maius eo, quia haberet tot spacia indivisibilia vel plura; quia cum mobile tardum acquirat in primo instanti motus continui ad minus indivisibile spacii, ergo quot sunt instantia in duobus diebus tot sunt partes indivisibiles spacii pertransiti secundum longitudinem pedalis quantitatis a mobili tardo in illis duabus diebus. Et non sunt plura, ymo tot, si continuum componitur ex indivisibilibus, in spacio bipedali pertransito uno die a velociori; quia velox in uno instanti diei acquirit plus quam indivisibile spacii, aliter instans esset divisibile sic quod in una eius parte acquireretur unum indivisibile spacii et in alia aliud. Et cum in fine diei per casum pertransiret spacium bipedale[m], ergo spacium bipedale non habet nisi medietatem tot partium quot probate sunt esse in spacio pedali; ergo et cetera."

15 The role of the inequality of infinites in Harclay, the support he finds for it in Grosseteste, and just how crucial this inequality was for Harclay are documented in the article on Harclay cited in n. 1.

16 Gerard of Odo (*Comm. Sent.*, I, dist. 37 MSS cit. [n. 8], fols. 239r & 123r): "Tertio arguitur sic: Si continuum componeretur ex indivisibilibus, sequeretur quod indivisibile divideretur. Consequens est impossibile, quia implicat contradictionem; ergo et antecedens. Probatio consequentie pro cuius evidentia permittuntur tres suppositiones. Prima est quod in omni tempore contingit aliquid moveri velocius et tardius. *Secunda quod mobile velocius plus pertransit de spatio in equali tempore movili tardiori.* Tertia quod contingit mobili tardo duplicem sesquialteram seu emioliam longitudinem pertransiri a velociori. Hiis premissis, fit talis deductio: Pertranseat mobile velocius equale spatium et dimidium (hoc est, sesquialterum seu emioliam) respectu spatii pertransiti a mobili tardiori in eodem tempore; et sit spatium pertransitum a velociori *AD*, spatium autem pertransitum a tardiori *EI;* et sit tempus utriusque *KN.* Si ergo continuum componatur ex indivisibilibus, dividantur magnitudines per quas moventur velocius et tardius in indivisibilia, et sit ita quod magnitudo velocioris dividatur in tria indivisibilia ut in *AB* et in *BC* et in *CD,* magnitudo autem tardioris dividatur in duo indivisibilia quorum unum sit *EZ* et aliud *ZI.* Cum ergo ita sit, oportet tempus in quo velocius pertransit suam magnitudinem dividi in tres partes indivisibiles, quia tempus et magnitudo sunt proportionaliter divisibilia. Et sit unum indivisibile *KL* et aliud *LM* et aliud *MN.* Et iterum oportet tempus dividi secundum divisionem magnitudinis pertransite a mobili tardiori. Sed illa magnitudo dividitur in duo indivisibilia et ita in duas partes equales. Sed impossibile est tempus dividi in duas partes equales nisi illud indivisibile quod est medium et signatum per *LM* dividatur. Ergo indivisibile dividetur. Istas tres rationes facit Philosophus et Commentator 6° *Phisicorum.* Ratio potest aliter sic formari: Moveatur aliquod mobile super magnitudinem compositam ex tribus indivisibilibus in tempore composito ex tribus instantibus. Datis duobus mobilibus quorum unum sit velocius primo mobili dato in duplo, reliquum tardius in duplo, velocius mobile movebitur super eandem magnitudinem in medietate temporis et sic in instanti cum dimidio, mobile tardius movebitur in eodem tempore (scilicet in tribus instantibus) per medietatem magnitudinis et sic in tribus instantibus temporis pertransibit unum indivisibile cum dimidio. Quare indivisibile dividetur. Et licet istud non sit eodem modo deductum, tamen est eadem difficultas."

17 Gerard of Odo, *Comm. Sent.*, I, dist. 37 MSS cit., fol. 241v–242r; 124r: "Ad aliud de proportione sesquialtera, cum dicitur quod indivisibile dividetur, consequentiam nego. Ad probationem respondeo, primo ad suppositiones: Nego eas omnes tres ad intellectum ad quem inducuntur. Quando enim dicitur in prima quod in omni tempore contingit velocius et tardius moveri, si intelligatur quod in omni tempore contingit velocius et tardius moveri, hoc est, plus et minus pertransiri de spatio, suppositio est falsa simpliciter. Quia possibile est aliqua duo moveri, semper unum velocius altero, et numquam mobile velocius pertransibit plus de spatio quam mobile tardius. Unde si motus primi mobilis duraret per imperpetuum, polus articus et polus antarticus moverentur continue, non tamen plus pertransirent de spatio quam poli orbis lune, dato quod orbis lune non revolveretur nisi semel in centum annis et quod etiam continue moveretur. Apparet etiam quod secunda suppositio est simpliciter falsa. Non enim necessario mobile velox plus pertransit de spatio in equali tempore quam mobile tardum, sicut dictum est de polis primi mobilis et polis orbis lune. Ex hoc apparet etiam quod tertia suppositio est simpli falsa. Non enim est necesse quod mobile velocius pertranseat duplicem sesquialteram seu emioliam longitudinem. Si autem contingat etiam, ut in pluribus, quod mobile velox plus pertranseat de spatio in equali tempore quam mobile tardum hoc est per accidens et extra rationem velocitatis et tarditatis, cum velocitas et tarditas possint reperiri sine maioritate et minoritate spatii pertransiti. Si autem prima suppositio intelligatur sic, scilicet quod in omni tempore contingit velocius et tardius non plus et minus pertransiri de spatio, sed velocius et tardius mutari secundum differentias situs et dispositiones loci, dico quod prima suppositio nichil facit ad propositum. Et cum hoc secunda et tertia semper remanent false. Hec autem responsio habetur ex duobus correllariis primi deffensivi. Defficit iterum ratio, quia supponit quod tempus et magnitudo et motus sint proportionaliter divisibilia; quia, ut dictum est superius, unum indivisibile posset semper moveri et tamen esset divisibile secundum divisionem temporis vel motus, sicut apparet de polo artico et antartico primi mobilis. Si vero dicatur quod polus distinguitur secundum partes temporis et motus per differentias respectivas loci – sicut apparet de puncto in medio rote, qua revoluta, potest terminare semidyametrum circuli a parte ante vel a parte post secundum diversos aspectus vel differentias loci – dico quod tunc ratio nihil facit contra me. Quia ratio supponit quod divisio magnitudinis fiat secundum divisionem temporis et motus secundum partes discretas, non solum secundum differentias loci. Istud autem simpliciter est falsum." The text cited in n. 8 follows directly after this.

18 Crathorn, *Comm. Sent.* (MS Basel B.V.30, p. 65): "Impossibile est aliqueno motum esse velociorem motu vere continue; licet enim unus motus vere unus et continuus, si aliquis talis est, sit velocior motibus illis qui non sunt vere continui et motus non continuus sit velocior alio motu non continuo, tamen si aliquid vere continue moveatur, non apparentur tantum, impossibile est, aliquem motum tali motu esse velociorem. Et hec conclusio patet sic: quando mobile movetur continue ita quod inter partes motus non sit aliqua quies intercepta, in quolibet instanti temporis mensurantis motum mobile est in alio et alio loco ita quod non est in aliqua parte temporis divisibilis in eodem loco. Suppono igitur quod aliquod moveatur continue vere in hora una et quod adquirat motu illo in illa hora mensurante motum mille loca; igitur in illa hora que mensurat precise motum illum sunt precise mille instantia, quia, si essent plura, illa hora non mensurat precise illum motum, sed pars illius hore; si essent pauciora instantia in tempore illo, oporteret quod corpus mobile in eodem instanti esset in pluribus locis, quod est impossibile. Si igitur aliquod mobile in eadem hora velocius moveretur, adquireret in eadem hora plura loca et per consequens esset in duobus locis in uno et eodem instanti, quod est impossibile. Igitur, si aliquod corpus moveatur sola continuitate, non est possibile aliud corpus in eodem tempore vel equali velocius moveri vel plura loca adquirere."

19 See Nicolas of Autrecourt as edited by J. R. O'Donnell, *Mediaeval Studies,* 1 (1939), p. 215; and Johannes Wyclif, *Tractatus de logica,* ed. M. H. Dziewicki (The Wyclif Society, 1894–9), cap. 9, pp. 38–40.

20 Crathorn, *Comm. Sent.* (MS Erfurt, Amplon. Q° 395A, fol. 24v): "Ad sextum principale contra conclusionem aliam dicendum quod punctum potest moveri continue et discontinue; sed si punctum moveatur super spacium trium punctorum, movebitur continue super illud spacium in tempore composito ex tribus indivisibilibus et, quantumcumque virtus motiva augmentatur, non potest moveri velocius, quia motus continuus est velocissimus. Si vero ponatur quod punctum motum moveatur per spacium trium punctorum discontinue, hoc potest contingere multipliciter: uno modo sic quod quiescat in quolibet puncto spacii per duo instantia, alio modo per 3, alio modo per 4, et sic de aliis numeris. Et talis motus potest sic velocitari. Unde si esset virtus motiva in tali gradu quod moveret punctum per spacium trium punctorum in 6 instantibus, alia virtus que esset dupla respectu prime moveret illud punctum per spacium trium punctorum in tribus instantibus; sed in minori tempore impossibile est punctum moveri per illud etsi virtus motiva esset infinita, quia motus continuus esset velocissimus. Ned potest unus motus esse velocior alio motu vere continuo, sed apparet nobis quod multi motus sunt continui, cum tamen non sint continui propter quietes interceptas quas non percipimus."

21 William of Alnwick, *Determinatio* 2 (MSS Vat. Pal. lat. 1805, fol. 15r–15v; Oxford, Bodl. Can. misc. 226, fol. 81r): "Item quarto sic: Accipiantur duo mobilia quorum unum velocius movetur ita quod in tempora unius diei pertranseat spacium unius miliaris, mobile tardum in equali tempore pertranseat medietatem illius spacii. Tunc sic: Mobile quod continue movetur in alico tempore in quolibet instanti illius temporis est in spacio sibi equali, ut dicit Philosophus in 6 *Phisicorum,* alioquin quiesceret et non moveretur; sed per positum mobile tardum continue movetur in uno tempore pertranseundo spacium medietatis unius miliaris, igitur in quolibet instanti illius temporis est in spacio sibi equali, et per consequens tot sunt partes in potentia in medietate spacii unius miliaris quot sunt instantia in potentia in tempore unius diei. Sic etiam arguitur de mobili velociori respectu maioris spacii (scilicet unius miliaris), scilicet quod tot sunt partes in potentia in spacio unius miliaris quot instantia in potentia in tempore unius diei. Tunc sic: Quecunque sunt equalia tertio, sunt equalia inter se. Sed iam probatum est quod quot sunt spacia in potentia in minori magnitudine tot sunt instantia in potentia in tempore unius diei, et non plura nec pauciora; quot etiam sunt spacia in potentia in maiori magnitudine tot sunt instantia in potentia in tempore unius diei, et non plura nec pauciora. Igitur inter se quot sunt spacia in potentia in maiori spacio tot sunt in minori, ita quod nec plura nec pauciora."

22 Chatton, *Comm. Sent.,* MSS cit. (n. 9), fols. 96r, 190r: "Item, accipiatur mobile tardum et moveatur duobus diebus super spacium unius pedis, et mobile velox moveatur super spacium duorum pedum uno die; tunc plura spacia equalia erunt in spacio pedis unius quam pedum duorum. Proba ut supra argumento septimo tertie opinionis. Istud est equaliter contra omnes, quia probaret, si sint infinita instantia in tempore mensurante motum, [probaret] quod essent in spacio infinita spacia equalia mobili, quia in quolibet instanti est in spacio sibi equali. Dico ad istud quod argumentum Zenonis de mobili tardo et veloci semper fuit difficile ad solvendum; set vadit ad hanc conclusionem: quod motus nichil sit; de indivisibilibus autem nichil ad rumbum."

23 Chatton, *Comm. Sent.,* MSS cit., fols. 90v, 185v: "Quod motus est aliqua res positiva preter res absolutas permanentes, scilicet motionis passive mobilis ad motorem."

24 Chatton, *Comm. Sent.,* MSS cit., fols. 95r, 189r: "Ad aliud: quod tunc tot vel plura spacio pedali quam bipedali, quia in duplo plura instantia sunt in duobus diebus quam uno. . . . Secundo, quero quid intelligis per instans. Si motionem alicam passivam raptim transeuntem subitam corporis celestis, qua posita corpus celeste sit in tali puncto magnitudinis, concedo quod tot sunt instantia in duobus diebus quot sunt spacia in pedali quanto sibi invicem equalia. Set ulteriorem conclusionem nego: ergo sunt scilicet in duplo plura spacia equalia in pedali spacio quam in bipedali; immo in bipedali tot spacia equalia sunt quot sunt instantia quattuor dierum ad minus. Et si accipias quod non sunt plura spacia quam motiones passive celi dum mobile velox transit spacium bipedale, nego. Non valet ergo argumentum volendo instantia esse motiones passivas celi. Immo tot sunt spacia non habentia partes in quolibet quanto super quod natus est fieri motus quot motiones passive celi possibiles dum mobile motu tardissimo transiret illud."

25 Chatton, ibid.: "Si autem intelligas per instans corpus celeste in pucto suo vel istud mobile in puncto suo, dicendum est consequenter. Dices dum mobile velox adquirit punctum, cum tardum non quiescat, aliquid adquirit, et non minus quam punctum; ergo tot puncta describet tardum motu suo quot velox suo motu. Concedo quod mobile velox per unam motionem efficacem quando habebit punctum secundum et erit in eo, tardum per motionem minus efficacem ambo tangit simul, et hoc correspondenter secundum aliam et aliam proportionem efficiencie motionum, et mobile motu efficaciori adquirit prius secundum punctum sui spacii et plus consequenter semper prioritate spacii, non temporis, quam minus efficacem motum sui spacii. Et ita cito adquirit plus de suo spacio sicud aliud de suo minus; et in eodem instanti quo illud adquirit punctum, minus efficax coexistit duobus simul."

26 Chatton, *Comm. Sent.*, MSS cit., fols. 96r, 190r: "Respondi ergo ad argumentum istud alio die, scilicet ante pascha, et illam teneo responsionem. Set modo pono etiam aliam: videtur michi quod simul stent quod motus sit continuus, et tamen quod mobile sit in eodem puncto spacii per tempus. Set suppono prius dicta, scilicet quod potes uno modo per motum intelligere quod mobile primo coexistat uni parti spacii, tum raptim et statim alteri, et sic deinceps. Et sic illa predicta includunt repugnantism et contradictionem." This first, "coexistence of the mobile with parts of space," view of motion is, in Chatton's eyes, Ockham's.

27 Chatton, ibid. (directly following text in previous note): "Alio modo, motionem passivam cuius partes raptim transeuntes continue succedunt sibi ita quod una motio passiva succedit continue alteri. Dico quod tamen hoc staret absque repugnantia et contradictione: quod mobile esset in eodem parte magnitudinis per tempus. Et hoc est rationale, quia per motum potest primo cadere oblique, secundo recte et tertio iterum oblique, et sic moveri continue, et tamen per tempus esse in eodem puncto spacii. Huic concordat antiqua distinctio quod motus habet duplicem divisibilitatem et duplices partes, scilicet unam a magnitudine et aliam ab agente. Id est, secundum quod esse, intelligo aliud est loqui de coexistentia mobilis primo huic puncto magnitudinis et statim post raptim in alia parte, et aliud de successione motionum passivarum ad invicem, ut dictum est. Hec est alia solutio ab illa quam prius dixi. Ad argumentum ergo, dico quod falsum est quod in spacio pedali sunt tot spacia quot instantia, set bene sunt tot motiones passive quot instantia. Dices in quolibet instanti est in alico [! *lege alio et alio*] spacio; nego, set dico quod natum // est sic esse, puta quando movetur motu velocissimo, et aliter non, set ubi partes spacii sunt pauciores quam motiones passive, propositio illa falsa est."

28 Wodeham, *Tractatus*, MSS cit. (n. 14), fols. 136v, 134r: "Sed hec responsio improbatur penitus, sicud prior. Nam eo ipso quod mobile secundum se et secundum quodlibet sui manet in eodem loco precise quo prius quiescit. Hoc enim et non aliud est mobile quiescere, ut notum est cuilibet, et Philosophus hoc diffuse declarat, ut prius. Sed impossibile est in eodem loco simul et semel precise existens per tempus secundum se et secundum quodlibet sui moveri et quiescere; ergo istud mobile tardum quod ponebatur continue moveri quiescebat et non movebatur, si per tempus remanet secundum se et secundum quodlibet sui in eodem loco."

29 See the article on Harclay cited in n. 1.

30 Harclay, "Utram mundus . . ." MSS cit (n. 12), fols. 91v–92v; 100r–100v: "Ad istud est dicendum quod tempus potest dupliciter considerari: vel in quantum continuum vel in quantum discretum. Secundum quod continuum dicitur longum vel breve, sed secundum quod discretum dicitur multum vel paucum et non longum vel breve secundum Aristotelem. Et Commentator hoc idem dicit, commento 23 quantum ad capitulum de tempore et commento 109 quantum ad totum 4^m *Phisicorum,* quod discretio est formale in tempore et continuatio materiale est in eo. Unde componitur ex continuo et discreto. . . . Dico igitur ad argumentum quod tempus mensurat in quantum discretum et in quantum continuum, sed diversimode, sicut alio modo mensurat discretum et alio modo continuum. Discretum enim mensurat quantitatem vel numerum per replicationem unitatis vel numeri; continuum mensurat aliud continuum in quantum continuum per applicationem et superpositionem unius ad alterum (unde communis animi conceptio in principio *Geometrice:* Si aliqua quantitas applicetur alteri ned excedit nec exceditur ab alia, ille sibi ad invicem sunt equales). Eodem modo de tempore, cum sit continuum et discretum, in quantum discretum est mensurat motum per replicationem mensure prime (puta motus primi, qui est motus diurnus uniformis circularis et continuus). Unde hoc modo motum mensurari tempore non est aliud nisi motum unum continere plures vel pauciores celi revolutiones, et unus motus dicitur diuturnior alio, quia pluries celum revolvitur durante uno quam revolvebatur durante alio. Et tunc maioritas et minoritas temporis et motus debet intelligi secundum rationem numeri, sicut multitudo et paucitas, non secundum rationem maioritatis et minoritatis in continuo. Alio modo mensurat tempus motum secundum quod unum continuum habet mensurare aliud continuum, ut unum applicetur alteri; hoc modo tempus dicitur longum vel breve sicut continuum, non multum vel paucum. . . . Et tunc dico breviter quod unum tempus est equale alteri et unus motus alteri, accipiendo tempus et motum ut sunt continua. Tempus enim, ut dicit Aristoteles, imitatur motum in continuitate et motus magnitudinem. Et ideo proprie realiter non est in motu alica extensio alia ab extensione magnitudinis super quam erit, nec est alia extensio temporis quam motus et magnitudinis nisi in ymaginatione et in anima. Vel si ponatur aliqua in motu et tempore preter illam que est in magnitudine, saltem una erit equalis alteri in quantum sunt continua, cuius ratio est quia una superponitur alteri nec excedit nec exceditur; et principium motus et magnitudinis sunt simul et terminus magnitudinum et motus simul, igitur necessario sunt equales ut considerantur ut continua quedam."

31 Harclay, "Utram mundus . . ." MSS cit., fols. 92v, 100v; "Et dico tunc breviter pro argumento quod accipiendo tempus et motum ut considerantur ut continua, non sunt plura in stantia in duobus diebus quam in uno die, supposito quod equale spacium mensuretur per duos dies et per unum diem. Sed accipiendo tempus ut discretum, sic sunt plura instantia in duobus diebus quam in uno."

32 Harclay, ibid.: "Et tunc secundum hoc patet responsio ad formam argumenti, quia non sunt plura instantia in tempore maximo quam in una hora, si accipiamus ut sunt mensure motuum equalium vel equalium magnitudinum; ymo hora una sic habet plura instantia quam 100 anni, ut intelligatur esse mensura motus continui super maiorem magnitudinem et tempus 100 annorum mensura motus super minorem propter dictam causam. . . . Primum ostendo sic: Nam tot sunt instantia in tempore quot sunt mutata esse in motu quem mensurat istud, non est dubium. Non enim contingit in tempore continuo signare diversa instantia nisi secundum diversa mutata esse que possunt inveniri in motu continuo. Sed, sicut probabo, duo motus, quorum unus, velocissimus, alter tardissimus (dum tamen sint super equales magnitudines), habent mutata esse equalia; nam mutari non est aliud quam aliter se habere nunc quam prius et mutatum est totiens quotiens habet aliter nunc quam prius. Modo totiens se habet unumquodque mobile, quocumque gradu motus moveatur, aliter quam prius, quotiens contingit assignare alium et alium gradum forme secundum quam est motus, quia non potest esse simul sub utroque gradu.

Igitur, aliter necessario se habet cum ⟨est sub⟩ uno gradu quam se habuit cum fuit sub alio et aliter quam se habebit cum erit sub tertio vel quarto. Quot igitur contingit assignare gradus novos in forma secundum quam est motus, quot contingit et possunt esse mutata esse in motu, et non plura neque pauciora. Ponamus modo exemplum de motu alterationis: Unum mobile alteratur de nigredine intensa ad albedinem intensam velocissime; aliud mobile alteratur tardissime a nigredine equali intensive ad albedinem equaliter intensam continue per omnia media. Certum est quod unum mobile in motu suo non pertransibit plures vel pauciores gradus quam aliud, ned igitur plura mutata esse in uno quam in alio, cum mutata esse non insint mobilibus nisi ratione diversorum graduum forme secundum quam est motus, licet sit unus velocissimus, alius tardissimus. Confirmatur istud: Nam motus secundum opinionem veriorem non est nisi forma fluens, ut dicit Commentator, 3-*Phisicorum,* ubi dicit quod est adquisitio partis post partem de termino ad quem vadit. Modo de motu locali de quo est mentio: Ibi spacium supra quod est motus est forma secundum quam movetur, de quo spacio semper novum aliquid acquiritur, et ideo quot sunt signabilia spacia in illo spacio toto, tot et non plura sunt mutata esse in motu toto, et hoc sive velocissime moveatur sive tardissime. Et ideo quicumque motus facti super equales magnitudines habent equaliter mutata esse et tempus per consequens mensurans unum motum et tempus mensurans alium motum habent equalia instantia, etiam si unum tempus esset 100 annorum et aliud tantum unius diei."

33 Harclay ibid., MSS cit., fols. 93r–93v; 100r–100v: "Si autem accipia ⟨n⟩ tur tempus et motus ut sunt discreta (nam motus est discretus sicut tempus, Aristoteles enim dicit quod motus mensurat tempus sicut tempus motum), sic plura sunt instantia in magno tempore quam parvo et plura mutata esse in tardo motu quam in veloci ceteris paribus, quod non est aliud dicere nisi quod motus primus, que est mensura aliorum, pluribus vicibus est in motu tardo quam veloci. Et ideo plus de tempore consumit motus tardus quam velox, sicut unus numerus plus est quam alius qui est plurificatus. . . . Sed tempus vel instans non mensurat velocitatem et tarditatem motus nisi secundum quod discretum est, non ut continuum est, ut dictum est supra."

34 Harclay, ibid., MSS cit., fols. 91v, 100r: "Modo ad propositum, accipiendo tempus ut mensura motuum inferiorum, sic tempus et motus inferiorum non sunt commensurabiles nec sunt eiusdem rationis, sicud nec numerus et quantitas continua (quia tempus ut est mensura sic est numerus, ut dictum est). Igitur sicut modo unus totus motus mensuratur aliquo tempore, ita totus ille posset mensurari tempore in duplo minori, quia mobile posset moveri tanto motu in medietate temporis; et sic quelibet pars aliquota temporis mensuraret duplam partem talem magnitudinis et eodem modo unum indivisibile temporis (puta instans) mensurat duo indivisibilia magnitudinis."

35 William of Alnwick, *Determinatio 2,* MS Vat. Pal. lat. 1805, fol. 14r: "Ad terciam rationem de tempore et instantibus temporis respondet distinguendo de tempore: Quod potest considerari ut est discretum, vel ut est continuum. Si ut est continuum, sic dicit instantia plurificari in tempore secundum mutata esse mobilis super magnitudinem. Et, quia in maiori magnitudine sunt plura puncta quam in minori, ideo, si in una hora pertransitur magnitudo duorum pedum, et in centum annis magnitudo unius pedis, plura erunt instantia in una hora quam in centum annis. Set econtra in tempore secundum quod est discretum."

36 Wodeham, *Tractatus* MSS cit. (n. 14), fols. 136r, 134r: "Ista opinio est irracionalis apud me, ut ipsam reputet indignam in probatione. Quia, si continuum vere et realiter integratur ex indivisibilibus, nichil ad rumbum quod continuum consideretur sic vel aliter; quoniam vere et realiter maius eiusdem rationis continebit tot indivisibilia quot minus et plura. Ergo una hora C annorum vere et realiter tot habet tanquam partes

indivisibilia quot una alia hora in qua movetur velocius mobile super maius spacium, et per consequens C anni incomparabiliter plura, quotiens scilicet ⟨C⟩ anni contineant unam horam. Unde, istam responsionem relinquo hiis qui in nullo studio laborant."

37 See pages 437–41 of the article on superposition and congruence cited in n. 1. Although Bradwardine draws on many "sciences" in his refutation of indivisibilism, at bottom his arguments are based on geometry.

3

"Something old, something new, Something borrowed, something blue" in Copernicus, Galileo, and Newton

JOSEPH T. CLARK
Canisius College

This essay (*herzlichst gewidmet*) is a study in the metahistory of science, a second-order discipline, philosophically perceptive but ideologically neutral, which examines (*von aussen hinein*) what it is, if anything at all, that our first-order researches (*von innen heraus*) in the conceptual history of scientific ideas jointly disclose about the phenomenon of human beings, or at least that subset of *Homo sapiens* each member of which – of either gender and at some cross section of time – risks everything to pose as a personal issue (*an sich selbst*) a decidable question about phenomena, and thereafter must somehow formulate a testable answer – or perish!

Such a supervening inquiry (the Newton that *was* is the only one there is) thus requires and assumes in both reader and writer a shared knowledge of a professional stock of reliably established matters of historical fact, either constantly available or readily accessible.

In this universe of discourse something x is construed as *old*, relative to some y, if and only if such x is anterior to such y. Something x is construed as *new*, if and only if, relative to some y, such x is posterior to such y. Something x is construed as *borrowed*, relative to some y, if and only if both x is old and y is new, and there exists a context z, such that both such x and such y are contemporaneous within the same z. And something x is construed as *blue* – in a psychic sense – if and only if there exists an x, such that for all y, if y is human, x produces pathos in y.

To identify something old or something borrowed or something blue, does not appear problematic. But to specify something *new* invites at once the shudder syndrome: "Shudder we must. I [Professor Cohen] do it all the time. And I shudder especially whenever I have to deal with the problems of the creative imagination on the part of any thinking scientist." For this malaise there is no cure. But a nostrum is available. It is the phenomenon of a *Gestalt* experience in which, for example, one *sees* what is in fact not there, or more precisely, not there just yet, but which can – and *must* – be made to supervene in order to complete the defective object in one's visual field, and thus satisfy the aesthetic demands of integrative intelligence. Such a *Gestalt* experience requires some antecedents of disturbing attention, and often produces recognizable consequents (such as a *new* conceptual scheme for the reinterpretation of old phenomenon), but is never reducible to categories of formal entailment or subject on demand to accurate recall. If so, then it is less difficult to comprehend how it sometimes happens that scientists, in autobiographical mood, either explain such gigantic *Gestalten* experiences in logical gibberish or in contrived fictions – or both.

COPERNICUS

To descry something *old* in Copernicus (or any other scientific figure) is not to indulge an antiquarian whim, but to identify one component, at least, in the total conceptual framework that delimited the horizon of his inquiries and restricted the range of his responses. I here select the orthodox set of mathematical devices for hollow-circle compositions, which Copernicus employed, as a geometrical pseudomechanism of mobile perimeters, in the astronomical part of his *De revolutionibus.* This choice remains valid even if Copernicus modified the practices of Ptolemy, in particular, by eschewing the equant, supplanting a calculus of chords with a calculus of sines, and preferring for tabular use the positional system of Arabic numerals. For despite these more modern touches the mathematical apparatus of Ptolemy and Copernicus remains basically the same over an interval of 1,400 years.

A prime clue to something *new* in Copernicus is to find what it was in contemporary astronomy that disturbed (*coepit me taedere*) his scientific conscience. The search area is narrowed when one recalls that Ptolemy's mathematical theory of the heavens in the *Syntaxis* became a working astronomer's portfolio of independent, even if kinematically comparable, unit solutions to independent, even if observationally similar, unit problems of angular displacements in longitude around a common optical center, but without other methods to determine relative linear distance intervals than what naked-eye parallactic measurements supplied for the sun and moon, and calculated lengths of retrograde arcs could suggest for reciprocating radii for deferent and epicyle pairs. Such a problem-oriented *Handbuch* presented to Copernicus, not a system of multiple parts, possessing a common unit of measure, but a *monstrum* whose grotesque disorder (and not mere complexity) caricatured his God of creation (*optimus et regularissimus opifex*) and principally (*res praecipua*) disturbed his intellectual conscience.

I therefore select as something *new* in Copernicus his suggestion that there exists some kind of common measure and thus an orderly proportion between (a) the average relative distances of the planets from their common center of rotation, and (b) the time periods respectively required for each such circumsolar revolution in a scalar model of the universe, measured by fractional or integral multiples of some now – and for the first time – available astronomical unit. This choice is confirmed by the witness of Kepler's indomitable conviction (itself a *Gestalt?*) that the vague proportion of Copernicus (*ut magnitudinem orbium multitudo temporis metiatur*) is not only mathematically specifiable, but also physically interpretable (*physica coelestis*) in a dynamical system of *true,* not fictive, celestial mechanics, of actual, not imaginary, trajectories, the respective perimeters of which enclose, not vacuous, but astronomically significant areas of celestial space.

What Copernicus *borrowed* from Ptolemy, and thus rendered in the new context contemporaneous, is the astronomical *apparatus technicus* that reappears in the *De revolutionibus.* For exactly as in the geostatic and heliokinetic cosmology of Ptolemy, the tables which accompany the heliostatic and geokinetic system of Copernicus yield geocentric planetary positions directly and from a single (not a double-entry combinatory) procedure. Hence, as equally representative specimens of positional astronomy, the basic techniques of both Ptolemy and Copernicus are observationally equivalent to the naked eye and empirically indistinguishable. Nor is such mimetic parallelism effectively diminished by the *circulorum metamorphosis* through which Copernicus tries to contrive such a combination of (a) eccentric circle and one epicycle or (b) concentric circle and two epicycles with reference to his own mean sun, as will be geometrically equivalent to Ptolemy's more candid combination of eccentric and equant circle with respect to the earth, and thus camouflage (but not eliminate) the intrasystematic abomination of variations in orbital velocity. For geometrically equivalent figures are, in effect, mathematical synonyms, and such transformations only paraphrase, as Viète saw and dubbed Copernicus *Ptolemaei paraphrastes.* The appellation is apropos. For while the *De revolutionibus* was still in progress, Rheticus narrated about his preceptor: *suum opus ad Ptolemaei imitationem instituere.* And the published text announces, and more than once, that *in caeteris Ptolemaeum sequemur paucis exceptis.* And after the event Reinhold refers to Copernicus as *novus Ptolemaeus.* For Copernicus borrows from the *Syntaxis* the format, the method, the structure, the chapter sequence, the theorems, the tables, the style, the language, the problematics, the solution techniques, the arrangement of materials, even the star catalogue that here, however, as comprising the absolutely fixed points in the reference frame for his reconstructed cosmology, Copernicus places before his discussion of the sun's (apparent) motions. So much, in sum, is borrowed and thus rendered in the new context contemporaneous that the authentic history of astronomy is more reverently respected by construing the *De revolutionibus* of Copernicus as the *zweite verbesserte Auflage* of Ptolemy's *Syntaxis,* than to pit one author against the other as Galileo appears to do in the dramatic title of his *Dialogo.* If, in brief, a new and provocative commensurability appeared in the cosmology of Copernicus, an *astronomia nova* arrived only with Kepler.

My clue to what is *blue* in Copernicus is found in the poignant epitaph that the same Kepler provided for his predecessor: *Copernicus, divitiarum suarum ipse ignarus, Ptolemaeum sibi exprimendum omnino sumpsit – non rerum naturam – ad quam tamen omnium proximus accesserat.* For an aura of pathos surrounds the portrait of any person whose intellectual reach exceeds his conceptual grasp.

To appreciate, first, the intellectual reach of Copernicus, one has only to ponder the first pages of the *De revolutionibus* (*in medio vero omnium residet sol*), and to gaze upon the diagram that, even if only a conventionally schematized representation of spherical zones, adorns the smoothly speculative text. In order to comprehend, second, the constraints on the conceptual grasp of Copernicus, one has only to plumb the later and technical parts of his treatise and come to realize – not without tears (*sunt lacrimae rerum*) – that the very same diagram is not an exact blueprint after all.

For in the technical astronomy of Copernicus not only is the linear distance interval of separation between the mobile sphere of Saturn and the immobile sphere of the fixed stars *paene infinitum,* but also that specific point that functions as the center of the earth's orbit as well as the momentary center of the other planetary revolutions is located, not in the sun, but in the solar neighborhood. The scandal here is not the absence of some physical pivot for such actual revolutions about any empty focus. For in the celestial revolutions of Copernicus uniform circular motion is natural to a sphere, a consequence of its geometrical form, and thus requires no dynamical support. The complaint is, rather, that if, as here, the sun functions as the cosmological (and photothermal) center of the universe of Copernicus, the same sun does not effectively function as the astronomical center of the circumsolar, indeed, but not heliocentric planetary orbits: *ut a principio diximus* ἀμφιβολικῶς *in sole vel circa ipsum esse centrum mundi.* For the bald fact of the matter is that uniform circular motion and a heliocentric astronomy are systematically incompatible.

It is no wonder, then, that continental astronomy developed after Copernicus only by revising his system, and sad – but true – to report that (a) the *De revolutionibus* is both the cradle and the coffin of the astronomical system it contains, and (b) the name of its author an eponym for a cultural revolution he did not intend to ignite or even envision as a consequence.

For Copernicus was persuaded that if career astronomers discreetly agreed (*mathemata mathematicis scribuntur*) to improve their discipline by conceding that our earth really rotated and revolved, everything else of consequence in human affairs would remain exactly the same as it always had been. But nothing did.

GALILEO

What I find very *old* in Galileo is his conviction that physics is a *demonstrative science,* equipped to produce *la verità assoluta* about (reasoned) matters of fact. If this ancient epistemological posture is thus preserved intact, it is not because Galileo credits as infallibly conclusive the (vacuously verbal) syllogistic of Aristotle's *Posterior Analytics.* The reason is rather that Galileo adopts from antiquity and then employs for the same demonstrative function in natural philosophy the unconditional validity of the (physically interpreted) geometric problematics that Euclid (geometry) and Eudoxus (proportionality of ratios) and Archimedes (compound ratio) jointly supplied for the kinematical analysis of *virtual* velocities in the venerable science of statics. For the same techniques could now be applied to a kindred kinematical analysis of *real* velocities in a new and hybrid science (mathematical physics) of *movimenti locali,* such as the historically problematic cases of *il grave cadente* and *projectilia.* It is, thus, the adventitious amalgam in one gifted person of these two ancient, but previously unrelated, philosophical and mathematical heirlooms that constitutes the idiosyncratic methodology in physics of the *Filosofo e Matematico primario* of the Grand Duke of Tuscany.

In this uniquely personal style of *philosophiae naturalis theoremata mathematica* I discern four principal stages. The first is *prior knowledge* (1604) of the empirical certainty, on the basis of careful computational measurements, of the *demonstrandum: gli spazii passati dal moto naturale esser in propozione doppia dei tempi.* The second is a painful, persistent, and systematic search for the *essential* definition (or *assioma*) of the motion under scrutiny: *per dimostrare li accidenti da me osservati, mi mancava principio totalmente indubitabile.* When that unique definition has been found, as Galileo successfully did (*post diuturnas mentis agitationes*) for *il grave cadente: motus qui, a quiete recedens, temporibus aequalibus aequalia celeritatis momenta sibi superaddit,* the philosophic task of correlating an essence with its *accidens proprium* ceases, and the mathematical task of its rigorous demonstration begins.

But Galileo's mathematical resources (without standard units of measure or physical constants) are systematically restricted to (1) proven proportions between ratios of geometric element pairs of the same kind (lines to lines, circles to circles, arcs to arcs), which are (2) themselves physically interpretable (*representetur per extensionem AB tempus*) as isomorphic counterparts to the same proportions between physical magnitude pairs of the same kind (*spatium MH ad spatium HL esse in duplicata ratione eius quam habet tempus EA ad tempus AD*). Step three in this quadripartite procedure thus requires the construction of a diagram, exhibiting the ingredients of the prior definition, and competent to demonstrate the desired conclusion: *ergo ratio spatiorum peractorum dupla est rationis temporum: quod erat demonstrandum.* The final step four, perfunctory for the author who already knows the result from prior experimental observations, but reassuring for a lay audience, discloses that the conclusion of the mathematical demonstration is, ceteris paribus, empirically verifiable to an acceptable degree of tolerance.

There is therefore much that is not new in this novel, but cumbersome and only syncretic, combination of *vetera et vetera* that constitutes the *Two New Sciences* of Galileo.

But what I find altogether *new* (some 200 years before Cantor's *Mannigfaltigkeitslehre*) and completely original (itself a *Gestalt?*) is Galileo's systematic understanding (*i numeri quadrati esser tanti quante sono le proprie radici, avvenga che ogni quadrato ha la sua radice, ogni radice il suo quadrato, nè quadrato alcuno ha più d' una sola radice, nè radice alcuna più d' un quadrato solo*), and effective use (in *Theorema I*) of set-theoretical equivalence – a dyadic relation as reflexive and symmetric and transitive as was the familiar (and paradigmatic?) relation of Eudoxian proportion between ratios of magnitude pairs.

For any set – such as the set S_1 of denumerably infinite instantaneous degrees of speed, from initial zero to final and maximum N, of a natural uniformly accelerated motion M_1 from the rest – is defined as equivalent to a set – such as the set S_2 of denumerably infinite instantaneous degrees of speed, at constant ½ N, of a uniform motion M_2 – if and only if there exists a third set, the members of which

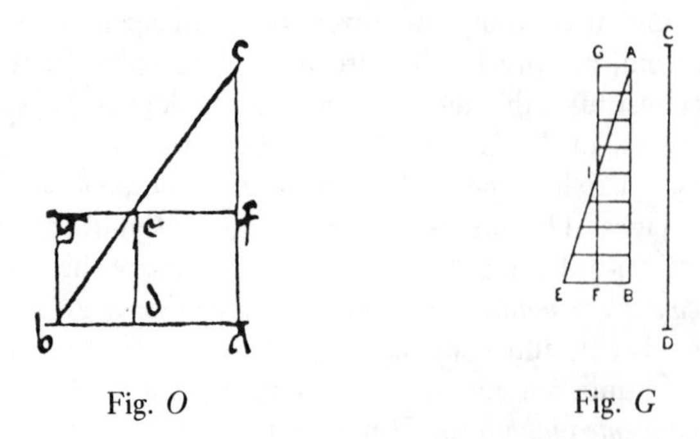

Fig. *O* Fig. *G*

are ordered pairs (and *not* sums of any sort), and such that (a) the first member of each such pair is an element of S_1 and the second member is an element of S_2, and (b) each element of S_1 occurs as a first member and each element of S_2 occurs as a second member of exactly one such matched pair.

If one now reviews the text of the *Discorsi* and there translates (but does not betray) *Theorema I* as:

If the spaces traversed are the same (*CD* in each case), and the speeds (S_1 of M_1 and S_2 of M_2) are equivalent, then the times elapsed are the same (*AB* in each case),

and its conclusion as:

If the times elapsed are the same (*AB* in each case), and the speeds (S_1 of M_1 and S_2 of M_2) are equivalent, then the spaces are the same (*CD* in each case),

one will instantly discern behind Galileo's retooled vocabulary (*aggregatum aequalem aggregatui; singulis et omnibus respondeant singula et omnia; totidem, itidem; totidem, ac*) how he proceeds to specify – but *not* total – the respective unit elements of the set S_1 of M_1 and the set S_2 of M_2, and then (2) establish a biunivocal correspondence between their respective members, some common to both and equal by construction, and the remainder equalized by compensatory line segments. If so, then *Theorema II,* which depends crucially on *Theorema I,* and what follows in *Corollarium I* proceed smoothly and easily.

If I here select such equivalence of sets S_1 and S_2 as something *new* in Galileo, I find that I echo the sentiments of Mersenne, who cites the *Discorsi* as *les nouvelles pensées* of Galileo and remarks in particular that the demonstrations therein contained about the proportions between local motions are *inconnues jusqu' à presént.*

Such novelty makes mandatory a critical review of precisely what it was, if anything at all, that Galileo is reported to have *borrowed* from his medieval predecessors in mechanics, and in particular the questionable claim that Galileo (Figure G) used the very same diagram that Oresme (Figure O) had employed.

I note first that Galileo (on the basis of empirical evidence: *li accidenti da me osservati*) already knew in 1604, and thus long before *G* was constructed, as an expository and *not* heuristic device, that *gli spazii passati dal moto naturale esser in proporzione doppia dei tempi,* which is the exact content of the conclusion (*Theorema II*) that such *G* was designed in part (*Theorema I*) to demonstrate: *ergo ratio spatiorum peractorum dupla est rationis temporum.* This space–time correlation in free fall Oresme never knew, neither before nor after his original construction of such *O.*

I note, second, that in the Eudoxian theory of proportions, which Galileo knew and Oresme did not, ratios exist if and only if the magnitudes they correlate are *eiusdem generis.* Hence the relevant quantities for the comparison of two such homogeneous magnitude pairs can be isomorphically represented by the counterpart components of a single geometrical figure *F,* if and only if such *F* is composite by superposition, such as a rectangle constructed upon a triangle. If therefore, as here, both *O* and *G* happen to be thus composite by superposition, there are independent and intrasystematic reasons, other than reproduction of *O,* why *G* should be such.

I note, third, that (a) whereas in *O* the hypotenuse *cb* is technically construed as a "line of summit" and theoretically composed of the adjacent upper extremities of the velocity lines, erected perpendicularly upon the extent of the total time base line *ab,* (b) in *G* the hypotenuse *AE* is drawn independently (*iunctaque AE*), and antecedently to the construction of the sample velocity lines (parallel to *BE*), and directed from *AB* (*ex singulis punctis lineae AB*) toward the target boundary line *AE,* already in place.

I note, fourth, that (a) whereas in *O* the point *d* bisects the base line *ab* of total time duration of the imaginary (*ymaginaciones*) motion in question, and the perpendicular *de,* erected thereon, represents the velocity line at that midpoint of time, (b) in *G* point *F* halves the arbitrary (*utcunque super AB*) maximum (*maximus*) velocity line *BE* at a measurable amount of speed, which is also final (*ultimus*) because the real free fall of *il grave cadente* ceases at whatever happens to be the moment of percussive impact on some terminating surface.

I note finally that (a) whereas in *O* in the equal areas of both component figures are respectively construed by modern scholars as the summation of all the velocity lines that each perimeter encloses, so that the total motion is in each case velocity through the whole time, and hence dimensionally equivalent to the equal distances respectively traversed, and (b) whereas Galileo in the earlier, polemical, and often sciolist propaganda *Dialogo* once tendered the idea, as at least *ben ragionevole e probabile,* that an area could represent *la massa e la summa di tutta la velocità,* (c) in the later, definitive, and deliberately demonstrative *Discorsi* where – to make a mathematical point – *G* occurs, and where unlike the successive degrees of some inherent *qualitas intensiva* that may be said to accumulate within a subject, Galileo says *totidem velocitatis momenta absumpta esse in motu accelerato ac in motu aequabili* – a distinct but related vertical line *CD* is also inserted into the total diagram to prevent a recurrence of this *Dialogo* misconception (*Repraesentetur per extensionem AB tempus in quo a mobili latione uniformiter accelerata ex quiete in C conficiatur spatium CD*). This latter linear maneuver – more mechanical than geometrical – exposes how hopelessly inadequate in fact were

Galileo's mathematical resources to handle effectively the physical problems of *real* velocity in progress, rather than, as here, in static retrospect. For such live velocity, although construed as a physical magnitude, lacked an appropriate geometrical counterpart, and was not yet defined as a derived quantity, a ratio of independent measurements of space and of time.

I therefore find it imperative to conclude that even if Galileo knew all the representative specimens so magisterially assembled in Clagett's *Mechanics* – and I know no documentary proof that he was personally acquainted with any of them – it is, I think, certain that Galileo did not here borrow anything at all. If this dour result makes Galileo the Melchizedek of seventeenth-century mechanics, sans immediately prior and identifiable parentage and sans certifiable progeny in the next generation of physicists, it may just be the case that such is exactly what Galileo *was* (and the only one there is).

If we now omit from this account several things that, it is sometimes said, Galileo *borrowed* (in the pejorative sense of misappropriated), such as the idea for the *perspicillum,* and several calendar dates for sunspots observation, the only thing that I am sure that Galileo *borrowed,* and thus rendered in the newer context contemporaneous, is the literary format of the *Dialogo,* and that – not directly from Plato – but from his father's *Dialogo* (1581) on ancient and modern music.

What I find *blue* in Galileo himself, as distinct from the altogether deplorable and completely indefensible Galileo affair, is the recurrence, as a mere character trait and in no way a moral fault, in correspondence and publications, of typical paranoid reaction states, four diagnostic symptoms of which are (1) abnormal self-reference; (2) persecution complex plus motivational projection; (3) delusion of grandeur plus masochistic self-depreciation; and (4) confabulation of disparate ideas into novel constellations.

Although these symptoms come packaged in the same person, they are displayed with variant emphases in different behavior configurations. I therefore subjoin several sample items from a more ample Galileo case file. Item 1 appears in the *Discorsi,* sandwiched between some confident (but incorrect) comments about impact and the essential definition of uniformly accelerated motion. The passage is a paradigm of motivational projection, including an enemies' list (*registro*), and poignant because such behavior is most often self-portraiture in reverse. Item 2 is the 1632 Cavalieri-Galileo-Marsili correspondence file, concerning priority of publication of the parabolic path of projectiles. These Galileo letters, although transparently understandable and altogether excusable under the distressing circumstances, nevertheless depict the classic clinical syndrome of a paranoid paroxysm from automatic arousal through eventual subsidence. The scenario is sad because it reveals not only the psychic pain (*il mio digusto*) that such reaction produces in its victim, but also in another who is, in turn, victimized and reduced to vassalage. Item 3 is the witness of Galileo's own demonstrative science of physics in which – happily in a constructive manner, like Newton (universal gravitation), rather than in a destructive way, like Hitler (biogenetic politics) and like Nixon (imperial presidency) – the *Filosofo e Mate-*

matico of Tuscany joined together in a novel constellation the previously disparate ideas and ideals of *mathesis* and *philosophia naturalis*.

Clinical practice further reveals that such adult psychic behavior patterns have an innocent history, frequently traceable to some inflicted (rejection) trauma in early life. Could it be that the abortive termination in July 1579, of the young (fifteen) novice Galileo's trial period of candidacy for acceptance as a (lifelong) member of the Vallombrosan Order of monks was, or at least was perceived by him to be, either an unjust dismissal (rejection) by the monastery authority figures, or an unwelcome intervention (recall) by paternal authority – or both? If so, then most analysts of my acquaintance would not hesitate to pursue seriously the research possibility, at least, that the real author of the *Dialogo* (*Mein Kampf*) was born, not at Pisa in 1564, but at the Monastero di Santa Maria in 1579. If so, then students of the conceptual history of scientific ideas – never disembodied as were the *Ideas* of Plato, but always incarnated in the psychosomatic mix of some human person – could well produce, if still possible, a reliable psychohistory of the Galileo that *was* (and the only one there is).

NEWTON

The magnificent and monumental *variorum* edition of the *Principia* discloses (almost cruelly) that the least suitable format for a living document that grows and matures with its author is the skeletal structure of an axiom system *S* in the construction of which the logical ideal is to select from a set of previously mastered materials and then employ as relatively primitive propositions such minimal statements as are, not only mutually consistent and reciprocally independent, but also jointly both necessary and sufficient to decide the validity of all other statements formulated in the idiom of the same such *S*. For a student of logic and of *axiomatisches Denken* in general quivers when he sees the historical Newton shift, almost at will, from *Hypotheses* to *Leges* to *Leges Motus* to *Axiomata sive Leges Motus*, and the same reader quakes when he observes that at least two such declared *axiomata* are implicitly contained in the prior *Definitiones*. The same reader, finally, who recalls *Definition VIII*, despairs of the deductive apparatus altogether, in any version from *M* through E_3, when he watches Newton (1) employ intuition (and not deduction) to extend *Law II* from *impulses* to *continuously acting forces*, and (2) interpolate – *Law III* is irrelevant, though cited, and *Corollary II* to the *Laws of Motion* lacks proof – into the entailment process itself a previously unstated rule for adding infinitely many forces.

The complaint here is not about the physics that Newton thus provides. For such intuition enables Newton to extend the concept of real forces, observed at work in impact studies, to the imagined forces at work in the measured motions of planets, comets, moons, and tides, and thus to construct the first successfully operational dynamics of the world system. And such interpolation allows Newton (a) to state that the force of attraction between two homogeneously layered spheres is directed along the line of centers and is independent of their respective diameters, and then (b) to employ a model in which such immense bodies as the sun, earth, and moon are represented as mathematical points. It is, therefore, a mistake to imagine that Newton's *Laws,* in any version, are logically sufficient for the demonstration of what he claimed to deduce from them and a delusion to suppose that they serve as an adequate logical resource for the science of rational mechanics as a whole. For the *Principia* is, at best, the most preposterous and the most proficient treatise on solar system mechanics theretofore produced, but presented behind the facade of a deductivist architectonic that is formalized only in part.

The fault, therefore, is not with the physics but with the format. It has, in this connection, been suggested that the author of the *Principia* was endowed with a natural penchant for constructing best available axiomatic systems. I am not so persuaded. For I find no trace of such a congenital noetic bias in any as yet available version of the *Opticks.* And I recall that Newton first composed the third book of the *Principia* in a popular method and with the intent that it might thus be read by many. But after considering that such readers as had not sufficiently comprehended his principles could not easily discern the strength of their consequences, or lay aside the prejudices to which they had been many years accustomed, and in order to prevent the disputes which might be raised upon such accounts, Newton chose to reduce the substance of that book into the form of propositions (in the mathematical way) which should be read by those only who had first made themselves masters of the principles established in the two preceding books. But not *everything* in *each.* For *Books I* and *II* abound with such matters as might cost too much time, even to readers of good mathematical learning, and it is enough, Newton advised, if one carefully reads the definitions, the laws of motion, and the first three sections of the first book, and then passes on to the third, consulting such of the remaining propositions of the first two books, as references thereto in the third and the reader's own occasions shall require.

Are we therefore to infer from this explicit *Book III* maneuver that Newton employed the deductivist style in the *Principia,* not because he thought that the ensemble of his multiple calculatory investigations and gigantic *Gestalten* experiences in mathematics and in mathematical physics formed one logically articulated unit system whole, but rather to render each and all of them less vulnerable to psychologically painful criticism? I do not know. But it is, at any rate, this ancient axiomatics that I find very *old* in Newton who, I surmise, very much wanted to do the preposterously novel unified and universal physics at which he excelled, but with as little abrasion to the scar tissue of his hypersensitive psyche as this always hazardous profession would then permit.

Printed words emit no sound. But even silent readers sometimes hear in the chambers of memory recognizable echoes of the same words when voiced, and even distinctly different tonal variations when semantical equivalents appear in the published texts of authors as diverse as Plato or Aristotle or Ptolemy or Oresme or Copernicus or Kepler or Galileo or Descartes. For all of them wrote in one way or another about gravity, orbits, and the moon, and thus provided variations upon the same basic sound of these key words, in a way somewhat comparable to the individually cadenced chords produced by different classical composers from the same type of keyboard instrument. But when I first read long ago in Rouse Ball, I believe, the following (and, although critically revalued, still accredited) words of Newton: "I . . . compared the force requisite to keep the Moon in her Orb with the force of gravity at the surface of the earth, and found them answer pretty nearly," I knew that I was hearing a *new* sound, never before produced upon this planet, just as, for example, the music devotee who has already listened time and time again to all the characteristically different sounds of all the masters knows instinctively that he is hearing for the first time a *new* sound from the electronic wizardry of the Moog synthesizer.

This choice of something new in Newton does not imply that the author of the *Principia* merely *synthesized,* as some say, such items as by a set of successively focused *Gestalten* experiences Newton selectively *borrowed* from the unfinished (and disturbingly defective) business of his predecessors, and thus rendered contemporaneous by *assimilation* into a new context, both unified in concept and universal in extent. Consider the case of Galileo's "law" of falling bodies. It reads in the *Discorsi: ratio spatiorum peractorum dupla est rationis temporum.* This statement is descriptive, not prescriptive. It in no way resembles the Stamp Act of 1765. The syntax is declarative, and in no way repeats the imperative idioms of the *Decalogue.* It is, of course, a historical fact that bodies fell before Galileo and fell as Galileo later found them to fall. But it is not a historical fact that the "law" of falling bodies, as a specimen statute from the *Codex Iuris Divini,* was always there, waiting to be discovered by Galileo. It is an anthropomorphic myth, prevalent in the scientific subcultures of the past, and from which few practitioners were altogether immune. This pervasive contagion is, indeed, a historical fact that it behooves historians of science to report, *but not to perpetuate by rote repetition.*

For it is systematically counterproductive to muffle the context of such "laws" while amplifying their content, or – in short – to expunge the text that alone endows an equation with significance for the scientist at work. For each such law comes packaged in history (if not in histories of science) with a portfolio of explanatory provisos, each of which functions in an important way to determine its field, its range, its scope, its relevance, and thus set the boundary conditions that alone preserve its algebraic formulation from complete inaccuracy. And it is always somewhere within the ambience of this *chiaroscuro* conceptual environment that the thinking scientist perceives the disturbing defects in the unfinished business of a predecessor that, in turn, provide the occasion for relief from such tension by a *Gestalt* experience in which integrative intelligence generates the elements of a new and more satisfactory conceptual scheme for a reinterpretation of old phenomena – but only until further notice!

For what science teachers, peripheral philosophers, and even historians of science perversely call laws are in fact nothing but functional correlations, always inaccurate and at best only approximate, between named (and measured) magnitudes, systematically connected by way of operational definitions to some one or other element in the reticular network of a theoretical conceptual scheme.

But if such functional linkages (not laws) are not sections of the original world blueprint, it does not follow that such formulas are fraudulent and feckless. For a theory that works (so far as it does) must somehow fit the universe, even though in an important sense it creates a world of its own. And in *that* world of nicknames – such as "electrons," "mesons," "cosmic rays," "cells," "asteroids," and on through the entire lexicon of the sciences – such linkages are not required to *do* anything other than to continue to provide its incurably inquisitive inhabitants with some provisional sketches that promote the cognitive enterprise of making sense out of sense.

Come with me now to alpha Centauri and from that cosmic perspective review, in briefest compass, some unfinished business (and disturbing defects) in the published natural philosophy of Galileo, Kepler, and Descartes. For the author of the *Dialogo,* so far at least as that work reveals, where it does not conceal, his authentic perspectives, the planets perform their circumsolar revolutions with a motion that is, quite naturally, natural. On the planet earth *il grave cadente* possesses an inherent *gravità* and falls quite naturally at the same rate at any distance, vertically or parabolically. For Kepler the amount of matter M determines the volume V of each planet in the laminar sequence, and possesses, quite naturally, an inherent tendency (*inertia*), proportional to M, to slow down and stop. Each such planet must therefore always be propelled, harmonically with its neighbor, along the length L of its elliptical path by the absorption A, proportional to V, of a magnetic species S, emitted from a fixed focal and rotating sun. The time period thus required for a planet–sun line to sweep out equal times and so complete one revolution is (a) directly proportional to the product of L and M, and (b) inversely proportional to the product of S and V. For Descartes our world is, quite naturally, the Lord God's private and patented royal automaton toy. But anyone can clearly and distinctively see that an operative mechanical universe is devoid of void, and funded at the start with a fixed and static amount of motion. Such motion is indeed conceived as a state, coequal with rest, but in such a way that no change in such a privileged rectilinear motion state can occur without the impact action of some other previously impacted body within the *plenum.* Hence all changes in direction and orbital variations in the speed of bodies within interplanetary space require involvement with cosmic vortexes, already in place, and difficult to analyze mathematically. I doubt that even God – and much less Newton – could structure our world from bodies that, quite naturally, either fall down when unsupported, or slow down when not propelled, or move right on until the next collision. And Newton said: "I . . . compared the force requisite to keep the Moon in her Orb with the force of gravity at the surface of the earth and found them answer pretty nearly."

When a fellow human being risks everything, as Newton did, to ask himself, as Newton did, a decidable question Q_1 about phenomena, and then proceeds, as Newton did, to exhaust his intellectual energies, as Newton did, in the formulation of a testable answer A_1 to Q_1, and when that A_1 passes every test to which it could then conceivably be put on trial, I find it *blue,* indeed, that the same triumphant A_1 breeds a further Q_2 to which the tired Newton not only did not devise in a further *Gestalt* experience an equally adequate A_2, but protested that such Q_2 was not even a legitimate scientific inquiry. But one does not demean the theoretical power of universal gravitation when one respectfully asks just what, *if anything at all,* a "mathematical force" can possibly be. And it is no help to argue, as Newton often does in public, that the controverted expression is completely meaningful. For glossaries available to historians of science teem with terms, such as "phlogiston," which are completely meaningful but just as completely devoid of reference, and remain names that fail to name anything. And if so, then Newtonian science may be less a physics than a faith – but one that worked wonders.

EPILOGUE

The complete version of the folklorist verse that has been pirated in part of the title of this essay closes with a third line: "and a lucky sixpence in her shoe." This sentiment is not alien to Kepler, who wrote: *non minus admirandae vindentur occasiones quibus homines in cognitionem rerum coelestium deveniunt quam ipsa natura rerum coelestium,* and thus reminds us all of the role of serendipity in the work of scientists who sometimes seem in restrospect – and especially to the moralistic methodologists among us – to have succeeded better than they had any right to do.

I have always though it lucky for Copernicus (and lucky for us) that Ptolemy followed the solar and not the lunar calendar year. And everyone knows how lucky it was for Galileo (and lucky for us) to hear about the existence of the new spyglass. And all who should know realize how lucky it was for Newton (and lucky for us) that Hooke wrote that letter of invitation in the autumn of 1679. And it will be lucky for everyone, especially all those idiosyncratic scientists who are what they are and do what has to be done (and whose autopsy time will come), if historians of science never forget that the muse of history is still Clio, and *not* Procrustes.

4

Conceptual revolutions and the history of mathematics

Two studies in the growth of knowledge

JOSEPH W. DAUBEN
City University of New York

> In most sciences one generation tears down what another has built, and what one has established another undoes. In mathematics alone each generation builds a new story to the old structure.
>
> Hermann Hankel

> Je le vois, mais je ne le crois pas.
>
> Georg Cantor

> Transformation, by presenting each anterior concept, theory, law, or principle as the *occasion* of an innovation, focuses attention on the *cause*, the possible reason why only one of the many scientists to whom the scientific idea was known produced the transformation in question.
>
> I. Bernard Cohen

It has often been argued that revolutions do not occur in the history of mathematics and that unlike the other sciences, mathematics accumulates positive knowledge without revolutionizing or rejecting its past.[1] But there are certain critical moments even in mathematics that suggest that revolutions do occur – that new orders are brought about and eventually serve to supplant an older mathematics. Although there are many important examples of such innovation in the history of mathematics, two are particularly instructive: the discovery by the ancient Greeks of incommensurable magnitudes and the creation of transfinite set theory by Georg Cantor in the nineteenth century. Both examples are as different in character as they are separated in time, and yet each provides a clear instance of a major transformation in mathematical thought. The Greeks' discovery of incommensurable magnitudes brought about changes that were no less significant than the revolutionary transformation mathematics experienced in the twentieth century as a result of Georg Cantor's set theory. Taking each of these as marking important transitional periods in mathematics, this essay is an attempt to investigate the character of such transformations.

Recently there has been considerable interest in the growth of mathematics,

This research was originally undertaken during 1977–78, when the author was a member of the Institute for Advanced Study, Princeton, under the auspices of a Herodotus Fellowship. Additional support has also been provided by grants from The City University of New York PSC-BHE Research Award Program. An early version of this paper was read at the New York Academy of Sciences on September 27, 1978.

the nature of that growth, and its relation to the development of knowledge generally. In the fall of 1974, at the fiftieth anniversary meeting of the History of Science Society, an entire session was devoted to the historiography of mathematics and to the relationship between the growth of mathematical knowledge and the patterns described in Thomas S. Kuhn's book, *The Structure of Scientific Revolutions.*[2] Naturally, the question of revolutions arose, and the problem of whether revolutions occur at all in the history of mathematics. When invited to consider the example of Cantorian set theory, I took the opportunity to suggest that revolutions did indeed occur in mathematics, although the example of transfinite set theory seemed to imply that Cantor's revolutionary work did not fit the framework of Professor Kuhn's model of anomaly-crisis-revolution.[3] Nor is there perhaps any reason to expect that a purely logicodeductive discipline like mathematics should undergo the same sort of transformations, or revolutions, as the natural sciences.

Similar interest in the nature of mathematical knowledge and its growth was evidenced at the Workshop of the Evolution of Modern Mathematics held at the American Academy of Arts and Sciences in Boston, August 7–9, 1974. Of all the participants at the workshop, no one questioned the phenomenon of revolutions in mathematics so directly as did Professor Michael Crowe of the University of Notre Dame. In a short paper prepared for the workshop and subsequently published in *Historia Mathematica,* he concluded emphatically with his "tenth law" that "revolutions never occur in mathematics."[4] My intention here, however, is to argue that revolutions can and *do* occur in the history of mathematics, and that the Greeks' discovery of incommensurable magnitudes and Georg Cantor's creation of transfinite set theory are especially appropriate examples of such revolutionary transformations.

REVOLUTIONS AND THE HISTORY OF MATHEMATICS

Whether one can discern revolutions in any discipline depends upon what one means by the term "revolution." In insisting that revolutions never occur in mathematics, Professor Crowe explains that his reason for asserting this "law" depends on his own definition of revolutions. As he puts it, "My denial of their existence is based on a somewhat restricted definition of 'revolution' which in my view entails the specification that a previously accepted entity *within* mathematics proper be rejected."[5] Having said this, however, Professor Crowe is willing to admit that non-Euclidean geometry, for example, "did lead to a revolutionary change in views as to the nature of mathematics, but not within mathematics itself."[6]

Certainly one can question the definition Professor Crowe adopts for "revolution." It is unnecessarily restrictive, and in the case of mathematics it defines revolutions in such a way that they are inherently impossible within his conceptual framework. Nevertheless, revolutionary moments have been identified, not only by historians but by mathematicians as well. Rather than dictate the meaning of revolution, there is no reason not to allow its use in legitimately describing certain penetrating changes in the evolution of mathematics. However, before challenging further the assertion that revolutions never occur in the history of mathematics, it will be helpful to consider briefly the meaning of revolution as a historical concept. Here we are fortunate in having a recent study by Professor Cohen to guide us. In fact, what follows is a very brief resumé of results owing largely to Professor Cohen's research on the subject of revolutions.[7]

The concept of revolution first made its appearance with reference to scientific and political events in the eighteenth century, although with considerable confusion and ambiguity as to the meaning of the term in such contexts. In general, the word was regarded in the eighteenth century as indicating a breach of continuity, a change of great magnitude, even though the old astronomical sense of revolution as a cyclical phenomenon persisted as well. But following the French Revolution, the new meaning gained currency, and thereafter revolution commonly came to imply a radical change or departure from traditional or acceptable modes of thought. Revolutions, then, may be visualized as a series of discontinuities of such magnitude as to constitute definite breaks with the past. After such episodes, one might say that there is no returning to an older order.

Bernard de Fontenelle may well have been the first author to apply the word "revolution" to the history of mathematics, and specifically to its evolution in the seventeenth century. In his *Éléments de la géométrie de l'infini* (1727), he was thinking of the infinitesimal calculus of Newton and Leibniz.[8] What Fontenelle perceived was a change of so great an order as to have altered completely the state of mathematics. In fact, Fontenelle went so far as to pinpoint the date at which this revolution had gathered such force that its effect was unmistakable. In his eulogy of the mathematician Rolle, published in the *Histoire de l'Académie Royale des Sciences* of 1719, Fontenelle referred to the work of the Marquis de l'Hôpital, his *Analyse des infiniment petits* (first published in 1696, with later editions in 1715, 1720, and 1768), as follows:

> In those days the book of the Marquis de l'Hôpital had appeared, and almost all the mathematicians began to turn to the side of the new geometry of the infinite, until then hardly known at all. The surpassing universality of its methods, the elegant brevity of its demonstrations, the finesse and directness of the most difficult solutions, its singular and unprecedented novelty, it all embellishes the spirit and has created, in the world of geometry, an unmistakable revolution.[9]

Clearly this revolution was qualitative, as all revolutions must be. It was a revolution that Fontenelle perceived in terms of character and magnitude, without invoking any displacement principle – any rejection of earlier mathematics – before the revolutionary nature of the new geometry of the infinite could be proclaimed. For Fontenelle, Euclid's geometry had been surpassed in a radical way by the new geometry in the form of the calculus, and this was undeniably revolutionary.

Traditionally, then, revolutions have been those episodes of history in which the authority of an older, accepted system has been undermined and a new, better authority appears in its stead. Such revolutions represent breaches in continuity, and are of such degree, as Fontenelle says, that they are unmistakable even to the casual observer. Fontenelle has aided us, in fact, by emphasizing the discovery of the calculus as one such event – and he even takes the work of l'Hôpital as the identifying marker, much as Newton's *Principia* of 1687 marked the scientific revolution in physics or the Glorious Revolution of the following year marked England's political revolution from the Stuart monarchy. The monarchy, we know, persisted, but under very different terms.

In much the same sense, revolutions have occurred in mathematics. However, because of the special nature of mathematics, it is not always the case that an older order is refuted or turned out. Although it may persist, the old order nevertheless does so under different terms, in radically altered or expanded contexts. Moreover, it is often clear that the new ideas would never have been permitted within a strictly construed interpretation of the old mathematics, even if the new mathematics finds it possible to accommodate the old discoveries in a compatible or consistent fashion. Often, many of the theorems and discoveries of the older mathematics are relegated to a significantly lesser position as a result of a conceptual revolution that brings an entirely new theory or mathematical discipline to the fore. This was certainly how Fontenelle regarded the calculus. Similarly, it is also possible to interpret the discovery of incommensurable magnitudes in antiquity as the occasion for the first great transformation in mathematics, namely, its transformation from a mathematics of discrete numbers and their ratios to a new theory of proportions as presented in Book V of Euclid's *Elements*.

THE PYTHAGOREAN DISCOVERY OF INCOMMENSURABLE MAGNITUDES

Aristotle reports the Pythagorean doctrine that all things were numbers and surmises that this view doubtless originated in several sorts of empirical observation.[10] For example, in terms of Pythagorean music theory the study of harmony had revealed the striking mathematical constancies of proportionality. When the ratios of string lengths or flute columns were compared, the harmonies produced by other, but proportionally similar lengths, were the same. The Pythagoreans also knew that any triangle with sides of length 3, 4, 5, whatever unit might be taken, was a *right* triangle. This too supported their belief that ratios of whole numbers reflected certain invariant and universal properties. In addition, Pythagorean astronomy linked such terrestrial harmonies with the motions of the planets, where the numerical harmony, or cyclic regularity of the daily, monthly, or yearly revolutions were as striking as the musical harmonies the planets were believed to create as they moved in their eternal cycles. All of these invariants gave substance to the Pythagorean doctrine that numbers – the whole numbers – and their ratios were responsible for the hidden structure of all nature. As Aristotle comments:

> The so-called Pythagoreans, having begun to do mathematical research and having made great progress in it, were led by these studies to assume that the principles used in mathematics apply to all existing things . . . they were more than ever disposed to say that the elements of all existing things are found in numbers.[11]

But what were these numbers? For the early Pythagoreans, Aristotle indicates that they were apparently something like physical "monads." In the *Metaphysics*, for example, one passage offers the following elaboration: "[The Pythagoreans] compose all heaven of numbers (ἐξ ἀριθμῶν), not of numbers in the purely arithmetical sense, though, but assuming that monads have size."[12]

Thus the Pythagoreans apparently came to regard the numbers themselves as providing the structure and form of the material universe, their ratios determining the shapes and harmonies of all symmetrical things. The Pythagoreans gave the word λόγοι to the groups of numbers determining the character of a given object, and later the meaning of this word was extended, as we shall see, from that of "word" to "ratio."[13]

This sort of arithmology found its realization in the Pythagorean's quest to associate numbers with all things, and to determine the internal properties, ratios, and relations between numbers themselves. Thus the number of stones needed to outline the figure of a man or a horse was taken by the Pythagorean Eurytus as the "number" for man or horse.[14] The essence of such things was expressed by a particular number. Moreover, some Pythagoreans sought to establish the number for justice, or for marriage. Others distinguished numbers that were perfect (the tetractys, for example, $1 + 2 + 3 + 4 = 10$), amicable, or friendly. Figured numbers, including pentagonal and solid numbers, were also subjects of great interest.[15] It is against this background of Pythagorean numerology in which the λόγος of all things was thought to be an invariant principle of the universe, expressible in terms of whole numbers and their ratios, that the discovery of incommensurable magnitudes must be viewed. The Pythagoreans' arithmology would doubtless have provided sufficient incentive for their search for the hidden numbers, the prevailing logos governing the most important objects of their mysticism, for example, the pentagon or the golden section. It is also possible that the discovery was made in less rarefied contexts, through study of the simplest of right triangles, the isosceles right triangle.

Exactly when incommensurable magnitudes were first discovered is not particularly relevant for the argument here.[16] Similarly, the details of the initial discovery are also of secondary importance, and we can dispense with the dilemma of whether the discovery was first made in the context that Aristotle reports it, by studying the ratio of the length of a square's edge with its diagonal, or whether, as has been argued by K. von Fritz (1945) and by S. Heller (1958), that Hippasus found incommensurability in considering the construction of the regular pentagon.[17] What concerns us is the discovery and its subsequent effect. Philosophically, it would certainly have represented a crisis for the Pythagoreans.[18] Having been tempted by the seductive harmony of generalization, some Pythagoreans had carried their universal principle that all things were number too far. The complete generalization was inadmissible, and this realization was a major blow to Pythagorean thought, if not to Greek mathematics. In fact, a scholium to Book X of Euclid's *Elements* reflects the gravity of the discovery of incommensurable magnitudes in the well-known fable of the shipwreck and the drowning of Hippasus:

It is well known that the man who first made public the theory of irrationals perished in a shipwreck in order that the inexpressible and unimaginable [Καὶ ἄλογον Καὶ ἀνεὶδεον] should ever remain veiled . . . and so the guilty man, who fortuitously touched on and revealed this aspect of living things, was taken to the place where he began and there is forever beaten by the waves.[19]

What deserves attention here are the words "inexpressible" and "unimaginable." It is difficult, if not impossible, for us to appreciate how hard it must have been to conceive of something one could not determine or name – the inconceivable – and this was exactly the name given to the diagonal: ἄλογον. This reflects the double meaning of the word *logos* as *word,* as the "utterable" or "namable," and now the irrational, the *alogon,* as the "unspeakable," the "unnamable." In this context, it is easy to understand the commentary: "Such fear had these men of the theory of irrationals," for it was literally the discovery of the "unthinkable."[20]

Ultimately, however, the Greeks regarded the discovery not as a crisis but as a great advance. Whether or not discovery of incommensurable magnitudes precipitated a crisis in Greek mathematics, and if so, whether it affected only the foundations of mathematics rather than the mathematics itself, the significant issue concerns the *response* mathematicians were forced to make once the existence of incommensurable magnitudes had been divulged and was a matter of general knowledge.[21]

What ultimate effect did this discovery have on the content and nature of Greek mathematics? Above all, the theories of proportion advanced by Theaetetus and Eudoxus in the early fourth century B.C. (390–350 B.C.) served to reverse the emphasis of earlier mathematics. Consider, for example, the statement of Archytas (an early Pythagorean and teacher of Eudoxus), who was emphatic that *arithmetic* was superior to geometry for supplying satisfactory proofs.[22] After the discovery of incommensurable magnitudes, such a statement would be virtually impossible to justify. In fact, the opposite was closer to the truth, as the subsequent development of Greek geometric algebra demonstrates.

Basically, the transformation from a simple theory of commensurable proportions (where geometry and arithmetic might be regarded as coextensive) to a new theory embracing incommensurable magnitudes (for which arithmetic was inadequate) centers on the contributions of Theaetetus and Eudoxus. However, we know from Plato's *Theaetetus* that a major step toward the better understanding of the irrational was taken by Theaetetus's teacher, Theodorus, who established the incommensurability of certain magnitudes up to (but not including) $\sqrt{17}$ by means of geometric constructions. Although Theodorus's achievements were limited owing to his lack of a sufficiently developed arithmetic theory, some historians have argued that he began to develop a metric geometry capable of handling arithmetic properties in much the form of propositions in Book II of Euclid's *Elements.*[23]

Following his teacher Theodorus, Theaetetus became interested in the general properties of incommensurables and produced the classification that so impressed Socrates in Plato's dialogue.[24] Also, Theaetetus realized that to treat incommensurables successfully, geometry had to embody more of the results of arithmetic theory, and so he sought to translate necessary algebraic results into geometric terms. Here he focused on the arithmetic properties of relative primes, using the process of determining greatest common factors by means of successive subtraction, or *anthyphairesis.*[25] This enabled Theaetetus to reformulate the theory of proportion to include certain incommensurable magnitudes that he classified as the *medial, binomial,* and *apotome,* and these were enough for the results in which he was interested. But Theaetetus apparently was not inspired to study the new theory of proportion itself–something his premature death certainly precluded.

Eudoxus, however, realized that the methods Theaetetus had brought to geometry from arithmetic for the purpose of studying incommensurables could actually provide the basis for an even more comprehensive theory of proportion. In studying the construction of the regular pentagon, dodecahedron, and icosahedron, Eudoxus seems to have realized that these, like segments divided into mean and extreme ratio, involved incommensurable magnitudes that were not included in the three classes treated by Theaetetus.[26] Because of his interest in a formal, more comprehensive theory of proportions, he transformed Theaetetus's methods involving *anthyphairesis* by focusing on the theory of proportion itself and producing in large measure the theorems elaborated in Book V of Euclid's *Elements,* where the concept of equal multiples made it possible to develop a theory of proportion that was generally applicable to incommensurables. The advantages of the new Eudoxean theory were considerable, and comparison with Theaetetus's anthyphairetic approach made clear the differences. Aristotle, in fact, contrasted the two on several occasions, and noted the superiority of Eudoxus's formulation explicitly.[27]

Having produced a comprehensive theory of proportion, however, Eudoxus and his followers, perhaps chief among them Hermotimus of Colophon, were also interested in providing a systematic development of the new theory that eventually provided the basic framework for Euclid's Book V of the *Elements,* a book a scholiast tentatively attributes to Eudoxus.[28] In dealing with incommensurable magnitudes, "unfamiliar and troublesome" concepts as Morris Kline has described them, the need to formulate axioms and to deduce consequences one by one so that no mistakes might be made was of special importance.[29] This emphasis, in fact, reflects Plato's interest in the dialectic certainty of mathematics and was epitomized in the great Euclidean synthesis, which sought to bring the full rigor of axiomatic argumentation to geometry. It was in this spirit that Eudoxus undertook to provide the precise logical basis for the incommensurable ratios, and in so doing, gave great momentum to the logical, axiomatic, a priori "revolution" identified by Kant as the great transformation wrought upon mathematics by the Greeks.[30]

In concluding this brief summary of Greek mathematics and the transformation caused by the discovery of incommensurable magnitudes, several aspects of that transformation deserve particular emphasis. Primarily, two things were unacceptable after the discovery of incommensurables: (1) the Pythagorean interpretation of ratio, and (2) the proofs they had given concerning commensurable magnitudes came into play. A new theory was needed to accommodate irrational magnitudes – and this was provided by Theaetetus and Eudoxus. The less dramatic transformation of the definition of the number concept was a lengthier process, but over the course of centuries, it eventually led to admission of irrational *numbers* as being as acceptable ontologically as natural numbers or fractions.[31]

Wholly apart from the slower, more subtle transformation of the number concept, however, was the dramatic, much quicker transformation of the character of Greek mathematics itself. Because Pythagorean arithmetic could not accommodate irrational magnitudes, geometric algebra (cumbersome though it was) developed in its stead. In the process, Greek mathematics was directly transformed into something more powerful, more general, more complete. Central to this transformation were auxiliary elements that reflected the transformation underway. A new interpretation of mathematics must have discarded as untenable the older Pythagorean doctrine that all things were number – there were now clearly things that did not have numbers in the Pythagorean sense of the word – and consequently their view of number was correspondingly inadequate. The older concept of number was severely limited, and in the realization of this inadequacy and the creation of a remedy to solve it came the revolution. New proofs replaced old ones.[32] Soon a new theory of proportion emerged, and as a result, after Eudoxus, no one could look at mathematics and think that it was the same as it had been for the Pythagoreans. Nor was it possible to assert that Eudoxus had merely added something to a theory that previously was perfectly all right. The lesson of the irrational was that everything was *not* all right. As a result of the new theory of proportion, the methods and content of Greek mathematics were vastly different, and comparison of Book V of Euclid with the Pythagorean books VII–IX (perhaps reflecting directly earlier arithmetics from the previous century) reveals the deep transformation that Eudoxus and his theory of proportion brought to Greek mathematics.[33] The old methods were supplanted, and eventually, although the same words, "number" or "proportion," might continue in use, their meaning, scope and content would not be the same.

In fact, the transformation in conceptualization from irrational magnitudes to irrational numbers represented a revolution of its own in the number concept, although this was not a transformation accomplished by the Greeks. Nor was it an upheaval of a few years, as are most political revolutions, but a basic, fundamental change. Even if the evolution was relatively slow, this does not alter the ultimate effect of the transformation. The old concept of number, although the word was retained, was gone, and in its place, numbers included irrationals as well.

This transformation of the concept of number, however, entailed more than just extending the old concept of number by adding on the irrationals – the entire concept of number was inherently changed, transmuted as it were, from a world view in which integers alone were numbers, to a view of number that was eventually related to the completeness of the entire system of real numbers.

In much the same way, Georg Cantor's creation of transfinite numbers in the nineteenth century transformed mathematics by enlarging its domain from finite to infinite numbers. Above all, the conceptual step from transfinite sets to transfinite numbers represents a shift that was in many ways the same as the shift from irrational magnitudes to irrational numbers. From the concrete to the abstract, the transformation in both cases revolutionized mathematics.

GEORG CANTOR'S DEVELOPMENT OF TRANSFINITE SET THEORY

Born in St. Petersburg (Leningrad) in 1845, Georg Cantor left Russia for Germany with his parents in 1856.[34] Following study at the Gymnasium in Wiesbaden, private schools in Frankfurt am Main and the Realschule in Darmstadt, he entered a *Höhere Gewerbeschule* (Trade School), also in Darmstadt, from which he graduated in 1862 with the endorsement that he was a "very gifted and highly industrious pupil."[35] But his interests in mathematics prompted him to go on to university, and with his parents' blessing he began his advanced studies in the fall of that same year at the *Polytechnicum* in Zürich. Unfortunately, his first year there was interrupted early in 1863 by the sudden death of his father, although within the year he resumed his studies at the university in Berlin. There he studied mathematics, physics, and philosophy and was greatly influenced by three of the greatest mathematicians of the day: Kummer, Weierstrass, and Kronecker.

After the summer semester of 1886, which he spent in Göttingen, Cantor returned to the University of Berlin from which he graduated in December with the distinction: Magna cum laude.[36] Following three years of local teaching and study as a member of the prestigious Schellbach seminar for teachers, Cantor left Berlin for Halle in 1869 to accept an appointment as a *Privatdozent* in the Department of Mathematics. There he came under the influence of one of his senior colleagues, Eduard Heine, who was just completing a study of trigonometric series. Heine urged Cantor to turn his talents to a particularly interesting but extremely difficult problem: that of establishing the uniqueness of the representations of arbitrary functions by means of trigonometric series.[37] Within the next three years Cantor published five papers on the subject. The most important of these was the last, published in 1872, in which he presented a remarkably general and innovative solution to the representation problem.

With impressive skill Cantor was able to show that any function represented by a trigonometric series was not only uniquely represented but that in the interval of representation an infinite number of points could be excepted provided only that the set of exceptional points be distributed in a specific way.[38] The condition was limited to sets Cantor described as point sets of the *first species.*[39] Given a set P, the collection of all limit points p in P defined its first derived set P'. Similarly, P'' represented the second derived set of P, and contained all limit points of P'. Proceeding analogously, for any set P Cantor was able to generate an entire sequence of derived sets P', P'',. . . . P was described as a point set of the first species if, for some index n, $P^n = \varnothing$.

As outlined in the paper of 1872, Cantor's elementary set theoretic concepts could not break away into a new autonomy of their own. Though he had the basic idea of the transfinite numbers in the sequence of derived sets P', P'', . . . ,P^∞, $P^{\infty+1}$, . . . , the basis for any articulate conceptual differentiation between P^n and P^∞ was lacking. As yet, Cantor had no precise basis for defining the first transfinite number ∞ following all finite natural numbers n.[40] A general framework within which to establish the meaning and utility of the transfinite numbers was lacking. The only guide Cantor could offer was the vague condition that $P \neq \varnothing$ for all n, which separated sets of the first species from those of the second. Cantor could not begin to make meaningful progress until he had realized that there were further distinctions yet to be made in orders of magnitude between discrete and continuous sets. Before the close of 1873 Cantor did not even suspect the possibility of such differences.

In order to argue his uniqueness theorem of 1872, Cantor discovered that he needed to present a careful analysis of limit points and the elementary properties of derived sets, as well as a rigorous theory of irrational numbers.[41] It was the problem of carefully and precisely defining the irrational numbers that forced Cantor to face the topological complexities of the real line and to consider seriously the structure of derived sets of the first species.

After the success of his paper of 1872, it was a natural step to search for properties that would distinguish the continuum of real numbers from other infinite sets like the totality of rational or algebraic numbers. What Cantor soon established was something most mathematicians had assumed, but which no one had been able to formulate precisely: that there were more real numbers than natural, rational, or algebraic numbers.[42] Cantor's discovery that the real numbers were nondenumerable was not in itself revolutionary, but it made possible the invention of new concepts and a radically new theory of the infinite. When coupled with the idea of one-to-one correspondences, it was possible to distinguish mathematically for the first time between different magnitudes, or powers, of infinity. In 1874 he was only able to identify denumerable and nondenumerable sets. But as his thinking advanced, he was eventually able to detach his theory from the specific examples of point sets, and in 1883 he was ready to publish his *Grundlagen einer allgemeinen Mannigfaltigkeitslehre,* in which he presented a completely general theory of transfinite numbers.[43] It was in the *Grundlagen* that Cantor introduced the entire hierarchy of infinite number classes in terms of the order types of well-ordered sets. More than twelve years later, in his last major publication, the *Beiträge* of 1895 and 1897, he formulated the most radical and powerful of his new ideas, the entire succession of his transfinite cardinal numbers:

$$\aleph_0, \aleph_1, \ldots .$$ [44]

Cantor's introduction of the actual infinite in the form of transfinite numbers was a radical departure from traditional mathematical practice, even dogma. This was especially true because mathematicians, philosophers, and theologians in general had repudiated the concept since the time of Aristotle.[45] Philosophers and mathematicians rejected completed infinities largely because of their alleged logical inconsistency. Theologians represented another tradition of opposition to the actual infinite, regarding it as a direct challenge to the unique and absolute infinite nature of God. Mathematicians, like philosophers, had been wary of the actual infinite because of the difficulties and paradoxes it seemed inevitably to introduce into the framework of mathematics. Gauss, in most authoritative terms, expressed his opposition to the use of such infinities in mathematics in a celebrated letter to Heinrich Schumacher:

> But concerning your proof, I protest above all against the use of an infinite quantity [*Grösse*] as a *completed* one, which in mathematics is never allowed. The infinite is only a *façon de parler*, in which one properly speaks of limits.[46]

Cantor believed, on the contrary, that on the basis of rigorous, mathematical distinctions between the potential and actual infinite, there was no reason to hold the old objections and that it was possible to overcome the objections of mathematicians like Gauss, philosophers like Aristotle, and theologians like Thomas Aquinas, and to do so in terms even they would find impossible to reject. In the process, Cantor was led to consider not only the epistemological problems his new transfinite numbers raised, but to formulate as well an accompanying metaphysics. In fact, he argued convincingly that the idea of the actual infinite was implicitly part of any view of the potential infinite and that the only reason mathematicians had avoided using the actual infinite was because they were unable to see how the well-known paradoxes of the infinite, celebrated from Zeno to Bolzano, could be understood and avoided. He argued that once the self-consistency of his transfinite numbers was recognized, they could not be refused a place alongside the other accepted but once disputed members of the mathematical family, including irrational and complex numbers.[47] In creating transfinite set theory, Cantor was making a significant contribution to the constellation of mathematical ideas.

Of central concern to Cantor's entire defense of transfinite set theory was the nature of mathematics and the question of what criteria determined the acceptability of mathematical concepts and arguments. He reinforced his support of transfinite set theory with a simple analysis of the familiar and accepted positive integers. Insofar as they were regarded as well defined in the mind, distinct and different from all other components of thought, they served in a connectional or relational sense, he said, to modify the substance of thought itself.[48] Cantor described this reality that the whole numbers consequently assumed as their intrasubjective or immanent reality. In contradistinction to the reality numbers could assume strictly in terms of mind, however, was the reality they could assume in terms of body, manifest in objects of the physical world. Cantor explained further that this second sort of reality arose from the use of numbers as expressions or images of processes in the world of natural phenomena. This aspect of the integers, be they finite or infinite, Cantor described as their transsubjective or transient reality.[49]

Cantor specifically claimed the reality of both the physical and ideal aspects of his approach to the number concept. The dual realities, in fact, were always found in a joined sense, insofar as a concept possessing an immanent reality always possessed a transient reality as well. Cantor believed that to determine the connections between the two kinds of reality was one of the most difficult problems of metaphysics.

In emphasizing the intrasubjective nature of mathematics, Cantor concluded that it was possible to study only the immanent realities, without having to confirm or conform to any subjective content. As noted earlier, this set mathematics apart from all other sciences and gave it an independence from the physical world that provided great freedom for mathematicians in the creation of mathematical concepts. It was on these grounds that Cantor offered his now-famous dictum that the essence of mathematics is its freedom. As he put it in the *Grundlagen:*

> Because of this extraordinary position which distinguishes mathematics from all other sciences, and which produces an explanation for the relatively free and easy way of pursuing it, it especially deserves the name of *free mathematics*, a designation which I, if I had the choice, would prefer to the now customary "pure" mathematics.[50]

Cantor was asserting the freedom within mathematics to allow the creation and application of new ideas on the basis of intellectual consistency alone. Mathematics was therefore absolutely free in its development and bound only to the requirement that its concepts permit no internal contradiction, but that they follow in definite relation to previously given definitions, axioms, and theorems. Mathematics, Cantor believed, was the one science that was justified in releasing itself from any metaphysical fetters. Its freedom, insisted Cantor, was its essence.

The detachment of mathematics from the constraints of an imposed structure imbedded in the natural world frees it from the metaphysical problems inherently part of any attempt to understand the ultimate status of the physical and life sciences. Mathematicians do not face the preoccupation of scientists who must try to make theory conform with some sort of given, external reality against which those theories may be tested, articulated, improved, revised, or rejected.[51] Mathematicians, if they worry at all, need do so only in terms of the internal consistency of their work. This effectively eliminates the possibility of later discrepancies. Thus the grounds do not seem present within mathematics for generating anomaly and crisis, or for displacing earlier theory with some incompatible new theory.

One important consequence, in fact, of the insistence on self-consistency within mathematics is that its advance is necessarily cumulative. New theories cannot displace the old, just as the calculus did not displace geometry. Though revolutionary, the calculus was not an incompatible advance requiring subsequent generations to reject Euclid, nor did Cantor's transfinite mathematics require displacement and rejection of previously established work in analysis, or in any other part of mathematics.

Advances in mathematics, therefore, are generally compatible and consistent with previously established theory; they do not confront and challenge the correctness or validity of earlier achievements and theory, but augment, articulate, and generalize what has been accepted before. Cantor's work managed to transform or to influence large parts of modern mathematics without requiring the displacement or rejection of previous mathematics.

REVOLUTIONARY ADVANCE IN MATHEMATICS

Does this mean then that mathematics, because it represents a form of knowledge in which progress is genuinely cumulative, cannot experience periods of legitimate revolution? Surely not. To say that mathematics grows by successive accumulation of knowledge, rather than by the displacement of discredited past theory by new theory, is not the same as to deny revolutionary advance. For example, Cantor's proof of the nondenumerability of the real numbers led to the creation of the transfinite numbers. This was conceptually impossible within the bounds of traditional mathematics, yet in no way did it contradict or compromise finite mathematics. Cantor's work did not displace, but it *did* augment the capacity of previous theory in a way that was revolutionary – that otherwise would have been impossible. It was revolutionary in breaking the bonds and limitations of earlier analysis, just as imaginary and complex numbers carried mathematics to new levels of generality and made solutions possible that otherwise would have been impossible to formulate. Moreover, the extensive revision due to transfinite set theory of large parts of mathematics, involving the rewriting of textbooks and precipitating debates over foundations, are all results of what Thomas Kuhn has diagnosed as companions to revolutions.[52] And all of these are reflected in the historical development of Cantorian set theory.

THE NATURE OF SCIENTIFIC RESOLUTION

I have deliberately juxtaposed the words "revolution" and "resolution" in order to emphasize what I take to be the nature of scientific advance reflected in the development of the history of mathematics – be it the Greek discovery of incommensurables and the concomitant creation of a theory of proportion to accommodate them, or Cantor's profound discovery of the nondenumerability of the real numbers and his subsequent creation of transfinite numbers and the development of a general, transfinite set theory. Because mathematics is restricted only by the limits imposed by consistency, the inherent structure of logic determines the structure of mathematical evolution. I have already suggested the way in which that evolution is necessarily cumulative. As theory develops, it provides more complete, more powerful, more comprehensive problem solutions, sometimes yielding entirely new and revolutionary theories in the process. But the fundamental character of such advance is embodied in the idea of resolution. Like the microscopist, moving from lower to higher levels of resolution, successive generations of mathematicians can claim to understand more, with a greater stockpile of results and increasingly refined techniques at their disposal. As mathematics becomes increasingly articulated, the process of resolution brings the areas of research and subjects for problem solving into greater focus, until solutions are obtained or new approaches developed to extend the boundaries of mathematical knowledge. Discoveries accumulate, and some inevitably lead to revolutionary new theories uniting entire branches of study, producing new points of view, sometimes wholly new disciplines that would have been impossible to produce within the bounds of previous theory.

This is as true of the discovery of incommensurable magnitudes as it of the advent of irrational, imaginary, and transfinite numbers, of the invention of the calculus, or the discovery of non-Euclidean geometries. None of these involved crisis or the rejection of earlier mathematics, although each represented a response to the failures and limitations of prevailing theory. New discoveries, particularly those of revolutionary import like those discussed here, provide new modes of thought within which more powerful and general results are possible than ever before. As Hermann Hankel once wrote, "In mathematics alone each generation builds a new story to the old structure."[53] This is the most obvious sense in which I mean that the nature of scientific advance can be understood directly, in terms of the logic of argument and mathematics, as one of increasingly powerful resolution.

RESISTANCE TO CHANGE

One last feature of the evolution of mathematics may help to corroborate further the fact that it does experience revolutionary transformations, for resistance to new discoveries may be taken as a strong measure of their revolutionary quality. One form of this resistance was reflected in the Greek's inability to conceive of anything as number except the integers – although eventually this prejudice was overcome, just as Cantor eventually overcame even his own discomfort with the actual infinite to support his transfinite numbers. Perhaps there is no better indication of the revolutionary quality of a new advance in mathematics than the extent to which it meets with opposition. The revolution, then, consists as much in overcoming establishment opposition as it does in the visionary quality of the new ideas themselves.

From the examples we have investigated here, it seems clear that mathematics may be revolutionized by the discovery of something entirely new and completely unexpected within the bounds of previous theory. Discovery of incommensurable magnitudes and the eventual creation of irrational numbers, the imaginary numbers, the calculus, non-Euclidean geometry, transfinite numbers, the paradoxes of set theory, even Gödel's incompleteness proof, are all revolutionary – they all have changed the content of mathematics and the ways in which mathematics is regarded.

They have each done more than simply add to mathematics – they have each transformed it. In each case the old mathematics is no longer what it seemed to be, perhaps no longer even of much interest when compared with the new and revolutionary ideas that supplant it.

NOTES

1 The most adamant statement that mathematics does not experience revolutions may be found in M. J. Crowe, "Ten 'Laws' Concerning Patterns of Change in the History of Mathematics," *Historia Mathematica, 2* (1975), pp. 161–6, esp. p. 165. The literature on the subject, however, is vast. Of authors who have claimed that mathematics grows by accumulation of results, without rejecting any of its past, the following sample is indicative: H. Hankel, *Die Entwicklung der Mathematik in den letzten Jahrhunderten* (*Antrittsvorlesung:* Tübingen, 1871; 2d ed., 1889), p. 25; G. D. Birkhoff, "Mathematics: Quantity and Order," *Science Today* (1934), pp. 293–317, esp. p. 302, and *Collected Mathematical Papers*, vol. III (New York: American Mathematical Society, 1950), p. 557; C. Truesdell: "While 'Imagination, fancy, and invention' are the soul of mathematical research, in mathematics there has never yet been a revolution," in *Essays in the History of Mechanics* (New York: Springer-Verlag, 1968), foreword.

2 T. S. Kuhn, *The Structure of Scientific Revolutions* (Chicago: University of Chicago Press, 1962; 2d ed., enlarged, 1970).

3 J. W. Dauben, "Set Theory and the Nature of Scientific Resolution," (MS) for the Colloquium *History of Mathematics and Recent Philosophies of Science* (at the semicentennial meeting of the History of Science Society, Burndy Library, Norwalk, Conn., October 27, 1974).

4 Crowe, "Ten 'Laws'," p. 165. See also M. J. Crowe, "Science a Century Ago," in *Science and Contemporary Society*, ed. F. J. Crosson (Notre Dame: University of Notre Dame Press, 1967), pp. 105–26, esp. pp. 123–4.

5 M. J. Crowe, "Ten 'Laws' Concerning Conceptual Change in Mathematics," *Historia Mathematica, 2* (1975), p. 470.

6 Ibid.

7 I. B. Cohen, "The Eighteenth-Century Origins of the Concept of Scientific Revolution," *Journal of the History of Ideas, 37* (1976), pp. 257–88. More recently Professor Cohen has also developed this material in a number of articles, as well as in *The Newtonian Revolution, with Illustrations of the Transformation of Scientific Ideas.* Cambridge: Cambridge University Press, 1980, esp. chap. 2, pp. 39–49.

8 Bernard de Fontenelle, *Éléments de la géométrie de l'infini* (Paris, 1727); refer in particular to the preface, which is also reprinted in Fontenelle, *Oeuvres de Fontenelle,* vol. VI (nouvelle éd., Paris, 1792), p. 43.

9 B. de Fontenelle, "Eloge de M. Rolle," *Histoire de l'Académie Royale des Sciences* (Paris, 1719), pp. 94–100, esp. p. 98. See also *Oeuvres,* vol. VII, p. 67.

10 For details of the background to Greek mathematics, and in particular to the history of incommensurability, see the recent works by W. R. Knorr, *The Evolution of the Euclidean Elements* (Dordrecht: Reidel, 1975); and H. J. Waschkies, *Von Eudoxos zu Aristotles (sic. Aristoteles). Das Fortwirken der Eudoxischen Proportionentheorie in der Aristotelischen Lehre vom Kontinuum,* vol. VIII in the series *Studien zur antiken Philosophie* (Amsterdam: Grüner, 1977). I am especially indebted to Wilbur Knorr for his comments on an early draft of this chapter. Our discussion of the many difficulties in dealing with pre-Socratic material has been of great help to me in clarifying many murky or puzzling aspects of the history of Greek mathematics, especially those connected with the history of the theory of incommensurable magnitudes and early Greek geometry.

11 Aristotle, *Metaphysics,* 985b23–986a3. Similarly, 1090a20–25. See the more direct interpretation that "things are numbers" and variations at 1080b16–21, 1083b11, 18.

12 Aristotle, *Metaphysics,* 1080b16–20; see also *De Caelo,* 300a16–19. The whole question of Pythagorean number theory and its character has been vigorously debated. For a general introduction that is careful to underscore the problems in reconstructing what the Pythagoreans may have believed, see J. A. Philip: "Pythagorean Number Theory," in *Pythagoras and Early Pythagoreanism,* supp. vol. VII to *Phoenix, Journal of the Classical Association of Canada* (Toronto: University of Toronto Press, 1966), pp. 76–109. Harold Cherniss has described the Pythagorean point of view as more "a materialization of number than a mathematization of nature," in H. F. Cherniss, "The Characteristics and Effects of Presocratic Philosophy," *Journal of the History of Ideas, 12* (1951), pp. 319–45, esp. p. 336. The source for number atomism in Pythagorean mathematics comes from Ecphantus of Syracuse, and as W. Knorr notes, this provides the basis for a "thesis long in fashion via P. Tannery and F. M. Cornford, but which seems more recently to have fallen into disrepute. Yet I believe a form of 'number-atomism' may be accepted as having been a doctrine of some Pythagoreans," in *Evolution,* p. 43. Gregory Vlastos, in a review of J. E. Raven's *Pythagoreans and Eleatics,* argued vigorously that "number atomism was not regarded by the tradition stemming from Theophrastus as an original feature of Pythagoreanism," in *Gnomon, 25* (1953), pp. 29–35, esp. p. 32. He carries this further by arguing that number atomism was surely not a feature of Pythagorean musical formulae, "nor could there be any question of number-atomism in the extensions of this theory to medicine, moral or psychological concepts." Fortunately, the question of number atomism is not crucial to the issues presented here. Whether the early Pythagoreans, or only some later Pythagoreans like Ecphantus, adopted a view of number as material monads, the significant feature of Pythagorean arithmetic for the present purposes was its emphasis on *ratio,* and its belief that all things could be expressed through ratios of whole numbers.

13 H. Vogt was among the first to attempt the reconstruction of the development of a theory of proportion in response to discovery of incommensurable magnitudes through transformations in terminology. See H. Vogt, "Die Entdeckungsgeschichte des Irrationalen nach Plato und anderen Quellen des 4. Jahrhunderts," *Bibliotheca Mathematica, 10* (1909–10), pp. 97–155, and "Zur Entdeckungsgeschichte des Irrationalen," *Bibliotheca Mathematica, 14* (1913–14), pp. 9–29. Later Kurt von Fritz developed a similar approach in his articles on "Theodoros" and "Theaitetos" for Pauly-Wissowa, *Paulys Real-Encyclopädie der classischen Altertumswissenschaft* (sec. ser., Stuttgart: Metzlersche Verlagsbuchhandlung, 1934), pp. 1811–31, 1351–72, respectively. See also K. von Fritz, "Discovery of Incommensurability by Hippasus of Metapontum," *Annals of Mathematics, 46* (1945), pp. 242–64.

14 Aristotle, *Metaphysics,* 1092b10. Aristotle reports that Eurytus decided what the number of man or of horse was, for example, "by imitating the figures of living things with pebbles." For commentaries on this passage by Alexander (*Metaphysics,* 827, 9) and Theophrastus (*Metaphysics,* 6a19), see G. S. Kirk and J. E. Raven, *The Presocratic Philosophers* (Cambridge: Cambridge University Press, 1957), p. 314. Wilbur Knorr maintains that Eurytus's approach was an attempt to modify Pythagorean number atomism in response to discovery of incommensurables. See Knorr, *Evolution,* p. 45.

15 For representative passages in Aristotle, *Metaphysics,* turn to 985b23–31, 986a2–8. See as well the discussion in Kirk and Raven, *Presocratic Philosophers,* pp. 236–62, esp. pp. 248–50. It should be noted that some writers minimize the significance of the Pythagoreans in the history of mathematics and science. See, for example, W. A. Heidel, "The Pythagoreans and Greek Mathematics," *American Journal of Philology, 61* (1940), pp. 1–33, esp. p. 31: "The role of the Pythagoreans must appear to have been much exaggerated." Even more emphatic is the view of W. Burkert: "The tradition of Pythagoras as a philosopher and scientist is, from the historical view, a mistake. . . . Thus after all, there lived on, in the image of Pythagoras, the great Wizard whom even an advanced age, though it be unwilling to admit the fact, cannot entirely dismiss," in *Lore and Science in Ancient Pythagoreanism,* trans. E. L. Minar, Jr. (Cambridge, Mass.: Harvard University Press, 1972), p. 482. As for the Pythagorean concept of a "perfect number," it must be remembered that their definition differed from that now standard in mathematics. For the Pythagoreans, the number 10 was perfect because it was the sum of the first four integers, i.e., $1 + 2 + 3 + 4 = 10$. Only after Aristotle did the sense of "perfect numbers," as used by Euclid, make its appearance. Then, as now, a perfect number is equal to the sum of its divisors. Consequently, $6 = 1 + 2 + 3$ and $28 = 1 + 2 + 4 + 7 + 14$ are both perfect numbers, but 10 is not, since $10 \neq 1 + 2 + 5$. For further information see Burkert, *Lore and Science,* p. 431.

16 This, too, is a question that has received much discussion but little agreement in the literature on the subject. For the most recent study of the problem, see W. Knorr, *Evolution,* pp. 36–49, esp. p. 40, where numerous arguments are presented to establish the discovery within a twenty-year span from 430–410 B.C.

17 For Aristotle's discussion of the incommensurability of the side and diagonal of a square, see *Prior Analytics,* 41a29. W. Knorr discusses this proof and its version in Euclid's Book X of the *Elements* at length, noting that "arguing for the antiquity of this version of the proof is its application of the even and the odd," *Evolution,* pp. 22–8, esp. p. 23. Arguing for discovery of incommensurability by Pythagoreans studying the method of *anthyphairesis,* discussed later (see n. 25), are Kurt von Fritz, "Discovery," p. 46, and S. Heller, "Ein Beitrag zur Deutung der Theodoros-Stelle in Platons Dialog 'Theaetet,' " *Centaurus, 5* (1956), pp. 1–58, and "Die Entdeckung der stetigen Teilung durch die Pythagoreer," *Abhandlungen der Deutschen Akademie der Wissenschaften zu Berlin, Klasse für Mathematik, Physik und Technik, 6* (1958), pp. 5–28. See also the discussion in W. Knorr, *Evolution,* pp. 29–36.

18 Although much debate has centered on the advisability of referring to the discovery as a "crisis," as did H. Hasse and H. Scholz, "Die Grundlagenkrisis der griechischen Mathematik," *Kant-Studien, 33* (1928), pp. 4–34, an important distinction must be made between the effect of the discovery of incommensurability upon mathematics as opposed to Pythagorean arithmology and its close connection with their cosmology or arithmological philosophy. For non-Pythagoreans and mathematicians in general, the ancient literature never mentions a "crisis" but refers instead to the discovery as an advance, or even as a great "wonder." This is precisely the attitude of Aristotle, *Metaphysics,* 983a13–20: "As we said, all men begin wondering that a thing should be so; the subject may be, for example, the automata in a peepshow, the solstices, or the incommensurability of the diagonal. For it must seem a matter for wonder, to all who have not studied the case, that there should be anything that cannot be measured by any measure, however small." For Pythagorean arithmology, on the other hand, the discovery must have posed a major problem, and in this context its effect can accurately be described as representing a "crisis."

G. E. L. Owen is even more emphatic in asserting that "discovery of incommensurables was a real crisis in mathematics," in "Zeno and the Mathematicians," *Proceedings of the Aristotelian Society, 58* (1957–8), pp. 199–222, esp. p. 214. For arguments that there was no such crisis, however, see K. Reidemeister, *Das exakte Denken der Griechen* (Leipzig: Classen & Goverts, 1949), p. 30, and H. Freudenthal, "Y avait-il une crise des fondements des mathématiques dans l'antiquité?" *Bulletin de la Société Mathématique de de Belgique, 18* (1966), pp. 43–55. Burkert comes to similar conclusions in *Lore and Science,* p. 462.

19 Scholium to Euclid, *Elementa,* X,1, in *Opera Omnia,* ed. J. L. Heiberg (Leipzig: Teubner, 1888), p. 417. For other accounts of the drowning episode, see Iamblichus, *De Vita Pythagorica Liber,* XXXIV, 247, and XVIII, 88, ed. Ludwig Deubner (Leipzig: Teubner, 1937), pp. 132 and 52, respectively, and Iamblichus, *De Communi Mathematica Scientia Liber,* XXV, ed. Nicola Festa (Leipzig: Teubner, 1891), pp. 76–78. Burkert writes that "the tradition of secrecy, betrayal, and divine punishment provided the occasion for the reconstruction of a veritable melodrama in intellectual history," *Lore and Science,* p. 455. Pappus, however, viewed the story of the drowning as a "parable," *The Commentary of Pappus on Book X of Euclid's Elements,* Book I, sec. 2, eds. G. Junge and W. Thomson. Harvard Semitic Series, VIII. (Cambridge, Mass.: Harvard University Press, 1930; reprinted, New York: Johnson Reprint Corp., 1968), p. 64: The story was "most probably a parable by which they sought to express their conviction that firstly, it is better to conceal (or veil) every surd, or irrational, or inconceivable in the universe, and, secondly, that the soul which by error or heedlessness discovers or reveals anything of this nature which is in it or in this world, wanders [thereafter] hither and thither on the sea of non-identity (i.e., lacking all similarity of quality or accident), immersed in the stream of the coming-to-be and the passing-away, where there is no standard of measurement."

20 Scholium to Euclid, *Elementa,* X, 1. For discussion of this passage, see M. Cantor, *Vorlesungen über Geschichte der Mathematik* (Leipzig: Teubner, 1894), vol. I, p. 175. As Burkert has pointed out, late commentators like Plutarch and Pappus might have been especially tempted to seize on the double entendre made possible by the multiple connotations of the word ἄρρητος as irrational and unspeakable: "In Plutarch it is clear that the word ἄρρητος, set in quotation marks, as it were, by λεγόμεναι, is to be understood in a double sense. The "ineffable because irrational" is at the same time the "unspeakable because secret" . . . The fascination of the ἄρρητον lies in the pretense to indicate the fundamental limitations of human expression, which are at the same time transcended by the initiate . . . This exciting double sense of the word ἄρρητος is what makes the story of the discovery and betrayal of the irrational an *exemplum* for Plutarch, and even more for Pappus, who is probably following some Platonic source," *Lore and Science,* p. 461. For additional discussion of these terminological transformations, refer to K. von Fritz, *Philosophie und sprachlicher Ausdruck bei Demokrit, Platon und Aristoteles* (New York: Steckert, 1939), p. 69, and "Die ARXAI in der griechischen Mathematik," *Archiv für Begriffsgeschichte, 1* (1955), pp. 13 – 103, esp. pp. 80–7, as well as to the articles by von Fritz and Vogt cited in n. 13. It should also be added that Mugler, in defining ἄρρητος, writes that "son sens étymologique était «indicible, inexprimable»; il était synonyme, à l'origine, de ἄλογλος au sens primitif . . . ," *Dictionnaire,* p. 83.

21 The position adopted by Michael Crowe, for one, maintains that "revolutions may occur in mathematical nomenclature, symbolism, metamathematics, methodology, and perhaps even in the historiography of mathematics," but *not* within mathematics itself. See Crowe, "Ten 'Laws',", pp. 165–6.

22 Archytas, *Fragment* B4 (Fragmente der Gespräche) in H. Diels, *Die Fragmente der Vorsokratiker,* vol. I (Berlin: Weidmannsche, 1922), p. 337: "Und die Arithmetik hat . . . einen recht betrachtlichen Vorrang, . . . besonders aber auch vor der Geometrie, da sie deutlicher als diese was sie will behandeln kann. . . . ⟨Denn die Geometrie beweist, wo die anderen Künste im Stiche lassen,⟩ und wo die Geometrie wiederum versagt, bringt die Arithmetik sowohl Beweise zustande wie auch die *Darlegung* der Formen [Prinzipien?], wenn es überhaupt irgend eine wissenschaftliche Behandlung der Formen gibt."

23 W. Knorr, "The Early Study of Incommensurability: Theodorus," *Evolution,* pp. 170–210, esp. pp. 199, 220–1. Here the recent research of D. Fowler is also relevant, above all his pair of articles, "Book II of Euclid's *Elements* and a pre-Eudoxian Theory of Ratio," *Archives for History of Exact Sciences, 22* (1980), pp. 5–36, and "Part 2: Sides and Diameters," *26* (1982), pp. 193–209. I am happy to acknowledge a very stimulating correspondence with David Fowler, covering a range of subjects including incommensurability, anthyphairesis, and Greek theories of ratio and proportion in general. Although our correspondence on these matters came after this essay was already in press, I am grateful for his very careful reading of my original paper, and his subsequent comments, only a few of which were possible to incorporate here. Readers should also note in particular D. Fowler, "Ratio in Early Greek Mathematics," *Bulletin of the American Mathematical Society* (NS), *1* (1979), pp. 807–46, and "Anthyphairetic Ratio and Eudoxean Proportion," *Archive for History of Exact Sciences, 24* (1981), pp. 69–72.

24 Plato, *Theatetus,* 147C–148B.

25 O. Becker, in analyzing the concept of ανθυφαίρεσις, reconstructed a pre-Eudoxean theory of proportion: "Eudoxus-Studien I. Eine voreudoxische Proportionenlehre und ihre Spuren bei Aristoteles und Euclid," *Quellen und Studien,* II (B) (1933), p. 311–33. For a detailed discussion of *anthyphairesis,* see W. Knorr, "*Anthyphairesis* and the Side and Diameter," *Evolution,* pp. 29–36; H. Waschkies, "Die anthyphairetische Proportionentheorie," *Von Eudoxus,* pp. 77–100. Mugler, *Dictionnaire,* p. 61, connects ἀνθυφαιρεῖν, the process of reciprocal subtraction, with study of the irrational magnitudes and the older, archaic term, *probablement d'origine pythagoricienne,* ἀνταναίρεσις, p. 65. See as well the commentary on Theatetus's demonstration and anthyphairesis by François Lasserre, *The Birth of Mathematics in the Age of Plato,* trans. H. Mortimer (London: Hutchinson, 1964), pp. 68–9.

26 W. Knorr, *Evolution,* pp. 286–8.

27 Aristotle, *Posterior Analytics,* 74a17–30, refers to the new, more general techniques of proof (ὃ καθόλου ὑποτίθενται ὑπαρχειυ). Moreover, scholia 1 and 3 to Book V of the *Elements* comment on the generality of the results obtained there. See Euclid, *Opera Omnia,* ed. Heiberg, vol. V, pp. 280, 282, respectively. In fact, the differences between the earliest theory of proportion, generally regarded as authentically Pythagorean and set forth in Book VII of Euclid's *Elements,* and Eudoxus' powerfully more general theory as represented in Euclid, Book V, may be seen in a comparison of several parallel definitions. For example:

Book VII, Definition 3: Μέρος ἐστὶν ἀριθμὸς ἀριθμοῦ ὁ ἐλάσσων τοῦ μείξονος ὅταν καταμετρῇ τὸν μείξονα.

Book V, Definition 1: Μέρος ἐστὶ μεγέθος μεγέθους τὸ ἔλασσον τοῦ μείζονος; ὅταν καταμετρῇ τὸ μεῖζον.
Book VII, Definition 5: πολλαπλάσιος δὲ ὁ μείξων τοῦ ἐλάσσονος, ὅταν καταμετρῆται ὑπὸ τοῦ ἐλάσσονος.
Book V, Definition 2: πολλαπλάσιον δὲ τὸ μεῖξον τοῦ ἐλάττονος, ὅταν καταμετρῆται ὑπὸ τοῦ ἐλάττονος.

Waschkies also underscores the significance of the term μεγαθος for magnitude in Book V by noting that it became a technical term in geometry directly as a result of Eudoxus's influence. See Waschkies, *Von Eudoxos,* p. 19.

28 Scholium 1 to Book V of Euclid's *Elements* in *Opera Omnia,* ed. Heiberg, vol. V, p. 280. As W. Knorr notes, "The fundamental conception of proportion in *Elements* V, if not the completion of the entire theory, is due to Eudoxus," *Evolution,* p. 274.

29 M. Kline, *Mathematical Thought from Ancient to Modern Times* (New York: Oxford University Press, 1972), p. 50.

30 I. Kant, *Kritik der reinen Vernunft,* ed. B. Erdmann (Hamburg: Voss, 1889), p. 7. Also in N. K. Smith (trans.), *Immanuel Kant's Critique of Pure Reason* (London: Macmillan, 1956, reprint of second impression with corrections of 1933), p. 19. See also the discussion by I. B. Cohen, "Concept of Scientific Revolution," pp. 283–4.

31 It should be stressed, however, that the Greeks never attained such a general concept of number. For them, ἀριθμοι, or numbers, were always defined, as in Euclid VII, Definition 2, as a sum of *units.* There were no rational or irrational *numbers,* only ratios of whole numbers and proportions defined as equal ratios. See B. L. van der Waerden, *Science Awakening,* trans. A. Dresden (New York: Oxford University Press, 1961, 2d English ed. with additions of the author), p. 125. Despite the conjectures of some historians (see, e.g., T. Heath, *A History of Greek Mathematics* [Oxford: Clarendon Press, 1921], vol. I, p. 327), the Greeks *never* had the concept of real numbers, Dedekind cuts, or even the set of rational numbers. For details, see F. Beckmann, "Neue Gesichtspunkte zum 5. Buch Euklids," *Archive for History of Exact Sciences, 4* (1967), pp. 1–144, esp. pp. 21 and 37–41. Knorr stresses that ἀριθμός (=number) and λόγος (=ratio) were *never* equated in the ancient tradition. See Knorr, *Evolution,* pp. 9–10.

32 Aristotle takes Theorem V, 16, on the *ennalax* property of proportions, as epitomizing the great transformation in proof techniques and capabilities brought about by Eudoxus's theory (see n. 27). On a simpler level, Book V duplicates propositions from Book II, where they were originally established for line segments only. Book V, of course, establishes similar theorems for all magnitudes in general. One may also compare, for example, specific propositions like the *di' isou* theorem for proportions, *Elements* V, 22, with the earlier version, VII, 14, where a different method was originally used employing the special properties of integers as opposed to magnitudes. Recently, Wilbur Knorr has argued that in Theorems X, 9–10, Euclid saw the unsuitability of the original pre-Eudoxean proofs of these propositions, and therefore gave them a new, if not very skillful version suitable to post-Eudoxean theory. See *Evolution,* p. 304.

33 By directly comparing the proofs of various Euclidean propositions in their pre- and post-Eudoxean forms, it is possible to make clear their comparative "advantages and limitations," as Knorr does in drawing direct comparisons where possible between theorems in Book V and their counterparts in Book VII. See his Appendix B, *Evolution,* pp. 332–44.

As Zeuthen observed, it is precisely at Theorem VII, 19, that the relation between Book V and Book VII is directly established, for in VII, 19, Euclid shows that the definition of proportion used in Book V is equivalent to definition VII, 20 when applied to numbers. It therefore follows that all theorems on proportion in Book V may be applied to any of the theorems dealing with proportions between numbers alone in Book VII. For details, refer to H.-G. Zeuthen, "Sur la constitution des livres arithmétiques des Éléments d'Euclide et leur rapport à la question de l'irrationalité," *Oversigt over det Kgl. Danske Videnskabernes Selskabs. Forhandlinger,* (1910), pp. 395–435, esp. p. 412: "l'importance logique du No. 19 consiste précisément en ce qu'on y établit que la définition d'une proportion donné dans le Ve livre a, si on l'applique à des nombres entiers, tout à fait la même portée que la définition donnée au VIIe livre."

Wholly apart from the significance of Eudoxus's theory of proportion for the development of Euclidean *Elements,* Kurt von Fritz has pointed out that Eudoxus was "the author of the method of exhaustion, of the theorem that the volume of a cone is one-third of the volume of a cylinder with the same base and altitude, and undoubtedly of other stereometric theorems which must have been used in the proof of that proposition. All this would have been impossible without the new definition of proportion invented by Eudoxus." See von Fritz, "Discovery," p. 264. Similarly, Wilbur Knorr has noted that "the renovation of proportion theory (Book V) was used to improve the foundations of geometry (Books VI and XI) and with the 'method of exhaustion' to effect the measurements in Book XIII," *Evolution,* p. 306.

34 For the details of Cantor's biography and the origins of transfinite set theory, sketched here only in the broadest outline, consult A. Fraenkel, "Georg Cantor," *Jahresbericht der Deutschen Mathematiker-Vereinigung, 39* (1930), pp. 189–266. This biography also appears separately as *Georg Cantor* (Leipzig: Teubner, 1930), and is reprinted in an abridged version in G. Cantor, *Gesammelte Abhandlungen mathematischen und philosophischen Inhalts,* ed. E. Zermelo (Berlin: Springer, 1932; reprinted, Hildesheim: Olms, 1966). For more recent studies, refer to H. Meschkowski, *Probleme des Unendlichen. Werk und Leben Georg Cantors* (Braunschweig: Vieweg, 1967); I. Gratten-Guinness, "Towards a Biography of Georg Cantor," *Annals of Science, 27* (1971), pp. 345–91 and plates xxv–xxviii; J. Dauben, *Georg Cantor. His Mathematics and Philosophy of the Infinite* (Cambridge, Mass.: Harvard University Press, 1979), and chap. V, "The Development of Cantorian Set Theory," in *From the Calculus to Set Theory, 1630–1910,* ed. I. Grattan-Guinness (London: Duckworths, 1980), pp. 181–219. I am grateful to Esther Phillips for her comments on an earlier version of this paper. Conversations with her on the subject of revolutions in mathematics have also greatly benefited the analysis that follows.

35 Fraenkel, "Georg Cantor," p. 192.

36 Ibid., p. 194.

37 See E. Heine, "Ueber trigonometrische Reihen," *Journal für die reine und angewandte Mathematik, 71* (1870), pp. 353–365, esp. p. 353. As Cantor noted in a footnote to his first paper on the subject, "Zu den folgenden Arbeiten bin ich durch Herrn *Heine* angeregt worden. Derselbe hat die Güte gehabt, mich mit seinen Untersuchungen über trigonometrische Reihen frühzeitig bekannt zu machen," in "Über einen die trigonometrischen Reihen betreffenden Lehrsatz," *Journal für die reine und angewandte Mathematik, 72* (1870), pp. 130–8, esp. p. 130. The paper is reprinted in *Gesammelte Abhandlungen,* pp. 71–9.

38 G. Cantor, "Über die Ausdehnung eines Satzes aus der Theorie der trigonometrischen Reihen," *Mathematische Annalen, 5* (1872), pp. 123–32, reprinted in *Gesammelte Abhandlungen,* pp. 92–102. A French translation of this paper appeared under the title "Extension d'un théorème de la théorie des séries trigonométriques," *Acta Mathematica, 2* (1883), pp. 336–48. For a discussion of the significance of this paper in the context of

Cantor's early work, consult J. Dauben, "The Trigonometric Background to Georg Cantor's Theory of Sets," *Archive for History of Exact Sciences, 7* (1971), pp. 181–216, and chap. 2, "The Origins of Cantorian Set Theory," in *Georg Cantor,* pp. 30–46.

39 See J. Dauben, *Georg Cantor,* pp. 41–2.

40 For a fuller discussion of Cantor's early conceptualization of derived sets and the distinction between sets of the first and second species, see J. Dauben, "Denumerability and Dimension: The Origins of Georg Cantor's Theory of Sets," *Rete, 2* (1974), pp. 105–34.

41 It should be noted that Richard Dedekind's famous theory of "cuts" used to define the real numbers was also published in the same year: *Stetigkeit und irrationale Zahlen,* 2d ed. (Braunschweig: Vieweg, 1892; translated into English by W. W. Beman as part of a collection of Dedekind's papers: *Essays on the Theory of Numbers, Continuity of Irrational Numbers, the Nature and Meaning of Numbers* (Chicago: Open Court, 1901; reprinted, New York: Dover, 1963). See also P. E. B. Jourdain, "The Development of the Theory of Transfinite Numbers (Part 3). Georg Cantor's Work on Trigonometrical Series and his Theory of Irrational Numbers (1870–1871). The Other Theories of Irrational Numbers," *Archiv der Mathematik und Physik, 22* (1910), pp. 1–21; and J. Cavaillès, *Philosophie mathematique* (Paris: Hermann, 1962), ep. pp. 35–44.

42 G. Cantor, "Über eine Eigenschaft des Inbegriffes aller reellen algebraischen Zahlen," *Journal für die reine und angewandte Mathematik, 77* (1874), pp. 258–62; reprinted in *Gesammelte Abhandlungen,* pp. 115–18; translated as "Sur une propriété du système de tous les nombres algébriques réels," *Acta Mathematica, 2* (1883), pp. 305–10.

43 G. Cantor, *Grundlagen einer allgemeinen Mannigfaltigkeitslehre. Ein mathematisch-philosophischer Versuch in der Lehre des Unendlichen* (Leipzig: Teubner, 1883); in *Gesammelte Abhandlungen,* pp. 165–208. Translated, in part, into French as "Fondements d'une théorie générale des ensembles," *Acta Mathematica, 2* (1883), pp. 381–408. An English translation has recently appeared by U. Parpart, "Foundations of the Theory of Manifolds," *The Campaigner* (The Theoretical Journal of the National Caucus of Labor Committees), *9* (January and February), pp. 69–96. The reader should be warned, however, that in addition to missing the distinction between *reellen* and *realen Zahlen* in translating the *Grundlagen,* Parpart also fails to distinguish between *Zahlen* and *Anzahlen,* translating both as "number" throughout without making clear the differences crucial to Cantor's introduction of the transfinite numbers. For fuller discussion of the significance of such terminological aspects of the *Grundlagen,* see J. Dauben, *Georg Cantor,* pp. 125–8.

44 G. Cantor, "Beiträge zur Begrundung der transfiniten Mengenlehre," part I, *Mathematischen Annalen, 46* (1895), pp. 481–512; part II, *Mathematischen Annalen, 49* (1897), pp. 207–46; in *Gesammelte Abhandlungen,* pp. 282–351. Part I was translated into Italian by F. Gerbaldi, "Contribuzione al fondamento della teoria degli inseimi transfinite," *Rivista di Matematica, 5* (1895), pp. 129–62. Both parts were translated into French by F. Marotte, *Sur les fondements de la théorie des ensembles transfinis* (Paris: Hermann, 1899), and into English by P. E. B. Jourdain, *Contributions to the Founding of the Theory of Transfinite Numbers* (Chicago: Open Court, 1915). For discussion of Cantor's terminology, and the remarkable fact that he only introduced the transfinite alephs in 1893, although he had introduced the ω for transfinite ordinal numbers in 1883, see J. Dauben, *Georg Cantor,* pp. 179–181.

45 See in particular Cantor's discussion in sec. 4 through 8 of the *Grundlagen,* in *Gesammelte Abhandlungen,* pp. 173–83. The following analysis presents, in its major outline, the views Cantor held on these matters.

46 Gauss wrote to Schumacher from Göttingen on July 12, 1831. See letter 396

(Gauss's letter 177) in K. F. Gauss, *Briefwechsel zwischen K. F. Gauss und H. C. Schumacher,* ed. C. A. F. Peters (Altona: Esch, 1860), vol. II, p. 269.

47 G. Cantor, *Grundlagen,* in *Gesammelte Abhandlungen,* p. 182.

48 Ibid., p. 181.

49 See Cantor's explanation of immanent and transient realities in sec. 8 of the *Grundlagen, Gesammelte Abhandlungen,* pp. 181–3.

50 Ibid., p 182.

51 This is exactly Cantor's point in sec. 8 of the *Grundlagen,* where he stresses that the natural sciences are always concerned with the "fit with facts," while mathematics need not be concerned with the conditions of natural phenomena as an ultimate arbiter of the truth or success of a given theory. In the natural sciences, however, historians and philosophers of science have been especially interested in the nature of the connections between observation, experiment, and theory. Among many works that might be cited that of Thomas Kuhn is perhaps the best known and will suffice here to give some sense of the connections that set the sciences in general apart from mathematics: "The decision to reject one paradigm is always simultaneously the decision to accept another, and the judgment leading to that decision involves the comparison of both paradigms with nature and with each other," *Revolutions,* p. 77. It was precisely its independence from nature that give mathematics, in Cantor's view, its "freedom" as characterized in n. 50.

52 See chap. XI: "The Invisibility of Revolutions," in T. S. Kuhn, *Revolutions,* pp. 135–42, esp. p. 136.

53 H. Hankel, *Entwicklung der Mathematik,* p. 25.

5

Cauchy and Bolzano

Tradition and transformation in the history of mathematics

JUDITH GRABINER
California State University, Dominguez Hills

Simultaneous discoveries, often accompanied by controversies over priority, have been frequent in the history of the sciences. The history of mathematics has been no exception. Newton, for instance, engaged in a priority controversy with Leibniz and his followers over the invention of the calculus. Although Newton's work was first in time, most scholars now hold that the discoveries were largely independent of each other. But the priority dispute in this case is less interesting than the determination of what made the simultaneous discovery possible. Scholars have reconstructed the common mathematical tradition that produced the invention of the calculus, identifying the predecessors of Newton and Leibniz – Fermat, Cavalieri, Pascal, Wallis, Barrow, Gregory – and documenting their influence.

Another celebrated example of simultaneous discovery in the history of mathematics is provided by the new, rigorous analysis of the nineteenth-century mathematicians Bernard Bolzano and Augustin-Louis Cauchy. This episode, too, has produced a priority controversy – though the controversy did not arise until a century and a half after the original publications. Even though the controversy may have been delayed, however, the circumstances that produced it are still of interest. The present essay will discuss the historical questions raised by the common achievement of Bolzano and Cauchy: what they did, what has been said about it recently, what made their accomplishments possible, and what can be learned from the whole episode.

BOLZANO, CAUCHY, AND THEIR COMMON ACHIEVEMENT

In 1817, Bolzano published a paper whose title indicates its subject: "Purely analytical proof of the theorem that between any two values that give results of opposite sign, there lies at least one root of the equation."[1] Bolzano's purpose was not simply to prove this theorem, but to exemplify his view of how rigorous mathematics ought to be done. His paper includes the following important features: essentially the modern definition of continuous function; a formal statement of the intermediate-value theorem for continuous functions; a "purely analytical" (*rein analytischer*) proof of the theorem; and a statement of what is now called the Cauchy criterion for convergence. Philosophically, Bolzano's paper advocates ridding the study of analysis of appeals to geometry, motion, and intuition, and using instead "purely analytical" means to work out the rigorous foundations of analysis. Both in this 1817 paper and in a number of other writings, Bolzano carried out particular details of his program of rigorization. In particular, instead of geometric intuition, his work is marked by rigorous arguments based on the algebra of inequalities.[2]

In 1821, Cauchy published his *Cours d'analyse.*[3] This course of lectures differs vastly in scope from Bolzano's paper, in part because of Cauchy's professional obligation to cover more topics. Nevertheless, Cauchy's book and Bolzano's paper have some features in common. The *Cours d'analyse,* like the *Rein analytischer Beweis,* includes: essentially the modern definition of continuous function; a formal statement of the intermediate-value theorem for continuous functions; a purely analytical (*par une méthode directe et purement analytique*) proof of that theorem; and a statement of the Cauchy criterion for convergence. Cauchy's 1821 book begins with a statement of his program for making analysis rigorous by avoiding overgeneralization from previously known results. Both in the *Cours d'analyse* and in other works, Cauchy developed the details of giving the calculus a foundation based on the algebra of inequalities.[4] The similarity between this list of Cauchy's achievements and the list of Bolzano's cannot possibly be a coincidence.

Surprisingly, this similarity was not widely recognized for many decades after the 1820s. In fact, Bolzano, though a gifted mathematician, had virtually no influence on the development of nineteenth-century analysis, because of his relative isolation in Prague, his lack of a major mathematics teaching position, the explicit and technical philosophical emphasis of his publications, and also because much of his significant work – for instance, the *Functionenlehre* – remained unpublished until the twentieth century. Bolzano's achievements were rediscovered only after the work of Weierstrass had shown the mathematical community the shape of rigorous analysis.[5] One way for post-Weierstrassian writers to give some indication of Bolzano's stature and the nature of his achievement was to compare him to Cauchy. So the statement that Bolzano and Cauchy accomplished essentially the same things – true up to a point – has been frequently repeated by historians.[6]

A new stage in the discussion of the similarity of the work of Bolzano and Cauchy has resulted from the recent writings of Ivor Grattan-Guinness. Struck by the similarity in the achievements of the two men, and struck also by the fact that Cauchy before 1821 seems to have shown considerably less interest in rigor than he did in the *Cours d'analyse,* Grattan-Guinness reasoned: Cauchy's achievement appears unprecedented; its similarity to Bolzano's work cannot be an accident; the similarity demands an explanation. What is the explanation? Grattan-Guinness concluded that Cauchy had read Bolzano's paper and appropriated the results without acknowledging their author. Grattan-Guinness buttressed this argument by showing that the Bibliothèque impériale in Paris began getting the journal in which Bolzano's paper appeared with precisely the issue containing it; by giving evidence that Cauchy could read German; and by claiming that Cauchy had been none too scrupulous in appropriating the results of others on other occasions.[7]

More recently, scholars unsympathetic to Grattan-Guinness's conclusions have argued against his view by convincingly demonstrating that there are real conceptual differences between the work of Bolzano in 1817 and Cauchy in 1821. Hans Freudenthal has shown this with full mathematical detail and has accompanied his argument with a spirited defense both of Cauchy's originality and of his character.[8] Hourya Sinaceur, concentrating on the work of Bolzano, has not only stressed the difference of conception, but has claimed Bolzano's work to be superior to Cauchy's.[9] Entirely independently of any Bolzano-Cauchy controversy, many scholars have studied the published and unpublished work of Bolzano and have given us a picture of him that his mathematical contemporaries unfortunately did not have, namely, a highly original, philosophically sophisticated mathematician whose writings anticipated not only some of Cauchy's work, but much in modern logic and set theory.[10] The work of these scholars, if we bring it to bear on the "simultaneous discovery or plagiarism?" question, strengthens the Freudenthal-Sinaceur observations about the differences between the outlook of Bolzano and that of Cauchy.

In one sense, however, Grattan-Guinness's question has not been adequately dealt with. He may have given the wrong answer, but he has asked the right question. The achievements of Cauchy and Bolzano are a real advance in the history of analysis. Though the similarities between the work of Cauchy and Bolzano are not exact, they exist, and they are striking. These similarities need to be explained. How is this to be done?

It is not enough simply to say that Bolzano and Cauchy did similar things, so they must have had common predecessors. This statement is true, as we will show, but it is too general to be historically interesting. The predecessors must be identified; their work must be described in some detail; the specific relevance of their work for the Cauchy-Bolzano achievement must be demonstrated; the access of Cauchy and Bolzano to this work must be shown; and the debt Cauchy and Bolzano owed it must be assessed. Only then will we have a picture of the mathematical tradition behind the work of Bolzano and Cauchy sufficiently clear to explain the similarities between their achievements.

In the pages that follow, we will discuss the common background of the Cauchy-Bolzano achievement of 1817–21 in detail. At this point let us state our principal conclusion: The most important common predecessor of Cauchy and Bolzano was Joseph-Louis Lagrange. Lagrange was not, of course, their only significant source, and Lagrange's own work crystallized and elaborated earlier eighteenth-century traditions as well as introducing ideas of his own. But whatever the sources of his ideas, we will argue that Lagrange's books constitute the greatest single prior common influence on the work of Cauchy and Bolzano. For each specific example we discuss, we will indicate the source and precise nature of his influence.

The reader may be skeptical of the conclusion just stated. After all, it is well known that Lagrange's proposed "algebraic" foundation for the calculus is wrong. Lagrange had defined the derivative $f'(x)$ as the coefficient of the linear term in the Taylor series expansion of $f(x + h)$. Expanding a function in a Taylor series is not, as Lagrange erroneously thought, an algebraic process; defining $f'(x)$ in terms of a Taylor series requires the existence of all the higher-order derivatives to define just the first. Furthermore, as Cauchy himself pointed out, a given Taylor series does not uniquely specify a function: For instance, e^{-x^2} and $e^{-1/x^2} + e^{-x^2}$ have the same Taylor series.[11] But there is more to Lagrange's books on the calculus than what is found in their initial chapters. As we will demonstrate, Lagrange did a great deal in the body of his works that Cauchy and Bolzano could use – and did use.

Of the general influence of Lagrange on Cauchy and Bolzano there can be little doubt. The works of both abound with references to Lagrange.[12] Bolzano's personal library contained German translations of both Lagrange's books on the calculus, the *Théorie des fonctions analytiques* and the *Leçons sur le calcul des fonctions.*[13] Cauchy's biographer, Valson, related that when Cauchy went on his first engineering job, Lagrange's *Théorie des fonctions analytiques* was one of the four books he took along.[14] Both Cauchy and Bolzano continually used Lagrangian innovations like the notation $f'(x)$, the term "derived function" (*fonction derivée, abgeleitete Function,* our "derivative"), and the Lagrange remainder of the Taylor series.[15] Thus, that Bolzano and Cauchy had access to Lagrange's work, and acknowledged this access, is indisputable; that there was some general influence is clear also. In addition, we will often be able to find explicit references in the work of Bolzano and Cauchy to the relevant prior work of Lagrange. Let us now turn to the specific accomplishments of Cauchy and Bolzano and consider the mathematical tradition out of which they came.

THE CAUCHY-BOLZANO PROGRAM FOR RIGORIZING ANALYSIS

The introduction to Cauchy's *Cours d'analyse* includes a clear statement of his desire to set a new standard for rigor in analysis, a standard he thought had previously been found only in the logically structured geometry of the Greeks. He attacked formalism (except as a heuristic device), especially the uncritical manipulation of infinite series without regard to convergence. He closed with a statement that once the algebraic foundation had been laid down *correctly,* all else would follow:

> As for methods, I have sought to give them all the rigor that exists in geometry, so as never to refer to reasons from the generalness of algebra. Reasons of this type, though often enough admitted, especially in passing from convergent series to divergent series, and from real quantities to imaginary expressions, can be considered only . . . as inductions, sometimes appropriate to suggest truth, but as having little accord with the much-praised exactness of the mathematical sciences. . . . Most formulas hold true only under

certain conditions, and for certain values of the quantities they contain. By determining these conditions and these values, and by fixing precisely the sense of all the notations I use, *I make all uncertainty disappear.*[16]

Besides this general statement, a number of specific accomplishments in Cauchy's 1821 book show that he tried to put this philosophy into practice: the careful investigation of convergence of series and the proof of criteria for convergence; algebraic proofs about limits; and, most important for our present purposes, his rigorous proof of the intermediate-value theorem for continuous functions.[17] Moreover, though Cauchy's language is sometimes not precise, this does not necessarily detract from the rigor of his work, because he clearly distinguished between a heuristically useful description and a rigorous proof; for instance, in the text of his *Cours d'analyse,* he was willing to settle for an intuitive geometric argument for the intermediate-value theorem for continuous functions, but he referred the reader to a *Note* for the "purely analytic and direct" proof that we will discuss at length later.

Similarly, Bolzano sought in 1817 to give a *rein analytischer Beweis* of the intermediate-value theorem. He advocated treating the problems of the calculus "without any consideration of the infinitely small . . . and without any suppositions not rigorously proved."[18] "It would be an unendurable offense against good method," he argued, "to derive truths of pure (or general) mathematics (i.e., arithmetic, algebra, or analysis) from considerations which belong to purely applied (or special) parts of mathematics, such as geometry," adding that "the concept of time, and even more that of motion, are just as foreign to general mathematics as is that of space."[19]

There are, of course, some differences between Bolzano's program as enunciated in 1817 and Cauchy's as given in the *Cours d'analyse* of 1821. A significant part of Bolzano's motivation was philosophical. Indeed, as Philip Kitcher has shown, there was strong Aristotelian influence on the way Bolzano sought to reduce less general or applied subjects like geometry to more general or pure ones like algebra.[20] Cauchy, by contrast, was already actively contributing to many mathematical fields. In 1821 he felt obliged not only to give a general account of the concepts of analysis, but to prove as many results – both old and new – as possible, so that mathematicians could use eighteenth-century results knowing precisely when they worked and when they did not. It is sometimes even observed that Cauchy's supposed rigor, unlike Bolzano's, is not consistently based on the algebra of inequalities, because he continued to use the language of infinitesimals.[21] But this criticism is not wholly justified, because Cauchy deliberately retained older terms and concepts because of their heuristic value. In the preface to his 1823 textbook on the infinitesimal calculus, Cauchy said, "My principal goal has been to reconcile the rigor that I made a law for myself in my *Cours d'analyse* with the simplicity that results from the direct consideration of infinitely small quantities."[22] Bolzano in 1817 had made no such compromises. Though he of course granted the heuristic value of intuition,[23] he barred it from the precise statement of theorems. To exemplify his new standards of rigor, he apparently felt he had to make a clear break with the past. Cauchy preferred to cover all topics, old and new, as well as he could, rather than cover only one with perfection.

Nevertheless, the similarity between the goals and methods of Cauchy and of Bolzano is great, and it needs explanation. The explanation lies in the common prior influence of a number of eighteenth-century Continental mathematicians who advocated turning from arguments based on geometry, motion, and intuition to the certainty of purely analytic demonstrations, most notably Euler, D'Alembert, and Lagrange.

Euler, for instance, in 1755 proudly said that his *Institutiones calculi differentialis* "remains throughout within the bounds of pure analysis" since there are no diagrams.[24] D'Alembert, objecting to Newtonian fluxions as a foundation for the calculus since they seemed to involve the idea of motion, argued that the calculus has "only algebraic quantities for its object."[25] Lagrange accepted and expanded on the views of these predecessors. In the preface to his *Mécanique analytique,* Lagrange boasted, "No diagrams are to be found in this work. The methods I expound here do not require constructions, geometric or mechanical reasoning, but only algebraic operations, proceeding by regular and uniform steps."[26] Lagrange renounced arguments based on space and time in his calculus as well, echoing D'Alembert's views that "motion is not a sufficiently clear idea on which to base the calculus."[27] In 1797, Lagrange taught a course in analysis at the *Ecole polytechnique* in which he tried to give a rigorous and purely analytic basis for the calculus. The title of the book Lagrange based on that course reveals his program: *Théorie des fonctions analytiques,*[28] *contenant les principes du calcul différentiel, degagés de toute considération d'infiniment petits, d'évanouissans, de limites*[29] *et de fluxions, et réduits à l'analyse algébrique des quantités finies.* The last phrase, "reduced to the algebraic analysis of finite quantities," is particularly characteristic of his approach.

Given this background, Bolzano's insistence on pure analysis, his rejection of geometry, and his rejection of the ideas of space, time, and motion sound quite Lagrangian. Further, Bolzano's knowledge of Lagrange's views on this subject is easy to document. In the *Rein analytischer Beweis,* for instance, Lagrange is cited twice in the first two paragraphs, and a specific argument from Lagrange's *Leçons sur le calcul des fonctions* is discussed at length in Bolzano's Section V.[30] In other works, too, Bolzano cited Lagrange with approval. Indeed, the influence of Lagrange's *Théorie des fonctions analytiques* may be reflected in the very title of Bolzano's *Functionenlehre* of 1830, as well as in the citations in the text of that work.

As for Cauchy's desire to firmly establish the received results of the calculus, Lagrange (along with the teaching requirements of the *Ecole polytechnique*)[31] prepared the way. Reading Lagrange's *Théorie des fonctions analytiques* in full demonstrates that Lagrange did not just suggest the things he advocated in his title: He tried to do them. After defining the derivative $f'(x)$ as the coefficient of the linear term in the Taylor series expansion of $f(x + h)$, he devoted the rest of his book to deducing the then-known results of the calculus from that "analytic" definition. The fact that his definition was inadequate for his purpose does not invalidate the purpose – which was Cauchy's purpose too.

At first glance, Cauchy's work appears to break sharply with Lagrange. Cauchy's attack on the assumption that algebraic arguments work for all values of the variables they contain is an attack on much eighteenth-century formal manipulation with series, including, at least implicitly, the work of Lagrange. The attack on Lagrange was made explicit in Cauchy's calculus lectures, published in 1823, in which Cauchy rejected Lagrange's Taylor series foundation for the calculus.[32]

Nevertheless, Cauchy accepted many of Lagrange's goals. Cauchy's *Cours d'analyse* is subtitled *analyse algébrique,* a phrase borrowed from the full title of Lagrange's *Théorie des fonctions analytiques;* and the program Cauchy pursued – giving a *cours d'analyse* that reduces the calculus to *analyse algébrique* – begins, historically, with Lagrange's book of 1797. Lagrange, like Cauchy, did not object to the heuristic use of infinitesimals, holding only that they had no place in rigorous foundations. As Lagrange observed in the preface to the second edition of his *Mécanique analytique,* "When one has convinced oneself of the exactness of [the] results [of the calculus] by the geometric method of first and last ratios, or by the analytic method of derived functions, one can use infinitely small quantities as a sure and easy instrument to abridge and simplify the demonstrations."[33] This was precisely the view expressed by Cauchy in the preface to his *Calcul infinitésimal,* which we have quoted. We have already indicated how Cauchy followed Lagrange's calculus in notation and terminology. And, as I have shown elsewhere, Cauchy owes much of his rigorous theory of derivatives to Lagrange's books on the calculus.[34]

Thus, both Cauchy and Bolzano accepted Lagrange's goal of reducing the calculus to the "algebraic analysis of finite quantities," even if they did not carry it out in the same way. Though Lagrange's vision was not one he himself could successfully carry out, it helped inspire Bolzano and Cauchy to their own versions of the general ideal of basing the calculus on purely analytic proofs. As we shall now see, Lagrange and his predecessors also helped inspire Bolzano and Cauchy in their analytic definitions of continuity and their purely analytic proofs of the intermediate-value theorem for continuous functions.

CONTINUOUS FUNCTIONS

In 1821, Cauchy defined continuous function as follows: "The function $f(x)$ will be, between . . . two given limits of the variable x, a continuous function of that variable, if, for every value of x included between these limits, the numerical [i.e., absolute] value of the difference $f(x + \alpha) = f(x)$ decreases indefinitely with that of α."[35] He added, by way of explanation, "In other words, the function $f(x)$ remains continuous with respect to x in the given limits, if between these limits an infinitely small increment in the variable always produces an infinitely small increment in the function itself." Cauchy had previously defined "infinitely small" as having zero as a limit, so the second version Cauchy gave means, to him, just what the first version means.[36] Similarly, Bolzano in 1817 said that if

$f(x)$ were a function with x included between certain limits, $f(x)$ "varies according to the law of continuity" between these limits when "the difference $f(x + \omega) - f(x)$ can be made smaller than any given magnitude, if ω can be taken as small as desired."[37]

P. E. B. Jourdain, who was interested in the question of Cauchy's sources, has suggested an eighteenth-century origin for Cauchy's concept of continuous function, arising out of the debate over the vibrating string.[38] Eighteenth-century mathematicians had sought a precise characterization of those functions F that would make physically acceptable solutions to the differential equation of the vibrating string; the problem goes back to D'Alembert, who stated the equation we now write as $\partial^2y/\partial x^2 = 1/c^2\ \partial^2y/\partial t^2$, and gave the solution of form $F(x + ct) - F(x - ct)$. We will not give a full account of these discussions here.[39] To get a feeling for the climate of mathematical opinion about continuous functions in which Bolzano and Cauchy worked, it will suffice to quote from the work of L. F. A. Arbogast, who won the St. Petersburg Academy's prize competition in 1787 by giving the best characterization of those functions that would be allowable solutions to the vibrating-string equation. Arbogast described these functions in several ways: The functions make "no jumps"; they have the intermediate-value property; and they increase in increments whose sizes correspond to the sizes of the increments in the variable. For instance, if the function is represented by two different formulas on adjacent intervals, "the last ordinate of the old form, and the first of the new, are equal to each other, or differ only by an infinitely small quantity." Again, "The ordinate y cannot pass brusquely from one value to another; there cannot be a jump from one ordinate to another which differs from it by an assignable quantity."[40] This "no-jumps" characterization, though it helps call attention to the crucial property of continuous function as defined by Cauchy and Bolzano, is not in itself an anticipation of that definition; it deals with functions that are piecewise continuous, and discusses the behavior of the function in detail only at the break. Thus a definition of the continuity of the "piece" is still lacking. But Arbogast was concerned about this question. He linked his no-jumps property to the intermediate-value property, saying that the functions had to obey what he called the "law of continuity": "A quantity cannot pass from one state to another without passing through all the intermediate states subject to the same law."[41] The closest Arbogast came to the Cauchy-Bolzano definition was to say "assuming that the variable increases continually, the function receives corresponding variations,"[42] though the language is not sufficiently precise to be a real anticipation of that definition.

Arbogast's paper, and the debate that stimulated it, could certainly have provided food for thought for mathematicians like Bolzano and Cauchy. Unfortunately, Bolzano and Cauchy cited neither Arbogast nor any of the other major contributions to the debate in connection with their definitions of continuous function. So, even though Arbogast's memoir won a major academy prize and can therefore be presumed to have been widely known, there is no evidence that it had a direct influence on Bolzano or Cauchy. Still, one result of the vibrating-string debate was that many descriptions of continuous function were current in the period around 1800. S. F. Lacroix, for instance, wrote in 1806 about such functions: "The smaller the increments of the independent variable, the closer the successive values of the function are to each other."[43] Bolzano, in 1830, actually cited another statement from the very page on which this passage appeared.[44] Jourdain is thus certainly correct in stating that the vibrating-string debate called attention to the various properties of continuous functions, even though he did not demonstrate direct influence.[45] Still, a more convincing candidate for the immediate source of the independent Bolzano-Cauchy definitions would be a more algebraic characterization of their defining property of continuity, one which we could prove both Bolzano and Cauchy had seen. In fact, there is not just one such characterization, there are two.

In 1798, Lagrange, in a book on approximating the solutions of algebraic equations, *Traité de la résolution des équations numériques de tous les degrés,* actually tried to prove the intermediate-value for continuous functions. We will return to Lagrange's proofs later on. Let us now look at the description Lagrange gave in his proof for the property we call the "continuity" of polynomial functions with positive terms. If P and Q are such polynomial functions of x defined between $x = p$ and $x = q$, then "when x is increased by all insensible degrees from p to q, they [P and Q] increase also by insensible degrees."[46] Here, then, is a characterization of continuity, resembling some of those arising from the vibrating-string debate, but that is in this case used in a proof of the intermediate-value theorem. And we know that Cauchy and Bolzano both read the work in which this version of "continuity" appears, because they explicitly cited Lagrange's book.[47]

Even so, Lagrange's 1798 description is verbal and thus not, as it stands, sufficiently precise to support valid proofs of theorems about continuous functions. But this verbal characterization of continuity by Lagrange was not his only one. Lagrange's *Théorie des fonctions analytiques* contains an algebraic characterization of the continuity of a function at zero. In the course of a proof, Lagrange needed to characterize the continuity of the function hP as h went to zero, where P was defined by $f(x + h) = f(x) + hP$.[48] If x is fixed, said Lagrange, and if P exists and remains finite, then hP vanishes with h. Considering the curve corresponding to that function, with h as abscissa and hP as ordinate, Lagrange observed that the curve cuts the axis at the origin, and "the course of the curve will necessarily be *continuous* from this point; thus it will, little by little, approach the axis . . . within a quantity less than any given quantity."[49] Lest this appear too geometric to satisfy Bolzano or Cauchy, here is the semialgebraic translation that follows: "So we can always find an abscissa h corresponding to an ordinate less than any given quantity; and then all smaller values of h correspond also to ordinates less than the given quantity."[50] For Lagrange, unlike Bolzano and Cauchy, this property of continuous function is not a definition. It is just one property of such functions, a characterization introduced on a one-time basis to help him prove a theorem. Nevertheless, we recognize in it what later became the essential defining property of continuous function.

The correspondence between Lagrange's view of continuity and the Cauchy-

Bolzano view was explicitly pointed out by Bolzano in 1830. Bolzano noted that the term "continuous function" had been used in several different senses: for instance, by Lacroix (on the page we mentioned earlier) to mean that the ratio $\Delta F(x)/\Delta x$ was bounded; by Kästner and Fries to mean that the function has the intermediate-value property; by Eytelwein to mean that the function is real and finite; and so on.[51] These properties are all important, said Bolzano, and need to be distinguished from one another. Nevertheless, he continued, the term "continuous function" should be reserved for the defining property. As he put it, "Thus it would be best, holding to the terminology introduced by *Lagrange, Cauchy,* and others, to understand by the *continuity* of a function only the property described in the previous section."[52]

We have already documented the acquaintance of both Cauchy in 1821 and Bolzano in 1817 with Lagrange's *Théorie des fonctions analytiques* and *Résolution des équations numériques.* Thus, on the basis of the similar concepts and explicit citations given so far, these two books are the most likely sources for both Cauchy's and Bolzano's definitions of continuous function. But the achievement of Cauchy and Bolzano about continuous functions did not end with the definition. Both used their definitions in proofs, particularly the proof of the intermediate-value theorem. Let us now turn to a discussion of this important theorem.

WHY PROVE THE INTERMEDIATE-VALUE THEOREM?

Surely the major common achievement of Bolzano in 1817 and Cauchy in 1821 was the proof of the intermediate-value theorem for continuous functions. We, of course, recognize the importance of this theorem in the foundations of analysis. Still, its truth might be said to be "obvious" on looking at a diagram. Why might a mathematician in the late eighteenth or early nineteenth century even want to prove it?[53] The key to answering this question is to look at the theorem as eighteenth-century mathematicians looked at it, not as a fundamental theorem in the calculus, but as part of the theory of equations. When we do this, we find that Lagrange, among others, had already tried to prove the theorem in 1767.

Lagrange's *Résolution des équations numériques* was, as we have said, a treatise on finding approximate solutions to equations. Obviously, approximate solutions are most urgently sought when the true root is not known. It would be well to be sure that there was a root before setting out to find it. Since approximation procedures so often begin with the finding of two numbers that bound the root above and below, the intermediate-value theorem assures the existence of a root in such procedures. Lagrange recognized this need: Accordingly, the first chapter of his treatise begins with this theorem.

Theorem I. If we have any equation, and if we know two numbers such that, if they are successively substituted for the unknown in that equation, they give results of opposite sign, the equation will necessarily have at least one real root whose value will be between those two numbers.[54]

This is the intermediate-value theorem, and the statement Lagrange gave here is not limited to polynomials.

In the main text of his *Résolution des équations numériques,* first published as a long article in 1767, Lagrange noted that this theorem is usually proved "by the theory of curved lines," but added – as one might expect him to add – that a proof based on the theory of equations would be preferable. He then gave an argument for this theorem – algebraic but incorrect – by breaking the original polynomial into linear factors and noting that each linear factor has the required property. Yet as he himself recognized, this proof does not apply to equations with complex roots.[55] So, in the book version of his treatise (1798; second edition, 1808), he added a *Note* in which he tried to supply a better proof. This *Note I* is the first of a long series of *Notes sur la théorie des équations algébriques,* and has the eye-catching title "On the demonstration of Theorem I." The reader turns eagerly to Lagrange's proof, which obtains the theorem from a result even stronger than the intermediate-value theorem for continuous functions: If Q and P are functions, and if for $x = p$, $Q < P$, and for $x = q$, $Q > P$, then there is a value of x between p and q such that $Q = P$. The reader then finds, however, an argument by Lagrange for this stronger theorem that cannot be called rigorous by Bolzano-Cauchy standards; it rests on a mental picture of two quantities approaching and passing each other, and is illustrated by the physical picture of two bodies moving along the same line while one overtakes the other.[56]

Cauchy and Bolzano, reading Lagrange's Theorem I and this *Note,* would have found three things: a stated need for the intermediate-value theorem as a fundamental result in the theory of equations; a call for a proof by algebraic methods; and then a proof clearly inadequate even by Lagrange's own standards. And we know beyond any doubt that Cauchy and Bolzano did read this, and with careful attention.

Bolzano began his *Rein analytischer Beweis* by saying that there are two fundamental results of equation theory. One is what is now called the fundamental theorem of algebra, the other, the intermediate-value theorem. The first, which Bolzano stated as "every algebraic entire function of one variable can be decomposed into real factors of the first or second degree," he said had now been rigorously proved by Gauss; the other, however, had yet to be given an adequate demonstration.[57] Bolzano observed that Gauss's first proof (1799) had been marred by resting on a geometrically based argument – in fact, it assumed the intermediate-value theorem itself – but that Gauss's new proofs (1816) were free from this defect.[58] Perhaps the immediate stimulus for Bolzano's 1817 paper was seeing Gauss's work of 1816; after seeing with satisfaction that an analytic proof now existed for one basic theorem, Bolzano might have renewed his attack on the other. In any case, the context of Bolzano's theorem in equation theory should now be obvious. And Bolzano explicitly criticized a proof of the intermediate-value theorem that is almost identical with Lagrange's (after having listed Lagrange's proof earlier in a footnote), pointing out quite properly that arguments of this type depend on the idea of motion.[59]

As for Cauchy, his own rigorous proof of the intermediate-value theorem has precisely the same logical role for him as Lagrange's had in the *Résolution des équations numériques;* it is Cauchy's Theorem I in a long *Note* (really a fair-sized treatise) at the end of the *Cours d'analyse,* entitled "Sur la résolution numérique des équations," which explicitly refers to Lagrange's treatise on the same subject.[60] Cauchy's *Note* develops many of the themes and methods of Lagrange's work. Cauchy had given a less rigorous argument for the intermediate-value theorem in the text of his *Cours d'analyse,* referring the reader to the *Note* for what he rightly called a "purely analytic and direct" proof.[61] Let us now turn to that proof and investigate its origin.

PROVING THE INTERMEDIATE-VALUE THEOREM

The first question to be asked is, Can Cauchy's proof be traced to Bolzano's? The answer is clearly no. Bolzano's proof is extremely involved and difficult, giving the theorem as one consequence of a chain of new and important results in analysis. Cauchy's proof, by contrast, is crystal clear and is based on a very simple procedure. Let us first indicate the contrast between the two proofs, and then return to the question of the origin of Cauchy's proof technique.

Bolzano began his proof by trying to show that every Cauchy sequence has a limit. He used this result to prove that a bounded set of real numbers has a least upper bound. He then used the least-upper-bound result to prove the same general theorem about continuous functions that Lagrange had used: If $f(\alpha) < \varnothing(\alpha)$ and $f(\beta) > \varnothing(\beta)$, then there is an x between α and β such that $f(x) = \varnothing(x)$. From this last theorem he, like Lagrange, obtained the intermediate-value theorem as a corollary.[62]

Cauchy's proof proceeds as follows. Let $f(x)$ be continuous between x_0 and X, let $f(x_0)$ and $f(X)$ have opposite sign, and let $X - x_0 = h$. Divide the interval $X - x_0$ into m parts each having length h/m, and examine the signs of $f(x_0)$, $f(x_0 + h/m)$, $f(x_0 + h/2m)$, . . . , $f(X - h/m)$, $f(X)$. In this set of functions, there will be found two values of $f(x)$ having opposite sign, whose arguments differ by h/m. Taking these, consider the new interval between them, of length h/m; and repeat the same process again and again. This procedure eventually produces two sequences of values of x, one increasing (which Cauchy designated by $x_0, x_1, x_2, \ldots$) and one decreasing (which he wrote $X, X', X'', \ldots$), such that the terms of one sequence get arbitrarily close to the terms of the other; the sequence of differences between the successive values is $X - x_0$, $(1/m)(X - x_0)$, $(1/m^2)(X - x_0)^2$, and so on, which can be made as small as desired. The two sequences $(x_0, x_1, x_2, \ldots)$ and $(X, X', X'', \ldots)$ must then converge to a common limit.[63] Cauchy designated that common limit by a. The continuity of $f(x)$ now insures that the corresponding sequences of functions $f(x_0)$, $f(x_1)$, $f(x_2)$, . . . and $f(X)$, $f(X')$, $f(X'')$, . . . converge to the common limit $f(a)$. Since the two sequences are of opposite sign, that common limit $f(a)$ must be zero.[64] It should now be apparent that this proof is not a version of Bolzano's;[65] it would be much easier to devise Cauchy's proof from scratch than to adapt it from Bolzano's.

As Cauchy himself pointed out, his proof constructs the intermediate value by a procedure that can also be used to approximate the roots of equations.[66] Indeed, the very technique used in the proof can be found in Lagrange's book on approximations. In the pages of the *Résolution des équations numériques* immediately following Lagrange's Theorem I, Lagrange suggested the technique of successively substituting, in the place of the unknown x in an equation, numbers "in arithmetic progression $0, \Delta, 2\Delta, 3\Delta, 4\Delta, \ldots$," adding that if Δ is equal to or less than the smallest difference between the roots of the equation, this method would identify *all* the real roots.[67] Though Lagrange did not suggest repeating this process, since the right choice of Δ was his immediate concern, many analogous approximation methods using iteration, such as repeated halving, were known in the eighteenth century.[68] Repetition of Lagrange's process would thus seem natural.

It is not hard to imagine Cauchy's interest in a proof being aroused by Lagrange's call for one; his disappointment in actually reading Lagrange's proof; and the resulting effect on him of reading, immediately after this disappointment, an approximation technique so well suited to construct the intermediate value. At least one of Cauchy's other existence theorems was also based on transforming an approximation technique into a proof of existence by using the approximation technique to construct the solution – the proof of the existence of a solution to a differential equation.[69] Since we have already demonstrated that Cauchy knew Lagrange's *Résolution des équations numériques* well, there seems little doubt that a plausible mathematical background for the idea of Cauchy's proof is provided by Section I and Note I of Lagrange's book.

A PROGRAM FOR FURTHER RESEARCH

Seeking the twin roles of tradition and transformation in scientific discovery in the work of Bolzano and Cauchy has brought us to a new level of understanding. We need no longer be content to marvel that there is a man – Bolzano – who independently stated and proved some of the same results as Cauchy. The fact that both these mathematicians made similar discoveries has led us to ask for a historical explanation, to ask what there was in the mathematical climate of opinion in general, and in the work of their immediate predecessors in particular, that made their discoveries possible. We have demonstrated here the existence of a great deal of previous work relevant to their achievements. Incidentally the results of our investigation considerably accentuate the importance of the work of Lagrange for the later rigorization of analysis. We have focused primarily on explaining the important common achievements of both Cauchy and Bolzano in 1817–21: their general programs for rigorizing analysis; their definitions of continuity; their intention and accomplishment in giving proofs of the intermediate-value theorem. But the differences between the philosophical approaches they took, the mathematical techniques they used, and the relative emphasis they gave to different topics mean that the answer to the question of who influenced them and how cannot be identical in both cases. There is, accordingly, more work to be done than we have done here. We have given a documented account of the background of Cauchy's proof of the intermediate-value theorem. A similar study seeking the background of the ideas in Bolzano's proof of the least-upper-bound property, and in his proof, using that property, of the intermediate-value theorem, would be extremely helpful in understanding the development of Bolzano's thought.

Another task that deserves attention is explaining the apparent coincidence of the discovery by Bolzano and Cauchy of the Cauchy criterion. To be sure, Cauchy himself made little real use of the Cauchy criterion, so its background has not been essential in our investigation. Nevertheless, the Cauchy criterion is the one version of the completeness property of the real numbers that both Cauchy and Bolzano explicitly stated. Both used, sometimes implicitly, sometimes explicitly, various forms of the completeness property. Given our present purpose, we must at least say enough about the possible genesis of Cauchy's version of the Cauchy criterion to acquit him of the charge of plagiarizing Bolzano.

Cauchy's criterion is perhaps best understood in the context of the many equivalent versions of the completeness property of the real numbers used in his *Cours d'analyse.* Since he assumed them implicitly, they seem for him to be "obvious" intuitions about the real numbers: (1) the monotone-sequence property (every bounded monotone sequence converges to a limit), used in his proof of the intermediate-value theorem; (2) comparison test (if a given series of positive terms is, term by term, bounded by a second, convergent series, then the given series converges also), used, with comparison to a convergent geometric progression, to derive his ratio test and root test for convergence; and (3) lim-sup (a bounded sequence has a limit superior, which Cauchy called "the greatest of the limits"), used in his proof of the root test for convergence.[70] As for the Cauchy criterion, he stated it explicitly, in part because it is less "obvious," in part because he needed an explicit statement of it for a proof.

Given the series $u_0 + u_1 + \dots + u_n + \dots$, Cauchy observed, it was necessary that the nth term of the series u_n go to zero for the series to converge, but this was not sufficient. "It is necessary also, for increasing values of n, that the different sums $u_n + u_{n+1}$, $u_n + u_{n+1} + u_{n+2}, \dots$, taken from the first, in whatever number we wish, finish by constantly having numerical [i.e., absolute] values less than any assignable limit. Conversely, *when these diverse conditions are fulfilled, the convergence of the series is assured.*"[71] this last statement gives the completeness of the real numbers in the form "every Cauchy sequence converges." Cauchy applied this criterion rarely, but he did use it to show that the alternating series $1 - \frac{1}{2} + \frac{1}{3} - \frac{1}{4} \dots$ converges; he added that his proof could be generalized to show the convergence of any alternating series whose terms go to zero.[72]

Bolzano, four years before Cauchy, not only had stated the "Cauchy" criterion, he had tried to prove it (though in fact he succeeded only in showing that the existence of the limit was noncontradictory).[73] He gave the criterion as follows:

If a sequence [*Reihe*] of magnitudes $F_1(x)$, $F_2(x)$, $F_3(x)$, . . . $F_n(x)$ . . . $F_{n+r}(x)$ . . . is subject to the condition that the difference between its nth member $F_n(x)$ and every later member $F_{n+r}(x)$, no matter how far beyond the nth term the latter may be, is less than any given magnitude if n is taken large enough; then, there is one and only one determined magnitude to which the members of the sequence approach closer, and to which they can get as close as desired, if the sequence is continued far enough.[74]

Though Bolzano stated the Cauchy criterion for sequences rather than for the sum of a series, his version is nevertheless a clear anticipation of Cauchy's criterion.

What predecessors are there for Cauchy's version other than Bolzano? There is no clear-cut answer, but let us give some suggestions. There is a natural motivation for Cauchy's formulation in the first illustration he gave of his criterion.[75] Cauchy showed that the convergent geometric progression $1 + x + x^2 + \ldots$ ($|x| < 1$) satisfies the Cauchy criterion because the finite expressions x^n, $x^n + x^{n+1}$, $x^n + x^{n+1} + x^{n+2}$, . . . , are always included between the bounds x^n and $x^n/(1 - x)$.[76].

Cauchy may have intended this as more than an illustration, since he often used the convergences of geometric progressions as the basis for proving other means of convergence (the ratio test, for instance) by term-by-term comparison. Cauchy proved the convergence of geometric progressions with ratio less than one by directly proving that the partial sums had a limit by explicitly computing the remainder. Thus, he may have reasoned, any criterion equivalent to the convergence of a geometric progression with $|x| < 1$ ought to be valid. If this interpretation of Cauchy's illustration is correct, he may have discussed the geometric series to increase the reader's confidence in the Cauchy criterion, and not merely to illustrate it.

A possible eighteenth-century source for Cauchy's criterion has been suggested by G. Eneström: Euler's 1734 paper on harmonic progressions.[77] In this paper, Euler showed the divergence of the harmonic series by showing that what we would write as $S_{nk} - S_n$ was bounded below by a precisely computable finite positive number. Cauchy was generally familiar with Euler's work, though I do not know of a specific reference to this 1734 paper. Of course, Euler did not describe this computation as a general criterion for divergence, let alone its contrary for convergence, but the specific computation might have been suggestive to a mathematician in search of general criteria for convergence.

Cauchy's motivating illustration and Euler's divergence criterion give possible, though by no means certain, explanations for his formulation of the Cauchy criterion. They do not, however, appear to me to be plausible predecessors for Bolzano. The way Bolzano stated the criterion does not sound as if it arose out of concern with proving the convergence of series, and his proof that the "Cauchy" criterion implies the least-upper-bound property suggests that his version originated in attempts to elucidate the properties of the real-number system, perhaps aided by the mental image of points about a cluster point. I do not know what sources in eighteenth-century work might have inspired Bolzano's conceptions of the completeness of the real numbers, and I think it would be of interest to find out. Further research on the origins of Cauchy's conceptions of completeness would also be of value. Additional studies of the origins of other aspects of Bolzano's philosophy and mathematics would further illuminate his thought. There is more work to be done on the different directions taken by Cauchy and Bolzano in executing their common program. I urge scholars to intensify their work on all these questions.

Meanwhile, we should not underestimate what we have already been able to explain. Cauchy and Bolzano pioneered the transformation of the work of eighteenth-century analysts into nineteenth-century rigor. Their common achievement, of course, calls attention to their stature as outstanding creative mathematicians. But the fact that their work is an example of simultaneous discovery should call attention also to the mathematical tradition that inspired their discoveries. We have accounted historically for the simultaneous announcement of their programs for rigorizing analysis, their definitions of continuous function, their common desire to prove the intermediate-value theorem, and the technique of Cauchy's proof of that theorem. Besides the intrinsic interest of an account of the origins of any important discovery, we hope our work can further the reconstruction of the full mathematical and philosophical tradition out of which the common achievements of Cauchy and Bolzano came. Completing the picture we have begun here would contribute to the general goal that has inspired this volume: understanding the relationship between great scientific thinkers and the tradition that their work transforms.

NOTES

1 Bernard Bolzano, *Rein analytischer Beweis des Lehrsatzes dass zwischen je zwey Werthen, die ein entgegengesetztes Resultat gewähren, wenigstens eine reele Worzel der Gleichung liege* (Prague, 1817). Published in *Abh. Gesell. Wiss. Prague* (3) *5* (1814–1817), pp. 1–60. Reprinted as Ostwalds Klassiker 153, ed. P. E. B. Jourdain, and readily available in French translation, *Revue d'histoire des sciences, 17* (1964), pp. 136–64. Page references will be made in this chapter, to the French translation, together with section numbers so the reference can be located in any edition.

2 Bernard Bolzano, *Functionenlehre* [1830], in *Bernard Bolzano's Schriften*, vol. I, ed. Karel Rychlik (Prague: Königlich Böhmische Gesellschaft der Wissenschaften, 1930), pp. 55, 52, 100–9.

3 A.-L. Cauchy, *Cours d'analyse de l'Ecole Royale Polytechnique. 1^{re}Partie: Analyse algébrique* [all published] (Paris, 1821). in *Oeuvres completes d'Augustin Cauchy* (2) 3. (Paris: Gauthier-Villars, 1897). All page references will be made to the *Oeuvres*.

4 *Cours d'analyse*, Note II, pp. 360–77, has the inequalities he proved for later use. See also J. V. Grabiner, "The Origins of Cauchy's Theory of the Derivative," *Historia Mathematica, 5* (1978), pp. 379–409.

5 Bolzano was rediscovered, and his work described, by men like Hermann Hankel (see H. Hankel, "Grenze," Ersch-Gruber *Allgemeine Encyclopädie der Wissenschaften u.*

Künste, 90 Teil [Leipzig, 1871], pp. 185–211), and O. Stolz, "B. Bolzanos Bedeutung in der Geschichte der Infinitesimalrechnung," *Math. Annalen, 18* (1881), pp. 255–79.

6 Following Hankel, "Grenze." See, e.g., Carl Boyer, *History of Mathematics* (New York: Wiley, 1968), p. 564.

7 Ivor Grattan-Guinness, "Bolzano, Cauchy, and the 'new analysis' of the early 19th century," *Arch. Hist. Ex. Sci, 6* (1970), pp. 372–400; compare I. Grattan-Guinness, *The Development of the Foundations of Mathematical Analysis from Euler to Reimann* (Cambridge, Mass.: MIT Press, 1970), pp. 76–7.

8 Hans Freudenthal, "Did Cauchy plagiarize Bolzano?" *Arch. Hist. Ex. Sci., 7* (1971), pp. 375–92.

9 H. Sinaceur, "Cauchy et Bolzano," *Revue d'histoire des sciences, 26* (1973), pp. 97–112. Sinaceur somewhat overstates his case by underestimating Cauchy's rigor. See especially his comparison, p. 106, of Bolzano's proof of the intermediate-value theorem with Cauchy's *first* (geometric) argument, *Cours d'analyse,* pp. 50–1, rather than with what Cauchy himself called his purely analytic proof, *Cours d'analyse,* pp. 378–80. For Sinaceur's claim that there is no "epsilonisation" in Cauchy, compare *Cours d'analyse,* pp. 54–7.

10 See, e.g., E. Kolman, *Bernard Bolzano* (Berlin: Akademie-Verlag, 1963); H. Wussing, "Bernard Bolzano und die Grundlegung der Infinitesimalrechnung," *Z. Gesch. Naturwiss. Tech. Med.* (1964), pp. 57–73; Philip Kitcher, "Bolzano's Ideal of Algebraic Analysis," *Stud. Hist. Phil. Sci., 6* (1975), pp. 229–69; B. van Rooteslaar, "Bolzano's Theory of Real Numbers," *Arch. Hist. Ex. Sci., 2* (1962–6), pp. 168–80; P. Funk, "Bolzano als Mathematiker," *Sitz. Ber. Öst. Akad. Wiss. Wien,* Phil.-Hist. Klasse, *252* (1967), pt. 5, pp. 121–34.

11 A.-L. Cauchy, *Résumé des leçons données a l'école royale polytechnique sur le Calcul Infinitésimal,* Tome Premier (all published), *Oeuvres* (2) 4, p. 230.

12 Bolzano, *Functionenlehre,* p. 170; *Rein analytischer Beweis,* Preface, p. 131; sec. VI, pp. 142–3; Cauchy, *Calcul infinitésimal,* pp. 9–10; *Cours d'analyse,* pp. 413, 429; e.g.

13 Karel Rychlik, "Anmerkungen," (separate pagination), p. 23, in Bolzano, *Functionenlehre.*

14 C. A. Valson, *La vie et les travaux du Baron Cauchy* (Paris, 1868), p. 27. The other three were Laplace's *Mécanique celeste,* Vergil, and Thomas à Kempis's *Imitation of Christ.*

15 For Cauchy, see, e.g., *Calcul infinitésimal,* pp. 22–3, 88–9, 217. For Bolzano, *Functionenlehre,* pp. 80, 155.

16 *Cours d'analyse,* pp. ii–iii. My italics.

17 For example, see, respectively, *Cours d'analyse,* pp. 121–8; p. 54, where he interprets "$\lim_{x \to \infty} f(x + 1) - f(x) = k$" as "we can give to h a value sufficiently large so that, x being equal to or greater than h, the difference in question is included between $k - \varepsilon$ and $k + \varepsilon$"; and pp. 378–80.

18 Bolzano, "Die drey Probleme der Rectification, der Complanation, und der Cubirung" (1817), *Schriften,* vol. 5, pp. 67–137; p. 67.

19 *Rein analytischer Beweis,* sec. I–II; pp. 137–8.

20 Kitcher, "Bolzano's Ideal of Algebraic Analysis."

21 For instance, Grattan-Guinness, *Foundations,* pp. 60–1; J. M. Dubbey, "Cauchy's Contribution to the Establishment of the Calculus," *Annals of Science, 22* (1966), pp. 61–7.

22 *Calcul infinitésimal,* "Avertissement," p. 9.

23 *Rein analytischer Beweis,* sec. II, p. 138.

24 Leonard Euler, *Institutiones calculi differentialis* (St. Petersburg: Imp. Acad. Sci., 1755); *Opera (1),* vol. X, p. 9.

25 J. D'Alembert et al., *Dictionnaire encyclopédique des mathématiques* (Paris, 1789), article entitled "Fluxion."

26 J.-L. Lagrange, *Mécanique analitique* [*sic*] (Paris: Desaint, 1788); "Preface." The preface to the first (1788) edition is reprinted in *Oeuvres de Lagrange,* XI, pp. xi–xii. The statement was reproduced in the Preface to the second edition, *Mécanique analytique* (Paris: Courcier, 1811–15), and may be found in *Oeuvres,* XI, p. xiii.

27 J.-L. Lagrange, *Théorie des fonctions analytiques* (Paris: Imprimérie de la république, An V [1797]); second edition (Paris: Courcier, 1813), reprinted in *Oeuvres,* IX; the quoted passages will all be from this edition. *Oeuvres,* IX, p. 17.

28 By "function," Lagrange meant any "expression de calcul" containing the variable, whether finite or infinite. See *Théorie des fonctions,* p. 15; compare J.-L. Lagrange, *Leçons sur le calcul des fonctions,* new ed. (Paris: Courcier, 1806), *Oeuvres,* X, pp. 11–12.

29 He viewed limits as insufficiently general, because "limit" in the eighteenth-century view had usually been restricted to a one-sided approach, *Calcul des fonctions,* p. 8; and as insufficiently precise because we have no "clear and precise idea" of a ratio when both terms simultaneously become zero, *Théorie des fonctions,* p. 18.

30 The *Calcul des fonctions* shares its philosophy with the *Théorie des fonctions,* seeking, for instance, "to lead back the differential calculus to a purely algebraic origin," *Oeuvres,* X, p. 9.

31 This circumstance – Cauchy's having to deal systematically with the subject because he had to teach a course – itself suffices to explain why the systematic rigor of the *Cours d'analyse* is not universally present in Cauchy's pre-1821 papers. Compare Freudenthal, "Did Cauchy plagiarize Bolzano?" p. 378.

32 *Calcul infinitésimal,* Preface, pp. 9–10. It was in the *Calcul infinitésimal,* as we have already noted, that he showed that two distinct functions could have the same Taylor series: pp. 229–30.

33 *Mécanique analytique, Oeuvres,* XI, p. xiv.

34 From $f(x + h) = f(x) + hf'(x) + h^2/2\, f''(x) + \ldots$, Lagrange obtained (∗) $f(x + h) = f(x) + hf'(x) + hH$, where H went to zero with h. (Lagrange used i for what is given here as h.) He used the inequalities based on that second equation to prove theorems about derivatives. For a full discussion of Lagrange's use of this property (∗) and his influence on Cauchy, see J. V. Grabiner, "The Origins of Cauchy's Theory of the Derivative."

35 *Cours d'analyse,* p. 43.

36 *Cours d'analyse,* p. 37. As we have said, Cauchy thought infinitesimals were heuristically valuable, so he used them. But he regarded the rigorous approach of the *Cours d'analyse,* which uses limits understood in terms of the algebra of inequalities, as logically prior to any such heuristic reliance on infinitesimals. See, e.g., the arguments cited in n. 17.

37 *Rein analytischer Beweis,* sec. II (a), p. 139.

38 P. E. B. Jourdain, "The origins of Cauchy's conception of the definite integral and of the continuity of a function," *Isis, I* (1914), pp. 661–703.

39 For such an account, see C. Truesdell, *The rational mechanics of flexible or elastic bodies,* 1638–1788, in Euler, *Opera* (II), vol. XI, pt. 2, pp. 237–300. Compare Grattan-Guinness, *Foundations,* chap. I. The main texts, with the original notation, can readily be consulted in D. J. Struik, ed., *A Source Book in Mathematics* (Cambridge, Mass.: Harvard University Press, 1969).

40 L. F. A. Arbogast, *Mémoire sur la nature des fonctions arbitraires qui entrent dans les intégrals des équations aux différences partielles* (St. Petersburg, 1791), pp. 5, 9.

41 Arbogast, *Mémoire,* p. 9.

42 Arbogast, *Mémoire,* p. 9.

43 S. F. Lacroix, *Traité élémentaire du calcul différentiel et du calcul intégral,* 2d ed. (Paris, 1806), p. 82.

44 *Functionenlehre,* p. 16, though in a different connection. This reference to Lacroix is cited by Grattan-Guinness, *Foundations,* p. 77, n. 23. We will discuss the purpose and content of Bolzano's reference to Lacroix later.

45 If one still wonders whether Cauchy took his definition from Bolzano, note that Cauchy was already thinking about the question in 1814, was familiar with characterizations of continuity like the Arbogast-Lacroix-Lagrange ones then, and used the Lagrangian phrase "insensible degrees" in his own characterization. In A.-L. Cauchy, "Mémoire sur les intégrales definies," *Mem. div. sav.* (*2*) 1, (1827) pp. 601–799; in *Oeuvres* (1), vol. 1, pp. 329–506, he defined the "Cauchy principal value" for the integral of a piecewise continuous function. He said that $_a\int^b f(x)dx = F(b) - F(a)$ (when $F'(x) = f(x)$) "only in the case of a function found increasing or decreasing in a continuous manner between the limits in question. If, when the variable increases by insensible degrees, the function found passes suddenly from one value to another . . . the differences between each of the brusque jumps that the function makes necessitate a correction." He later gave an algebraic version: for small ζ, when two values of a function $\varnothing$, $\varnothing(z + \zeta)$ and $\varnothing(z - \zeta)$, are "sensibly different" from each other, there is a finite fixed Δ such that $\varnothing(z + \zeta) - \varnothing(z - \zeta) = \Delta$. *Oeuvres* (1) 1, p. 332; pp. 402–3.

46 J.-L. Lagrange, *Traité de la résolution des équations numériques de tous les degrés, Oeuvres,* VIII, p. 134 (Note I).

47 For Cauchy, see, e.g., *Cours d'analyse,* p. 413. Compare his use of the phrase "insensible degrees," quoted in n. 45. For Bolzano, see, e.g., *Rein analytischer Beweis,* preface, p. 137 n.

48 The theorem being proved, in modern notation, was that if $f(x + h) = f(x) + hp(x) + h^2q(x) + h^3r(x) + \ldots$, there is an h sufficiently small to that $|f(x)| < |hp + h^2q + h^3r + \ldots|$, $|hp| < |h^2q + h^3r + \ldots|$, etc.

49 *Théorie des fonctions, Oeuvres* IX, p. 28. My italics.

50 Ibid. The *Calcul des fonctions* has a closely similar characterization; *Oeuvres,* X, p. 87.

51 *Functionenlehre,* pp. 15–16.

52 *Functionenlehre,* p. 16; his italics. Among the "others" who used this definition, Bolzano on p. 15 cited one: Klügel. He gave essentially the Bolzano-Cauchy definition in his mathematical dictionary. G. S. Klügel, *Mathematisches Wörterbuch,* vol. 4 (Leipzig: Schwickert, 1823), p. 550. (Note that this was not published until after the *Cours d'analyse.*) The "previous section" to which Bolzano referred gives the 1830 version of his own definition of continuous function, where the absolute value of the difference $F(x + \Delta x) - F(x)$ "becomes and remains smaller than any given fraction $1/N$ if Δx is taken sufficiently small." *Functionenlehre,* p. 12.

53 That Cauchy may have been stimulated to prove the intermediate-value theorem by seeing the title of Bolzano's 1817 paper is, incidentally, one of the few points Freudenthal conceded in his response to Grattan-Guinness, "Did Cauchy plagiarize Bolzano?" p. 384. We will show that even this supposition is entirely unnecessary.

54 *Résolution des équations numériques,* Theorem I, *Oeuvres,* VIII, pp. 19–20.

55 Ibid., Note I, p. 133.

56 The proof begins by representing, in general, the proposed equation by $P - Q = 0$, P being the sum of terms with plus sign, $-Q$ being the sum of all those with minus sign. "From the form of the quantities P and Q . . . it is clear that these quantities increase necessarily as x increases, and that, making x increase by all insensible degrees from p to q, they increase also by insensible degrees; but in such a way that P increases more than Q, since the smaller it was the larger it becomes. Thus there will necessarily be a term between the two values p and q where P will equal Q, just as two moving bodies which are supposed to traverse the same line in the same direction and which, beginning simultaneously from two different points, arrive in the same time at two other points in such a way that the one which was formerly behind is later found ahead of the other, must meet on their paths." Ibid., p. 134.

57 *Rein analytischer Beweis,* preface, p. 136. His statement of the fundamental theorem of algebra is just a German translation of the title of Gauss's 1799 paper.

58 One example of the "defect" in Gauss's proof may be found in C. F. Gauss, *Demonstratio nova theorematis omnem functionem algebraicam rationalem integram unius variabilis in factores reales primi vel secundi gradus resolvi posse* (1799), in *Werke* III, pp. 3–56; see sec. 19, part II. Cited by D. J. Struik, *Source Book,* p. 119, n. 4. For Bolzano's assessment, see the place cited in n. 57, this chapter.

59 *Rein analytischer Beweis,* II, p. 138; the citation of Lagrange is in the preface, p. 137.

60 *Cours d'analyse,* pp. 378–425. For references to Lagrange's treatise, see p. 413, e.g.

61 Ibid., pp. 50–1.

62 See *Rein analytischer Beweis,* II (c), p. 140.

63 This conclusion implicitly assumes the completeness of the real-number system in the form of the monotone-sequence property: A bounded monotone sequence converges to a limit.

64 *Cours d'analyse,* pp. 378–80.

65 As claimed by Grattan-Guinness; quoted by Freudenthal, "Did Cauchy plagiarize Bolzano?" p. 383, in the context of his dėmolition of this claim, pp. 382–4.

66 *Cours' d'analyse,* p. 51.

67 *Résolution des équations numériques,* sec. 6, *Oeuvres,* VIII pp. 22–23. Lagrange also tried to find ways of calculating such a Δ; see note IV, pp. 146–58. He used the same approximation method elsewhere; compare J.-L. Lagrange, *Leçons élémentaires sur les mathématiques données a l'école normale en 1795* (Paris, An III [1794–5]), *Oeuvres,* VII, pp. 181–288; see pp. 260–1.

68 Compare, for instance, Colin Maclaurin, *A Treatise of Algebra,* 2d ed. (London: A. Miller and J. Nourse, 1756), p. 230.

69 A.-L. Cauchy, *Exercises d'analyse* (Prague, 1840), in *Oeuvres* (2) XI, pp. 399ff. The approximation method is due to Euler. See P. Painleve, "Gewönliche Differentialgleichungen: Existenz der Lösungen," *Enc. der Math. Wiss.,* II B, 1, 1, sec. IIA 4a, pp. 189–293; see p. 193, n. 5. Indeed, Cauchy's Note III itself contains a proof drawn from an approximation method of A.-M. Legendre; see *Cours d'analyse,* p. 381, for an acknowledgment of this by Cauchy.

70 For (1), see *Cours d'analyse,* p. 379; for (2), see pp. 121–2; for (3), see p. 121.

71 *Cours d'analyse,* pp. 115–16. My italics.

72 Ibid., pp. 130–1.

73 Often noted; see, e.g., Sinaceur, "Cauchy et Bolzano," p. 102.

74 *Rein analytischer Beweis,* sec. 7; p. 150.

75 As pointed out by Freudenthal, "Did Cauchy plagiarize Bolzano?" pp. 380–1.

76 *Cours d'analyse,* pp. 116–17.

77 G. Eneström, "Ueber eine von Euler aufgestellte allgemeine Konvergenzbedingung," *Bib. Math.* (3) 6 (1905), pp. 186–9. Leonhard Euler, "De progressionibus harmonices observationes," *Comm. Petrop.* 7 (1734–5), *Opera* (1) XIV, pp. 87–100.

6

Idolatry, automorphic functions, and conceptual change

Reflections on the historiography of nineteenth-century mathematics

UTA C. MERZBACH
Smithsonian Institution

Nineteenth-century mathematics is one of the great underdeveloped territories of historic inquiry. It is also one of the most rewarding areas of research for the student of conceptual change. Only recently has an apparent trend been established that brings an increasing number of both mathematicians trained in history and historians trained in mathematics to explore this rich field. Many of these contemporary historians of mathematics work independently of established institutions in the history of science. They encounter a vast mass of untapped source materials, numerous studies conducted mostly by mathematicians and philosophers who had a personal involvement in the subject, and a small quantity of frequently cited primary sources. Their attempts to delve into the intricate network of nineteenth-century mathematical ideas are made difficult by a variety of factors, the most obvious of which is the scarcity of related studies carried out by historians. Other obstacles, however, threaten to affect the quality of future work. Of the handicaps the present-day student of conceptual change in nineteenth-century mathematics must face, few are more insidious than adherence to the idols of *rhizofiliation, nomination, dehumanization,* and *axiomatization.*

Rhizofiliation (root tracing) has been the source of most of our knowledge about nineteenth-century mathematics. The few surveys of the history of mathematics that include the nineteenth century are either "highlight" chronologies or, like many of the specialized studies on which they are based, set out primarily to trace the roots of specific branches of contemporary mathematics. As we know from analogous situations in other disciplines, root tracing can bring together much valuable information; but it is distortive. The distortion is twofold: The subject matter treated depends on what is considered significant in mathematics at the time of writing, or, more specifically, by the writer; and the concepts or problems singled out for study are those that appear similar over a period of time. The first distortion accounts for such things as the fact that until recently we knew far less about number theory or numerical analysis or even topology in the nineteenth century than non-Euclidean geometry or complex function theory; the latter two were of greater interest to the writers who gave us our first influential historic accounts of the nineteenth century.[1] The second distortion especially affects the study of conceptual change. For, as I. B. Cohen has noted,

By stressing differences between successive generations of concepts (through "transformation"), rather than emphasizing their apparent and undifferentiated similarities, the point of view I am advocating will reveal the fine structure of the creative enterprise of science where today we may tend to see only the gross features of scientific change.[2]

The distortive effect is intensified when the mathematical language and notation of an earlier period are translated into present-day terms.

Nomination leads to preoccupation with established theories and explicitly defined concepts at the expense of needed concern with suggestive techniques or implicitly accepted ideas. One frequent result is that a topic or concept is studied at a stage past that at which the major creative transformations have occurred. For example, nomination accounts for the efforts to assess the rise and fall of the very specific topic that came to be known as "invariant theory" and is frequently assigned a lifetime of fifty years (1841–90) without contemplating the wider issue of the emergence to prominence during the early nineteenth century of the concept of invariance in all branches of mathematics, and its role since that time. Similarly, nomination brings us a variety of discussions of Riemann's function theory without exploration of the interesting and enlightening question: What was it in the lectures and publications of his teachers Dirichlet and Jacobi, as well as Gauss, that caused Riemann to propagate a "revolutionary" point of view concerning the nature of functions? A combination of rhizofiliation and nomination has led writers of historical accounts to explain the importance of elliptic function theory in terms of its having served as a "model" or "testing ground" for subsequent theories of functions. This has forced them to deal, albeit unwillingly, with a vast body of specific results and computations, resulting in bewilderment at so many nineteenth-century mathematicians' having engaged in such cumbersome work, and leaving the impression that Jacobi was primarily a master of algorithm who loved to while away his hours in calculation. What is missed in such evaluations is Jacobi's recognition through elliptic functions of the relationship binding various arithmetic, algebraic, analytic, and geometric problems and his conscious exploitation of elliptic functions as a unifying tool.[3] This perception had a profound influence not only on his work but also on the orientation of his contemporaries, although few of them expressed it as directly as the admiring Hermite, who noted:

Mais c'est surtout l'oeuvre propre de l'immortel auteur des *Fundamenta* d'avoir reconnu ces rapports si remarquables des nouvelles transcendántés avec l'Algèbre et les propriétés des nombres.[4]

Dehumanization refers to the tendency to divorce studies of conceptual change from "external" factors affecting those who produced that change. This deprives the investigator of important clues. In mathematics, as in other disciplines, straightforward matters such as student–teacher relationships, the identity of colleagues on a faculty, presence at a meeting, or attitude toward a contemporary or predecessor are significant keys to the study of change. This is so, not only because these factors contribute to the acceptance or rejection of a "new" concept, but also because they may predispose the individual to consider or overlook a particular notion in the first place. Thus, knowledge of Felix Klein's geometric training under Plücker and Clebsch, of his association with Lie, of their stay in Paris during 1870, of their resulting exposure to Darboux and to Jordan's *Traité des Substitutions* provides a clearer guide to the formative stages of Klein's conceptualization of automorphic functions than do his later writings.

Axiomatization causes commonly accepted statements to be elevated to the level of self-evident truths, where they become a framework for future studies. For example, a survey of histories of mathematics currently in print that deal with the nineteenth century will show that most conform to the following set of statements:

1. Gauss [was] the greatest mathematician at least since Newton.
2. In the nineteenth century . . . generalization and abstraction began to become the order of the day.
3. Riemann . . . more than any other has influenced the course of modern mathematics.
4. The most stimulating and important nineteenth century event pertaining to the principles of geometry was the discovery of non-Euclidian geometry by Gauss, Bolyai and Lobachevski.
5. The most stimulating and important nineteenth century event pertaining to the principles of analysis was the arithmetic conceptualization of the continuum in the works of Cauchy, Bolzano and Lobachevski.
6. In the numerical epsilons and deltas of rigorous Weierstrassian analysis the calculus of the nineteenth century attained its classic perfection.
7. From the technical standpoint, the most original creation of the nineteenth century was the theory of functions of a complex variable.
8. In algebra, the two most revolutionary contributions were made by . . . Sir William Rowan Hamilton [and] George Boole. The most prolific contributors to nineteenth century algebra were . . . Arthur Cayley and J. J. Sylvester and it was chiefly from their alma mater, Cambridge, that the rise of modern algebra stemmed.[5]

Since these statements tacitly have been assigned axiomatic status, the student of the subject rarely feels called upon even to investigate in what sense or to what extent they are true. The problem with the statements is not so much their validity as their emphasis, however. Through rhizofiliation they grew out of historic analyses conducted by men whose chief preoccupation lay with classical analysis or axiomatics. The statements were reinforced by, and reinforced, philosophical concerns about foundational problems and mathematical involvement with turn-of-the-century issues. They are allied with the nominative tendency to explore in detail post-1850 theories but to limit consideration of work done during the earlier decades of the century to episodic treatments and flashbacks that credit Gauss with numerous unpublished anticipations; Cauchy with the "rigorous" definition of a half-dozen fundamental analytic concepts; and Hamilton, Galois, Abel, Bolyai, and Lobachevski with being geniuses who made discoveries not duly appreciated in their lifetime. This does not aid the study of conceptual change.

In studying the transformation that mathematics underwent in the nineteenth century, it is useful to focus on the genesis and utilization of concepts that emerged during that period. An example of such a notion is that of "automorphic function," to which several of the illustrations given earlier relate. An automorphic function is a complex function staying invariant under a group of substitutions of the form $w = (az + b)/(cz + d)$.[6] The concept is commonly associated with the theory of automorphic functions that was established in the last quarter of the nineteenth century. The story of the "creation" of that theory is well known because the two central figures it involves, Felix Klein and Henry Poincaré, wrote about it repeatedly.[7] Poincaré stressed that in fashioning his theory he had used the theory of elliptic functions as a model; his motivation was the need for a tool to solve linear differential equations, and he had been inspired by the work of Lazarus Fuchs. Klein emphasized his own geometric orientation; he had first come upon the importance of the function in his work on the icosahedron; he credited Riemann as the source of his outlook and of the basic concepts of geometric function theory he utilized.

Historic surveys of automorphic functions tend to follow the outline established by Klein's student, Fricke, in his article in the *Encyklopaedie der mathematischen Wissenschaften.*[8] It is noted that elliptic modular functions (those where a, b, c, and d are integers such that $ad - bc = 1$) were studied earlier in the nineteenth century. Passages from Gauss's unpublished writings are cited that show that he recognized certain modular functions, performed related conformal mappings that demonstrate his recognition of "Schwarz's" symmetry principle, understood the relationship of these matters to number theory and entertained the idea of generalizing his results. A variety of results concerning modular functions by Abel, Jacobi, Hermite, Eisenstein, and Weierstrass are listed; it is usually emphasized that none of these authors understood the implications of their results in terms of automorphic function theory. Next, Riemann's influence is emphasized; singled out are his inaugural dissertation with its new concept of complex functions, especially his method of defining them by conformal mapping, and his work on the hypergeometric function. To this are usually added references to his posthumously published related lectures and contributions to minimal surface theory and other matters. Finally, the resulting work in the 1870s by Schwarz, Fuchs, Schottky, Dedekind, and H. J. Smith is cited, emphasis being placed on their specialized interests that contrast with the general conception of Klein and Poincaré whereby each founded the theory of automorphic functions.[9]

The dominance of our four idols in even the best of these accounts is startling. There is some variation, of course. In particular, attempts to be rigorous and objective, and to go back to the original sources, have caused the most recent accounts to be freer of axiomatization and nomination, but subject to heavier dehumanization and rhizofiliation.[10]

A number of approaches could be taken that would reveal more than these historic summaries do about the process of change surrounding the emergence of the automorphic function concept. One that is proving rather effective in studying the conceptual transformations in terms of both individual and group thought processes may be outlined as follows.

Let the idea or concept to be studied be denoted by C_1. We shall assume that any concept C_1 can be expressed as the result of an operation or "transformation" t on a concept C_2, so that $C_1 = t(C_2)$. An *event* E_1 is said to occur when an individual H_1 performs an operation t_1 on an idea C_1. We denote this as $E_1 = (H_1, t_1, C_1)$. If $t_1(C_1) = C_2$, we shall say that the individual H_1 *generated* the concept C_2 at the event E_1.[11] Certain relations may exist between an event E_r and an individual H_s. One such relation leads us to say that H_s *has knowledge* of E_r; this relationship may be denoted by $H_s = H_s/E_r$. Two events E_r and E_s are said to be *historically related* if there exists a sequence of events E_i, where $i = 1, 2, \ldots, n$, $1 \leq r \leq n$, $1 \leq s \leq n$, such that $E_i = (H_i/E_{i-1}, t_i, C_i)$. If $r \leq s$, then the event E_r *occurred before* the event E_s. Two ideas C_p and C_q are said to be *conceptually related* if there exists a sequence of ideas C_j, where $j = 1, 2, \ldots, n$, $1 \leq p \leq n$, $1 \leq q \leq n$, such that $C_j = t_j(C_{jf-1})$.

To study conceptual change in the history of mathematics we are concerned with the establishment of a network of sequences consisting of historically related events generating conceptually related ideas. For example, if we wish to determine whether there is a historic relationship between the generation of two conceptually related ideas C_r and C_s, we select events E_j at which these ideas were generated and see whether a sequence of events of the form $E_i = (H_i/E_{i-1}, t_i, C_i)$ can be established that contains at least two of the E_j.

Just as family relationship is accepted as significant only within a rather small number of generations and branches, so the historical and conceptual relationships established by our sequences should not extend too far if they are to be meaningful. To cut down on the density of sequences it is useful, for example, to identify certain events as being *critical.* Critical events may mark the first known appearance or explicit definition of a concept or its initial use in a major problem area, branch of mathematics, or technique; others may mark significant unions formed by linking use of the same concept in two areas or by linking two concepts.

To illustrate some of these points we turn again to the history of the automorphic function concept. Noting that the definitions of automorphic functions consistent with late nineteenth-century usage all involve the concepts of (linear) transformation, group, invariance, and complex function, it appears useful to study their interrelationship during the century. This makes it easy to identify a number of critical events. For example, Galois participated in several such events, not only because of his linking of the concepts of group (substitution group) and invariance, but also because of his applying the concept to Jacobi's modular equation.[12] By paying special attention to sequences of concept transformations that include unions of these notions in number-theoretic, analytic, and geometric applications, one arrives at several sequences of nineteenth-century events leading to those in which Klein and Poincaré participated and generated the automorphic function concept. What is particularly striking in these sequences is the way in which they illustrate the different stages of conceptual junctures that led to one of the more famous cases of "simultaneous discovery." Establishing such sequences over a period extending into the twentieth century, one perceives an interesting pattern reflecting the relative influence of number-theoretic and geometric approaches on the course of automorphic function theory. Participants in critical events of this period include Koebe, Hecke, Petersson, and Siegel.[13]

Another interesting problem area that is easily analyzed within our framework relates to unpublished discoveries. For instance, suppose that the Gauss *Nachlass* indeed establishes that an event occurred at which Gauss generated the modular function concept. That establishes a conceptual relationship to other events involving modular functions and to the broader concept of general automorphic function. It also establishes a historic relationship to other ideas generated by Gauss. But it is historically related to the study of the modular or automorphic function concept prior to 1900 only if a sequence of events exists such that each participating individual had knowledge of the preceding event; such that at least one had knowledge of Gauss's generation of the modular function concept; and such that at least one participated in an event generating the automorphic function concept.[14] The evidence so far suggests that if such a sequence exists, the first event in it in which someone other than Gauss participated will fall after 1855, the year of Gauss's death; for the allusions to the subject in his correspondence and publications are too vague to have been recognized, that is, to have enabled an individual to have knowledge of the event being considered. This state of affairs makes all the more interesting the question of Riemann's familiarity with the *Nachlass.*[15] That question could be raised for Felix Klein as well; but the nature and unusually dense cluster of conceptually related critical events in his writings from 1870 on make it unlikely that he would have had knowledge of the material in the *Nachlass* even had he seen it during his first stay at Göttingen.[16]

Yet this does not mean that Gauss's work is historically unrelated to the automorphic function concept. Other events, further removed in conceptual relationship, in which Gauss participated enable us to establish appropriate sequences; apparently, they all include events in which Abel, Jacobi, or Galois participated; the participation of others, such as Kummer, Hermite, Kronecker, Betti, Brioschi, or later contributors, varies with the sequence branch one pursues.

The scheme outlined here is a device that may be useful for the historian of mathematics in establishing historic and conceptual relationships, and in keeping in mind the distinction between the two. It also helps avoid the four idols discussed earlier; it does this more effectively for dehumanization and nomination than the other two, which tend to influence the choice of critical events. It does not provide a formula for appropriately choosing critical events, or optimal sequences of events or of concept transformations. Events can be listed by the chronicler, historic associations established by the archivist, conceptual linkages by the mathematician. The diversity of the relationships among these three factors remains the special challenge to the student of the history of mathematics. Glimpses of a structure provide the reward for the explorer of conceptual change.

NOTES

1 For examples of such works note the major references in the publications cited in n. 5.

2 "History and the Philosopher of Science," in *The Structure of Scientific Theories*, ed. by Frederick Suppe (Urbana: University of Illinois Press, 1974), p. 322.

3 See, e.g., C. G. J. Jacobi, *Gesammelte Werke* 1 (Berlin: G. Reimer), p. 263.

4 C. Hermite, "Sur la théorie des équations modulaires," *Oeuvres* 2 (Paris: Gauthier-Villars, 1905–1917), p. 81.

5 Statements 4 and 5 are derived from Hilbert's introduction to his address on mathematical problems at the 1900 Paris Congress; see P. S. Alexandrov, ed., *Die Hilbertschen Probleme* (Leipzig: Akademische Verlagsgesellschaft, 1971), p. 22. For the other statements, see E. T. Bell, *The Development of Mathematics* (New York: McGraw-Hill, 1952), p. 294 (6); Carl B. Boyer, *A History of Mathematics* (New York: Wiley, 1968), p. 620 (8); Howard Eves, *An Introduction to the History of Mathematics*, rev. ed. (New York: Holt, Rinehart and Winston, 1964), p. 366 (2); Morris Kline, *Mathematical Thought from Ancient to Modern Times* (New York: Oxford University Press, 1972), p. 871 (1, 7); Dirk J. Struik, *A Concise History of Mathematics*, 2d rev. ed. (New York: Dover, 1948), p. 232 (3).

6 The term was used by F. Klein in 1890 to denote "single-valued functions of w which reproduce under infinitely many linear substitutions of w." "Zur Theorie der allgemeinen Laméschen Funktion," *Gesammelte mathematische Abhandlungen* 2 (Berlin: Julius Springer, 1922), p. 549.

7 Felix Klein, *Lectures on the Icosahedron*, trans. G. G. Morrice (New York: Dover, 1956); *Vorlesungen über die Entwicklung der Mathematik im 19. Jahrhundert* 1 (New York: Chelsea, 1956), pp. 345–81; *Gesammelte mathematische Abhandlungen* 3(Berlin: Julius Springer, 1923), pp. 577–86. H. Poincaré, "Analyse des Travaux scientifiques . . . faite par lui-meme," *Acta Mathematica* 38(1921), pp. 42–50.

8 IIB4: "Automorphe Funktionen und Modulfunktionen" (Leipzig: Teubner, 1913).

9 For variations of this approach see Joseph Lehner, *Discontinuous Groups and Automorphic Functions* (Providence, R.I.: American Mathematical Society, 1964), chap. I, and Jean Dieudonné, *Abrege d'histoire des mathematiques 1700–1900* 1 (Paris: Hermann, 1978), chap. 7.

10 For example see Dieudonné, *Abrege d'histoire des mathematiques*.

11 H_1 is then said to *participate* in E_1.

12 Letter to Auguste Chevalier of May 29, 1832. E. Galois, *Oeuvres* (Paris: Gauthier-Villars, 1951), p. 29.

13 For a summary of such events see Lehner, *Discontinuous Groups*.

14 The question is of chief interest for the period prior to 1900 because that is the year Gauss's notes on the subject were publicized through the editorial comments accompanying their publication in vol. 8 of the *Werke*.

15 This was a subject of controversy between Schering and Hattendorff in the 1870s. See Gauss, *Werke* 8.

16 These critical events are reflected in the *Erlanger Programm* and the papers collected in vols. 2 and 3 of Klein's *Abhandlungen*.

7

The Andalusian revolt against Ptolemaic astronomy

Averroes and al-Biṭrūjī

A. I. SABRA
Harvard University

I

The episode referred to in the title of this chapter as "the Andalusian revolt" is the well-known anti-Ptolemaic program of research that was conceived and defended by twelfth-century scholars in Muslim Spain. This was not a widely characteristic or long-lasting phenomenon of Arabic science, being definitely limited both geographically and in time. Nor was it in any way representative of the high degree of mathematical accomplishment already reached and subsequently maintained in the Islamic world. It was, nonetheless, an intriguing phenomenon that was associated with towering figures such as Averroes and Maimonides, and its very confinement to Andalusia under one rule, that of the Almohads, gives rise to rather interesting historical questions. I shall argue that there existed in twelfth-century Spain a certain cultural situation without reference to which the surprising position taken by Averroes and al-Biṭrūjī would be difficult if not impossible to explain.

The Middle Ages witnessed no revolutions in science, at least not in the sense we have come to associate with certain features of the combined achievement of men like Copernicus, Kepler, Galileo, Descartes, and Newton. But this does not mean that the ideal of innovation was beyond the imagination of medieval scholars or that they lacked a critical attitude toward their admittedly respected predecessors. In Medieval Islam the concept of innovation in intellectual endeavor found expression in terms like *istikhrāj* or *istinbāṭ (discovery)*, which denoted accomplishments that went beyond merely elucidating, emending, or completing an earlier contribution to knowledge;[1] and a critical attitude clearly revealed itself in the not infrequent composition of *shukūk* (*aporiai, dubitationes*), a form of argument in which difficulties or objections were raised against ancient authorities. Indeed, it would be impossible to explain the high quality of much of Islamic scientific writings without noting the intellectual ambition and independence of mind their authors often possessed to a remarkable degree.

One example, which is of particular relevance to the subject of this chapter, is provided by the mathematician Ibn al-Haytham, who flourished in Cairo some

This research was completed during tenure of grants from the National Science Foundation and the National Endowment for the Humanities. Grateful acknowledgment is made of this support.

150 years before Averroes. Ibn al-Haytham wrote a series of *shukūk* or objections mainly directed against certain aspects of Ptolemaic astronomy, in particular what (rightly) appeared to him as an inconsistency generated by Ptolemy's introduction of the equant hypothesis.[2] It is not that he (or the later Islamic critics of Ptolemy) failed to appreciate the predictive function of that hypothesis. But, being convinced (as Ptolemy had been) that planetary motions must ultimately be understood in terms of the motions of real spherical bodies in which the planets were embedded,[3] he saw that the equant hypothesis would make it necessary to attribute a nonuniform motion to the deferent sphere that carried the planet's epicycle around. According to Ibn al-Haytham this was unacceptable because it contradicted Ptolemy's position as exhibited jointly in the *Almagest* (where the uniformity principle is repeatedly asserted) and in the *Planetary Hypotheses* (where a system of the universe is outlined in terms of nested spheres that produced the motions already described in the *Almagest*). It is to Ibn al-Haytham's credit that he had the courage to draw and boldly state what he believed to be an inevitable conclusion: that the arrangements proposed for planetary motions in the *Almagest* were "*false*" (his own word) and that the true arrangements were yet to be discovered.[4] It is now known that Ibn al-Haytham's criticisms played an important part in stimulating the research at thirteenth-century Marāgha that led to the construction of "non-Ptolemaic" models the purpose of which was to preserve the uniformity principle *without* sacrificing the mathematical effect of the equant.[5] This research may be characterized as an attempt to save the principles of Ptolemaic astronomy (uniformity and circularity of motion, eccentrics, and epicycles) as well as the phenomena the models of the *Almagest* had been designed to account for. In other words, the Marāgha astronomers were not aiming to overthrow Ptolemaic astronomy but only to reform it. And I would therefore venture to say that the results reached at Marāgha (and later at Damascus), insofar as they were successful, would have been perfectly acceptable to Ptolemy himself (and to Ibn al-Haytham).

II

It is important to distinguish clearly between this line of research and what took place in Andalusia in the century separating Ibn al-Haytham and the Marāgha astronomers. The final outcome of the Andalusian endeavor was a book on *The Principles of Astronomy* in which its author – al Biṭrūjī (or Alpetragius) – gave sample illustrations of how the apparent motions of the planets could in his opinion be produced by means of concentric spheres, without the use of eccenters or epicycles.[6] Thus whereas the Marāgha astronomers of a later century would aim to straighten out Ptolemaic astronomy by bringing it into line with its own principles, the goal of al-Biṭrūjī was to get rid of two of Ptolomy's basic principles altogether. Al-Biṭrūjī wrote his book toward the end of the twelfth century (probably around 1200 A.D.), but he tells us that he received his inspiration from the philosopher and court physician Ibn Ṭufayl (d. 1185), author of the well-known philosophical narrative of *Ḥayy ibn Yaqẓān,* the only written work of his that has survived. According to al-Biṭrūjī, Ibn Ṭufayl claimed that he had "come upon" (or found) an arrangement that brings about the motions of the planets without assuming eccenters or epicycles and that he had promised to write on this subject.[7] Apparently that promise was never made good, but it was these remarks that set al-Biṭrūjī on the long path that finally led him to write his own book. As we shall see, the writings of Averroes (who died in 1198) and Maimonides (who died in 1204) exhibit a similar, negative attitude toward Ptolemaic astronomy. Since Averroes was very close to Ibn Ṭufayl (who had introduced him as a young man to the Almohad ruler, probably in 1168 or 1169), we may assume that he, too, was acquainted with Ibn Ṭufayl's views.

There is nothing in the available sources to indicate a direct connection between Averroes and al-Biṭrūjī, but at least their common philosophical parentage in the person of Ibn Ṭufayl is beyond doubt. Intellectual relations between Averroes and Maimonides are more difficult to ascertain. Maimonides was about thirty years old when he left the Maghrib (northwest Africa) for Egypt where, in his mature age, he wrote the *Guide for the Perplexed,* the book in which he deprecated the Ptolemaic system.[8] It is known that he received copies of Averroes's commentaries on the Aristotelian corpus before he completed the *Guide.*[9] But whether or not he was directly influenced by Averroes's arguments against Ptolemy, it is at least clear that the two thinkers shared a common culture and a common philosophical background. Despite his long sojourn in Egypt Maimonides was in matters of philosophy much closer to the Spanish brand of Aristotelianism than to the Eastern version that had been forged by Avicenna. To quote the words of Shlomo Pines, "When . . . [Maimonides] wrote the *Guide,* he was, in the domain of philosophy and philosophic theology, still almost exclusively involved with the problems with which he must have been familiar in his youth in Spain and the Maghrib."[10]

From this brief account it is clear that when we consider the movement of thought that culminated in the formulation of a new astronomical theory by al-Biṭrūjī, we are dealing with a compact situation in which a small number of individuals were bound together by a distinctive intellectual milieu and, in some decisive cases, by direct personal ties. Sometimes, the ideas of these individuals, particularly their commitment to certain Aristotelian doctrines, are linked to a slightly earlier philosopher, Ibn Bājja (d. 1138), frequently called by modern scholars "the founder of Spanish Aristotelianism," and who is known, on the authority of Maimonides, to have criticized the use of epicycles in planetary theory.[11] But although it is fairly certain that such a link existed, a line must be drawn between partial criticisms such as those of Ibn Bājja (and Jābir ibn Aflaḥ) and the radical rejectionism associated with Ibn Ṭufayl, Averroes, al-Biṭrūjī, and Maimonides.[12]

According to al-Biṭrūjī the universe consists of nested spheres (or spherical shells) that have the earth as their common center. Each of these spheres transfers to the one immediately below it a portion of the motion it receives from the one immediately above, the ultimate source of motion being a *primum mobile* that lies above the sphere of the fixed stars and that rotates with a constant motion from east to west. As a result of this mode of transference of energy, velocities diminish and variations of motion increase as one proceeds from the outermost sphere toward the center.

The planets and the fixed stars, being rigidly fixed in their appropriate spheres, maintain a constant distance from the earth, which, being at the center, is itself at rest. With only constant and concentric motions available, the apparent irregularities of the sun, the moon, and the planets, and the movement of precession, were to be explained in terms of a spiral motion, a rotation of poles around poles; and by imposing polar motions upon one another al-Biṭrūjī hoped to explain planetary stations and retrogressions as well as motions in longitude, latitude, and anomaly. All this was presented as conformable to the solid principles of Aristotelian physics.

To give just one or two examples. The annual motion of the sun is governed by the motion of what is called the pole of the sun, a point intended to maintain a quadrant's distance from the sun. The pole of the sun moves on a small circle in the sphere of the sun about the pole of the celestial equator. Taking the radius of this small circle to be equal to the inclination of the ecliptic, and assuming the pole of the sun to move on this circle (from west to east) twice as fast as the mean motion of the sun, al-Biṭrūjī manages to show that the pole will be a quadrant away from the sun at the equinoctial and solsticial points. (Unfortunately, however, the sun will not move on the ecliptic between these points although not departing too far from it.)

The case of the planets is more difficult and even less successfully dealt with. Again, the motion of the planets in the vicinity of the ecliptic is governed by the rotation of poles near the poles of the equator. With a radius equal to the inclination of the ecliptic, a circle is drawn about the north pole of the equator. Call this circle the polar deferent, the name given to it by B. Goldstein, whose interpretation of the text I am following here. Then, with a radius equal to the maximum latitude of the planet, another circle (smaller than the first) is drawn about a point on the polar deferent. Call this circle "polar epicycle," a name al-Biṭrūjī would probably have objected to.[13] It represents the path of a point called the pole of the planet. The planet itself, which initially may lie on the ecliptic, is supposed to be a quadrant's distance from its pole. The idea is that the motion of the planet is governed by the rotation of its pole about the center of the "polar epicycle" as this center slides on the polar deferent. The two rotations combine to produce the motion of the planet in longitude and in latitude.

Recent analysis of al-Biṭrūjī's system has revealed many shortcomings and many unanswered questions. For example, the planetary model as it stands would require Saturn to depart sometimes from the ecliptic by more than 26°, while a divergence of only 3°3′ (the mean of the Ptolemaic extreme latitude) is intended (Kennedy).[14] The planet's pole will not always be a quadrant's distance from the planet (Kennedy and Goldstein), again contrary to what is intended. And al-Biṭrūjī makes no accommodation for Ptolemy's equant hypothesis, nor does he refer to it (Goldstein).

The solar model is equally inadequate. Even on the hypothesis of a polar motion twice that of the sun, the latter will not always be found on the ecliptic, as already pointed out; nor is it clear why the hypothesis itself should be accepted (Kennedy). Then, in the attempt to arrange for this model to exhibit an annual irregularity, al-Biṭrūjī "heaps chaos upon confusion" (in the words of Kennedy) by introducing elements required by the incompatible system of Ptolemy.[15]

Thus, though ingenious and inventive, al-Biṭrūjī's system does not stand up to close astronomical examination, and its appeal to some medieval minds contrasts greatly with the judgment passed on it by modern scholars. Dreyer calls it "quaint,"[16] and Carmody describes it as a "delusion."[17] Goldstein concludes his analysis of the planetary model with these words: "The attempt to find satisfactory philosophical principles for the description of planetary motion was perhaps noble, but its success in this instance was quite meager."[18] Kennedy concludes a review of Goldstein's study of al-Biṭrūjī with the statement that "serious planetary theory was beyond al-Biṭrūjī."[19] And, in an earlier review of Carmody's edition of Michael Scot's translation, he wrote:

[al-Biṭrūjī's] basic device, that of rotating poles on concentric spheres, was highly ingenious and he exploited it to the full. But the resulting system represented an improvement only to minds more firmly attached to the concept of concentric spheres than they were influenced by observable facts. As an interpretation of the real universe it is vastly inferior to the Ptolemaic one. Perhaps the situation is best summed up by the statement that al-Biṭrūjī was a philosopher.[20]

There is no doubt that al-Biṭrūjī deserves to be flunked. Like Humpty Dumpty he had a great fall, and no one can put the pieces together again. We may simply bury his remains and forget about the poor fellow, or we may choose to meditate further on the nature and circumstances of his failure. In what follows I shall choose the latter course, on the advice of those historians who believe that failures as well as successes can be instructive in the study of scientific endeavor. And, first, let us look at how al-Biṭrūjī viewed his own task and achievement. This is what he wrote at the end of his discussion of planetary motions:

Neither time nor good fortune has enabled me to complete the inquiry into the details of planetary motions, or to pursue all that belongs to the planets in respect of their risings and settings and the times of their visibility, or to learn the conditions of their conjunctions and eclipses and all that is contained in the *Almagest* regarding them. These are matters that require a long time and the collaboration of those who are expert in them. Indeed the remainder of [our] life would not be enough if the ability is not there. Our aim was merely to draw attention to the way in which the ascertained [east to west] motion brings about the different and diversified (*al-mukhtalifa al-mutafannina*) motions, and to make known (reading "*wa al-taʿrīf*" in place of "*waal-taʿriḍ*") a possible arrangement for the heavens.[21]

Al-Biṭrūjī, it is clear, believed in the truth of a program he knew he had not accomplished. He did not intend the models presented in his book as constituting final solutions to the problems involved but rather as illustrations of the *kind* of solution that must be sought. He doubted his own ability to fulfill the task in hand, but his successes, limited though they were, would have enhanced his confidence in and hope for the general program. Such an attitude is not unheard of in the history of science. Al-Biṭrūjī calls to mind another philosopher, Descartes, who offered a plan for the whole of physics with nothing to support its validity but a number of a priori arguments and a few problematic applications. It will be remembered that the Cartesian models for gravitational, optical, and other phenomena were not only inadequate, but sometimes contradictory. They, too, were meant as illustrations of a type of explanation, not as final results. The comparison with Descartes may, however, be objected to on the grounds that Descartes lived before Newton whereas al-Biṭrūjī ventured on his project a thousand years after Ptolemy, whose powerful system of the universe would have been well known to him. So, probably, it would be more to the point to compare al-Biṭrūjī with the Bernoullis, Johan and Daniel, who, in the eighteenth century, were diligently trying to reconcile Newton's law of gravitation with the Cartesian theory of vortices.[22] The fact that al-Biṭrūjī was not mathematically strong is, I think, irrelevant to this comparison. He and his eighteenth-century counterparts were motivated by the same desire, which was to harmonize some empirically confirmed results with a rationally satisfying idea, and both his program and theirs were doomed to failure.

It may be noted that both al-Biṭrūjī and the Marāgha astronomers were driven by the same sort of theoretical concerns. But again it must be emphasized that their respective commitments and aims were not identical. The Marāgha astronomers were committed to Ptolemaic astronomy, and theirs was the limited aim of reconciling certain features of the Ptolemaic system with Ptolemaic principles. Al-Biṭrūjī, on the other hand, had inherited a stronger and much more rigid program that, as it turned out, was impossible to execute. It is this rigid character of his program that I should now like to consider.

III

It is often asserted that al-Biṭrūjī belonged to an Aristotelian school of philosophy that had been initiated in Spain by Ibn Bājja and of which Averroes was the strongest representative. The assertion is justified inasmuch as the authority and views of Aristotle are expressly invoked and defended in the writing of so-called members of this school, including al-Biṭrūjī. What we have to do with here, however, is a certain attitude to Aristotle and a certain interpretation of his doctrines that I find quite puzzling. Aristotle had always been an authority, indeed the foremost authority, for Islamic philosophers. But it was only Averroes in twelfth-century Spain who raised Aristotle to the status of a perfect human being, an individual in whom the intellectual faculty had reached its highest possible degree of human perfection.[23] Again, it would not be difficult to point to examples of an earlier critical attitude on the part of Islamic scientists toward certain aspects of Ptolemaic astronomy, but it was only in twelfth-century Spain that Islamic thinkers in the Greek philosophical tradition went as far as to reject Ptolemaic astronomy in toto. Al-Biṭrūjī's position may have been simply a consequence of such a belief in Aristotle's infallibility, but this belief itself needs some explanation.

In order to persuade the reader that the attitude of Averroes to Ptolemaic astronomy is something of a puzzle, I shall quote a few passages from two of his works that were concerned with Aristotle's *Metaphysics*. The first of these works is a paraphrase (*talkhīṣ*) of the *Metaphysics* that Averroes is supposed to have written in 1174, when he was forty-eight years old.[24] The second and better known work, composed at a later date (after 1186), is the large commentary (*tafsīr*) on the *Metaphysics*.

The following is what Averroes wrote in the *Talkhīṣ* regarding the question of number of celestial motions:

As for the number of these motions, it must be received from the mathematical science of astronomy. Let us lay down what is most widely held regarding them here [in al-Andalus] and in our own time, and what has not been subject to dispute among practitioners of this science [*ṣināʿa*] from Ptolemy to the present time; and let us leave out disputable matters to the experts in that science. Moreover, it is not possible to determine many of these motions without using the generally accepted premises [*muqaddamāt mashhūra*], given that a period of many life-times would be required to determine many of these motions; and generally accepted premises in a given science are those which practitioners of that science do not dispute. For this reason, then, we have here adopted such premises; and we therefore say: that the motions agreed upon for the heavenly bodies are thirty-eight [*sic*]: five for each of the three superior planets – Saturn, Jupiter and Mars, five for the moon, eight for Mercury, seven for Venus, one for the sun (provided that it is imagined to move in an eccentric sphere only and not in an epicyclic sphere), and one for the all-embracing sphere, i.e., the sphere of the fixed stars.[25]

Averroes departs here from the Aristotelian theory, derived from Eudoxus and Callippus, which had led Aristotle to posit fifty-five or forty-seven movements.[26] It is to be noted that he justifies this departure by appealing to Aristotle's view that the metaphysician must accept the number of such motions from the astronomer who may introduce modifications in the light of new observations.[27] It is as if Averroes were saying that just as Aristotle had been obliged to adopt the number of motions determined by the astronomy of his day, our duty requires us to adopt the motions established by later astronomers, including those of our own time. In other words, Ptolemy and his successors were to be preferred to Eudoxus, Callippus, and, consequently, Aristotle.[28]

After briefly discussing and rejecting the assumption of a ninth sphere as a debatable and doubtful doctrine, Averroes goes on to argue that the number of movers must then be the same as those accepted movements, provided it is assumed that a single mover is responsible for the daily motion of all the planets – an opinion Averroes accepts, following (as he implies) Alexander. The idea of regarding the daily motion of the planets as essentially one and therefore proceeding from one mover, leads Averroes to adopt a view Ptolemy had argued for in the *Planetary Hypotheses*. This is the view in which the whole celestial sphere is regarded as a unique celestial living being. These are Averroes's words:

If all that is as we described (or assumed), then a motion that is essentially one must be attached to a single thing in motion. But a single thing in motion must be moved by a single mover. We therefore ought to think of the whole [celestial] sphere as a single spherical animal whose convex [surface] is that of the sphere of the fixed stars and whose concave [surface] touches the sphere of fire; and that its motion is one and universal; and that the motions that exist in it for each one of the planets are particular motions; and that the great motion resembles the locomotion of an animal while the particular motions resemble those of the members of an animal.[29]

Averroes goes on:

And it is for this reason that these motions do not require centers about which they move, like the earth in relation to the great motion, for most of these motions have been shown in [the science of] mathematics to have centers other than the center of the world, and that their distances from the earth are not the same. Accordingly, we do not need to imagine a multiplicity of spheres whose centers are the same as the center of the world and whose poles are those of the world but which are distinct from one another. Let us rather imagine that between the spheres proper to each of the planets there exist certain bodies [which behave], not separately from one another, nor as [individually] endowed with a proper motion, but which [move] by virtue of being parts of the whole, and that the planets perform their daily motion on these bodies.[30]

These words clearly show that it was not only with regard to the number of motions that Averroes was willing (at the time of writing the *Talkhīṣ*) to differ from Aristotle. As he argues for the validity of the picture that he derives from the *Planetary Hypotheses* he makes it quite clear that he adopts the Ptolemaic principle of nonconcentric spheres. For in that picture, he says, such spheres do not require the existence of a stationary body at the center of their motions. One may thus adopt the Ptolemaic assumption of eccentric and epicyclic spheres without abandoning a fundamental feature of Aristotelian cosmology, a unique earth at rest representing the absolute center of the world. As we shall immediately see, Averroes's position here is in contradiction with the view he later took in the *Tafsīr.* But before we turn to that view let us insist again that, for the Averroes of 1174, to be an Aristotelian did not mean that one was shut up in an absolutely closed system, at least as far as the science of astronomy was concerned.

IV

In the large commentary on Aristotle's *Metaphysics* (the *Tafsīr*), Averroes comments on the same passage with which the words quoted from the *Talkhīṣ* were concerned. Here he compares the doctrines derived by Aristotle from the mathematicians of his own age with those developed by later mathematicians, especially Ptolemy. He repeats and endorses the Aristotelian view that the metaphysician must accept the number of such motions from the astronomer, whose job it is to add (if necessary) new motions in the light of new observations. As an example he mentions two new motions ascribed by Ptolemy to the moon.[31] He draws the lesson that in such matters, which by their difficulty preclude the possibility of indubitable premises, one is obliged to rely on generally accepted propositions (*al-mashhūrāt*) as long as they are not disputed (?by astronomers). He then reports the opinion of Eudoxus, who attributed three spheres each to the sun and moon: one for the daily east-to-west motion, a second for the motion in longitude, and a third for the motion in latitude. "As for the mathematicians of our own time," Averroes adds, "they have assumed only one motion" (in place of the last two), "namely that of the planet in its inclined sphere, thus giving rise to a motion in longitude and another in latitude with respect to the ecliptic."[32] Thus, he concludes, "the sun will have only two motions, *unless it is necessary to introduce a third on account of its observed acceleration.*"[33]

It is clear from these last words that Averroes was not thinking of a second *eccentric* motion (or sphere) for the sun. He explains his reason in the following passage, which I shall translate in full because of its historical importance.

[a] For to assert the existence of an eccentric sphere or an epicyclic sphere is contrary to nature. As for the epicyclic sphere, this is not at all possible; for a body that moves in a circle must move about the center of the universe (*al-kull*), not aside from it; for it is the revolving thing itself (*al-mutaḥarriku dawran*) that produces (*yafʿal*) the center. Thus if a revolution about a center other than this center were to take place, then a center would exist other than this center, and there would exist an earth other than this earth. But all this has been shown to be impossible in natural science.

[b] It is similarly the case with the eccentric sphere proposed (*yaḍaʿuhu*) by Ptolemy. For if many centers existed, we should have a multitude of heavy bodies outside the place of the earth, and the center would cease to be unique, and it would be extended and divisible. But all this is not possible.

[c] Moreover, if eccentric spheres existed, then certain parts of the heavenly bodies would be redundant, their use being restricted to filling in an empty space, as is thought to be so in the case of the bodies of animals. But nothing of what appears of the motions of these planets makes it necessary to assume the existence of an epicyclic or an eccentric sphere.

[d] It may be possible to replace these two things by the spiral motions (*al-ḥarakāt al-lawlabiyya*) assumed by Aristotle in this astronomy (*hādhihi l-hayʾa*) in imitation of (*ḥikāyatan ʿan*) those who came before him. For it appears that astronomers before Hipparchus and Ptolemy assumed no epicyclic or eccentric spheres. Ptolemy stated this in his book on *Planetary Hypotheses,* and he claimed that Aristotle and his predecessors had assumed instead spiral motions, thereby increasing, as he claimed, the number of motions. Those who came after them, however, found a simpler way – that is, they were able to account for the phenomena (*amkanahum an yaḍaʿū mā yaẓharu*) by reference to fewer bodies, by which he meant the epicyclic and the eccentric sphere. He also claimed that this way was better inasmuch as it was accepted that nature does nothing redundant and therefore would not employ more means than it needs to bring about the motion of something.

[e] Ptolemy was not, however, aware of what had obliged the ancients to appeal to

spiral motions, namely the impossibility of epicyclic and eccentric spheres. But when people came to believe that this [new] astronomy was simpler and easier for [explaining] the revolutions (*ʿawd al-ḥarakāt*) now recorded in Ptolemy's book, they neglected the ancient astronomy until it became so obsolete that people are not now able to understand what Aristotle says in this place [in the *Metaphysics*] about those [ancient] people. This has been admitted by Alexander and Themistius but without their being aware of the reason we have mentioned.

[f] We should therefore embark on a new search for this ancient astronomy, for it is the true astronomy that is possible from the standpoint of physical principles. It is in my view based on the motion of one and the same sphere about one center and different poles, which may be two or more in accordance with the phenomena. For such motions can give rise to the acceleration, retardation, accession, and recession [*iqbāl wa idbār*] of a planet, and other motions for which Ptolemy failed to produce an arrangement [*hayʾa*]. Such motions would also be the approaching and receding of a planet, as in the case of the moon. In my youth [*fī shabābī*] I had hoped to accomplish this investigation, but now in my old age [*fī shaykhūkhatī*] I have despaired of that, having been impeded by obstacles. But let this discourse spur someone else to inquire into these matters [further]. For nothing of the [true] science of astronomy exists in our time, the astronomy of our time being only in agreement with calculations [*al-ḥusbān*] and not with what exists.[34]

This long and rich passage raises more questions than can be examined here. What is interesting from the point of view of our present problem is that in it Averroes reveals an attitude that is entirely different from that which he had expressed at the time of writing the *Talkhīṣ*. Averroes still maintains that in matters relating to celestial motions the philosopher or metaphysician must rely on the results of mathematical astronomy. He clearly recognizes the explanatory power of the Ptolemaic system by his unequivocal admission that the astronomy of his own time "agrees with calculations." And yet he finds himself obliged to reject the whole basis on which this system stands. In doing so he completely ignores the arguments he had himself produced in the *Talkhīṣ* and simply puts in their place an a priori argument that eliminates eccenters and epicycles as a matter of principle. Can we convincingly explain Averroes's new attitude by merely describing it as Aristotelian? Did he not write the *Talkhīṣ* also as an Aristotelian? Did he not once think that he was following Aristotelian methodology when he preferred the Ptolemaic system to the earlier one of Eudoxus? But now, instead of characterizing the later theory as more informed, he postulates a perfect astronomy that antedated Aristotle, that Aristotle only hinted at in his writings, and that perished as a result of the triumph of Ptolemaic astronomy. That triumph, Averroes explains, was due to a misguided application of the principle of economy – misguided because it ignored the demonstrable requirements of natural philosophy; and therefore, according to him, it is the duty of the astronomer who is interested in truth as well as computation to try to discover anew that ancient and forgotten astronomy. That is the task al-Biṭrūjī set for himself a little later, and we may accordingly look at his and Averroes's deliberations as a chapter in the history of the myth of ancient wisdom that is known to have continued to play a part in the later development of Western thought.[35]

V

The "Aristotelianism" revealed in the passage from the *Tafsīr* requires explanation because of the unexpectedly extreme position it exhibits. An explanation must be sought outside of the text itself and, as it turns out, in cultural and, perhaps, even psychological rather than strictly cognitive and "rational" terms. Here I can only outline such an explanation briefly and somewhat dogmatically. It consists in looking at Averroes's (and al-Biṭrūjī's) position as part of a more general phenomenon, an intellectual trend that prevailed in Andalusia under the Almohads among scholars working in such diverse fields as law, grammar, medicine, and philosophy. And this trend may itself be related to a noticeable and often expressed Andalusian self-assertiveness vis-à-vis the rest of the Islamic world. On one level this attitude shows itself in the many essays composed by Andalusian scholars at different times on the distinctive and superior virtues of their land (*faḍā'il al-Andalus*). But the Andalusian sense of identity went further than self-praise and actually expressed itself in the creation of systems of ideas that were distinctly Andalusian and consciously directed against intellectual authorities in the Eastern part of Islam. Already in the eleventh century one of the most original thinkers of Muslim Spain, Ibn Ḥazm of Cordova (d. 1064), developed a literalist doctrine of law that he set against all other recognized *sunnī* doctrines. By equating religion (*dīn*) exclusively with what can be found explicitly stated in the Qur'ān and in the *Sunna* (Traditions of the Prophet), and by denying religious merit to all efforts on the part of legists to form inferences or opinions or preferences of any kind he was undermining the authority of the legal schools and their followers. It is to be remembered that the Almohad rule in North Africa and Spain came into being as a result of the forceful implementation of an articulated ideology. The founder of the Almohad movement, Ibn Tūmart (d. 1130), may not have been a thoroughgoing literalist, but his writings seem to exhibit literalist tendencies.[36] In any case such tendencies have been attributed to the second Almohad ruler Abū Ya'qūb Yūsuf (r. 1163–84) by Spanish historians of the period, and it is known that his son Ya'qūb al-Manṣūr (r. 1184–99) openly called for a literalist approach in the practice of law. He went as far as to order the burning of Mālikite books on *furū'*, or recipes for detailed applications of religious law as developed by the followers of Mālik who dominated the theological field in Andalusia and the Maghrib. Averroes, who served both Yūsuf and Ya'qūb as court physician and as a judge, appears to have had some such approach to law and theology in mind when, at the end of his famous treatise on the harmony of religion and philosophy, he wrote that through the "triumphant rule" of the Almohads the masses had been summoned to a "middle way . . . which is raised above the low level of the followers of authority but is below the turbulence of the theologians."[37] It was also during the reign of Ya'qūb al-Manṣūr that Ibn Maḍā' of Cordova, another protégé of the Almohad court who had been appointed chief judge by al-Manṣūr's father, wrote his attack on the theory of the agent, until then the almost universally accepted basis of Arabic grammar. This attack shared certain features with Ibn

Ḥazm's literalism: It rejected the grammatical equivalent of legal reasons (*ʿilal*) and it was aimed at freeing the study of language from the clutches of the established authorities of Arabic grammar.[38]

As for Averroes, his negative attitude toward the Muslim philosophers of the East and the pride he took in his own country and culture should be well known and can be easily documented. It was he who wrote the most vigorous and most detailed reply to al-Ghazālī's resounding attack on philosophy, some seventy years after al-Ghazālī's death.[39] Anyone who looks at this reply will see that it is a criticism not only of al-Ghazālī, but also of Avicenna, the leading philosophical authority in Islam up to the time of Averroes. One of al-Ghazālī's mistakes, Averroes argued, was to attribute to the ancient philosophers doctrines that Avicenna had invented or borrowed from the discredited Mutakallimūn (dialectical theologians).[40] Time and again Averroes castigates Avicenna, and also al-Fārābī,[41] for having corrupted or at least departed from the true teachings of the first, and indeed perfect, philosopher. The reply to al-Ghazālī is only one example. When Averroes finds occasion in his other writings to compare the views of Aristotle with opposing views of later Islamic thinkers, such as al-Kindī[42] and Ibn al-Haytham,[43] it is Aristotle that he judges to be right and the others wrong. It is difficult not to regard this attitude of Averroes's and the commentatorial style he adopted in most of his philosophical writings as a literalism in philosophy that paralleled the theological literalism of Ibn Ḥazm. Thus Averroes's glorification of Aristotle and his rejection of the authority of Muslim peripatetics in the East can be seen as two aspects of the same attitude.

If Averroes considers himself closer to Aristotle than any of his Islamic predecessors, then it should not come as a surprise that he should be concerned to find an explanation of that privileged position. The explanation he develops in several of his writings is interesting as it relates to that Andalusian self-consciousness to which I referred. In the *Kulliyyāt,* Averroes has a few words about climatology, a theory that accounted not only for the physical properties of the various regions of the earth but also for the intellectual as well as physical features of their inhabitants. Averroes reports Galen's view that the most temperate climate is the fifth, where Greece, the country of Hippocrates, is located, and pointedly adds that "this land of ours, namely al-Andalus," lies at the beginning of the fifth climate.[44] But whereas in his commentary on the *Republic,* Averroes is content to remark, against Plato, that "individuals" may excel in the sciences outside of Greece, as is found, for example, in Iraq, Egypt, and al-Andalus,[45] he takes an exclusive position in the middle commentary on Aristotle's *Meteorology.*[46] Here he repeats Galen's opinion that it is the fifth climate, not the fourth (as others think), that is the most temperate, and, having decided that geographical latitude (and proximity to the sea) is the crucial factor, he groups Andalusia together with Greece and ignores Iraq. Of course, the Arabs and the Berbers came to Andalusia from elsewhere, but they had been acclimatized over the centuries, a process that, we are led to assume, had reached maturity in the time of Averroes. That, Averroes tells us, is the reason why the sciences have flourished among the Andalusians.

NOTES

1 The idea of levels or orders of achievement in writing or composition (*marātib al-sharaf fī al-tawālīef*) is mentioned, for example, by Ibn Ḥazm of Cordova (eleventh century A.D.), who lists seven such orders headed by that in which an author puts forward something that has not been previously discovered (*shay'lam nusbaq ilā istikhrājih fa-nastakhrijuh*); see his *al-Taqrīb li-ḥadd al-manṭiq . . .*, ed. Iḥsān ʿAbbās, Beirut: Dār al-Ḥayāh, ?1959. The same idea already occurs in the opening sentences of al-Khwārizmī's *Kitāb al-Jabr wa al-muqābala,* also using the term *istikhrāj* for unprecedented discovery. A marginal note in the Bodleian MS of this work gives some examples of authors in this highest category: Archimedes, Abū Ḥanīfa (founder of a Muslim legal school), and al-Khwārizmī himself, see the edition by ʿA. M. Musharrafa and M. M. Aḥmad, *Kitāb al-J. wa al-m.,* Cairo: The Egyptian University, 1939, p. 15 and the facing facsimile.

2 An edition of Ibn al-Haytham's *al-Shukūk ʿalā Baṭlamyūs* (*Dubitationes in Ptolemaeum*) has been published by A. I. Sabra and N. Shehaby, Cairo: Dār al-Kutub, 1971. An English translation of Ibn al-Haytham's criticisms of Ptolemy's equant hypothesis is included in A. I. Sabra, "An eleventh-century refutation of Ptolemy's planetary theory," in *Science and History: Studies in Honor of Edward Rosen* (Studia Copernicana XVI), Wrocław etc.: Ossolineum (The Polish Academy of Sciences Press), 1978, pp. 117–31.

3 It is sometimes forgotten or ignored that, in view of the cosmology that prevailed in ancient and medieval times, such a conviction was in fact inevitable. The stars are physical bodies that can be seen to move across the sky. They did not of course make their journey in empty space (for the void did not exist), and of course they did not swim like fish in some lowly matter that opened up before them and closed in behind them (for the substance of the heavens was impassable). They must therefore be embedded in immense, solid, and transparent spheres or spherical shells that rotate inside one another about their own centers, each unimpeded by its neighbor. That was the conclusion reached in classical antiquity and universally accepted by medieval astronomers. The history of theoretical astronomy from Ptolemy to Copernicus cannot be understood without realizing that astronomers during that period were not merely interested in saving the appearances by means of purely mathematical devices. There was no such thing as a mathematical astronomy consciously conceived as a discipline entirely independent of physics or natural philosophy. It is true that Ptolemy's *Almagest,* like Newton's *Principia,* was not itself a work on natural philosophy, but rather laid down certain mathematical principles that govern celestial phenomena. But Ptolemy himself had been no more positivistic in his outlook than was Newton. Like Newton he was interested in questions of nature and causes, and his *Planetary Hypotheses* may in some respects be seen as having somewhat the same relationship to the earlier *Almagest* as Newton's *Scholia* and *Quaeries* had to the *Principia.* The two works, *Almagest* and *Planetary Hypotheses,* differed greatly in character but they were parts of one and the same enterprise, which was to save the principles as well as the phenomena of the heavens. Some of these principles were mathematical, but others were physical and had to do with the nature of celestial bodies. In the thirteenth century, Naṣīr al-Dīn al-Ṭūsī expressed the general understanding of the astronomers of his time when he wrote the following in his famous and influential *Tadhkira:* "Every science must possess the following: [a] an object which this science investigates, [b] principles which are either self-evident or need to be proved in another science but are taken for granted in this science, and [c] problems (*masāʾil*) which are proved in this science. Now the objects of astronomy (*hayʾa*) are the simple bodies, both

superior and inferior, in respect of their quantities, positions and inseparable motions. The principles of astronomy that need proof are demonstrated in three sciences: metaphysics, geometry, and physics. The problems of astronomy are aimed at gaining knowledge of these bodies themselves, and their quantities and distances and the reasons (*ʿilal*) for their varying positions" (*Tadhkira,* MS. Leiden Or. 905, fols. 1b–2a). The physical principles of astronomy, as al-Ṭūsī also explained, included statements that asserted the spherical shape of the universe and of all bodies in the celestial part of it, the circular and uniform motion of all such bodies, the impossibility of the void and incorruptibility of the heavens, the impossibility for any celestial body of combining contrary motions or tendencies toward such motions (Ibid., fols. 4b–5a). These, and other principles, constituted boundary conditions that a wholly successful solution of a theoretical problem in astronomy was desired to satisfy. Sometimes we find them dogmatically repeated in philosophical writings whose authors were not active or successful in astronomical research. In the case of al-Ṭūsī, however, they formed the basis of inquiries that went beyond paying lip service to a dominant natural philosophy. And similar statements to those of al-Ṭūsī are also found in the *Nihāyat al-sūl* of Ibn al-Shāṭir in the fourteenth century.

4 See the article by the author referred to in n. 2, pp. 121–2. Also idem, "Ibn al-Haytham's treatise: Solution of difficulties concerning the movement of *iltifāf,*" *Journal for the History of Arabic Science,* 3 (1979), Arabic text, pp. 183–210, esp. pp. 206–7; English summary, pp. 388–92.

5 On the Marāgha astronomers and Ibn al-Shāṭir see E. S. Kennedy, "Late medieval planetary theory," *Isis,* 57 (1966), pp. 365–75. In addition to the articles listed by Kennedy in this article (p. 365, n. 1 and p. 369, n. 9), see Willy Hartner, "Naṣīr al-Dīn al-Ṭūsī's lunar theory," *Physis,* 11 (1969), pp. 287–304; idem, "Copernicus, the man, the work, and its history," *Proceedings of the American Philosophical Society,* 117 (1973), pp. 413–22; idem, "Ptolemy, Azarquiel, Ibn al-Shāṭir, and Copernicus on Mercury: A study of parameters," *Archives Internationales d'Histoire des Sciences,* 24 (1974), pp. 5–25; Bernard R. Goldstein, "Remarks on Ptolemy's equant model in Islamic astronomy," in *Prismata* (Festschrift für Willy Hartner), ed. Y. Maeyama and W. G. Saltzer, Wiesbaden: Franz Steiner Verlag, 1977, pp. 165–8; George Saliba, "The original source of Quṭb al-Dīn al-Shīrāzī's planetary model," *Journal for the History of Arabic Science,* 3 (1979), pp. 3–18 (includes an extract from the Arabic text of a work by al-ʿUrḍī); idem, "The first non-Ptolemaic astronomy at the Marāghah School," *Isis,* 70 (1979), pp. 571–6; idem, "A Damascene astronomer proposes a non-Ptolemaic astronomy," *Journal for the History of Arabic Science,* 4 (1980), Arabic text pp. 3–17; English summary, pp. 97–8. A volume edited by E. S. Kennedy and ʿImād Ghanem, *The Life and Works of Ibn al-Shāṭir,* Aleppo: Institute for the History of Arabic Science, 1976, includes many of these articles. To this must be added "Les sphères célestes selon Nasîr-Eddîn Attûsî" [which includes a French translation of chap. 11 of al-Ṭūsī's *Tadhkira,* made from the Paris MS Bibl. Nat. ar. 2509], published as Appendix VI in P. Tannery, *Recherches sur l'histoire de l'astronomie ancienne,* Paris: Gautier-Villars & Fils, 1893, pp. 337–61.

6 See Bernard R. Goldstein, *Al-Biṭrūjī: On the Principles of Astronomy,* an edition of the Arabic and Hebrew versions with translation, analysis, and an Arabic-Hebrew-English glossary, vol. I (analysis and translation), vol. II (the Arabic and Hebrew versions), New Haven, Conn.: Yale University Press, 1971. Also Francis J. Carmody, Al-Biṭrûjî: *De motibus celorum,* Critical edition of the Latin translation of Michael Scot, Berkeley: University of California Press, 1952.

7 These are al-Biṭrūjī's words: "As you know, brother, Abū Bakr ibn al-Ṭufayl, may

the exalted God have mercy on him, used to tell us that he had come upon (*'athara 'alā*) an arrangement (*hay'a*) and principles (*uṣūl*) for these [planetary] motions other than those two principles which Ptolemy had laid down, that is without in any way assuming an eccentric or an epicyclic sphere, while maintaining at the same time all these motions without entailing anything impossible. He promised to write on this matter, and his rank in learning is such that ignorance cannot be attributed to him [reading: *lā yujahhal*]. I have continued to think about this from the time I heard it from him, searching the statements of those who came before us, but without finding anything that relates to it other than a few hints, such as the statement of the Philosopher in the Second Treatise of the book *On the Heavens* – [namely:] We say further that two motions [properly] belong to the spherical body (*al-jism al-mustadīr*), one circular [?or rotational] [*mustadīra*) and the other a turning round in the manner of a spiral (*idāra lawlabiyya*) [*kulisis kai dinēsis*] (my translation: see Goldstein's edition referred to in n. 6, vol. II, p. 49, and Aristotle's *De caelo*, II, 8, 290^{a}10). The Arabic translation of *De caelo* (?by Yūḥannā al-Biṭrīq) published by 'A. Badawī has *mutaqalliba* (rolling) for *mustadīra* in al-Biṭrūjī's quotation (Aristotelis *De Caelo et Meteorologica*, Cairo: Maktabat al-Nahḍa, 1961, pp. 259–60). In *De caelo*, II, 8, Aristotle is concerned to argue that the stars and the planets have neither a rolling (*kulisis*). nor a rotational motion (*dinēsis*). Al-Biṭrīq's translation refers to the second of these motions by a number of confusing expressions: *idāra lawlabiyya* (turning in the manner of a spiral, or whirling), *dawriyya* (circular or rotational), *multawiya mudawwara* (twisting and rotating), *multawiya mutalaffifa* (twisting and winding). Rolling motion (*kulisis*) is, however, happily rendered by *mutaqalliba* and *mudaḥraja.*

Al-Biṭrūjī quotes only the opening lines in Aristotle's passage, but his text calls one of the two proper motions for a sphere *mustadīra* (circular) and the other *idāra lawlabiyya* (turning in the manner of a spiral). *If* we assume that the two motions are referred to in the same order in all three texts (i.e., Aristotle's Greek text, al-Biṭrjīq's Arabic translation, and the translation as quoted by al-Biṭrūjī), then the same motion would be called rolling (*mutaqalliba*) in al-Biṭrīq's text and circular or rotational (*mustadīra*) in al-Biṭrūjī's quotation. Both texts, however, would call the second motion (*dinēsis*) a "turning in the manner of a spiral." In spite of what has been written on this subject (I have in mind particularly the article by F. J. Carmody, "The planetary theory of Ibn Rushd," *Osiris*, 10 [1952], pp. 556–86), the confusion surrounding the transmission of Aristotle's passage to al-Biṭrūjī and Averroes has yet to be sorted out. The confusion appears to have set in already at the first stages of the transmission, in the process of translating the passage into Arabic, and it is there that further research must begin.

8 Maimonides' critique and ultimate rejection of Ptolemaic astronomy *as a representation of the true arrangement of the heavenly spheres and of their motions* are set forth in chap. 24 of book II of his *Guide for the Perplexed* (see translation by S. Pines, University of Chicago Press, 1963, pp. 322–7). The opening two paragraphs, puzzling though they are at some points, outline some of the main reasons for this rejection: "You know of astronomical matters what you have read under my guidance and understood from the contents of the 'Almagest.' But there was not enough time to begin another speculative study with you. What you know already is that as far as the action of ordering the motions and making the course of the stars conform to what is seen is concerned, everything depends on two principles: either that of the epicycles or that of the eccentric spheres or on both of them. Now I shall draw your attention to the fact that both those principles are entirely outside the bounds of reasoning and opposed to all that has been made clear in natural science. In the first place, if one affirms as true the existence of an epicycle revolving round a certain sphere, positing at the same time that that revolution is not around the center of

the sphere carrying the epicycles – and this has been supposed with regard to the moon and to the five planets – it follows necessarily that there is rolling, that is, that the epicycle rolls and changes its place completely. Now this is the impossibility that was to be avoided, namely, the assumption that there should be something in the heavens that changes its place. For this reason Abū Bakr Ibn al-Ṣā'igh [Ibn Bājja] states in his extant discourse on astronomy that the existence of epicycles is impossible. He points out the necessary inference already mentioned. In addition to this impossibility necessarily following from the assumption of the existence of epicycles, he sets forth there other impossibilities that also follow from that assumption. I shall explain them to you now.

The revolution of the epicycles is not around the center of the world. Now it is a fundamental principle of this world that there are three motions: a motion from the midmost point of the world, a motion toward that point, and a motion around that point. But if an epicycle existed, its motion would be neither from that point nor toward it nor around it" (ibid., pp. 322–3).

Maimonides (as the first lines of this quotation show) was not unaware of the predictive power of Ptolemaic astronomy, but in his view the Ptolemaic system did not conform with the accepted principles of natural philosophy. But, unlike Ibn al-Haytham, Averroes, the Marāgha astronomers, and Ibn al-Shāṭir, he did not think it incumbent upon (or even possible for) the astronomer to discover *the true* arrangement: "However, I have already explained to you by word of mouth that all this does not affect the astronomer. For his purpose is not to tell us in which way the spheres truly are, but to posit an astronomical system in which it would be possible for the motions to be circular and uniform and to correspond to what is apprehended through sight, regardless of whether or not things are thus in fact . . . However, regarding all that is in the heavens, man grasps nothing but a small measure of what is mathematical; and you know what is in it. I shall accordingly say in the manner of poetical preciousness: *The heavens are the heavens of the Lord, but the earth hath He given to the sons of man.* I mean thereby that the deity alone fully knows the true reality, the nature, the substance, the form, the motions, and the causes of the heavens. But He has enabled man to have knowledge of what is beneath the heavens, for that is his world and his dwelling-place in which he has been placed and of which he himself is a part. This is the truth. For it is impossible for us to accede to the points starting from which conclusions may be drawn about the heavens; for the latter are too far away from us and too high in place and in rank" (ibid., pp. 326–7). It is seen that Maimonides' position here has much in common with that of Osiander, who, in a well-known preface to Copernicus's *De revolutionibus,* maintained that truth was beyond the reach of the astronomer and should be of no concern to him, the only valid aim of the science of astronomy being to provide tools for accurate predictions.

9 See Pines's introduction to his translation of the *Guide,* (n. 8), p. cviii.

10 Ibid., p. cix.

11 Maimonides' words: "I have heard that Abū Bakr [Ibn Bājja] has stated that he had invented [*awjada*] an astronomical system in which no epicycles figured, but only eccentric circles. However, I have not heard this from his pupils. And even if this were truly accomplished by him, he would not gain much thereby. For eccentricity also necessitates going outside the limits posed by the principles established by Aristotle, those principles to which nothing can be added" (ibid., p. 323).

12 Jābir ibn Aflaḥ (who flourished at Seville in the first half of the twelfth century) wrote an *Iṣlāḥ* or emendation of the *Almagest* that was known to Maimonides and to al-Biṭrūjī and was epitomized by Quṭb al-Dīn al-Shīrāzī in the thirteenth century (see the article on Jābir by R. P. Lorch in *Dictionary of Scientific Biography,* VII [1973], pp. 37–

9). Ibn al-Shāṭir in the fourteenth century also knew some of Jābir's views (such as placing Venus above the sun – contrary to Ptolemy) either directly through being acquainted with the *Iṣlāḥ* or indirectly through the writings of Mu'ayyad al-Dīn al-ʿUrḍī and al-Shīrāzī (see Ibn al-Shāṭir's *Nihāyat al-sūl,* Bodleian MS Marsh 139 [dated A.H. 768], fol. 4^b). Jābir's criticisms of Ptolemy did not, however, extend to the principles of eccenters and epicycles as such.

13 See E. S. Kennedy's review of Goldstein's edition and analysis of al-Biṭrūjī's *Principles;* in *Journal for the History of Astronomy,* 4 (1973), pp. 134–6, esp. p. 134: "The reviewer would occasionally have preferred a more literal terminology. For instance, the translation frequently cites the 'polar epicycle' . . . However, the text has *dāʾirat mamarr,* literally 'passage circle.' Here the translation, to some degree, begs the question of interpretation."

14 E. S. Kennedy, review of Francis J. Carmody's edition of Michael Scot's translation of al-Biṭrūjī's work; in *Speculum,* 29 (1954), pp. 246–51, esp. p. 248.

15 Review cited in n. 13, p. 136.

16 J. L. E. Dreyer, *A History of Astronomy from Thales to Kepler,* Dover, 1953, p. 267.

17 See introduction to Carmody's edition cited in n. 6, p. 11.

18 Goldstein, *al-Biṭrūjī,* I, p. 14.

19 Kennedy, *Jour. Hist. Astronomy,* 4 (1973), p. 136.

20 *Speculum,* 29 (1954), p. 251

21 Goldstein, *al-Biṭrūjī* pp. 427–9 (my translation).

22 See A. J. Aiton, *The Vortex Theory of Planetary Motions,* London: Macdonald, and New York: American Elsevier, 1972, esp. chap. 9.

23 Some of the strongest expressions of admiration for Aristotle occur in the *Middle Commentary* on the *Meteorology,* where Averroes lavishes on the Greek philosopher such descriptions as infallible, singled out by God for perfection, always correct regardless of the occasional criticism of his commentators. Aristotle is also said here to be aided by divine power and to have abrogated the philosophy of his predecessors. This attitude found good reception among Jewish philosophers, some of whom went one step further and claimed that Aristotelian science had been derived from the ancient Hebrews. See Irving Maurice Levey, "The Middle Commentary of Averroes on Aristotle's Meteorologica (Ph.D. diss. Harvard University, 1947), pp. 36ff. See later remarks on ancient wisdom.

24 Cf. S. Munk, *Mélanges de philosophie juive et arabe,* Paris: J. Vrin, 1955, p. 423. The *Middle Commentary on the Meteorology* was written after 566/1170–1 (ibid., p. 422).

25 *Talkhīṣ Mā Baʿd al-Ṭabīʿa,* ed. ʿUthmān Amīn, Cairo: Al-Bābī al-Ḥalabī, 1958, pp. 130–1. The numbers add up to 37, not 38. The larger number would be obtained if an epicyclic model for the sun were adopted. There is a German translation of the *Talkhīṣ* by Simon van den Bergh, *Die Epitome der Metaphysik des Averroes,* Leiden: Brill, 1924; and an earlier Spanish translation by Carlos Quirós Rodríguez: Averroes, *Compendio de metaphysica,* texto arabe con traducción y notas, Madrid, 1919.

26 Regarding Aristotle's calculation (or miscalculation) of the number of agent and reagent spheres, see article by G. E. L. Owen, "Aristotle: Method, Physics and Cosmology," in *Dictionary of Scientific Biography,* I (1970), esp. pp. 257–8.

27 Aristotle's text, as quoted by Averroes in his larger and later Commentary (*Tafsīr*), may be translated as follows: "As for the number of motions (*kathrat al-ḥarakāt*), we ought to infer it (*nastadillu ʿalayhā*) from what is said regarding the motions of the planets in [that part of] philosophy that is especially concerned with the mathematical sciences" (Bouyges's edition, vol. III, p. 1646). And again, "As for the number of these [motions] we shall now state [it] ourselves in accordance with what some mathematicians have said, so that it may be imagined in some way (*yutawahhamu bi-nawʿin mā*) and in order that our mind (*fikrunā*) may receive a determinate number (*kathra maḥdūda*). As for what remains of that [inquiry], we may either search for it ourselves or base our investigation on those who searched for [those motions] if something other than what has just been said becomes manifest to those who deal with these matters. We must, however, appreciate (*nuḥibb*) both parties but follow those whose investigation is more thorough" (ibid., vol. III, pp. 1657–8; compare *Metaphysics* XII, viii, 7–8, Loeb edition, vol. 2, London, 1962, pp. 155–7).

28 It may be said that a departure from Aristotle with regard to the number of celestial motions is not the same thing as a departure from Aristotelian cosmology. True. But it should be observed that in the passage quoted from the *Talkhīṣ* Averroes shows no qualms about eccenters or epicycles: The motion associated with the sun may be one (if the sun moves in an eccentric circle or sphere), or two (if the sun moves in an epicycle carried around by a deferent).

29 *Talkhīṣ* (n. 25), p. 133. The psychobiological view of the universe Averroes adopts here from Ptolemy's *Planetary Hypotheses* is so little known and so frequently overlooked that it deserves to be quoted at some length. Ptolemy wrote in the second part (lost in Greek but extant in Arabic) of that work: "Physical reasoning (*al-qiyās al-ṭabīʿī*) leads us to say that the ethereal bodies do not admit of being acted upon nor suffer change, though they vary [their positions] throughout time, inasmuch as this befits their wonderful substance and accords with the power of the stars (*al-kawākib*) which are [embedded] in them, [stars] whose light clearly goes through all these things that spread around them (*al-mabthūtha ḥawlahā*) without hindrance or affection – in the same way that [those bodies] are penetrated by what is in us that is homogeneous with them (*mimmā yujānisuhā*), such as sight and understanding (*al-fahm*). We are also led to maintain the unchangeability of ethereal bodies by our doctrine that their shapes are circular and that their actions are the actions of things whose parts are uniform. Now for each one of these movements that differ in respect of quantity or kind there exists a body that moves about poles in its proper place with a volitional movement (*ḥaraka irādiyya*) in accordance with the power of each one of the planets from which there originates the movement which proceeds from the governing faculties (*al-quwā al-raʾīsa* = *?ta hēgemonika*) that resemble the faculties residing in us, and which [movement] imparts as much motion to the bodies which are homogeneous with them [i.e., the planets] and which are like parts of the universal animal, as is proportionately suitable to each of them." (B. Goldstein, ed., *The Arabic Version of Ptolemy's Planetary Hypotheses, Transactions of the American Philosophical Society*, new ser., vol. 57, pt. 4, 1967, p. 36).

Ptolemy continues a few pages later: "Now if someone imagines that the earth is at the center and that the air and fire revolve along with the revolution of what envelops them and forces them to move; further, if he takes the observable behavior of birds as an analogy (*mithāl*) for the motion of what exists in the heavens; then such an analogy (*mā maththala* [or *muththila*] *min hālik*) would not be objectionable. For the proper motion of birds proceeds first from the psychic faculty (*al-quwwa al-nafsāniyya*) that resides in them, then the impulsion (*inbiʿāth: ? hormē*) produced by this faculty goes forth to the nerve, then – in [our] example – to the legs or hands or wings, and there the process of being passed on from one thing to another comes to a stop. . . . We should similarly imagine the situation with respect to the celestial animal (*al-ḥayawān al-falakī*). We should, that is, consider that each of the planets has the rank of a governing power (*quwwa siyāsiyya*), and that it possesses a psychic faculty and that it moves itself and naturally confers motion upon the bodies adjacent to it: first to that which lies close to it, then to the next; for example, it conveys motion first to the epicyclic sphere, then to the eccentric sphere, then to the sphere whose center is the center of the world. This motion conferred by the [planet] varies in the several places, just as in our [human] case the motion of the muscle (*al*ˣ*aḍal*) is not similar to that of the impulsion (*inbiʿāth*) nor the latter to that of the nerve nor this to that of the leg; but [all] vary somewhat as they proceed outward" (ibid., pp. 40–1).

30 *Talkhiṣ* (n. 25), pp. 133–4.

31 A reference to double elongation and prosneusis, which Averroes calls, respectively, *al-ḥaraka al-muḍāʿfa* and *ḥarakat al-muḥādhāh. Tafsīr* (n. 27), vol. III, p. 1659.

32 *Tafsīr* (n. 27), vol. III, p. 1661.

33 Ibid. Emphasis added.

34 Ibid., vol. III, pp. 1661–4. I have divided the text into paragraphs and supplied them with letters in square brackets. In connection with Averroes's reference to *Planetary Hypotheses* in [d] see my article cited in n. 4.

35 See F. A. Yates, *Giordano Bruno and the Hermetic Tradition,* London: Routledge & Kegan Paul, 1964; D. P. Walker, *The Ancient Theology,* Ithaca, N. Y.: Cornell University Press, 1972; J. E. McGuire and P. M. Rattansi, "Newton and the 'pipes of Pan,' " *Notes and Records of the Royal Society of London,* 21 (1966), pp. 108–43.

36 I. Goldziher, *Le livre de Mohammed ibn Toumart, mahdi des Almohades. Texte arabe accompagné de notices biographiques et d'une introduction.* Alger: Imprimerie Orientale Pierre Fontana, 1903. See introduction, pp. 1–101.

37 G. E. Hourani (trans.), *Averroes on the Harmony of Religion and Philosophy,* London: Luzac & Co., 1976 (first printed in 1961), pp 70–71 and n. 197, pp. 116–117. In the same passage Averroes also acknowledges his rulers' "many benefits, especially to the class of persons who have trodden the path of study and sought to know the truth" – obviously a reference to such scholars as Ibn Ṭufayl and Averroes himself. Averroes was here envisaging nothing less than an alliance between himself and the ruling authorities against the "turbulent" dialectical theologians (or practitioners of *kalām*) and the slavish followers of legal authorities. All this, of course, was in keeping with his well-known elitist approach to the study of philosophy.

38 An edition by Shawqī Ḍayf of Ibn Maḍā''s "Refutation of the Grammarians," *Kitāb al-Radd ʿalā al-nuḥāh,* was published in 1947 (Cairo: Dār al-Fikr al-ʿArabī). See editor's introduction.

39 Ibn Rushd, *Tahafot al-Tahafot,* texte arabe établi par M. Bouyges, S. J., Beirut: Imprimerie Catholique, 1930. English translation and extensive notes by Simon van den Bergh as Averroes's *Tahafut al-Tahafut* (*The Incoherence of the Incoherence*), 2 vols., London: Luzac & Co., 1969.

40 Averroes's criticisms of the earlier Islamic philosophers are not, of course, confined to the *Incoherence of the Incoherence.* Reference may be made here to his *Sermo de substantia orbis* (dated 1178 – Munk, *Mélanges,* n. 24, p. 423) where, in the words of Arthur Hyman, he "refutes what he considers the erroneous opinions of earlier philosophers, especially those of Avicenna, and reestablishes what in his opinion are the pure Aristotelian doctrines" (Arthur Hyman, Ph.D. diss., Harvard University, 1953, p. iii). The following passage from the same work (Hyman's translation from the Hebrew version) reveals the exaggerated view of Aristotle's position in the history of thought that was mentioned earlier (see n. 23): "The starting point of the investigation is what we have gathered from Aristotle concerning these matters. For concerning existent things no opinion has reached us from the ancients which is truer than his, or less subject to doubt or presented in better order. Therefore we take his opinion to be that human opinion which man can attain by nature, that is it is the most advanced of those opinions which man, insofar as he is man, man by his own knowledge and intellect obtain. Thus, as Alexander puts it, "It is Aristotle on whom we are to rely in the sciences' " (Hyman, p. 146).

41 Of the seven references to al-Fārābī in the *Incoherence of the Incoherence,* five are critical. In the last two, Averroes draws support from al-Fārābī against Avicenna, but in general the two Eastern philosophers are bundled together as misguided interpreters of Aristotle who had misled al-Ghazālī from the true doctrines of the ancients. See indexes of Bouyges's edition of the *Tahafot* and van den Bergh's translation, both cited in n. 39.

42 In *Kitāb al-Kulliyyāt* (Latin *Colliget*), a compendium on the generalities of medicine composed before 1162, Averroes accused "the man known as al-Kindī" of having misled people by having introduced into medicine (a natural science) considerations that properly belonged to arithmetic and harmonics (*ṣināʿat al-ʿadad wa ṣināʿat al-mūsīqā; artem alhabachi vel algorismi et musice*), a reference to al-Kindī's treatise on compound medicines. See Abu el Ualid . . . ben Roxd . . . (Averroës), *Quitab el Culiat* (*Libro de las Generalidades,* Larache (Morocco): Publicationes de Instituto General Franco para la Investigación Hispano-Árabe, 1939, p. 168 (this publication being a facsimile of the Arabic text from a MS dated 583/1187, not 1186). For the Latin translation of the *Colliget,* see Averrois Cordubensis *Colliget Libri VII,* in *Aristotelis Opera cum Averrois commentariis,* supp. I (Venice, 1562; repr. Frankfurt am Main, 1962). For a new edition of the passage cited here, see M. R. McVaugh, *Arnaldi de Villanova Opera Medica Omnia II: Aphorismi de Gradibus,* Granada-Barcelona, 1975, esp. p. 323, lines 30ff. The Arabic text and a French translation of al-Kindī's treatise have been published by Léon Gauthier, *Antécédents greco-arabes de la psycho-physique,* Beirut, 1939.

43 Averroes's criticism of Ibn al-Haytham, like his criticism of al-Kindī, concerns the question of the relation of physics to mathematics. In the *Middle Commentary* (*Expositio Media*) *on Aristotle's Meteorology,* Averroes contrasts Aristotle's idea of subordination of sciences (when the conclusions of one science are taken as hypotheses in another) with Ibn al-Haytham's view of optical inquiry as "composed of" physics and mathematics, which view Averroes takes to be a wrong departure from Aristotle. See A. I. Sabra, "The physical and the mathematical in Ibn al-Haytham's theory of light and vision," in *Commemoration Volume of Bīrūnī International Congress in Tehran,* Tehran: High Council of Culture and Art, 1976, pp. 439–78, esp. pp. 448–50.

44 "The most temperate lands (*al-bilād*) are those in which autumn is short and springtime is long, and these are the lands that lie in the fifth climate, especially those among them that are close to the sea. Now autumn in these lands of ours, namely the lands of al-Andalus, is about two months and they lie at the beginning of the fifth climate. There is no temperate time (*zamān muʿtadil*) below the equator, as is claimed by many people, and we have shown this in what we have written elsewhere; nor is [any] part of the fourth climate better in any way than the fifth (*wa lā ayḍan yafḍḍulu* [ms. *yfṣl*] *min baʿḍ al-iqlīm al-rābiʿ ʿalā al-khāmis bi-shayʾ*). And Galen is of the opinion (*yarā*) that the most temperate places are the lands of the Greeks (*bilād yūnān*) among which is the land of Hippocrates." (*K. al-Kulliyyāt,* n. 42, p. 42; *Colliget,* Venice ed. cited, supp. I, fol. 32ᵛ.)

45 However, even if we accept that they [the Greeks] are the most disposed by nature to receive wisdom, we cannot disregard [the fact] that individuals like these – i.e. those disposed to wisdom – are frequently to be found. You find in this the land of the Greeks and its vicinity, such as this land of ours, namely Andalus, and Syria and Iraq and Egypt, albeit this existed more frequently in the land of the Greeks" (*Averroes on Plato's "Repub-*

lic", trans. Ralph Lerner, Ithaca, N.Y.: Cornell University Press, 1974, p. 13; first bracket added by the present writer).

46 "Et ideo nos videmus hanc terram nostram magis propinquam in sua natura terris Graecorum quam terrae Babyloniae. Clima autem aequale seu contemperatum est quintum, ut inquit Galenus, non quartum, ut crediderunt multi homines. Signum autem huius est quod illic inveniuntur complexiones aequales magis. Signum autem forte super complexiones est color et capilli. Color autem aequalis est albus et clarus; capillus autem aequalis est magis propinquus ad illum qui est quasi medius inter planitiem et crispitudinem seu ad planum quam ad crispitudinem. Esse autem huius coloris et capilli parum invenitur in terra Arabum, et ideo vocant album rubeum; terrae autem Babyloniae sunt mediae ad terras Arabum, scilicet quod color brunus dominatur super eos homines, sicut est dispositio Arabum; iste autem color in capillis invenitur naturaliter, scilicet magis in hominibus climatis quinti, quando non coniunguntur cum gentibus extraneis, sed cum habitantibus extraneis quae habitant illic de propinquo. Sed, cum prolongaverit tempus in istis hominibus, tunc redit natura eorum ad naturam illorum hominum illius climatis, sicut accidit filiis Arabum et Barbarorum in terra Andalosiae, scilicet quod ipsi conversi sunt ad naturam gentis propriae illi terrae, ideo multiplicatae sunt in eis scientiae . . ." (*Aristotelis Opera cum Averrois Commentariis,* n. 42, V. fols. 435^v–436^r).

8

"Success sanctifies the means"

Heisenberg, Oppenheimer, and the transition to modern physics

GERALD HOLTON
Harvard University

I

It is likely that no science had a more turbulent and fruitful three decades than did physics between the turn of the century and the end of the 1920s – a period to which Professor I. B. Cohen has turned his attention in several of his works. Notions that were unquestioned at the beginning became barely remembered vestiges of a distant "classical" period. The very process of approaching scientific problems was dramatically altered. Even the names of some of the main actors – the Curies, Rutherford, Planck, Einstein, Minkowski, Schrödinger, Bohr, Born, and their generation – commonly stand for "revolutionary" innovations.

And yet, an argument can be made that for all their hard-won advances, those physicists were at heart still closer to the classical tradition than to the conceptions most physical scientists today carry in their very bones. Even Bohr had to wean himself painfully from his correspondence-principle approach, with its base in mechanistic explanations; and his complementarity point of view did not turn its back on that tradition, but tried to incorporate it.

To find those who, in that transition period, dedicated themselves without visible ambivalence to the fashioning of a new physics, to find the real "radicals," one has to turn to the pupils of that older generation. Among these, none is a more appropriate candidate than Werner Heisenberg. Here was a young person who appeared to have no ties to old ideas that might impede or delay him. One probably has to agree with the point that "there is simply no ultimate *logical* connection between classical and quantum mechanics";[1] it was therefore perhaps unavoidable that a crucial intervention in the transition from one to the other came from one who seemed to have no *psychological* or *presuppositional* connection with the old physics, either.

Heisenberg's frank letters and interviews[2] greatly enrich our understanding of his many scientific and nonscientific publications, and so are important sources

I am happy to acknowledge support from the National Endowment for the Humanities and from the National Science Foundation for research on which this paper is based in part. An early version of this paper was part of a presentation made at the first meeting of the Council of Scholars at the Library of Congress, November 1980.

for identifying the personal responses of a modern scientist as he tried to deal with contrary demands and loyalties. He has given us a revealing picture of the context in which he discovered himself as a physicist, just as he was turning twenty, while a third-semester student at the University of Munich. Only four weeks after beginning Arnold Sommerfeld's seminar on "Theory of Spectral Lines on the Basis of the Bohr Atomic Model," the young man is given the honor of being allowed to try his hand at a problem. In fact, Sommerfeld himself had been defeated by it: the explanation of the experimental values of the anomalous Zeeman effect. In those days, before the discovery of electron spin and of spin-orbit coupling, it was not possible to associate the observed splitting of the doublet and triplet spectra in a weak magnetic field with the mechanisms of the otherwise successful Bohr-Sommerfeld quantized model of the atom. The separation between the components of the split spectral lines could be rendered by rational fractions (e.g., ⅔, ⅓). But how could these fractions be related to some atomic model governed by the familiar orbital mechanics, where whole quantum numbers characterized the initial and final states of the emitting atom?

In an interview in 1963, Heisenberg recalled this first test as a research scientist:

> Sommerfeld gave me the experimental values of the anomalous Zeeman effect. He had just published a minor paper which he told me was not important at all. He said, "Since we supposed from Bohr's work that every frequency is a difference between two energies, one should expect that in these funny numbers – which one had in the anomalous Zeeman effect for the splitting – "that the denominator should be the product of two denominators belonging to the two states," . . . and he wanted me to find out what the initial and final states were, using the selection rules . . .
>
> So after a very short time, I would say perhaps one or two weeks, I came back to Sommerfeld, and I had a complete level scheme. Then I came up with a statement which I almost didn't dare to say, and he was, of course, completely shocked. I said, "Well, the whole thing works only if one uses *half* quantum numbers." Because at that time nobody ever spoke about half quantum numbers; the quantum number was an entire number, you know, an integer. "Well," he said, "that must be wrong. That is absolutely impossible; the only thing we know about the quantum theory is that we have integral numbers, and not half-numbers; that's impossible." . . . Since I was a complete dilettante and amateur, and didn't know anything, I thought, "Well, why not try half quantum numbers?"[3]

There is no indication that the week or ten days of work had cost the young man any deep anguish or soul searching. On the contrary; when Alfred Landé wrote Heisenberg not long afterward, to warn that this approach challenged the old quantum theory at its foundation, Heisenberg responded simply (October 26, 1921) that "one must give up much of the previous mechanics and physics if one wants to arrive at the Zeeman effect." Even the iconoclastic Wolfgang Pauli was worried; but he got the reply from Heisenberg (November 19, 1921), which announced a leitmotif: *der Erfolg heiligt die Mittel,* success sanctifies the means. When Heisenberg's paper—his first—appeared early in 1922, the reaction was the same as to many of his papers during the next few years: Its brilliance and daring challenged and perplexed his elders.[4]

Perhaps the readiness to give up the past came more easily to Heisenberg because, as he explained in the interview of 1963, his idiosyncratic, quite irregular education had brought him to quantum theory even before he had a good grounding in classical mechanics. But there is also some irony in the fact that this radical innovator was born in 1901 as the son of a scholar of Greek philosophy of the old German school. If we trace his personal trajectory we find that his development was affected by two very different forces: the vicissitudes of German political history and the breathtaking developments in physics. In both of these, Heisenberg saw himself as having to choose between the themes of tradition and innovation, or to put it in terms of their nearest thematic equivalents, continuity and discontinuity.

On the political side, the choice was predominantly for conservatism; he confessed once that he was attracted by the "principles of Prussian life – the subordination of individual ambition to the common cause, modesty in private life, honesty and incorruptibility, gallantry and punctuality." One of the earliest glimpses Heisenberg has given us of his youth in Munich shows him at the age of about 17.[5] Germany is in the grips of the spring 1919 revolution. The streets of Munich are battlefields. Far from engaging in youthful rebellion, young Werner and his friends do military service on behalf of the new Bavarian government's troops that are recapturing the city. Between guard-duty shifts, the youth retires to the roof of a nearby building, a theological college. He lies down in the wide gutter to soak up the warm sun and to catch up on his neglected studies, specifically on his Greek school edition of Plato's dialogues, as Kepler himself might well have advised.

And there he encounters, in the *Timaeus,* Plato's discussion of the theory of matter. Each small particle is said to be formed and determined by the mathematical properties of four regular solids. Heisenberg is fascinated by the ancient idea that matter in its chaotic diversity is explained by a few examples of mathematical form, and that thereby an orderliness is discerned behind all the seeming infinity of different behaviors of different materials. Going further, he finds himself speculating whether there can be found some similar sense of order to deal with the turbulent events of the day in the streets below. He asks himself, Does one have to discard the old order of traditional Europe and the "bourgeois virtues" taught by one's parents, or should the new order be built on the old? Strangely, more than most others, Heisenberg was to be confronted with such questions all his life, in physics as well as in politics.

II

By the time Heisenberg was twenty-one, the quality of his mind was unmistakable. In 1922, Niels Bohr had come to give some lectures at Göttingen, not long before going to pick up his Nobel prize. Heisenberg was now there for a term. In the discussion period, this young student asked questions so trenchant that Bohr invited him to continue their talk on a long *Spaziergang.* Their three-hour-long discussion was the first in a long series of collaborations.

In essence, the problem that first brought Bohr and Heisenberg together was this. Bohr's classic paper of 1913 had served splendidly to show a connection to exist among a number of separate findings. For example, the atom, such as that of hydrogen, can be perturbed by high-speed collisions with other atoms, or by chemical processes, or by radiation; but it always returns to its original, "normal" or stable state. Another curious fact is that the light emitted by a hydrogen source has many, very specific frequencies – but always the same ones, the rest being somehow forbidden for that atom.

Bohr had dealt with the puzzle by proposing the well-known model of the atom as a nucleus surrounded by electrons. The electrons, Bohr held, would normally be in "discrete stationary states" during which there is no radiation. But when an orbiting electron descends between stationary states, radiation is emitted, and shows up as a line in the characteristic spectrum. While in the stationary states, the atom obeys classical laws of mechanics, those going back to Newton; while the atom emits such radiation, however, it chiefly exhibits quantum behavior, by laws first proposed by Planck in 1900. Thus Bohr's atom of 1913 was really a kind of mermaid – the improbable grafting together of disparate parts, rather than a new creation incorporating quantum theory at its core.

In its favor was that the model worked excellently to explain a number of phenomena, for example, the known lines of the hydrogen spectrum. But almost everyone, including Bohr himself, felt from the beginning that this mixed model was only the first step on a long road. Some were repelled by the way in which the discrete quantum rules were being imposed on the continuous laws of dynamics, in violation of classical physics. Otto Stern, after reading Bohr's paper soon after it appeared, had turned to a friend and said, "If that nonsense is correct which Bohr has just published, then I will give up being a physicist." Ernest Rutherford, writing to his protégé Bohr on March 20, 1913, had gently scolded him: "Your ideas as to the mode of origin of spectra in hydrogen are very ingenious and seem to work well; but the mixture of Planck's ideas [quantization] with the old mechanics make it very difficult to form a physical idea of what is the basis of it."

It also became suspect that a number of ideas basic to Bohr's atom could be imagined or visualized, but not measured. Thus one could not, of course, actually see or test the presumed orbits of the electrons around the nucleus. Nor did the frequency of the assumed orbital motions turn up as anything observable, the frequency of the actually emitted light being connected only with the *differences* of energy in the electron's transition between stationary states.

Now during Bohr's Göttingen lecture in 1922, Heisenberg thought he detected that Bohr was inclining to question the suitability of his own atomic model. As Heisenberg recalled, it "was felt even by Bohr himself to be an intolerable contradiction, which he tried merely to patch over in desperation."[6] During their long walk, Heisenberg learned for the first time how "well-nigh hopeless these problems of atom dynamics then appeared."

Bohr knew already that the problem was one involving not only physics but also epistemology. Human language is simply inadequate to describe processes within the atom, to which our experience is connected in only very indirect ways. But since understanding and discussion depend on the available language, this deficiency made any solution difficult for the time being. Bohr confessed that originally he had not worked out his complex atomic models by classical mechanics; "they had come to him intuitively . . . as pictures," representing events within the atom. Thinking and talking about electron paths in the atom were easy, but the imagery had really been borrowed from macroscopic phenomena, such as watching actual electron tracks in cloud chambers, all on a scale billions of times larger than the atom itself. Similarly the properties of light allowed one to talk about it by analogy either with water waves or, in other experiments, in terms of energetic bullets (quanta). But how could light as such be both? Visualizability and model-based intuitability (*Anschaulichkeit*) of physical conceptions had always been a help in the past; by 1922 in this new realm it was beginning to appear to be a trap. Yet, how could one do without it?

In 1921–2 there began a long journey through the wilderness for most of the physicists – even for Heisenberg, who has given a description of that period.[7] These were several years of "continuous discussion, and we always saw that we got into trouble, because we got into contradictions and into difficulties. And we just could not resolve these difficulties by rational means. . . . So we actually reached a state of despair, even when we had the mathematical scheme – which was to every one of us to begin with a kind of miracle . . . Out of this state of despair finally came this change of mind. All of a sudden [we said] well, we simply have to remember that our usual language does not work any more, that we are in the realm of physics where our words don't mean much." Wolfgang Pauli declared in 1925, "Physics is decidedly confused at the moment; in any event it is much too difficult for me and I wish I . . . had never heard of it." But by 1928, the struggle had yielded the matrix form of quantum mechanics and the equivalent wave mechanics, and P. A. M. Dirac could write confidently: "The general theory of quantum mechanics is now almost complete. The underlying physical laws necessary for the mathematical theory of a large part of physics, and of all chemistry, at last are completely known." Indeed, although at first only the most adventurous spirits followed this road (and perhaps the majority of European physicists was not converted for several years more), physics had reached a new maturity, with quantum mechanics furnishing the tools for attacking a whole range of obstinate puzzles, from spectroscopy and magnetism to physical chemistry and nuclear physics.

A key event in this transition from despair to euphoria was Heisenberg's own work done in the spring of 1925.[8] He had asked Max Born, for whom he was now acting as assistant, for two weeks of leave. Heisenberg was having a bad case of hay fever and wanted to recover on the barren island of Heligoland.

In this lonely retreat, he worked out a formalistic approach to the understanding of atomic spectra, using a mathematics that more experienced scientists knew as matrix algebra. Heisenberg had totally eliminated the concept of electron orbits, or indeed of any "picture" of the atom. (When Born later saw it, he found this at first "disconcerting.") In its place, Heisenberg put a mathematical schema based on the laws of quantum physics, adjusted by introducing the data (e.g., the frequencies and intensities of the spectral lines) that long ago had been established by observation. In his revealing article "Quantenmechanik,"[9] he declared that the customary *Anschauung* – derived from ordinary space–time conceptions, which are in principle "continuous" – has to give way in the atomic realm to the apparent *Unanschaulichkeit* of "discontinuous" elements. He advised that the grasping of those elements, and of a "kind of reality" appropriate to them, was the real problem of atomic physics.

After the initial consternation among physicists about this approach, it turned out to give results that fit splendidly with the work of others, among them Bohr, Pauli, Dirac, and Schrödinger. Although a small number of physicists reject it to this day as nothing but a halfway house – and although any meaningful version of it has yet to penetrate beyond the walls of science faculties – the goal of a true quantum mechanics had been reached. Even the issue of visualizability, so important in German circles that it found its way into the very title of basic research papers,[10] seemed resolvable when a new kind of *Anschaulichkeit* was reached.[11]

The new quantum mechanics was the product of many hands in direct or indirect collaboration, of individuals and groups, of debates formal and informal. The week-by-week sequence of events, and the apportionment of specific credit, have been the subject of many published and unpublished reminiscences by most of the main actors, as well as of detailed treatises and occasional disputes among historians of science during the last two decades. It does not detract from the usefulness of such research into collegial relations and the operation of the "invisible college" if I focus instead on the individual, and ask here a question prompted by my interest in comparing aspects of the work of two brilliant physicists of about the same age, both trained by the same patron, active at about the same time in quantum mechanics, but with very different careers. The question of interest here is: In what consisted the novelty of Heisenberg's own contribution during the crucial period of 1925–7?

Heisenberg's claim to a key position in the transition to modern physics is hardly challengeable. Thus Max Jammer writes flatly: "The development of modern quantum mechanics had its beginning in the early summer of 1925, when Werner Heisenberg . . . conceived the idea of representing physical quantities by sets of time-dependent complex numbers."[12] Such evaluations draw attention to a set of closely related methodological elements in Heisenberg's approach during those crucial years.

a. The theory for the description of quantum-physical events eliminates the immediately pictorial conceptions and interpretations (e.g., orbits, and later the electron itself) that had generally been associated with basic physical quantities. A typical Heisenberg dictum was "The program of quantum mechanics has to be free itself first of all from these intuitive pictures. . .The new theory ought above all to give up visualizability [*Anschaulichkeit*] totally." [13]

b. As signaled by the discrete nature of the observed spectral lines, the physics of the atomic realm is intrinsically discontinuous, requiring the elimination of notions based on continuous representations characteristic of macrophysical kinematics and mechanics. As Heisenberg wrote to Pauli, on November 23, 1926, "That the world is continuous I consider more than ever totally unacceptable," and Born and Jordan in 1925 referred to Heisenberg's work as "a true discontinuum theory."

c. The statistical interpretation as the way to think about nature is to be adhered to fully. Probability is a fundamental feature of phenomena in the submicroscopic regime and need not be imposed on quantum mechanics, as Heisenberg thought Born had done when introducing the probabilistic interpretation of the wave function in 1926. (Born had confessed to be following Einstein in interpreting probability to be a property of a sort of "phantom field.")[14]

Although other views (e.g., Bohr's complementarity, Einstein's realism) persisted as minority opinions, this set of three interpenetrating elements, developed in Heisenberg's work, came to define a position that was identified with the majority point of view on how to think about modern physics. The young man had shown the path by which to cross the Continental Divide.

III

Heisenberg had given us a number of discussions of Einstein and his work; none is more revealing than the account of an occasion in 1926 when, having attracted Einstein's attention, he was invited after a lecture in Berlin to walk home with him. In the flush of his new successes, Heisenberg reported he now believed one must build the atomic model only on the direct results of experimental observations. Moreover, he confessed that in this he thought of himself as following just the philosophy Einstein had used in fashioning the theory of relativity in 1905. But to his consternation, Einstein answered, "This may have been my philosophy, but it is nonsense all the same. It is never possible to introduce only observable quantities in a theory. It is the theory which decides what can be observed."[15] In experiments of any sophistication, we just cannot separate the empirical processes of observation from the mathematical and other theoretical constructs and concepts.

When Heisenberg told me this story, some years before he published it, he added: "Einstein was of course right. Indeed, I myself showed in the paper on the uncertainty relation, written soon afterwards, that the theory even decides what we *cannot* observe." It was a fair point. But Einstein's objection really touched also on another distinction. Unlike Einstein, Heisenberg had never been interested in firsthand, direct experimentation. He trusted that his theory would be safely built on experimental results obtained behind the walls of some other building. For the frequencies of spectral lines, this was (by the 1920s) safe enough. But with respect to the more sophisticated and ambiguous experiments at the frontier of physics and technology, it could be a dubious policy, as we shall see later.

In his last book, *Tradition in Science*, Heisenberg retells the Einstein story and uses it to open the question whether the new physics, in whose growth he had played such a crucial role, had really overthrown "the traditional method in science." Is quantum mechanics, often referred to as a "revolution" in physics, really outside the tradition of work inaugurated in the time of Galileo? Is one right in seeing Heisenberg as the heroic conqueror of qualitatively new territory? Heisenberg did not think so. He said, "The fundamental method has always been the same"; and elsewhere; "I think science always has more or less the same structure. Science changes considerably in its philosophical aspect, but this I would not consider a crisis of science. It is not a revolution, it is evolution of science. . . . [Science actually proceeds by] a more gradual change, which afterwards could be called a revolution."[16]

The changes Heisenberg himself introduced involved nothing as fundamentally different from ongoing science as, for example, Goethe had demanded when he called for a return to a descriptive science, one based not on experiments that draw out new effects "artificially," but on the directly visible, natural phenomena themselves. Rather, Heisenberg held, the tradition of physics in the twentieth century has been continuous in three essential respects: in the *types of problems* scientists select, in the *methods* they employ, and most strongly in the *type of concepts* they use to deal with phenomena.

The tool kit of specific concepts changes in time, of course; thus, luminiferous ether had to be given up after 1905, and with it the dream of reducing electromagnetism to mechanics. Space and time turned out to be more dependent on each other than the Newtonians had thought. The absolutely "objective description" of a physical system turned out to be impossible. At each of these steps, tradition was eventually perceived to have been a hindrance to the progressive development because, as it is usually put, it "filled the minds of scientists with prejudices." But, said Heisenberg at the end of his career, this is a false view of the role of tradition in science, and "prejudice" is a derogatory label for a profound, more positive, and more functional mechanism in the scientific imagination. If we wish to study phenomena, we need a language. At the start of a deep investigation, the new words are not yet available, so we must use old, traditional (and therefore intuitable) ones, to think about the problem and to ask the initial questions in the first place. But these words are tied to the so-called prejudices that form a necessary part of the old language. It may be painful to realize that a natural, traditional conception like the orbit or path of an electron has no meaning when applied to the atom. But during the very act of coming to this seminal realization, the tradition against which the advance pushes is both a hindrance and a necessity.

Moreover, there is no escape from tradition in yet another sense: The opposite of a currently reigning tradition is another, perhaps equally ancient tradition. To illustrate that, we need only look at the topic that most preoccupied Heisenberg in his final years: the concept of the elementary particle in modern physics. He traces the modern theme of elementary entities as explanatory devices back 2,500 years, to Democritus. But he notes that our elementary particles, such as the proton, are in fact not uncuttable as the Greek atom was thought to be; on the contrary, a proton has a finite size, and it can be divided or transformed. When energetic protons collide they can give rise to many pieces with sizes like those of the proton. Hence the proton is not truly elementary, but "consists of any number and kind of particles." Even the current candidate for the ultimate unit of matter, the quark, is really no better. In some theories it can be divided – into two quarks plus one antiquark, or similar combinations. Thus, in principle every particle needs for its full explanation all other particles.

Out of this vicious circle, as Heisenberg saw it at the end, there is only one way: to abandon the philosophy of Democritus and the concept of fundamental elementary particles. Instead, one should go back to precisely the antithesis of the Greek atomistic materialists, namely, the concepts derived from Plato, specifically his ideas of symmetry – thereby exchanging allegiance from one old thema to its old opposite, difficult though it now might be to do so: "We have returned to the age-old problem whether the Idea is more real than its material realization." It is not the embodiment of physical discontinuity, the material elementary particle, that beckons as fundamental explanatory bedrock, but the mathematical symmetry properties by which matter on the larger scale is construed.

This thought, Heisenberg confessed, is "unfortunately . . . very far from intuitable, and hardly intelligible to the mathematically untutored reader." He might have added that few physicists today are tempted to follow him in this direction. But this is where his pilgrimage had taken him, he who formerly had been the Democritus of our age, who had helped to build the modern atom and describe its essential discontinuities. In his final years he returned to Plato, as if pulled back to the *Timaeus* that had captivated him in his youth.

IV

Let us return now to the crucial period in the 1920s when quantum mechanics was fashioned. I leave Heisenberg and Germany to trace, rather briefly, a parallel view of work in progress on the same grand project, but in North America. The transition is most appropriately introduced by Heisenberg's own account of his visit to the United States in 1929. He came to give a series of lectures on the principles of quantum theory at the University of Chicago, and several other campuses in many parts of the country. It was shortly after he had assumed his new post at the University of Leipzig, where he had, in his first seminar on atomic theory, only a single student. Contrary to what he might have expected, his experience in the United States was very different. His lectures were well attended, interest was keen, and his audience seemed well informed. As Heisenberg recorded in his autobiographical account, *Physics and Beyond:* "The New World cast its Spell on me right from the start. The carefree attitude of the young, their straightforward warmth and hospitality, their gay optimism – all this made me feel as if a great weight had been lifted from my shoulders." As if to support this rather rosy picture, Heisenberg adds a revealing interchange with a young experimental physicist that hints at a characteristic style of research at that time, one that helped many a young American scientist to forge ahead with remarkable speed in very unfamiliar territory:

I told him of a strange feeling I had acquired during this lecture tour: while Europeans were generally averse and often overtly hostile to the abstract, nonrepresentational aspects of the new atomic theory, to the wave–corpuscle duality and the purely statistical character of natural laws, most American physicists seemed prepared to accept the novel approach without too many reservations. I asked Barton how he explained the difference, and this is what he said: "You Europeans, and particularly you Germans, are inclined to treat such new ideas as matters of principle. We take a much simpler view. . . . Perhaps you make the mistake of treating the laws of nature as absolute, and you are therefore surprised when they have to be changed. To my mind, even the 'natural law' is a glorification or sanctification of what is basically nothing but a practical prescription for dealing with nature in a particular domain. I believe that once all absolutist claims are dropped, the difficulties will disappear by themselves."

"Then are you not at all surprised," I asked, "that an electron should appear as a particle on one occasion and as a wave on another? As far as you are concerned, the whole thing is merely an extension of the older physics, perhaps in unexpected form?"

[Barton replied]: " 'Oh, no, I am surprised; but, after all, I can see that it happens in nature, and that's that.' "

Unlike Heisenberg's colleagues in Germany, whether friends or opponents, virtually nobody in the United States seemed to be caught up in those fierce debates on *Anschaulichkeit* that had been raging among Heisenberg's colleagues. What he claims to have encountered here was a pragmatism that brought with it a hospitality to new ideas and to those who could convincingly present them. Perhaps the best summary of this essentially antimetaphysical approach to the new physics was the characteristically American philosophy of P. W. Bridgman, with its stress on operational demarcation criteria expressed in his widely circulated *Logic of Modern Physics,* published a year or two earlier.

Heisenberg may well have noted another, more institutional difference: In the 1920s, when some of the younger scientists in the United States pounced on the new quantum physics, they evidently did not have to worry too much about the incredulity or displeasure of their older mentors with respect to those strange new ideas. Since the older generation in the United States was trained not only in classical physics but in addition predominantly in experimental physics, young scientists who wanted to do their Ph.D. thesis on theoretical quantum physics were pretty much on their own. But emotional and financial support was made available to them, in a way that would have been more difficult in a more hierarchical system. The tendency was to put one's money on the younger people, as R. A. Millikan was fond of expressing it, whereas in Europe it was more likely that the young candidate or *Dozent* worked on a problem put to him by the *Ordinarius.*

Something had happened during the decade of the 1920s – the period Robert Oppenheimer later called the "heroic time" – to which few Europeans had paid attention. In convenient shorthand terminology, the period of catching up with Europe has been called "the coming of age" of physics in the United States.

Although the issue of national differences is usually overplayed, in this case the existence of an identifiably American effort during those years has been documented by the work of many scholars.[17]

It all happened in a remarkably short time. To be sure, in earlier years, the work of a few outstanding contributors in the United States, mainly experimentalists, had achieved world renown, and that of some others, such as the theorist J. W. Gibbs, had been unjustly neglected. On the whole, however, America at the beginning of the 1920s seemed, with respect to theoretical physics, an "underdeveloped country," far from ready to play a major role on the world stage of physics. The atmosphere has been caught in a few lines by John H. Van Vleck:

> The American Physics Society was a comparatively small organization, with only 1,400 members in 1921 . . . and only a small number of the communications were theoretical. Very few physicists in this country were trying to understand the current developments in quantum theory . . .
>
> The problem I worked on was trying to explain the binding energy of the helium atom by a model of crossed orbits which [Professor E. C.] Kemble proposed independently of the great Danish physicist, Niels Bohr, who suggested it a little later. In those days the calculations of the orbits were made by means of classical mechanics, similar to what an astronomer uses in a three-body problem. The Physics Department at Harvard did not have any computing equipment of any sort, and to get the use of a small hand-cranked Monroe desk calculator, I had to go to the Business School. I felt very blue when the results of my calculation did not agree with experiment."[18]

But by the mid-1920s this picture had begun to change rather dramatically. "Although we did not start the orgy of quantum mechanics, our young theorists joined it promptly."[19] It was as if they had been waiting in the wings: Carl Eckart, Robert S. Mulliken, Robert Oppenheimer, Linus Pauling, John Slater, Van Vleck himself, and a rapidly increasing number of others came forward with widely noted contributions, on the very stage where Europeans such as Heisenberg and Pauli, working at well-established centers, had so recently been near despair.

By the end of the decade there was in the United States in physics – as had begun to be achieved in chemistry a decade earlier – an adequate balance between experimental and theoretical work, adequate provision for training at all levels, a much-strengthened professional society, and a spectrum of well-run research publications. The interplay between academic and industrial science, between "pure" and "applied" research, was well launched. R. A. Millikan had made a brash prophecy in 1919: "In a few years we shall be in a new place as a scientific nation, and shall see men coming from the ends of the earth to catch the inspiration of our leaders and to share in the results [of] our developments."[20] It would have been quite sufficient to predict that after a long period of intellectual and institutional subservience to European contributions and styles, American physicists, after absorbing the necessary research attitudes, would be joining the Europeans in the front ranks in terms of the quantity and quality of their contributions.

V

These points, and more, can be illuminated by a sketch of another individual career, that of Robert Oppenheimer.[21] Born in New York in 1904, thus three years younger than Heisenberg, young Oppenheimer, the son of a well-to-do businessman and of a trained artist, received a fine education in scientific and humanistic studies (he later called himself "a properly educated esthete"). In school, writing stories and poetry, learning everything easily, he developed a baroque and exaggerated style that shows up in many of his early letters, and which may even have penetrated his early work in physics. He later confessed that he "probably still had a fascination with formalism and complication," before he fully realized the need for "simplicity and clarity."

It is significant that by his own account Oppenheimer's aim on entering college was to study chemistry or mineralogy, "with an idea of becoming a mining engineer." He had come to science first of all by becoming fascinated, as a child of five or seven, when on a visit to his grandfather in Germany he was given a "perfectly conventional tiny collection of minerals." But although the center of gravity of his college life was clearly in chemistry, mathematics and physics, he also stretched in other directions: philosophy, literature, continuing his study of languages: "I labor, and write innumerable theses, notes, poems, stories, and junk. . . . I make stenches in three different labs, listen to Allard gossip about Racine, serve tea and talk learnedly to a few lost souls, go off for the weekend to distill the low grade energy into laughter and exhaustion, read Greek, commit faux pas, search my desk for letters, and wish I were dead." At an age when Heisenberg walked in Göttingen with the likes of Niels Bohr, the various fragments of Oppenheimer's soul were still waiting for a magnetic field that would line them up.

That magnet came in his encounter with P. W. Bridgman, the experimentalist who later won the Nobel Prize for his researches on high-pressure physics and who was then also developing his operationalist philosophy of science. Looking back, Oppenheimer recalled that as far as science goes, Bridgman's course presented "the great point of my time" at college; for Bridgman "didn't articulate a philosophic point of view, but he lived it, both in the way he worked in the laboratory which, as you know, was very special, and in the way he talked. He was a man to whom one wanted to be an apprentice."

There is, however, important asymmetry. While young Heisenberg shocked Arnold Sommerfeld with his confident innovations in theoretical physics, Oppenheimer was heading for a terrible disappointment with the type of physics he had planned as his life's work. Bridgman later reminisced with some amusement that Oppenheimer had come to work on an experimental problem in his laboratory and at the beginning "didn't know one end of the soldering iron from the other." The delicate suspensions of galvanometers had to be replaced constantly when this apprentice was using them. But he stuck with it and did a quite respectable job of measuring the pressure coefficient of the electric conductivity of alloys to about 15,000 atmospheres – a kind of apotheosis of his initial interest in mineralogy and crystal physics.

Yet for a would-be physicist in the early 1920s, to be not very good in the laboratory, a test Heisenberg never had to face, raised in an American university some doubts about his ultimate promise. In Bridgman's letter of recommendation, intended to help Oppenheimer to go on to Ernest Rutherford's laboratory at Cambridge, he had to confess Oppenheimer's "weakness is on the experimental side. His type of mind is analytical, rather than physical, and he is not at home in the manipulations of the laboratory." Oppenheimer himself, in a letter of about that time, had discovered that "my genre, whatever it is, is not experimental science." That realization may help to explain, in some part, the depression and identity confusion that overtook Robert when, after continued poor success in laboratory matters, now at the Cavendish, he realized that he still had not made the central discovery that every young person must make: to "find and obey the demon who holds the fibers of one's very life," to quote the splendid words of Max Weber at the end of his essay, "Science as a Vocation."

Like many creative persons in their youth, Oppenheimer faced a terrible problem: He felt himself to have the ability to look for central questions, but neither saw them yet with clarity nor had the ability to survive psychologically, without deep anguish, the disappointment of remaining still outside the center. A few years later, in one of his beautiful letters to his brother Frank, he hints at the internal resolution that, to a large degree, came to him upon passing over that threshold: "I know very well, surely, that physics has a beauty which no other science can match, a rigor, an austerity and depth." Its study induces, and demands, a kind of mental discipline that at its best helps one to "achieve serenity, and a certain small but precious measure of freedom from the accidents of incarnation, and charity, and that detachment which preserved the world which it renounces. . . . We come a little to see the world without the gross distortion of personal desire."

Indeed, there is considerable evidence that it was both his good luck and his bad that this labile, vulnerable, and brilliant person came to maturity at the very time when physics itself was passing through a period of despair. The difficulty and fast pace of the science at that time of great excitement clearly exacerbated what can be called at its best a serious problem of temperament; and at the same time, it put to a test what one of his friends of that time called Robert's "ability to bring himself up, to figure out what his trouble was and to deal with it." Clearly, once he had made the decision not to go back into a laboratory but to throw himself into the grand theoretical problems agitating European physicists, he had essentially embarked on a course of self-therapy. Those of us who as teachers and mentors of young would-be scientists watch them over the period of a few crucial years not infrequently see some version of this process of self-identification taking place before our very eyes.

Great things were afoot out there; and joining in, "coming into physics," helped to resolve his dilemmas once Oppenheimer had his first sense of success. Of course, some things were left over, not least a vulnerability that ran through his personality like a geological fault, to be revealed when an earthquake came along. Thus when Oppenheimer was put on secret trial by the AEC Hearing Board in 1954, the record that was later made public shows clearly how ineffective Oppenheimer was under fire, acquiescing in his own destruction. As he explained later wearily, "I had very little sense of self."

But the letters of the young Oppenheimer, and reminiscenses recorded later both by Oppenheimer and his friends of that period, show that the years 1926–7, spent at the University of Göttingen, produced that inner alignment that formed the strong armature of his mature self. He met like-minded, challenging young people such as Courant, Heisenberg, Wentzel, and Pauli – a support group with which, as he later said elegantly, "I began to have some conversations." The formal and informal education proceeded at an incredible pace. Thus, a few months after arriving he could write (perhaps with much satisfaction) a letter to his former professor, Bridgman, explaining why Bridgman's rather classical theory of metallic conduction – on the basis of which Oppenheimer had embarked on his ill-fated laboratory work – had to be changed to take the new quantum mechanics into account. At the age of twenty-three, he had essentially achieved his goal of "making myself for a career."

Oppenheimer's mentor at Göttingen was, as it happened, Max Born – no easy taskmaster; Enrico Fermi had been so intimidated by him that he had toyed with the idea of giving up physics, before being rescued by finding sympathetic tutelage under Paul Ehrenfest in Holland. But Born had encountered Robert on a visit to the Cavendish Laboratory, and they discovered that they both were interested in the same problem, the light spectrum radiated from excited molecules. Within a few months, Oppenheimer had finished his Ph.D. thesis, in an article on the quantum theory of continuous spectra, received at the *Zeitschrift für Physik* on Christmas Eve, 1926. Before a year was over, he had written four more articles, one of which, with Max Born, was a quantum-theoretical treatment of the behavior of molecules that remains in use, six decades later ("The Born-Oppenheimer Approximation").[22]

Appropriately enough, having passed through his ordeal, Oppenheimer comes back to the United States as a kind of young guru of the new physics. He chooses for his main appointment a university in which there is no one working in the new quantum mechanics, where he can begin a career of leadership of a major school of theoretical physics. But it is most significant that he also cultivates genuine working relationships with experimental physicists, the happy harvest of his earlier struggles in laboratories, and also a characteristic of the possibilities in American universities, which rarely separated the theoretical and experimental divisions in the manner that was still rather usual at the time in European instruction and research.

VI

The personal trajectories of Heisenberg and Oppenheimer, when set forth in more detail than would be appropriate here, show remarkable symmetries and asymmetries, intersections, and divergences. Both came to intellectual maturity about the same period, anguishing about the same problems of physics. They were touched by many of the same people (thus a chance meeting with Niels Bohr, while Oppenheimer was still struggling at the Cavendish Laboratory, was fateful to Oppenheimer too; he said later: "At that point I forgot about beryllium and films and decided to try to learn the trade of being a theoretical physicist"). And it is rather uncanny that each ended up being in charge of the project that aimed to produce a nuclear weapon for his respective country during the 1940s.

The outcomes of those projects were, of course, of historic importance, and the subsequent fate of the world would have been very different if the Americans had failed and the Germans succeeded, or both succeeded, or both failed. Consequently, these wartime projects have been much analyzed and many reasons have been found for accounting for the actual results. But one element in the eventual outcome of these "applied science" projects, not negligible but not much remarked on, may well be the difference of scientific styles characterizing the heads of the two projects and the sociology of the scientific communities within which they worked.

In barest outline, the point in question is this. In Heisenberg's project, the top responsibility and many of the essential ideas were those of a theoretician, who, in the tradition of his country's university system, had been able to keep a distance from the experimental side of physics. The evidence points strongly to the possibility that the design of the early uranium pile on the German side, strongly influenced by Heisenberg's own ideas, was quite impractical. Also, at least two crucial experimental measurements – of the diffusion of neutrons in the graphite lattice of the pile, and the number of neutrons freed during fission – seem to have been botched at an early stage, and not effectively challenged for a long time or at all, possibly because the hierarchical system, transferred from the university to the weapons project, made such challenges more difficult there.

On the other side, one of Oppenheimer's first acts on assuming leadership – even a condition he made on accepting the assignment – was that an experimental physicist should be provided to assist him. In addition, of course, Oppenheimer's access to, and knowledgeable interaction with, other experimentalists was not a problem for him at all. The interlacing of the theoretical and experimental aspects was complete under Oppenheimer's influence and natural for all who worked with him. In a letter to John H. Manley (the experimental physicist who assisted him, often by mail until the actual focusing of the project, in the spring of 1943, at the Los Alamos site), Oppenheimer writes, "We are up to our ears in every kind of work," especially in a careful determination of the fission-neutron yields obtained from uranium bombarded with neutrons.[23]

Before it was all over, Oppenheimer had to dig deep into his own reservoir of knowledge, both of theory and of experiment, had to stretch and supplement his command of physics ranging from theoretical nuclear physics to metallurgy and ballistics, and, by assignment and example, had to do the same for many others. Those who worked with him then seem to agree that his taste and skills, across this spectrum, provided the essential glue that held the whole enterprise together despite great strains. Without implying a theory of causation, one has no great difficulties in seeing some of the roots of later skills and interests in the documented early period.

VII

I have sketched the outlines of what a "dense account" would be like that focuses on two persons caught up in the outburst of creativity in the physics of the mid-twenties. If we had a fully constructed, dense account, how would it be best used toward the goal of studying the mechanisms that propel the creative work in the sciences? In closing, a brief personal remark on methodology seems appropriate here. Following the assembly of a dense account, the second stage is to select for attention a determinate, small portion of it, a portion that on the basis of that account suggests itself as a promising keyhole to the laboratory of the mind under study.

The third stage is to dive into that segment and, typically, select one or more particular documentary artifacts, be it a publication, a laboratory notebook page, a letter, a sketch, a photograph, a piece of apparatus, or nowadays a computer printout or voice record – an artifact prepared during the nascent moment or phase so chosen. A document of this sort might be one of Heisenberg's early letters written while he formulated his approach to quantum mechanics,[24] or Oppenheimer's first paper of lasting scientific quality.[25] Depending on one's preference, the more conventional dimensions of scholarship lead one to establish (a) the historic state of science at the time of the production of the document, (b) the time trajectory of the state of public scientific knowledge leading up to (and, if possible, beyond) the time chosen, (c) the sociological, (d) the cultural developments outside science and the ideological or political events that may have influenced the work of the scientist, and (e) the epistemological or logical structure of the work.

Recently, we have seen also more serious attempts to deal with the personal aspects of the scientific activity, and with the psychobiographical development of the scientist under study. The discussion of Oppenheimer's self-therapy of an evident identity disturbance in the mid-1920s, through his progressive attainment of mastery in quantum physics, may well deserve a place in the full description of the forces at work in the production of one of his early papers.[26] Similarly, young Heisenberg's readiness to make radical departures is made more complex if we also hear the ticking of the time bomb of his early Platonic enchantment.

Last but not least, another tool for the analysis of a scientific work is what I have termed thematic analysis. Here I would point only to the previously noted example, in Heisenberg's work, to establish a physics that is thematically fully built on discontinuity, thereby freeing physics from what he called, in a letter to Pauli (January 1925), the "swindle" of allowing physics to work with a mixture of quantum rules and classical physics, as Bohr and Sommerfeld still tolerated.

When these various dimensions of modern analysis, originating from different directions, are brought to bear on a specific case in the history of science, one can hope to see emerging from the dense account – akin to the ethnographer's "thick description" – an entity with an orderly structure of its own. At the very least, the striking diversity we have seen in the responses of a Heisenberg and an Oppenheimer when both are caught in the same system of tensions, indicates the riches that still wait to be discovered in the continuing study of that turbulent period in the making of modern physics.

NOTES

1 Max Jammer, *The Conceptual Development of Quantum Mechanics* (New York: McGraw-Hill, 1966), p. 219.

2 E.g., in the *Archives of the Sources for the History of Quantum Physics,* twelve sessions, November 30, 1962–July 12, 1963.

3 *Archives,* pp. 5–6.

4 Heisenberg's paper, "Zur Quantentheorie der Linienstruktur und der anomalen Zeemaneffekte," appeared in *Zeitschrift für Physik 8* (1922), 273–97. For further details, including a good discussion of Heisenberg's "unique style" and the reaction to it, see David C. Cassidy, "Heisenberg's First Paper," *Physics Today 31* (1978), pp. 23–8, and David C. Cassidy, "Heisenberg's First Core Model of the Atom: The Formation of a Professional Style," *Historical Studies in the Physical Sciences 10* (1979), pp. 187–224. As Cassidy points out, half-integral quantum numbers had been used by A. Landé and others, but other of Heisenberg's conscious derivations from accepted principles such as half-integral momenta and a magnetic core had not. "Not only had Heisenberg introduced real non-integral momenta, but he had also violated [mostly without explicitly drawing attention to it] the Sommerfeld quantum conditions, the angular-momentum selection rules, space quantization, classical radiation theory, the Larmor precession theorem, and the semi-classical criterion of perceptual clarity (*Anschaulichkeit*) in model interpretation" (ibid., pp. 190–1).

Among other useful articles touching on these points, see Paul Forman, "The Doublet Riddle and Atomic Physics *circa* 1924," *Isis 59* (1968), pp. 156–74; Edward MacKinnon, "Heisenberg, Models, and the Rise of Matrix Mechanics," *Historical Studies in the Physical Sciences 8* (1977), pp. 137–88; Daniel Serwer, "*Unmechanischer Zwang:* Pauli, Heisenberg, and the Rejection of the Mechanical Atom, 1923–1925," *Historical Studies in the Physical Sciences 8* (1977), 189–256; David Bohm, "Heisenberg's Contribution to Physics," in W. C. Price and S. S. Chissick, (eds.) *The Uncertainty Principle and Foundations of Quantum Mechanics* (New York: Wiley, 1977), pp. 559–63 (on Heisenberg's "radically new mathematical and physical account of the facts as a whole"); and A. Hermann, K. V. Meyenn, and V. F. Weisskopf (eds.), *Wolfgang Pauli: Wissenschaftlicher Briefwechsel mit Bohr, Einstein, Heisenberg, u.a.,* vol. I, 1919–29 (New York: Springer-Verlag, 1979).

5 W. Heisenberg, *Physics and Beyond* (New York: Harper & Row, 1971), ch. 1; see also ch. 20.

6 W. Heisenberg, *Tradition in Science* (New York: Seabury Press: in press); a collection of the essays of the late period. See also his articles, "Development of Concepts in the History of Quantum Theory," *American Journal of Physics 43* (1975), pp. 389–94, and "The Nature of Elementary Particles," *Physics Today 29,* (1976), pp. 32–9.

7 "Discussion with Professor Heisenberg," in O. Gingerich (ed.), *The Nature of Scientific Discovery* (Washington, D.C.: Smithsonian Institution Press, 1975), 556–73, esp. pp. 560 and 565.

8 W. Heisenberg, "Uber quantentheoretische Umdeutung kinematischer und mechanischer Beziehungen," *Zeitschrift für Physik 33* (1925), pp. 879–93.

9 W. Heisenberg, "Quantenmechanik," *Die Naturwissenschaften 14* (1926), pp. 989–94.

10 W. Heisenberg, "Über den anschaulichen Inhalt der quantentheoretischen Kinematik und Mechanik," *Zeitschrift für Physik 43* (1927), pp. 172–98.

11 See A. I. Miller, "Vizualization Lost and Regained: The Genesis of the Quantum Theory in the Period 1913–1927," in Judith Wechsler (ed.), *On Aesthetics in Science* (Cambridge, Mass.: MIT Press, 1978), pp. 73–102. Also, A. I. Miller, " Redefining Anschaulichkeit," in *Festschrift for Laszlo Tisza* (Cambridge, Mass.: MIT Press, in press), which carefully treats the complex of meanings and transitions associated with the conception of *Anschauung* and *Anschaulichkeit* in German physics research in the 1920s.

12 Max Jammer, *The Philosophy of Quantum Mechanics* (New York: Wiley, 1974), p. 21. Jammer assigns Max Born a relatively secondary role in the further development: "As Max Born soon recognized, the 'sets' in terms of which Heisenberg had solved the problem of the anharmonic oscillator" were entities known from the theory of matrices. "Within a few months Heisenberg's new approach was *elaborated* by Born, Jordan, and Heisenberg himself." A similar estimate is given in Jammer, *Quantum Mechanics,* p. 197.

13 Heisenberg, "Quantenmechanik."

14 See M. Born, *Zeitschrift für Physik 37* (1926), p. 863 and *38* (1926), p. 803; also Born's letter to Einstein, Nov. 11, 1926.

Heisenberg was upset that Born imposed the notion of probability on the quantum mechanics that Heisenberg thought of as a closed theory; moreover, Born was using wave functions taken from Schrödinger's wave mechanics that purported to be based on a continuum interpretation of the atomic regime, replete with pictures, i.e., *Anschauungen.* Heisenberg immediately submitted a paper in which Born's probabilistic interpretation of the wave function is not mentioned and Schrödinger is roundly criticized (W. Heisenberg, "Schwankungserscheinungen und Quantenmechanik," *Zeitschrift für Physik 40* (1926), pp. 501–6. There Heisenberg demonstrated that a probability interpretation emerges naturally from quantum mechanics and is properly understood in terms of essential discontinuity (quantum jumps).

Later, Heisenberg, distancing himself further from Born, went so far as to declare Born's statistical interpretation of the wave function to be the result of "developing and elaborating an idea previously [1924] expressed by Born, Kramers, and Slater." Heisenberg, in M. Fierz and V. F. Weisskopf (eds.) *Theoretical Physics in the Twentieth Century* (New York: Interscience, 1960), pp. 40–7. Moreover, Born's original probabilistic interpretation of 1926, "Zur Quantenmechanik der Stossvorgänge," *Zeitschrift für Physik 37* (1926), pp. 863–7, with all its successes, failed when applied to electron diffraction. Again, it was Heisenberg's role here to make a crucial step, by changing the interpretation of the ψ waves, "not to regard them as merely a mathematical fiction but to ascribe to them some kind of physical reality." Jammer, *Philosophy of Quantum Mechanics,* p. 44.

15 Heisenberg, *Physics and Beyond.*

16 "Discussion with Professor Heisenberg," in Gingerich (ed.).

17 See, for example: Stanley Coben, "Scientific Establishment and the Transmission of Quantum Mechanics to the United States, 1919–1932," *American Historical Review 76* (1971); Laura Fermi, *Illustrious Immigrants,* 2d ed. (Chicago: University of Chicago Press, 1971); Daniel J. Kevles, *The Physicists: The History of a Scientific Community in Modern America* (New York: Knopf, 1978) (see esp. chap. 14–16 and his bibliographical essay on Resources, pp. 450–457 for further primary sources); Katherine R. Sopka, *Quantum Physics in America, 1920–1935* (New York: Arno Press, 1980); J. H. Van Vleck, "American

Physics Comes of Age," *Physics Today 17* (1964); Spencer R. Weart, "The Physics Business in America, 1919–1940: A Statistical Reconnaissance," in Nathan Reingold (ed.), *The Sciences in the United States: A Bicentennial Perspective* (Princeton, N. J.: Princeton University Press, 1979); and Charles Weiner, "A New Site for the Seminar: the Refugees and American Physics in the 1930s," in Donald H. Fleming and Bernard Bailyn (eds.), *The Intellectual Migration: Europe and America, 1930–1960* (Cambridge, Mass.: Harvard University Press, 1969).

18 J. H. Van Vleck, "American Physics," p. 25.

19 Ibid., p. 24.

20 Quoted in Kevles, *The Physicists,* p. 169.

21 Of excellent use for the following passages are the newly published letters, in Alice Kimball Smith and Charles Weiner (eds.), *Robert Oppenheimer, Letters and Recollections* (Cambridge, Mass.: Harvard University Press, 1980). The quotations that follow are from the letters in that book.

22 M. Born and R. Oppenheimer, "Zur Quantentheorie der Molekeln," *Annalen der Physik 84* (1927), pp. 457–484. The first footnote reference in the paper is to a collaborative paper by Born and Heisenberg, published three years earlier.

23 Smith and Weiner (eds.), *Robert Oppenheimer,* p. 227.

24 E.g., in A. Hermann, et al. (eds.), *Wolfgang Pauli.*

25 Born and Oppenheimer, "Zur Quantentheorie der Molekeln."

26 Smith and Weiner (eds.), *Robert Oppenheimer;* see also G. Holton, "Young Man Oppenheimer," *Partisan Review XLVIII,* (July 1981), pp. 380–8.

9

Einstein's image of himself as a philosopher of science

ERWIN HIEBERT
Harvard University

Since antiquity, natural philosophers and scientists have expressed the conviction that the observational and experimental study of nature brings with it a good measure of intellectual and aesthetic satisfaction. Indeed, scientists on the whole claim to derive considerable personal pleasure from their work. I believe these claims are true. Now it seems plausible to assert that the machinery of human perception and cognition is both biologically structured and socially motivated to accentuate certain characteristic benchmarks of excellence in human performance. These distinctive characteristics are by no means the property of scientists. They certainly are seen to be prominent as well in the arts and humanities. Still, they are glaringly visible in the work of scientists.

To be more specific, we might mention in this context a number of criteria of excellence: structure, order, and symmetry; the power of metaphor and analogical reasoning; comprehensiveness; simplicity – or a move in the direction of efficiency and economy of thought and expression; prediction into the unknown; logical rigor and internal consistency; and elegance of conception and formulation. All of these criteria obviously occupy a position of high priority among scientists, but, of course, as already intimated, scientists by no means have a corner on them.

In view of the fact that the sciences have been conspicuously successful, from the standpoint of the criteria just mentioned, it comes as no surprise to discover that scientists and their accomplishments and methods have provided the subject matter for perennial analysis by philosophers, historians, psychologists, anthropologists, theologians, and sociologists. In spite of all the commendable analyses, dissections, and reconstructions that scientists and their methods have been subjected to, we may find it shocking to see how inadequately they deem themselves to have been analyzed by others external to their intellectual and professional framework. Indeed, scientists frequently are at a loss even to identify themselves in the analyses of outsiders who purport to be examining what goes on in their own special disciplines. Although this may be a commonly encountered phenomenon, viz., the questioning or rejection by insiders of the analyses of outsiders, perhaps it tells us that the insiders are (or think they are) playing quite a different game. At minimum they are analyzing that game in another way. In any case, we have here a number of competing perspectives.

Let me express this observation in another way. At least since the middle of the nineteenth century, scientists have become increasingly more confident that professional, working scientists, on their own, can provide the most meaningful philosophy of science that is feasible. Whether this is true or not, historians of science are increasingly anxious to understand what kind of game scientists believe they are playing. Thus, this essay focuses almost exclusively on the question of the scientist's self-image.

If, in fact, there is widespread consensus among scientists that they do not really recognize themselves in treatises devoted to analyzing their methods and their work and behavior as scientists, then it is appropriate to suggest that there may be some positive merit in studying and evaluating their own self-reflections, namely, those in which they attempt to tell us what they do and how and why. Conceivably, some of the most pertinent questions to be explored in the philosophy of science touch on these aspects of the scientific enterprise that characterize the distinctive ways in which scientists see and understand themselves and their work.

Quite simply, what do scientists *claim* to be doing when they allege that they are engaged in scientific activity? What are the motivations, what methods (if any) are consciously cultivated; and what constitutes evidence to buttress an argument or support an explanatory hypothesis? Such questions have dimensions that can be explored profitably for the perspectives of history, psychology, the logic of science, conceptual frameworks, and various environmental contexts. But no *n*-dimensional analysis will help us here. We must adopt a more modest objective.

Against a specific conception of intention and outlook, viz., one that attempts to look at the scientist in the role of philosopher of science, I have undertaken a study of the self-image analysis of a number of scientists in hopes that some intrinsic pattern will emerge, if not for scientists in general, perhaps for scientists within a given discipline; and, at least within the more specialized domain of a particular science.

What I propose to do here is to examine one particular scientist's attempt to play the role of philosopher of science, namely, Albert Einstein. The rationale for this choice is that we recently celebrated Einstein's 100th birthday. Birthdays aside, the case for Einstein, his own self-image, and his view of himself as a philosopher of science provides a splendid example of the genre of questions this study purports to illuminate.

First let us mention a few landmarks in Einstein's career. In 1902, after completion of his studies at the Polytechnic in Zurich, Einstein became a Swiss citizen and worked for six years in the patent office in Bern. It was there between 1905 and 1906, when he was in his mid-twenties, that Einstein published four papers that contributed conspicuously to establishing the direction of twentieth-century theoretical physics. As is well known, these papers are models of originality, clarity, and elegance. They deal with totally diverse topics: the light quantum hypothesis, a theory of Brownian motion, an analysis of the electrodynamics of moving bodies that incorporates new views on the structure of space and time into a special theory of relativity, and a paper on the relation of the inertia and energy, or the general equivalence of the mass and energy, of a body. In one way or another, this early work of Einstein – each paper a landmark in its own right – sets the stage not only for much of his subsequent scientific work, but also for the direction of his philosophical reflections.

In 1913, after short intervals at the University of Zurich, the University of Prague, and the Polytechnic in Zurich, Einstein moved to Berlin. There, three years later, in 1916, he essentially completed his first enunciation of the general theory of relativity. The theory received its first confirmation in 1919 with the observation of the deflection of light in a gravitational field. Einstein certainly regarded his general theory of relativity as his true lifework. He said of his other contributions that they were *Gelegenheitsarbeit,* that is, performed as the occasion arose. But Max Born wrote that Einstein "would be one of the greatest physicists of all times even if he had not written a single line on relativity." In fact, Einstein received the Nobel Prize for 1921 for such *Gelegenheitsarbeit,* namely, the theory of the photoelectric effect.

In his later years Einstein turned his attention more and more to the object, methods, and limits of science. In exercising these rights, namely to pursue the philosophy of science as a scientist, Einstein was completely in step with the trends that had been set by late nineteenth-century investigators and that were being perpetuated with vigor, if not always with logical rigor, by the scientists who belonged to his generation. In his 1936 essay on physics and reality, Einstein tells us why it is not right for the physicist to let the philosopher take over the philosophy of science, especially at a time when the very foundations of science are problematic. He says: "The physicist cannot simply surrender to the philosopher the critical contemplation of the theoretical foundations: For he himself knows best, and feels more surely, where the shoe pinches." For Einstein, the philosophy of science definitely called for an in-depth knowledge of science.

On the other hand, Einstein by no means assumed that the narrow scientific specialist was qualified as a philosopher of science:

> The whole of science is nothing more than a refinement of everyday thinking. It is for this reason that the critical thinking of the physicist cannot possibly be restricted to the examination of the concepts of his own field. He cannot proceed without considering critically a much more difficult problem, the problem of analyzing the nature of everyday thinking.

Certainly here is a viewpoint that mirrors one of the central themes of Mach and the nineteenth-century critical positivists.

As already stated, our main objective is to search out such self-reflective aspects of Einstein's career as may shed light on the conception he had of himself as a philosopher of science. Before doing so, however, I would like to offer an explanation for approaching the subject in the way here indicated. It is advisable to be open and honest about one's methodology and specifically to mention at this point that the self-image study this approach entails has its own intrinsic complexities. We cannot take the time to outline them here. Suffice it to say that not the least vexatious of the difficulties encountered is that scientists, including Einstein, in analyzing their own motivations and methodological directives, are apt to construct self-images that conform to what their scientific communities expect of them. Thus, in an attempt to become philosophical and sophisticated about these matters, scientists are prone to fulfill the prophecy of philosophical climates of opinion. These may relate to such factors as a hierarchy within the sciences vis-à-vis theory and experiment, master–pupil relationships, schools of thought, centers of research activity, and so on.

So we might as well acknowlege explicitly and candidly that there are some severe limitations imposed upon the investigator who chooses this approach, namely, to focus on what scientists *say* they are doing when they claim to be engaged in science, rather than analyzing more single-mindedly their published scientific contributions in order to discover *what they do* when they claim to be engaged in science. In relation to this issue, Einstein once said:

> If you want to find out anything from the theoretical physicists about the methods they use, I advise you to stick closely to one principle: don't listen to their words, fix your attention on their deeds. To him who is a discoverer in this field, the products of his imagination appear so necessary and natural that he regards them, and would like to have them regarded by others, not as creations of thought but as given realities.

If Einstein has suggested here that one should not listen to what scientists *say* they do, but rather look at their *works* in order to learn *what* they do, he also confessed in his autobiography (or in his obituary as he called it): "The essential in the being of a man of my type lies precisely in what he thinks, not in what he does or suffers." We discover that over the years, Einstein, as so many other scientists, surrendered to the temptation to reify his own methodological preferences into a credo that guided him in all of his work – at least that is what he seems to want to tell us. But it is not that simple.

The point that needs to be stressed in advance, with these remarks, is that what a scientist *really* does, if we may speak that way, is not revealed to the historian of science unambiguously, either by an analysis of the scientist's reflective and retrospective account of what is going on, or by an examination of the finished, formal, published, product. In my opinion, anything that contributes to the clarification of the methodological question about how science advances, or retrogresses, is fair game for the historian.

Suffice it to say that one way to search out the self-image of Einstein as a philosopher of science, and to discover the way in which he conceives of his own work and thought within the context of the scientific currents of his times, is to listen seriously to what he has to say as he reflects on these matters in so many of his essays and lectures. Besides, the historian can take advantage of Einstein, so to speak, by invading his more unbuttoned, private, and internal life, to examine the uninhibited outpourings of his soul as revealed in the correspondence and informal interchanges with his most intimate friends and invisible opponents. Although this invasion may not be quite fair to a man like Einstein, since he undoubtedly never intended to add these documents to the historical record, they in fact substantially help to answer the questions that have been posed here.

I want to assert that Einstein had two self-images of himself and his work. The self-image that dominated his early career may be characterized by an attraction to critical positivism and the empirical status of theories advocated by Ernst Mach. The other more mature, more consciously worked out self-image of Einstein, and the one I want to talk about here, was one in which Mach's sensationalism and pluralism were abandoned and replaced by a realistic, unitary, and deterministic world view that lays claim to the intuitive recognition, or near-recognition, of rock-bottom truths about nature. Concerning his mature position Einstein wrote: "My epistemological credo . . . actually evolved only much later (in life), and very slowly, and does not correspond with the point of view I held in younger years."

We might mention here, parenthetically, that Einstein's dualistic image confronts us with a paradox: Virtually all of his most creative and lasting achievements were made while he was under the influence of a philosophy that he later categorically rejected. Or had he not thought through the consequences of his philosophy for his science? In fact, why does Einstein not struggle with the question of the influence of his own philosophy of science on his scientific work?

To analyze with psychological insight and historical credibility the many reflective accounts of Einstein that reveal something substantive about his self-image as a philosopher of science is an undertaking that would be far too ambitious on this occasion. Therefore we are confronted with the more modest objective of examining the way in which Einstein was prodded into explaining his philosophical position by two of his closest colleagues and critics, namely, Arnold Sommerfeld and Max Born. In both cases we have at our disposal a substantial portion of correspondence and intellectual interchange that covers a period of almost forty years.

Sommerfeld and Einstein both were enthusiastically committed to the technical mastery and critical evaluation of everything that transpired in the intellectual realm of relativity and quantum mechanics during the revolutionary era of physics from 1900 to 1930. However, no two persons could have followed the shifting scientific scenario from more diverse perspectives. We learn that Einstein, the philosopher, with cool detachment, was attracted to general overarching unitary principles and through the years became increasingly impatient with, and even hostile toward, quantum mechanics with all of its outlandish baggage of indeterminacy, statistical and probability functions, and discontinuity: He simply felt that the future of physics lay more in geometry, and therefore in continuum theory, than in particles. Intellectually independent, he continued for decades to puzzle deeply about scientific questions that most physicists had accepted as self-evident. Despite his tremendous scientific contributions, he had no school or pupils or close disciples.

By contrast, Sommerfeld, the unphilosophically disposed master of broad domains in theoretical physics, ten years older than Einstein, surrounded by an energetic and productive school of disciples in Munich, became a staunch supporter of the revolutionary quantum trends. He managed, with his unique mathematical dexterity, and his facility with intuitively clever mechanical models, to squeeze out and exploit subtle implications hidden beneath the basic principles that had been laid down by other investigators. We may add, that in the process, he formulated new problems eminently worthy of being explored on their own merits. Sommerfeld was an early enthusiast for both relativity and quantum theory. We shall concentrate on the Sommerfeld-Einstein discussions about quantum theory, because they demonstrate most convincingly the distinctive philosophy that Einstein generated over the years. Einstein became increasingly confident that the failure to provide a unitary continuum field theory that would encompass both macro and microphenomena provided proof positive that the quantum theorists were on the wrong track. All this, in spite of his own tremendous contributions to early quantum theory.

It was one of Einstein's early papers, the revolutionary 1905 hypotheses on light quanta, that brought him in contact with Sommerfeld. They first met in Salzburg in 1909 at the Society for Natural Scientists and Physicians, where Einstein lectured on the new quantum ideas. The next year Sommerfeld traveled to Zürich to spend a week in discussions with Einstein. At the first Solvay conference in 1911, Sommerfeld explored the theoretically exciting idea that the existence of the molecule was to be taken as a function and result of the elementary quantum of action *h*, and not vice versa, as had been argued.

Sommerfeld early on was stirred to action by Einstein's deduction from quantum principles about vanishing heat capacities at the absolute zero of temperature. He also was encouraged by the experimental support for the quantum theory being provided by the low-temperature heat capacity measurements conducted by Nernst and his colleagues in Berlin. Sommerfeld did his best to get into the act in 1912 by requesting from Einstein an in-principle clarification of quantum ideas. Unfortunately for Sommerfeld, Einstein was largely preoccupied with gravitational theory, in spite of the fact that he did not manage to attract much attention to this work from his colleagues. It was *not* Einstein's views on relativity, but rather quantum mechanics, that was the topic of lusty debates.

In 1916, Einstein wrote to Sommerfeld;

> You must not be angry with me that I have not answered your interesting and friendly letter until now. During the last month I have experienced one of the most exciting and trying, and certainly one of the most successful times of my life.

What follows in the letter is a discussion of some of the germinal ideas and consequences of Einstein's general theory of relativity. Somewhat late, in 1916, while commenting favorably on Sommerfeld's spectral investigations and successful extension of Bohr's theory of the atom, Einstein remarked: "If only I could know which little screws the Lord God is using here." This remark I interpret to mean something like this: It is rather inconceivable that the real world is like that, namely, that atoms are quantized; but if it should turn out that the world *is* so constructed, then I must ask, is it not a bit undignified for God to have to use little screws to run the world that way?

Neither disturbed by Einstein's cavalier disregard of what was going on among quantum theorists, nor overly sensitive about the fundamental theoretical or philosophical rationale behind it all, Sommerfeld continued courageously to work out the mathematical formalism of the modified Bohr theory with great finesse and virtuousity. Obviously impressed, Einstein responded in 1918: "If only it were possible to clarify the principles about quanta! But my hope in being able to experience that is steadily diminishing." What Einstein had been trying to show, but unsuccessfully, was that particles can be treated as stable regions of high concentration of the field.

Dubious about the direction in which quantum theory was moving, Einstein believed, by 1918, that general relativity, by contrast, was an accomplished theory. Thus he wrote:

> Behind general relativity henceforth there is nothing new to be found. In principle all has been said: Identity of inertia and mass; the metrical proportion of matter (geometry and kinematics) determined by the mutual action of bodies: and the nonexistence of independent properties of space. In principle, thereby, all has been said.

In this domain, Einstein was very certain that he had uncovered real physical truth about nature. In a letter to Sommerfeld in 1921, concerning a small supplementary addition to relativity theory that he and Hermann Weyl had published, he wrote: "I have my doubts about whether this thing has any physical worth. God makes it as he wills, and does not allow something to be put over on him." When asked to lecture on relativity, Einstein remarked that he had nothing new to say, and added: "The old stuff already is whistled by all the younger sparrows from the roof tops better than I can do it."

In 1920 Sommerfeld succeeded in explaining the multiplicity of many of the spectral lines by introducing an inner quantum number that had no physical meaning for him. "I can only further the technique of quanta," he wrote to Einstein, "you must construct their philosophy." Beginning with the work of Sommerfeld's pupil Heisenberg, in the summer of 1925, and promoted by the dramatic and ingenious contributions of Born, Jordan, Dirac, Schrödinger, Bohr, and Pauli, the elaboration of quantum theory was approached from quite different directions. It was given a formalism and mathematical structure that represents one of the most magnificent theoretical and practical accomplishments in the history of science. Much has been written about this subject and I only will mention here that in the outcome two opposing camps were created that divorced the enthusiasts for the Heisenberg-Born matrix mechanics – Born, Jordan, Dirac, Hund, and Pauli – from the supporters of the Schrödinger wave mechanics, for example, de Broglie, Planck, and Einstein. We have here the physics of the discrete versus the physics of continua.

Actually, Einstein essentially alienated himself from the direction in which quantum mechanics was headed, but felt moved now and then to take an occasional pot shot at the whole enterprise. Sommerfeld, typically engrossed in anything that would result in a practically useful and theoretically sound outcome, and philosophically uncommitted, stood outside the debate, but continued to elicit reactions from Einstein that at times revealed more about his (Einstein's) native intuitions and deep convictions that can be learned from studying his scientific papers.

In 1926 Einstein wrote to Sommerfeld:

> I have worried a great deal about searching out the relationship between gravitation and electromagnetism, but now am convinced that everything that has been done in this direction by me and others has been sterile . . . The theories of Heisenberg and Dirac, in fact, force me to admiration, but they do not smell of reality.

Or again, in another letter:

> The results of Schrödinger's theory make a great impression, and yet I do not know whether it deals with anything more than the old quantum rule, i.e., about something with an aspect of real phenomena.

Concerning Sommerfeld's monograph of 1930 on wave mechanics, Einstein said, in the same vein, that it was very nice, but that in spite of tremendous successes accomplished, the whole development and the prevailing trends did not satisfy him.

After 1930, as we well know, scientific investigations and communications suffered miserably in Germany. Research and discussion groups were splintered so severely that Sommerfeld in a reminiscent mood in 1937, wrote to Einstein (by then in Princeton) that he was consoling himself for having been able to experience personally the golden age of physics from 1905 to 1930. A decade later Sommerfeld was curious to know whether Einstein had changed his views about quantum theory.

> Perhaps you will tell me what you *now* think about continua and discontinua. Or do you take the situation to be hopeless?

Einstein replied:

> I still believe in all earnesty that the clarification of the basis of physics will come forth from the continuum [i.e., not quantum mechanics] because the discontinuum provides no possibility for a relativistic representation of action at a distance.

It is a fact that physicists more and more came to be preoccupied with the problems of quantum theory. It promised a better immediate yield. This did not deter Einstein (and others like Weyl and Eddington) from regarding reality as a continuous singularity-free manifold and from exploring a unified field theory on the model of general relativity. It was to include the laws of electromagnetism as well as those of gravitational fields.

In 1949 Einstein was seventy years old. In that year he believed that he had found the solution for which he strove for thirty years. The work was published by Princeton University Press as a new edition of *The Meaning of Relativity*.

In a tribute for Einstein's seventieth birthday, Sommerfeld, commenting about Einstein's outstanding contributions to the field of atomic theory, remarked:

> In spite of all this, in the old question "continuum versus discontinuity" Einstein has taken his position most decisively on the side of the continuum. Everything of the nature of quanta – to which, in the final analysis, the material atoms and the elementary particles

belong also – he would like to derive from the continuum physics by means of methods which relate to his general theory of relativity. . . .

His unceasing efforts, since he resides in America have been directed toward this end. Until now, however, they have led to no tangible success. . . . By far the most of today's physicists consider Einstein's aims as unachievable, and consequently aim to get along with the dualism: wave-corpuscle, which he himself first clearly uncovered.

In December of 1951 Einstein wrote to his lifelong friend, Michele Besso:

All these fifty years of conscious brooding have brought me no nearer to the question: What are light quanta? Nowadays every Tom, Dick and Harry thinks he knows it, but he is mistaken.

I feel that one of the most significant aspects of Einstein's attempts to formulate a unified field theory was the ease with which he rejected his own theories that did not work out. When that happened he blithely took up another approach. He did this until he died in 1955.

In 1951 Sommerfeld died at the age of eighty-three, thus terminating the discussions between the philosopher-physicist Einstein and the no-nonsense master of physics, Sommerfeld, who claimed no expertise at all in the philosophy of science but who had been anxious to exchange ideas with a colleague whose philosophy he respected.

In contrast to the picture we have sketched of Sommerfeld in Munich, as the philosophically neutral correspondent of Einstein, we have at our disposal the long-standing scientific interchange of ideas between Einstein and another physicist who was himself passionately inclined to philosophize about relativity theory and quantum mechanics at the slightest provocation. This was Max Born, in Göttingen, the physicist whose completion of Einstein's statistical interpretation of quantum theory earned him the Nobel Prize twenty-eight years after it was presented. So here in Born and Einstein we have two would-be philosophers of science wrestling intellectually with one another.

The philosophical views of Einstein and Born invariably were 180 degrees out of phase on the subject of quantum mechanics. Accordingly, an examination of their intellectual debates is all the more important because of Born's relentless efforts to entice Einstein, the independent and relatively isolated thinker, to explain and defend his position as he moved around the world and took up new positions in Prague, Zürich, Berlin, and Princeton.

Leopold Infeld tells us how Max Born first learned about Einstein's revolutionary paper – the early one on special relativity – a thirty-page work that bore the modest title, "On the Electrodynamics of Moving Bodies."

When Professor Loria met Professor Max Born at a physics meeting in 1908, he told him about Einstein and asked Born if he had read the paper. It turned out that neither Born nor anyone else there had heard about Einstein. They went to the library, took from the bookshelves the seventeenth volume of *Annalen der Physik* and started to read Einstein's article. Immediately Max Born recognized its greatness and also the necessity for formal generalizations. Later, Born's own work on relativity theory became one of the most important early contributions to this field of science. Thus it was not before 1908 or 1909 that the attention of great numbers of scientists were drawn to Einstein's results.

As in the case of Sommerfeld, Born first met Einstein in Salzburg in 1909. Born characterizes the young Einstein, up through the early 1920s, as an empiricist and enthusiast for the philosophy of Hume, Mach, and Schlick. But, already in 1919, when Einstein was first ruminating about a unitary field theory that would bring gravitation and electromagnetic theory together, he was expressing a degree of discomfort about the developments in quantum mechanics. The theorists operate, he wrote, as though "the one hand is not allowed to know what the other does."

Basically at odds with the upsurge of the idea of discontinuity in physics, Einstein wrote to Born in 1920: "I do not believe that the quantum can be detached from the continuum. By analogy one could have supposed that general relativity should be forced to abandon its co-ordinate system." Einstein also was unhappy about what seemed to him the failure of the strict law of causality in quantum mechanics and the simultaneous encroachment of statistical and probability arguments. There is no doubt that by 1920, Einstein sought to hold tenaciously to continuum theory, in hopes that quantum phenomena would be absorbed somehow into the differential equations of a unified field theory.

In the 1920s and 1930s, Einstein was preoccupied mostly with general relativity. He was clarifying and perfecting its theoretical exposition and pursuing its practical consequences with great determination. But he wrote to Born that in his spare time he was "brooding . . . over the quantum problem from the point of view of relativity" because, as he said "I do not believe that (quantum) theory will be able to dispense with the continuum."

In February of 1929 in newspapers all over the world it was announced that Einstein had formulated his unified field theory in which the phenomena of electricity and magnetism were combined in a single set of equations. These ideas were reformulated and refined for twenty-five years. But during all of those years, until the last major attempt in 1949, Einstein was compelled, periodically, to admit that he was getting nowhere with his *Lieblingsidee* (viz.; the continuum) despite all attempts to analyze the issues. Now and again over the years, he felt, and announced, that he had achieved at least the glimpse of a reconciliation between relativity and quantum theory under the umbrella of continuum ideas, but these hopes were shattered one after another either by himself or others. In his letters to Born we come to see how often and how deeply Einstein was distressed about the conception of a wave–particle duality for radiation – a view he could not embrace except as a temporary crutch devoid of physical reality. In 1924 Einstein confided to Born:

> My attempts to give the quantum a tangible form . . . have been wrecked time and again, but I am nowhere close to giving up hope. And if nothing works, there still remains the consolation that the failure is my fault.

In truth, the state of quantum theory in the early 1920s was one of considerable confusion. For example, there were the negative correlations with the Bohr-Sommerfeld rules. Attempts to connect quantum theory with classical mechanics were not successful. Qualitatively things worked out tolerably well, but the quantitative predictions were not impressive. Max Born certainly recognized very clearly that many technical difficulties simply escaped resolution. He referred to this state of affairs as *das Quatenrätzel* and in 1921 wrote to Einstein: "The quanta are a hopeless Schweinerei" – as he expressed himself. As already mentioned, from 1925 to 1930 we witness a series of dramatic and bold moves that reveal that the negative results of current quantum theory pushed investigators in the direction of making a sharper break with classical mechanics. This simultaneously provided a new quantum mechanics, much to the consternation of Einstein.

After 1925, Einstein and Born carried on a running commentary characterized by hard arguments in which neither could convince the other to change perspectives. Commenting about this interchange some forty years later, Born wrote:

> Einstein was fairly convinced that physics provides knowledge about the objective existence of the external world. But I, along with many other physicists gradually was converted, by experience in the domain of atomic quantum, to realize that it is not so – but rather that at every point in time we have no more than a rough approximate knowledge of the external world and that from this, according to specified rules of the probability laws of quantum mechanics, we can draw some conclusions about the unknown world.

In response to singular achievements in quantum mechanics by Born's Göttingen group (Heisenberg, Jordan, and Hund) Einstein could only respond: "Your quantum mechanics commands much attention, but an inner voice tells me that it is not yet the true Jacob. The theory offers much, but it brings us no closer to the secrets of the old one [der Alte]. In any case, I am convinced that he does not play dice." As Einstein said it: "Gott würfelt nicht." When Einstein spelled out some of the details of his attempt to establish a quantum field theory [i.e., continuum theory], Born wrote back politely that it was very interesting but not convincing. This was tit for tat.

In a letter of 1944 to Born we have a compelling illustration of Einstein's mature image of his own philosophy of science. It demonstrates convincingly how two talented scientists can be worlds apart in their interpretations of the same cognitive subject matter. Einstein writes:

> In our scientific expectations you and I have reached antipodal positions. You believe in a God who throws dice, and I believe in complete lawfulness, *viz.* in a world of something that exists objectively and that I have attempted to snatch in a wild speculative way. I believe firmly, but I hope that a more realistic way, and especially that a more tangible evidence will be found than *I* was able to discover. The great initial success of the quantum theory cannot bring me to believe in the fundamental nature of a dice-throwing God, even if I know that my younger colleagues interpret this position of mine as the result of calcification. Someday it will be known which instinctive conception was the right one.

Born responded by saying that Einstein's expression about a dice-throwing God was totally inadequate:

> In your determined world, you must throw dice too – that is not the difference . . . First of all, you underestimate the empirical basis of quantum theory . . . and second, you have a

philosophy that somehow brings the automaton of dead things in accord with the existence of responsibility and conscience.

Einstein at this point could do no better than say (1947) that he was sorry to discover that "I just cannot manage to express my position so that you will find it to be intelligible"; and then he adds the comment that the mathematical difficulties involved in trying to reach his objectives of a comprehensive unitary theory are so severe that,

I will bite the dust before I get there . . . But concerning this I am convinced – that eventually we shall land a theory in which law-like things will not be probabilities for facts – facts such as formerly were just taken for granted. But to prove this conclusion I have no logical reasons.

And so the debate wore on and on. Einstein called Born a positivist. Born said that that was the last thing he wanted to be called by anyone. Einstein to Born: Don't you believe in the reality of the external world? Born to Einstein: Don't dodge the real issue: Do you really maintain the quantum mechanics is a fraudulent affair? Einstein to Born: You talk about the philosophy of quantum mechanics, but your remarks in essence are not philosophy at all but the manipulation under the cloak of indeterminacy to a hidden machinery of reasoning. Born to Einstein: Your position is one of metaphysics and not philosophy. That was the tone of the intellectual interchange.

We see that Einstein had formulated his own image of what the philosophy of science should be and what it should accomplish, and so had Born. Nevertheless, neither Einstein nor Born felt that they were being successful in communicating what that image was. Or were they just stubborn? At one point Wolfgang Pauli entered the debate and managed to convince Born that they had not so much disagreed, as argued from basically different premises. But it is clear to me that Pauli, in fact, also had constructed his own image of the philosophical positions that Einstein and Born represented.

I want to suggest that when Einstein died in 1955, he was holding in firm grasp essentially the world view that he had formulated in the 1920s and 1930s. What was this world view? In his lecture on the theory of relativity at King's College, London, in 1921, Einstein said:

The theory of relativity may indeed be said to have put a sort of finishing touch to the mighty intellectual edifice of Maxwell and Lorentz, inasmuch as it seems to extend field physics to all phenomena, gravitation included . . . I am anxious to draw attention to the fact that . . . (the theory of relativity) is not speculative in origin; it owes its invention entirely to the desire to make physical theory fit observed fact as well as possible. We have here no revolutionary act but the natural continuation of a line that can be traced through centuries. The abandonment of certain notions connected with space, time, and motion hitherto treated as fundamentals, must not be regarded as arbitrary, but only as conditioned by observed facts.

Here is a plug for the theoretical soundness and fertility of the great accomplishments of the nineteenth century.

A decade later, in an essay on the problems of space, ether, and fields, Einstein wrote:

The theory of relativity is a fine example of the fundamental character of the modern development of theoretical science. The initial hypothesis becomes steadily more abstract and more remote from experience. On the other hand, it gets nearer to the grand aim of science, which is to cover the greatest possible number of empirical facts by logical deduction from the smallest number of hypotheses or axioms. Meanwhile, the train of thought leading from the axioms to the empirical facts or verifiable consequences gets steadily longer and more subtle.

According to Einstein the theoretical scientist is compelled in an increasing degree to be guided by purely mathematical, formal considerations in the search for a theory, because the physical experience of the experimenter cannot lead him up to the regions of highest abstraction. This is the line of thought, he says, that led from the special to the general theory of relativity and hence to its latest offshoot, the unified field theory. That unified theory, however, never came within Einstein's grasp, as he freely admitted toward the end of his life.

In the early 1930s, in an essay on the methods of theoretical physics, Einstein raised the question whether we can ever hope to find the *right* way – seeing as he believed that the axiomatic basis of theoretical physics cannot be extracted from experience but must be freely invented. In other words, has this *right* way any existence outside our illusions? Einstein's position is unequivocal:

I answer without hesitation that there is in my opinion, a *right* way, and that we are capable of finding it. Our experience hitherto justifies us in believing that nature is the realization of the simplest conceivable mathematical ideas. I am convinced that we can discover by means of purely mathematical constructions the concepts and the laws connecting them with each other, which furnish the key to the understanding of natural phenomena. Experience may suggest the appropriate mathematical concepts, but they most certainly cannot be deduced from it. Experience remains, of course, the sole criterion of the physical utility of a mathematical construction. But the creative principle resides in mathematics. In a certain sense, therefore, I hold it true that pure thought can grasp reality, as the ancients dreamed.

I believe it fair to say that Einstein's native epistemological credo comes through with remarkable consistency over the period of his last thirty to thirty-five years. I would mention first his characterization of God as a mathematician; or since Einstein did not believe in a personal God, we might better express his position by saying that natural phenomena can only be understood in depth, and natural laws can only be formulated successfully in the language of mathematics. Perhaps it is appropriate to mention in this context that Einstein's attitude changed considerably as he became increasingly preoccupied with relativity theory. Early in his career he displayed a far more skeptical attitude toward the role of mathematics in physics. Like Mach, who influenced him at that time, he must have felt that the abstract formalisms of mathematics were too closely allied with metaphysics and thus might disguise the deep *physical* significance of natural phenomena. He left such views behind when he came to recognize that his goal of achieving a more general and unitary representation of the world necessarily rested far more on sophisticated formal mathematical models than on physical terms and intuition.

Second, we see that Einstein had placed himself firmly on the side of those investigators of the classical period of physics who had expressed an unshaken faith in the ultimate conceptual unity of the physical world. In fact, in this regard he was exploring an old theme expressed cogently by d'Alembert in 1715 when he wrote: "To someone who can grasp the universe from one unified viewpoint, the entire creation would appear as a unique fact and a great truth." According to Einstein, a correct or right unitary theory of natural phenomena was conceivable and feasible, and scientists, he believed, were making steady progress in achieving that right unitary theory. Imbedded in this conception of a right theory is the belief that unambiguous progress had been achieved in moving toward the goal of constructing (discovering) a real picture, or physical representation of phenomena, that corresponds with the way things really are in nature. The right theory was equated with existence. Implied, of course, was also the conviction that the right theory is unique and not merely one of a plurality of alternative theories that might be constructed to do the job equally well.

For Einstein there was not only a right, correct, unique theory to explain the cosmos, but this theory was seen to be within our grasp. He said: "The Lord God is subtle but he is not malicious or vicious." ("Raffiniert ist der Herr Gott aber boshaft ist er nicht.") His God was a cosmic God, the God of Spinoza on a sublimated plane.

> It was not a personal God, who makes notes on whether a person behaves or misbehaves, but a cosmic God, who represents the all pervading intellect which manifests itself so marvelously in the Creation.

It was man's special mission to lay bare the marvels of that Creation.

A third point. The right unitary theory upon which Einstein placed all of his stakes, is seen to rest on a mathematical foundation that deals with fields (continua), and *not* quanta, that is, not discontinua. He felt, as he once put it,

> so long as no one has new concepts, which have sufficient constructive power, mere doubt remains; this is unfortunately my own position. Adhering to the continuum originates with me not as a prejudice, but arises out of the fact that I have been unable to think up anything organic to take its place. How is one to conserve four-dimensionality in essence (or in near approximation) and [at the same time] surrender the continuum?

For the same reasons that Einstein rejected quanta, he also put aside statistical or probabilistic arguments – because he believed that strict determinism was not to be sacrificed or even weakened.

In Einstein's fifty-year-long battle over the interpretation of quantum mechanics, one theme recurred again and again: his instinctive dislike of the idea of a probabilistic universe in which the behavior of individual atoms depends on chance. Was it likely that God would have created a probabilistic universe? Einstein felt that the answer must be no. If God was capable of creating a universe in which scientists could discern scientific laws, then God was capable of creating a universe wholly governed by such laws. He would not have created a universe in which he had to make chance-like decisions at every moment regarding the behavior of every individual particle. This was not something that Einstein could prove. It was a matter of faith and feeling and intuition. Perhaps it seems naive. But it was deep-rooted, and Einstein's physical intuition, though not infallible, had certainly stood him in good stead.

So Einstein's overall aim was a field theory that would encompass macro and micromechanics, or gravitation, electromagnetism, radiation, and atomistics including all aspects of science that pertain to the ultimate constituents of matter and their interactions at all levels. This grandiose, ambitious, and all-encompassing *Weltbild* was not one that Einstein was able to achieve despite more than twenty years of writing on the subject. Louis de Broglie summarized the position that most theoretical physicists took toward the end of Einstein's life when he commented that there exists

> a fundamental difference between gravitational and electromagnetic fields which does not allow an extension to the latter of the geometrical interpretation which succeeded in respect to the former. . . . Moreover, the nature of the electromagnetic field is so intimately bound to the existence of quantum phenomena that any non-quantum unified theory is necessarily incomplete. These are problems of formidable complexity whose solution is still in the lap of the gods.

Finally, Einstein believed, that however abstract and remote from experience the mathematical formalism of theory turned out to be, the investigator nevertheless could use experience to suggest appropriate concepts. Although the concepts themselves could not be deduced from experience, experience was still acknowledged as the sole criterion of the physical utility of the theory. That is, in the end, it was absolutely crucial that the physical theory fit the empirical facts.

To these four landmarks of Einstein's image of what he considered to be a correct philosophy of science, namely, his own, I suppose we might want to add that he obviously held in high regard the importance of philosophizing about the foundational analysis, critique, and reformation of the basic concepts that lie at the heart of a right and realistic scientific world view. How closely Einstein's image of his own views about physics borders on philosophy may be seen in the way that philosophers of science have continued to discuss his philosophy perhaps even more than physicists. I therefore would not hesitate for one moment to assert that Einstein, in a special sense, was a truly important modern philosopher and that his highly individualistic philosophical ideas and self-image have exerted an influence that, as Professor Dirac has emphasized, has changed the course of history.

We have done no more here than sketch some facets of Einstein's self-image and his conception of himself as a philosopher of science. I suggest that a parallel examination of, say, Niels Bohr or Boltzmann would land us in an entirely different ballpark. But then, there should be no doubt that Niels Bohr's views and approach to physics as well as Boltzmann's changed the course of history.

Einstein was neither systematic philosopher nor analytical philosopher. He was not given over to concerns about symbolic logic or the syntax of language or the construction of a metalanguage. He was a scientist's philosopher in the tradition of Helmholtz, Mach, Duhem, Planck, Poincaré, and Boltzmann. In spite of being worthy of being called the most revolutionary physicist of the twentieth century, he was at heart very much at home with the great scientific systems of the nineteenth century. He was, in fact, the natural philosopher par excellence. His revolutionary ideas are deeply imbedded in the three magnificent theoretical monuments of classic nineteenth-century thought, namely, mechanics, thermodynamics, and electromagnetic theory.

I do not believe that we know today whether nature is fundamentally simple and governed deep down by overarching simple general laws. Personally, philosophically, intuitively, I doubt it. That is, I doubt that scientists will ever reach rock bottom in such a way that no deeper digging is possible. To me, simplicity seems a myth and the lure of completeness deceptive. But having said this, I would want to add that whether nature is fundamentally simple or not, it probably is wise for investigators (if they want to get on with their work) to act as if it is. It is in this sense that I see Einstein's work as a guide and stimulus to scientific productivity and inventiveness.

Let me conclude with a quote from an article that Einstein wrote in 1918 for Planck's sixtieth birthday. It rather captures the spirit of Einstein's philosophy of science.

> The supreme task of the physicist is to arrive at those universal elementary laws from which the cosmos can be built up by pure deduction. There is no logical path to these laws; only intuition, resting on sympathetic understanding can lead to them. . . . The state of mind that enables a man to do work of this kind is akin to that of the religious worshiper or the lover; the daily effort comes from no deliberate intention or program, but straight from the heart.

BIBLIOGRAPHY

Born, Max. *Ausgewählte Abhandlungen,* 2 Bände, hrsg. von der Akademie der Wissenschaften in Göttingen (Göttingen: Vandenhoek & Ruprecht, 1963).

Einstein, Albert. *Mein Weltbild,* hrsg. Carl Seelig (Vienna: Ullstein, 1980). First published in 1934.

Einstein, Albert. *Out of My Later Years* (New York: Philosophical Library, 1950). German translation: Stuttgart: Deutsche Verlags-Anstalt, 1952.

Einstein, Albert/Arnold Sommerfeld, *Briefwechsel: Sechzig Briefe aus dem goldenen Zeitalter der Physik,* hrsg. und kommentiert von Armin Hermann (Basel: Schwabe Verlag, 1968).

Einstein, Albert/Hedwig und Max Born. *Briefwechsel 1916–1955: Kommentiert von Max Born* (Munich: Nymphenburger Verlagshandlung, 1969).

Schilpp, Paul Arthur (ed.). *Albert Einstein: Philosopher-Scientist* (Evanston, Ill.: Library of Living Philosophers, 1949).

Seelig, Carl. *Albert Einstein: Eine dokumentärische Biographie* (Zürich: Europa Verlag, 1954).

PART II

The eighteenth-century tradition

10

The Paracelsians in eighteenth-century France

A Renaissance tradition in the Age of the Enlightenment

ALLEN G. DEBUS
University of Chicago

In the past two decades there has been an ever-increasing interest in the role played by the followers of Paracelsus during the Scientific Revolution. Their chemical philosophy combined alchemical, chemical, and medical concepts in a universal scheme of nature that seemed to many to be a viable "new science" or "new philosophy." These Paracelsians carried on a widely trumpeted and spirited debate with Aristotelians, Galenists, and mechanists alike until well into the third quarter of the seventeenth century. But what happened after that time? Since they were seldom referred to in the eighteenth-century scientific journals it is sometimes suggested that they faded away with the triumph of the mechanical philosophy. In this essay we hope to show that this was not the case.

For the most part historical research on eighteenth-century chemistry has focused on Lavoisier and the background to the chemical revolution. Such an approach may be warranted if we confine ourselves to the origins of modern chemical theory and to the views of those authors who were associated with the established scientific academies. However, once we venture beyond their works, we find other chemical texts that confirm the existence of a persistent interest in alchemy, natural magic, and Paracelsian medical chemistry. Books on these subjects were published throughout the century. Indeed, if we examine the two major bibliographies of alchemical and early chemical texts (John Ferguson, *Bibliotheca Chemica* [2 vols., Glasgow: Maclehose and Sons, 1906]; Denis I. Duveen, *Bibliotheca Alchemica et Chemica* [London: Dawsons, 1949]) we find well over five hundred eighteenth-century titles of this type in the one and three hundred in the other. Many of these were new editions of texts dating from earlier centuries and we are indebted to eighteenth-century editors and their publishers for some of the most important collected editions of the alchemical classics. But there were also an impressive number of new works written, and

This paper was written for the present festschrift in 1977 and 1978. Due to unavoidable delays in the publication of this volume the author published it first in *Ambix, 28* (1981), 36–54. The present text is an updated version of the *Ambix* paper and is reprinted by permission. The research for this paper was supported in part by NIH Grant LM 03014 from the National Library of Medicine. The author is grateful also for the help received from John Neu in the Department of Rare Books at the Memorial Library of the University of Wisconsin (Madison) during a visit in the final stages of research.

these testify to the industry of authors who sought to identify themselves with the chemical philosophy of the Renaissance.

The overwhelming majority of these eighteenth-century alchemical texts derive from Central Europe (approximately 81 percent of the Ferguson titles and 57 percent of the Duveen titles are in German), but a preliminary study of the Paracelsian and mystical chemical works of this period is best directed at the French texts both because of the emphasis that has always been placed upon the French Enlightenment and also because of a new interest among scholars in a connection between the growth of occultism and the end of the "Age of Reason."[1]

Because of the number of works available this cannot claim to be an exhaustive study of the eighteenth-century French Paracelsian, iatrochemical, and alchemical texts. However, an attempt has been made to cover the century chronologically and to indicate with selected texts something of the breadth of interests expressed by a group of chemical authors whose works have not yet been integrated into the history of science or the history of medicine. It will be seen that this important Renaissance interpretation of nature survived throughout the century down to the Romantic period.

A PARACELSIAN VIEW OF THE BATTLE BETWEEN THE ANCIENTS AND THE MODERNS

Nowhere had the followers of Paracelsus caused more debate than in France.[2] As early as 1566 the Medical Faculty in Paris had condemned the internal use of antimony, but this did not stop the preparation of numerous Paracelsian translations and syntheses. The earliest were prepared by Pierre Hassard, who translated the *Grossen Wundartzney* (1567), and Jacques Gohory, who compiled a *Compendium* of Paracelsian philosophy and medicine (1567), but these were only a taste of what was to come. A strong critique of the views of Paracelsus on chemical remedies and the origin of metals (1575) offered Joseph Duchesne the opportunity to prepare a *Responsio* (1575) that proved to be the first of many works by him written in defense of Hermetic and Paracelsian medicine. Roch le Baillif's Paracelsian *Le demosterion* (1578) resulted in a trial in which the author sought to defend his unorthodox medical views. This was to no avail since he was ordered to leave the capital and return to his native Brittany (1579).

The outcome of the trial of le Baillif was surely a Galenist victory, but it did nothing to halt the growing interest in the medical views of the Paracelsian chemists. Numerous new works were made available by French publishers in the following decades, and when Joseph Duchesne penned a lengthy defense of Hermetic medicine in 1603 he caused a confrontation with the Parisian medical establishment that soon spread beyond the confines of France. His book was immediately condemned by Jean Riolan and other Galenists. Duchesne and his colleague, Theodore Turquet de Mayerne, then replied in the opening salvos of a battle of polemical works that was to last for decades. So fundamental did this dispute seem that summaries and histories appeared early for the benefit of the European physicians. A number of the works were translated into German, French, and English.

Within France there were far-reaching overtones. One of these involved a growing tension between the medical schools of Montpellier and Paris. Most of the chemical physicians with formal medical training came from Montpellier while the Galenist stronghold was surely Paris. Again, this debate led to a conflict between the followers of the Hermetic philosophy and the early French mechanists. This is best seen in the official condemnation of fourteen alchemical theses by the doctors of the Sorbonne in 1624,[3] and in the relentless attack on the alchemical cosmology of Robert Fludd by Marin Mersenne and Pierre Gassendi.[4] Nor was this debate devoid of political overtones. The chemist, Théophraste Renaudot, a graduate of Montpellier, was supported in his medical projects by Cardinal Richelieu, who saw in them an opportunity to diminish the power of the Galenic Parisian faculty of medicine (c. 1638–42).[5]

These debates testify to the intense interest in alchemy and the chemical philosophy to be found in France in the first half of the seventeenth century. Nor did this lessen in the last half of the century, a period we are accustomed to think of as dominated by Cartesian mechanism. In addition to many new works on chemistry and chemical medicine published by French authors, there were translations of the latest texts by Jean Baptiste van Helmont (1670) and Johann Rudolph Glauber (1659, 1674).[6] Even Descartes had early been attracted to the Rosicrucians and some of his late seventeenth-century adherents sought to interpret his work in mystical and chemical contexts rather than in mechanical terms.[7] Surely the establishment of the Académie des Sciences and the ever-increasing interest in the works of Descartes seemed to do little to lessen the publication of more traditional works on chemistry and chemical medicine.

With this background we need not be too surprised to learn that the familiar late seventeenth-century "battle of the ancients and the moderns" is much too complex to interpret simply as a struggle between Aristotelians and mechanists. In England the debate between John Webster and Seth Ward is understood best as a confrontation of the chemical and the mechanical philosophies.[8] This interpretation of the "battle" was also expressed in France. Evidence for this may be found in an anonymous text of 1697, *Le Parnasse assiegé ou La guerre declarée entre les Philosophes Anciens & Modernes.* The author's intent is clearly expressed in the preface:

> Dans l'idée que l'on s'est proposé de faire assieger le Parnasse par les Philosophes, on prétend demontrer la realité de la Science d'Hermes, & la verité de la Médicine de Paracelce.[9]

The plot is simple; Apollo, god of the sun and of the healing arts, has died on Mount Parnassus. This event seems to each philosopher to be an opportunity to assert his primacy over all others.[10] The mountain need only be climbed and the throne seized. But lack of success on the part of any one philosopher (or sect) to dominate the others leads to the abandonment of this civil war and the philosophers join together to assault the mountain in unison. There follows a list of the various philosophers and their place in this unusual army. At one side are the academicians with units commanded by Plato, Epicurus, and the various Ionic philosophers.[11] Closer to the mountain are to be found the followers of Gassendi and Descartes, who discover roads that seem to lead to the top.[12] Even Confucius and other Chinese philosophers are present and demand a proper place for the attack.[13]

Dissension arises within the ranks when Aristotle is appointed the Prince of Philosophers. Diogenes and Descartes object loudly, but this does not delay the continuing preparations for the assault.[14] Galileo is placed in charge of the cavalry, and Cardan and Porta are appointed to lead the artillery, while the infantry is to be commanded by Parmenides, Heraclitus, Democritus, and others.[15] Descartes commands the dragoons and his lieutenants include Mersenne, Regius, le Grand, Rohault, Boyle, De la Boë (Sylvius), and other friends. Surprisingly we find that the chemical physicians Daniel Sennert and Jean Baptiste van Helmont have been placed in charge of the baggage.[16]

But now four spies (the alchemists Jean Chortolasse [Johann Grasshof], Arnald of Villanova, Hortulanus, and Basil Valentine) inform the assembled army that the mountain is nearly inaccessible and open only to philosophers of the school of Hermes. The officers of Hermes carry a standard marked FRC (Fraternity of the Rosy Cross). After a lengthy discussion the spies disappear.[17]

In a new strategem various groups unsuccessfully try to penetrate the mists leading to the summit. Among a group of vivandiers we have a glimpse of *Harvée porta des oeufs*[18] while a group of chemists (including Libavius and Glauber) are forced to return to camp after losing their way.[19]

Another spy (Geber) is caught. He informs the leaders that there are many defenders at the top, philosophers who are guided by reason and truth. These are men who have been taught by Hermes, the father of all knowledge.[20] This statement enrages Galen,[21] but he is stilled by the announcement that a prisoner of great consequence has just been captured. This is no less than Paracelsus, who had injured the members of his escort and was using such strong language that some called for his immediate death. Aristotle then called for his interrogation thus permitting Paracelsus to defend himself.[22]

It was from Paracelsus that the leaders of the attacking army were to learn the names of the principal defenders of the summit. In the first rank were to be found Moses, Solomon, Roger Bacon, Nicholas Flamel, Hippocrates, Basil Valentine, Ramon Lull, Arnald of Villanova, and others while behind them stood such worthy figures as Joseph Duchesne, Gerhard Dorn, Roch le Baillif, Agrippa von Nettesheym, Oswald Crollius, Robert Fludd, Heinrich Khunrath, and Michael Maier.[23]

Unexpectedly a copy of Paracelsus's own *Archidoxes magica* is found and condemned to the fire. Paracelsus is to be given a reprieve from his own fate only if he agrees to show the philosopher-warriors the road by which they might avoid the mists and clouds that shield the summit from those below. He agrees to this and begins to lead Andreas Laurentius (du Laurens) up the slope by hand. But the latter is not worthy of his charge and he falls to the ground in the darkness that symbolizes his own ignorance. Paracelsus, a true champion of truth, continues on to join his comrades at the top, leaving behind the bickering philosophers, who represent every modern and ancient philosophical sect except the true one.[24]

PARACELSIAN MEDICINE IN A NEW CENTURY

The new fantasy outlined in *Le Parnasse* may serve as a point of entry to the eighteenth-century literature. Surely the author of *Le Parnasse* could still view Paracelsus and the alchemists as the primary guides to scientific and medieval advance. He was not alone in his conviction for we find that the French publishers offered their public numerous chemical texts that reflected the entire spectrum of views characterized by the Renaissance chemical philosophy. So great was the interest in this subject at this time that the reknowned chemist, E. F. Geoffroy, presented a paper to the Royal Academy of Sciences on the deceits of the alchemists (1722).[25]

Among the older texts to be reprinted early in the century were Jean Colleson's *L'Idée Parfaite de la Philosophie Hermétique* (first edition 1630; reprinted 1719) and Joachim Polemann's *Novum lumen Medicum de Mysterio Sulphuris Philosophorum* (first edition 1647; first edition in French, 1721). The former discussed the philosopher's stone and the principles of nature, the latter expounded van Helmont's views on the sulphur of the philosophers.

French authors also produced new works in the same vein. Thus Jean Le Pelletier revealed the secret of *L'Alkaest ou le dissolvant universel de Van-Helmont* (1704) whereas Francois Pousse offered an *Examen des principes des Alchymistes sur la Pierre Philosophale* (1711). Alexandre Le Crom discussued the salt of the philosophers, which he identified as the true quintessence in his alchemical text, *Plusiers Experiences utiles et curieuses. Concernant la Medecine, la Metallique, l'Oeconomique & autres Curiosités . . .* (1718). An anonymous author writing from London complained of Ben Jonson's *The Alchemist,* which made Ramon Lull and Basil Valentine appear ridiculous.[26] He argued that the search for a noncorrosive dissolvant for gold and silver was as essential for the advance of medicine as was the search among mathematicians for a correct method for the determination of longitudes.[27]

Works that bridged the gap between the world of the theoretical alchemist and the practical chemist include Charles Le Breton's *Les Clefs de la Philosophie Spagyrique . . .* (1722) and an anonymous *Traité de Chymie, Philosophique et Hermetique* (1725). Both emphasize the preparation of real chemicals having medical value. This was in the tradition of the seventeenth-century chemical textbooks of Beguin, Le Fèvre, Davisson, Glaser, and Lemery. Indeed, Le Fèvre's popular *Cours de Chymie* (1660) was reprinted at Paris as late as 1751 with much new material[28] and Nicolas Lemery's *Cours de Chymie* (1675) witnessed many greatly expanded eighteenth-century editions including sixteen editions in French printed between 1701 and 1757.

This continued interest in chemical tradition is also evident at the university level. The identification of the University of Montpellier with chemical medicine may be traced back to the sixteenth century and throughout the seventeenth century French Paracelsists and Helmontians were associated with this institution. For this reason alone we need not be unduly surprised to find the vitalistic school of French medicine originating at Montpellier rather than elsewhere in the mid-eighteenth century.[29]

At Montpellier the chemical textbook tradition was carried on and defended by enthusiastic physicians. Thus, when Paul Jacques Malouin (1701–77), physician to the queen and professor of chemistry at the Jardin du roi, was criticized for his interpretation of the history of this science, he was defended by an anonymous author identified only as a "physician of Montpellier."[30] Similar to Malouin's very practical book of chemical preparations were the lectures given by A. Fizes at Montpellier.[31]

A key figure in the development of chemical medicine at Montpellier in the late seventeenth century had been Raymond Vieussens (1635 – 1715) who had been taught by the Galenist, Lazarus Riverius, but who had also assimilated the iatrochemical views of Sylvius and Willis. Vieussens's *Tractatus duo* (1688) had presented a chemical interpretation of physiology based on fermentation as a fundamental explanatory device.[32] But although Vieussens later moved away from chemical means of explanation toward a mechanical interpretation of bodily processes, he continued to support others who favored chemical medicine.

Antoine Deidier (d. 1746), physician to the king and Royal Professor of Chemistry at Montpellier, was one who received a complimentary letter from Vieussens.[33] He expressed his gratitude to him in his *Chimie raissonée* (1715), a collected volume of Deidier's lectures on chemistry that had been given at Montpellier. The names of Paracelsus, Glauber, and other seventeenth-century figures appear frequently, and, in addition to the inevitable chemical preparations there is to be found an extensive discussion of the chemical principles that Deidier contrasted with the Cartesian principles of matter.[34] His discussion of the particulate nature of matter reflects the late seventeenth-century chemical interest in the shape of atoms as a cause for their characteristic chemical action. Of special interest is Deidier's reference to the aerial niter, which had been described first by Paracelsus and then by late seventeenth-century authors such as Robert Boyle, Robert Hooke, and John Mayow. After noting the healing properties of aerial niter, Deider referred to those who explained that the animal spirits were formed through the volatilization of this substance by fermentation in the blood.[35] A second work, his *Institutiones Medicinae* (1711) presented a chemical introduction to physiology. Here, too, Deider began his account with a lengthy discussion of the chemical principles of matter.[36]

THE PARACELSIAN MEDICINE OF JOSEPH CHAMBON AND FRANÇOIS MARIE POMPÉE COLONNE

Although it has been customary for historians of chemistry and medicine to emphasize the work of Stahl, Boerhaave, and Geoffroy in the early eighteenth century, it is evident that French chemists also had available to them a broad spectrum of books relating to alchemy and to chemical medicine. Here it is of special importance to note that several authors were primarily influenced by Paracelsus. It is in the medical publications of Joseph Chambon (1647–c.1733) and François Colonne (c. 1649–1726) that we may best witness the survival of Paracelsian medicine into the eighteenth century. For if Vieussens, Deidier, Lemery, and Malouin reflect the late seventeenth-century chemistry of Willis, Boyle, and Sylvius, Chambon and Colonne urged the adoption of an earlier form of iatrochemistry.

Even the life of Chambon is reminiscent of the wandering antiestablishment followers of Paracelsus of earlier centuries.[37] Born at Grignon in 1647, he studied medicine at Aix, where he received his doctorate. He practiced at Marseilles until a quarrel forced him to leave the city; he then traveled to Italy, Germany, and Poland, where he became physician to the king, Jan Sobieski. At the siege of Vienna (1683) Chambon left the royal service to confer with Paracelsian and Helmontian physicians in the Low Countries. From there he traveled to Paris where he was well received by Fagon, the physician to Louis XIV, but not by the Faculty of Medicine, whose members objected both to his practice and to his choice of medicines. Hoping to bypass the medical establishment, Chambon obtained an *arrêt du parlement'* which authorized him to practice with the grade of *licencié.* He proceeded to build a successful practice until his involvement in politics, which resulted in imprisonment in the Bastille for two years. After his release he returned to Grignon for the remainder of his long life.

Chambon's reputation ultimately rested on two publications, his *Principes de physique* (1711 [new edition], reprinted 1714 and 1750) and his *Traité des metaux* (1714, reprinted 1750), works that have been largely ignored by historians of medicine and science. Chambon was distressed by the fact that the level of medical science was so low compared to that of the other sciences.[38] The slow progress of medicine was all the more evident in a period when astronomy and mathematics were rapidly changing. He understood that this was partially due to the difficult nature of medicine, but even after making this allowance, the fact remained that the basic principles of medicine had not yet been discovered. Part of the blame was to be ascribed to the emphasis in medical schools on the study of human anatomy through the dissection of cadavers. Chambon argued that we must replace this with the study of the human body as a living whole.[39]

We know that the rules of mathematics are infallible; the rules of medicine should be as well.[40] But how should one proceed? The ancients said that one must travel in order to learn and Chambon agreed that this was essential. He had no desire to rest on book learning alone so for "eight years I went to study medicine in foreign countries."[41] The result was his firm conviction that the advance of medicine was dependent on its connection with chemistry. For Chambon, Paracelsus was "le plus grand de tous les hommes" and it would have been much better if people believed that he had spoken seriously.[42]

True religion, a knowledge of nature, and the healing art are all interrelated.

This truth is unknown to the academic physicians who follow the books of Galen and Hippocrates and their commentators. Therefore,

Il faut renoncer à Hypocrate & Galien, il faut renoncer à ces Philosophes spéculatifs, qui n'ont acquis ce nom, que par des sophismes, des argumens, & dont les spéculations imaginaires nous écartent du bon chemin, mais il ne faut jamais renoncer à soi-même.[43]

The true physician should know that it is not the books of the ancients that teach us, but rather that

La prudence & la simplicité sont les véritables guides, pour développer les mysteres de la nature; c'est par elles, que nous devons nous laisser conduire: ce ne sera même qu'avec ces secours, que nous parviendrons à la connoissance de la nature, & par la nature, que nous deviendrons profés en Médecine.[44]

For Chambon there was a very special importance to be found in the study of chemistry.

Je me suis donc attaché à ces Sciences & principalement à la Chymie. Si l'Anatomie nous apprend à connoître les ressorts de la machine qui fait l'objet de la Medecine, la Chymie nous donne beaucoup de lumieres sur les liqueurs qui nourissent ces ressorts. D'ailleurs, ce que ne fait point l'Anatomie, la Chymie nous administre des remedes, qui sont seuls capables de guerir une infinité de maladies.[45]

Even van Helmont would have applauded Chambon's insistence that to properly "pénétrer dans les véritables connoissances de la nature, le Philosophe & le bon Medicin n'ont besoin que du feu; ils naissent du feu, ils se perfectionnent avec le feu, & pratiquent le feu: *In igne, cum igne & per ignem.*"[46] With this conviction we need not be surprised to find that Chambon explained physiological processes in chemical terms. Thus "la digestion, ou la transmutation qui se fait dans l'estomach, est le premier des ouvrages du petit monde."[47]

Chambon's emphasis on the need for personal experience conflicted somewhat with his desire to understand the difficult works of the alchemical authorities whom he frequently cited. Indeed, Chambon felt the need to present to the reader a set of *Regles naturelles* that must be understood prior to delving further into the study of medicine. The chemical orientation of these rules is evident from the outset.

La Nature fait tous ses ouvrages en dissolvant & en coagulant.
Lorsqu'elle dissout, elle réincrude, & lorsqu'elle coagulle, elle cuit & murit.
Il y a un esprit ou un feu caché dans tous les corps de la nature.
Cet esprit ou ce feu est, pour ainsi dire, l'ame de chaque corps, qui est toujours en mouvement.[48]

The spiritual cause of movement tends either to formation or destruction and it is this that one calls fermentation, a process that leads to the separation of the pure from the impure.[49] The reader is reminded of van Helmont in Chambon's belief that each body perfects its seed and that this is required for its own generation. This is as true for the mineral world as it is for animals and vegetables.[50]

Above all Chambon insisted on the essential nature of the three Paracelsian principles, salt, sulphur, and mercury, of which all bodies are composed.[51] Each of the principles operates differently in the matter in which it resides. The chemist is directed to the properties of color, odor, taste, liquidity, solidity, and weight since these are signs by which they may be distinguished. For the physician it is essential to know that since there are only three principles there can only be three fundamental sicknesses: there is a "maladie du sel, du souphre, & du mercure." How different this is from the teachings of the schools, "Quel echec pour les Bibliotheques de Médecine . . ."[52]

But beyond the principles, Chambon believed in a fundamental *prima materia.* With van Helmont he was convinced that this was water. I believe

avec quelques Philosophes, que l'eau est le principe de toutes choses, & que toutes choses sont faites d'eau; que le soleil est la source & le center des eaux, & que tout le Monde est composeé de cette même matiere; le mouvement & le brillant qui paroît dans la matiere dont le soleil est composé, ressemble si fort à un or qui se purifie, qu'un Philosophe l'a regardé comme un or en coupelle, & les parties qui composent le soleil, ou cette même matiere dans son tout, comme une cire que sert à former tous les differens ouvrages de la Nature, qui ne different entre-eux qu'en ce que les parties des uns sont en repos, & les autres en mouvement . . . [53]

There was an essential unity to be found in nature: "L'homme, le ciel & la terre étant une même chose, faut-il estre surpris, s'il y a de l'accord entre'eux: *Coelum est moderator omnis sanitatis, morbi, veneni, boni & mali usque ad mortem.*"[54]

The books of Chambon are lengthy, but they read well. His frequent digressions often recount personal experiences. Thus the story of his meeting with a group of Spanish monks leads to a discussion of the differences between true religion and superstition.[55] Again, his critique of Descartes sheds light on the way in which a Paracelsian viewed the then dominant form of the mechanical philosophy.[56] And, again as a Paracelsian, Chambon discussed and compared at length the growth of metals with the stony deposits in the body resulting from "tartaric" diseases.[57] In the tradition of the alchemists he discoursed on the similarity between the calcination of metals and their subsequent recovery through reduction with the death and resurrection of Jesus Christ.[58] But above all, Joseph Chambon was presenting to a new century an approach to chemical medicine that differed little from that proposed by earlier followers of van Helmont.

The fact that these works were to be reprinted as late as 1750 attests to the interest in this approach to the subject in the midst of the Enlightenment.

The Paracelsian interests we find in the work of Chambon are even more explicit in the work of François Marie Pompée Colonne (c. 1649–1726). We know little of the life of this author beyond the fact that he died in the flames of his house and that a student (Gosmond) prepared several of his manuscripts for the press and answered one of his master's critics, the Reverend Father Castel.

Like Joseph Chambon, Colonne published very little until late in life. There is an *Introduction à la philosophie des anciens, par un amateur de la verité* that appeared in 1698, but then there was complete silence for a quarter century. His *Les Secrets les plus cachés* (1722, reprinted 1762), *Les Principes de la nature* (1725), *Suite des Experiences utiles* (1725), a work on geomancy (1726), and possibly a few additional texts appearing under the name of Le Crom were published shortly prior to his death. There appeared posthumously his *Principes de la nature ou de la génération* (1731) and a multivolume *Histoire naturelle de l'univers* (1734).

Although all of these texts are of importance for a fuller understanding of French chemistry in the early eighteenth century, the work of most concern for us is Colonne's *Abrégé de la Doctrine de la Paracelse et de ses Archidoxes* (1724). In this substantial volume of five hundred pages he is unequivocal in his praise: "Je dirai donc que parmi les Modernes, Paracelse semble avoir surpassé tous ses Prédécesseurs; & qu'avec raison il s'est attribué le titre, de *Monarque des Arcanes.*"[59] For Colonne this title is well deserved since Paracelsus established his doctrine on "raisons phisique & palpables sans se servir de ces énigmes inintelligibles qui font tourner la tête plûtôt que d'instruire."[60] But Paracelsus was also a true physician and he has presented to his readers the rules for the preparation of all sorts of medicines, "lesquels remedes, ou dumoins une grande partie, sont également bons soit pour la santé, soit pour la perfection des métaux."[61] The key to this great accomplishment was to be found in Paracelsus's *Archidoxes,*[62] which Colonne prepared for the reader through an abridgment that reflected his own interpretation.

Colonne began with his own introductory explanation of the principles of chemistry. Chemistry is described as the art of resolving mixed substances and of separating the pure from the impure.[63] In his discussion of the principles, Colonne accepted the five most commonly accepted in the late seventeenth century, the Paracelsian salt, sulphur, and mercury, plus phlegm (water) and caput mortuum (earth) derived from the four elements of Aristotle.[64] These elements, he argued, are true and certain since they are based upon laboratory experience.

We may not be too surprised to find that Colonne proceeded to discuss the traditional four qualities; heat, dryness, cold, and humidity. However, for him these differ from the visible elements. Rather they are the small invisible parts of them. Thus river water

> n'est pas proprement ce qu'on appele *la qualite humide;* mais il faut comprehende, que ce qu'un appelle *qualite* c'est la vapeur la plus subtile, ou si vous voulez la plus petite particule d'icelle, & dont un nombre innombrables de ces particules jointes ensemble forment les gouttes de l'eau sensible.[65]

Similarly, fiery flame is very different from the pure quality of heat. The latter consists of most subtle and mobile ethereal particles.[66]

It did not seem useless to Colonne to speculate on the shapes of the elementary particles. In so doing one might compare the views of the Cartesians and the chemists.

> Et on peut, si l'on veut, imaginer les figures que l'on voudra dans ces particules qui composent les qualités, & au lieu de trois sortes d'élemens que les Cartesiens supposent l'une trés-subtile, l'autre trés grossier, & un autre moyen, ou peut mettre quatre degrez differens étant au fond la même chose; puisque les trois élemens des Cartesiens & leurs particules ne sont pas absolument égalles, ni en substance, ni en figure, ni en vitesse de mouvement.[67]

Having prepared the reader through this primer of chemical theory, Colonne then proceeded to discuss the *Archidoxes*. He presented to the reader what might be termed a "commentary-abridgment" of the *Archidoxes* and then added two additional works on alchemy. For him, as for earlier alchemists, the true chemist should be able to apply his knowledge no less to the imperfect metals than to human ills. The macrocosm–microcosm universe assured the operator that a cure for the one would succeed also for the other. One could be assured that the slow natural transmutational growth of the the imperfect metals leading to gold could be hastened. For this reason Colonne discussed at great length the growth process of metals from the seed,[68] and he was clearly convinced of the possibility of transmutation in the laboratory.

FRENCH ALCHEMY AT THE HEIGHT OF THE ENLIGHTENMENT

The middle decades of the eighteenth century were to witness a series of important French publications on alchemy.[69] Pierre Brodin de la Jutais argued that the traditional texts actually described the fertilization of the soil rather than the transmutation of the metals (*L'Abondance, ou Véritable Pierre Philosophale* [1752]).[70] This agricultural interpretation may be contrasted with a more customary proof of *L'Existence de la Pierre Merveilleuse Des Philosophes* by Etienne César Rigaud (1765).[71] These and other monographic texts are interesting, but even more important are the multivolume studies that appeared during these years.

The three-volume *Histoire de la Philosophie Hermétique* by the Abbé Nicholas Lenglet du Fresnoy (1674–1752/5) appeared first in 1742 and was then reprinted twice in 1744.[72] Lenglet du Fresnoy thought it necessary to write such a history because it had not been done earlier.[73] However, this did not mean that he believed in the truth of the claims of the alchemists.

> Il faut remarquer qu'il y a deux sortes de *Chimie;* l'une sage, raisonnable, nécessaire même pour tirer des remedes utile de tous les êtres de la nature . . . : l'autre est cette Chimie folle & insensée, & cependant la plus ancienne des deux. . . . La premier a conservé la nom de *Chimie,* & l'on a donné à la seconde celui d'*Alchimie.*[74]

Accordingly, Lenglet du Fresnoy devoted himself to the composition of a history "de la plus grande folie, & de la plus grande sagesse, dont les hommes soient capables."[75] The first volume is a history of alchemy; the second contains a group of French translations; and the work closes with a catalog of book titles including a list of chemical papers to be found in the *Memoires* of the French Academy and the *Transactions* of the Royal Society of London.

Lenglet du Fresnoy's *Histoire* appeared at the same time as the most extensive collection of alchemical texts ever printed in French, Jean Maugin de Richebourg's *Bibliotheque des Philosophes Chimiques* (3 vols., 1740; reprinted 1741; a fourth volume published in 1754). Here are to be found thirty-three French texts plus an abridged dictionary of alchemical terms. The first two volumes had been edited in the seventeenth century (1672, 1678) by William Salmon,[76] but the fact that this enlarged edition appeared in the mid-eighteenth century attests to the interest in this subject at that time. We are assured here that this subject forms no part of the history of human error as Lenglet du Fresnoy suggested. Rather, in a 145-page preface, Richebourg (or Salmon) explained that his task was "établir la vérité de las Science, ou de l'Art, qui enseigne à faire Transmutation, & d'en faire voir la possibilité."[77] Indeed, his definition of chemistry was "qui enseigne à faire la Transmutation." A history of the art was, in essence, a recital of past transmutations.[78]

The midcentury French adept was soon able to supplement these works with the publications of Antoine-Joseph Pernety (1716–1800/1), the royal librarian of Frederick II at Berlin. His *Les Fables égyptiennes et grecques devoilées* (2 vols.; 1758; reprinted, 1758, 1786, 1795) maintained an earlier tradition in its thesis that the whole of ancient mythology was an alchemical allegory. But an additional work seemed necessary.

> Mon Traité des Fables Egyptiennes & Grecques développe une partie de ces mysteres. De l'obligation dans laquelle j'étois de parler le langage des Philosophes, il en est résulté une obscurité qu'on ne peut dissiper que par une explication particulere des termes qu'ils employent, & des métaphores qui leur sont si familieres. La forme de Dictionnaire m'a paru la meilleure, avec d'autant plus de raison qu'il y peut servir de Table raisonnée, par les renvois que j'ai eu soin d'inférer, quand il a été question d'éclairer des fables déja expliqées.[79]

For this reason Pernety prepared his *Dictionnaire Mytho-Hermétique* (1758; reprinted 1787), which explained the chemical significance of the allegorical terms used in the art in a volume nearly six hundred pages in length. Nor did he stop at the traditional terminology. Since

> beaucoup de gens regardent la Médecine Paracelsique comme une branche de la Science Hermétique; & Paracelse son auteur ayant, comme les Disciples d'Hermès, fait usage de termes barbares, ou pris des autres langues, j'ai cru rendre service au Public d'en donner l'explication suivant le sens dans lequel ils ont été entendus par Martin Ruland, Johnson, Planiscampi, Becker, Blanchard & plusiers autres.[80]

The first edition (1758) of Pernety's extensive alchemical dictionary was published only eight years prior to the first edition of Pierre Joseph Macquer's *Dictionnaire de Chymie* (1766), which has been praised as "the first scientific work of its class." The second edition (1787) appeared the same year as Lavoisier's fundamental revision of chemical nomenclature.

The man of the Enlightenment may have been confused by the simultaneous presentation of the alchemical classics simultaneously with chemical papers that were to lead to the chemical revolution, and he may have been made more uncertain of the goals of the chemists if he read the article on "Chymie" in Diderot's *Encyclopédie* (1753). Although unsigned, this lengthy paper was prepared by Gabriel François Venel (1723–75), who became professor of chemistry at Montpellier in 1759. Venel deplored the fact that chemistry was so little studied by scientists in his day. "Les Chimistes forment encore un peuple distinct, trés-peu nombreu, ayant sa langue, ses lois, ses mysteres, & vivant presque isolé au milieu d'un grand peuple peu curieux de son commerce, n'attendant presque rien de son industrie."[81]

If there were alchemical charlatans and those who busied themselves only with the preparation of chemical medicines, there were also great systematizers such as Johann Joachim Becher and Georg Ernst Stahl. Indeed, Venel felt that the study of chemistry was essential for all of the sciences. It was necessary that a

nouveau Paracelse vienne avancer courageusement, que *toutes les erreurs qui ont défiguré la Physique sont provenues de cette unique source; savoir que des hommes ignorant la* Chimie, *se sont donné les airs de philosopher & de rendre raison des choses naturelles, que la* Chimie, *unique fondement de toute la Physique, étoit seule en droit d'expliquer,* &c. comme Jean Keill l'a dit en propres termes de la Géométrie, & comme M. Desaguliers vient de le repétér dans la *préface* de son cours de Physique expérimentale.[82]

Venel sought to rectify the neglect of the greatest chemists through a history of chemistry and this forms an important part of his article. Here he pointed to the achievements of van Helmont, Glauber, Becher, and Stahl. The mechanists were clearly of less interest to him. Rather, we note once again his fascination with Paracelsus.

Paracelse est un des plus singuliers personnages que nous présente l'histoire littéraire: visionnaire, superstitieux, crédule, crapuleux, entêté des chimeres de l'Astrologie, de la cabale, de la magie, de toutes les sciences occultes; mais hardi, présomptueux, enthousiaste, fanatique, extraordinaire en tout, ayant sû se donner, éminemment le relief d'homme passionné pour l'étude de son art (il avoit voyagé à ce dessein, consultant les savans, les ignorans, les femmelettes, les barbiers, &c.) & s'arrogeant le singulier titre de prince de la Medecine, & de Monarque des Arcanes, &c. Il a été l'auteur de la plus grande révolution qui ait changé la face de la Medecine . . . & il a fait en *Chimie* la même figure qu'Aristote a fait en Philosophie.[83]

And even though his writings are "absolument inintelligibles," "quel que soit le mérite réel de Paracelse, il est évident que c'est à lui qu'est dûe la propagation & la perpétuité de la *Chimie.*"[84] Here Venel was referring once more to the chemical medicines.

JOYAND'S PARACELSIAN CRITIQUE OF MESMER

The closing decades of the century witnessed an increasing interest in the occult sciences. André Charles Cailleau claimed that he wrote not for the "treasure seekers," but for those who believed that God would aid them in their quest.[85] Of special interest is his discussion of the books (all French translations) that convinced him of the truth of alchemy.[86] An anonymous *L'Art Hermetique . . .* discoursed on the wonders of magical mysteries and alchemical wonders;[87] *Le Grand Livre de la Nature* explained occult philosophy, the Rosicrucians, the transmutation of the base metals, and the communication of man with God through angelic intermediaries. We are assured that tout ce qu'on lit dans *Paracelse, Van-Helmont, Raimond, Lulle, Glauber, Trevisan, Swedenborg, &c.*, n'est point un effet de leur erreur, ni de l'imposture: c'est donc dans ces écrivains qu'il faut chercher les préceptes des sciences occultes.[88]

Also deriving from Renaissance tradition is Pierre Joseph Buc'hoz's book of secrets in which he devoted much space to chemical information.[89]

However, there is no doubt that the three most famous figures associated with alchemy and the occult sciences were the Comte de Saint Germain (d. 1784); Guiseppe Balsamo, Comte de Cagliostro (1743–95); and Franz Anton Mesmer (1734–1815). Robert Darnton notes that the interest in the pseudosciences

carried Parisians into the territory of occultism, which has bordered on science since the Middle Ages. Cagliostro was only the most famous of the many alchemists (L.-C.) Mercier found in Paris. Street vendors hawked engravings of the Comte de Saint Germain, "célébre alchimiste," and booksellers displayed alchemist works like *Discours philosophiques sur les trois principes animal, végétal & minéral; ou la suite de la clef qui ouvre les portes du sanctuaire philosophique* by Claude Chevalier.[90]

An eyewitness described the alchemical activities of Cagliostro in Poland in 1780. In a day-by-day account we read of the count's deceptions. On June 12 the author exclaimed: "Ma patience est à bout; chaque jour de nouvelles bêtises & impostures." Finally exposed, Cagliostro fled Warsaw the night of June 27.[91]

The mysterious Comte de Saint Germain had no more devoted a disciple than the occultist Alliette, who signed his works Etteilla. He disputed a rumor of Saint Germain's death and announced that the Count would be in Paris in July 1785.[92] Alliette was an inexhaustible pamphleteer who wrote on tarot cards, divination, talismans, numerology, free masonry, medicine, astrology, and alchemy. Saint Germain had been an authority on all of these, but Alliette stated that he too had devoted thirty-three years to their study and he believed that he would not die without proving to all Europe that the universal medicine, the transmutation of copper into gold, and other alchemical wonders were true and possible for all adepts.[93]

The increased interest in the occult was accompanied by the growth – or revived interest in – numerous secret societies, the most notable of which were the Freemasons and the Rosicrucians. Several religious and prophetical works ascribed to Paracelsus were published in the final decades of the century,[94] but his name appears more prominently – and in a scientific context – in the controversy over mesmerism. Franz Anton Mesmer was convinced that the key to the sciences was to be found in the existence of an ethereal fluid that existed throughout the universe. This he thought could adequately explain the phenomena of magnetism, light, heat, electricity, and gravity. But he went far beyond this in the application of his "magnetic" fluid to medicine. Here he argued that all sickness was due to obstructions in the flow of the fluid within the human body. The body itself acted like a magnet and it was the physician's task to reinforce a natural polarity so that the flow of this fluid would be returned to normal. The proper flow resulted in the restoration of health and harmony between the patient and nature. Mesmer was capable of producing convulsive "crises" in his Parisian salons and these were convincing to many contemporary observers.

Seeking scientific and medical approval for his discovery, Mesmer approached both the Academy of Sciences and the Royal Society of Medicine. He gained a few prominent adherents (above all, Charles Deslon), but the end result was disastrous. A royal commission was appointed to investigate mesmerism. Including scientists as well known as Franklin, Lavoisier, and Bailly, the members of the commission made a thorough study and concluded that "Mesmer's fluid did not exist, the convulsions and other effects of mesmerizing could be attributed to the overheated imaginations of the mesmerists" (1784).[95]

This result led to a torrent of publications both defending and attacking Mesmer. Of these we will turn only to the works of M. Joyand, who signed his works as "Dr. en Médecine de la Faculté de Besançon, Médecine de l'Hôspital militaire de Brest." His major work is the *Précis du Siècle de Paracelse* (Paris, 1787).[96] This is the first volume (and the only one published) of a two-volume work that was planned to show the Paracelsian origin of Mesmer's thought. Nearly seven hundred and fifty pages in length, the *Précis* represents the most detailed study of its sort surviving from the eighteenth century.

Rather than analyze Joyand's major work, we will touch here only on his short *Lettre sur le siècle de Paracelse* (Paris, 1786), a pamphlet in which he discussed the reasons why he prepared the larger study. In effect, this *Lettre* served as an advertisement for his forthcoming *Précis.*

Joyand's account reflects mixed admiration and disappointment in Mesmer. He was clearly impressed by Mesmer's description of the universal fluid and animal magnetism. He was aware of the great discoveries of the past century in the physical sciences due to the work of Descartes, Newton, and Leibniz. The work of Mesmer seemed to him to be of the same order of magnitude, but the Viennese physician had not published a proper explanation of his theory and he had in effect denied himself the honor he deserved.[97]

Joyand had been in Paris when Mesmer arrived from Vienna in 1778, but he had left for Besançon in 1780 before this new system of medicine had become well known. Two years leater he learned of Deslon's expulsion from the Faculty of Medicine because of his conversion to mesmerism and of Deslon's subsequent challenge to the medical establishment.[98] It was at that time that Joyand sought to learn more of Mesmer's theories and system of cure. As yet he had had no correspondence with Mesmer or with any of his disciples, but he was already engaged in research of Paracelsus and he soon noticed strong similarities between the two systems.

By 1784 Joyand had completed his own work on Paracelsus and he visited Paris where he had a conversation with Mesmer. He clearly expected to be named a member of Mesmer's Société de l'Harmonie Universelle, but he was not accepted because he did not offer to pay the required initiation fee of one hundred louis.[99] Clearly distressed by this, it seemed to him that a pecuniary interest had been attached to a scientific discovery. "Je me propose aussitôt de publier le Siècle de Paracelse, et en montrant les découvertes qui appartiennent à Paracelse, d'obliger M. Mesmer à ne plus tenir secret ce que. les progrès de la physique."[100] But he hesitated since he realized that if he did publish his work he might appear envious of Mesmer.

At this point Joyand contacted Deslon, who encouraged him and invited him to observe the effects of animal magnetism.[101] Deslon promised that he would be immediately initiated into the Société but he soon saw that Deslon could add little by way of explanation:

Je n'entendes que la théorie fait vague *d'un fluide qui agit en se communiquant d'un animal à un autre,* comme le dit M. Mesmer dans ses propositions; et cette idée est la moindre de toutes celles de Paracelse.[102]

The seance Joyand attended was occupied almost entirely by practical matters related to various articles and conditions imposed on the membership by Deslon. None of the secrets Joyand sought were divulged, and he departed wondering whether Deslon actually knew anything more than the little Mesmer had published.[103]

In June 1784, a friend urged Joyand to publish his work on Paracelsus and the following month he lectured on the *Précis* at Besançon to the doyen of the university and other physicians of the city.[104] Here he discussed the views of the ancient philosophers on the system of the universe and their doctrine of life. He then went on to the doctrines of the alchemists and particularly the work of Paracelsus "où, d'l'occasion de l'un des deux agens magnétiques désignés par lui, je dis: *Si c'est-là la vertu opposée positive dont a voulu parler M. Mesmer, il ne s'est pas trompé.*"[105] He also presented a new theory of chronic illness and of other maladies characterized by acute pain, shivering, and fever as well as a mechanism of contagion.

Joyand's *Lettre* concluded with a summary of the two volumes of the forthcoming *Précis.* The unpublished second volume is of special interest since it was to cover those "monumens que attestent que tous ces principes étoient connus des anciens. Ils ont été la base de l'alchimie. Développés principalement dans Paracelse; excepte la loi des révolutions célestes, dont il s'est fort peu occupé."[106]

CONCLUSION

Historians have ignored Joyand's critique of Mesmer and the names of Chambon, Colonne, Pernety, and Lenglet du Fresnoy seldom appear outside histories of the occult sciences. Even the chemists associated with Montpellier have never been studied in detail. Perhaps the only exception here is Venel and the reason for this is understandable: The "Paracelsian" orientation of his article on "Chymie" seems quite out of place with the rest of the *Encyclopédie.*

But in fact, Venel was not alone. His hope for a "new Paracelsus" reflects the views of many eighteenth-century chemists. To relegate this to "popular science" or to an undercurrent of occultism is a gross oversimplification. In fact, this was a vigorous continuation of the Renaissance chemical philosophy. The fact that these authors and their works played such a small role in the establishment science of the period may be at least partially explained by the effect of the organization of science in the seventeenth century. Surely one result of the triumph of the mechanical philosophy had been to exclude the proponents of rival sects from the new scientific academies.

But if contemporary academicians ignored them, should we also? I think not. There is no doubt that alchemy and Paracelsian medicine continued to attract adherents throughout the eighteenth century. The increased interest in a mystical explanation of nature at the end of the century was based on a long tradition and did not arise suddenly as a phoenix from its ashes. These works are also of interest for our understanding of the background to early nineteenth-century science. At that time there was a critical debate between the proponents of the rather mystical *Naturphilosophie* and a science that was becoming ever more dominated by mathematics. This debate might be studied profitably in terms of the similarities it shows with the seventeenth-century confrontation of chemists and mechanists.

Perhaps we may look for a still greater significance. The persistent eighteenth-century inquiries into alchemy and Paracelsian chemical medicine may well lead us to a new model for the development of the sciences. Rather than seeking a continued march of progress to the present we might do better to unravel the continuing debate between the mathematical, the observational, and the experimental components on the one hand, and the spiritual, the mystical, and the religious on the other.

NOTES

1 A provocative discussion of the French scene is to be found in Robert Darnton, *Mesmerism and the End of the Enlightenment in France* (Cambridge, Mass.: Harvard University Press, 1968). For Central Europe see the recently translated edition of Henri Brunschwig's *Enlightenment and Romanticism in Eighteenth Century Prussia,* trans. Frank Jellinek (1st ed. in French, 1947; Chicago: University of Chicago Press, 1974), esp. pp. 190–204; Dietlinde Goltz, "Alchemie und Aufklärung: Ein Beitrag zur Naturwissenschaftsgeschictsschreibung der Aufklärung," *Medizin historisches Journal, 7* (1972), pp. 31–48, and Hermann Kopp, *Die Alchemie in älterer und neurer Zeit: Ein Beitrag zur Kultergeschichte,* 2 parts, (Heidelberg, 1886; reprinted, Hildesheim: Olms, 1962). Also essential is August Viatte, *Les Sources occultes du Romanticisme – Illuminisme – Théosophie 1770–1820,* 2 vols. (Paris: Champion, 1928; reprinted, Paris: Champion, 1969) and Contantin Bila, *La croyance à la magie au XVIIIe siècle en France dans les contes romans & traités* (Paris: Gamber, 1925).

2 On the early French Paracelsians see Allen G. Debus, *The Chemical Philosophy: Paracelsian Science and Medicine in the Sixteenth and Seventeenth Centuries,* 2 vols. (New York: Science History Publications, 1977), 1, pp. 145–173.

3 Ibid., pp. 262–5.

4 Ibid., pp. 265–79.

5 Howard M. Solomon, *Public Welfare, Science and Propaganda in Seventeenth Century*

France: The Innovations of Théophraste Renaudot (Princeton; N.J.: Princeton University Press, 1972), see esp. pp. 162–200.

6 The French translation of van Helmont by Jean Le Conte *Les Oevvres de Iean Baptiste Van Helmont, traittant des Principes de Medecine et Physique, pour la guérison assurée des Maladies* [Lyon: Hvgvetan and Barbier, 1670]) is extensive, but incomplete. Glauber's *La Description des Novveavx Fovrneavx Philosophives ov Art Distillatoire* was translated by Le Sieur dv Teil and published at Paris (Thomas Iolly) in 1659 along with the *Oeuvre minerale,* the *De l'or Potable, La Medecine universelle* and *La Consolation des Navigants.* Separate French editions of the work on furnaces appeared both in Paris and Brussels in 1674.

7 See Steven Blankaart, *Cartesianische Academie, oder Grund-lere der Arzney-Kunst, worinnen die völlige Arzney-lere auf den naturgemässen Grunden des welt-berümten Cartesii aufgefuret wird* (Leipzig: T. Fritsch, 1699) and Blankaart, *Die neue heutiges Tages gebräuchliche Scheide-Kunst, oder Chimia nach den Gründen des fürtreflichen Cartesii und des Alcali und Acidi engerichtet . . .* (Hannover: G. H. Grentz, 1689). See also Kurt Sprengel, *Versuch einer pragmatischen Geschichte der Arzneikunde,* 5 vols. (Halle, 1799–1803), 4, pp. 321–469 (367–86) for a discussion of Cartesian thought in seventeenth-century iatrochemistry.

8 The texts are presented with an introduction by Allen G. Debus in *Science and Education in the Seventeenth Century. The Webster-Ward Debate* (London: Macdonald, and New York: American Elsevier, 1970).

9 Anon., *Le Parnasse assiegé ou La guerre declarée entre les Philosophes Anciens & Modernes* (Lyon: Antoine Boudet, 1697), sig. Aii^{v}. John J. Renaldo has recently described Gaetano Tremigliozzi's *Nuova stafetta da Parnaso* (1700), "A Note on the History of Medicine and a Debate on Parnassus," *Bulletin of the History of Medicine, 53* (1979), pp. 606–13. There is evidently no connection between this work and *Le Parnasse assiegé.*

10 *Le Parnasse assiegé,* p. 1.

11 Ibid., p. 4.

12 Ibid., p. 6.

13 Ibid.

14 Ibid., p. 11.

15 Ibid., p. 12.

16 Ibid., p. 13.

17 Ibid., pp. 14–15.

18 Ibid., p. 100.

19 Ibid., p. 103.

20 Ibid., pp. 109–110.

21 Ibid., p. 115.

22 Ibid., p. 116.

23 Ibid., pp. 127–8.

24 Ibid., pp. 136–8.

25 E. F. Geoffroy, "Des supercheries concernant la pierre philosophale," *Histoire de l'Academie Royale des sciences (1722) Avec les Memoires de Mathematique & de Physique pour la meme Année* (Amsterdam: Pierre de Coup, 1727); *Memoires,* pp. 81–94.

26 Anon., *Lettre a un ami, touchant La Dissolution Radicale & Philosophicale de l'Or, & de l'Argent, sans corosifs. Avec des Remarques Sur l'opinion général, qu'il ne faut point chercher de Reméde à la Goutte* (London: Pierre Dunoyer, 1719), pp. 3–4.

27 Ibid., sig. A2^{r}.

28 Nicolas Le Fevre, *Cours de Chymie, pour Servir d'Introduction à cette Science,* edited and augmented by M. Du Monstier, 5th ed., 5 vols. (Paris: Jean Noel Leloup, 1751). The work of Joseph Chambon is cited frequently in the new material added on minerals (vol. 3). This edition comes to nearly 2,300 pages. Du Monstier is a pseudonym for the Abbé Nicolas Lenglet du Fresnoy. See J. R. Partington, *A History of Chemistry*, vol. 3 (London: Macmillan Press, 1962), p. 18.

29 On the rise of French vitalistic medicine see Elizabeth Haigh, *Xavier Bichat and Medical Theory of the Eighteenth Century* (New York: Franklin, 1979). Jacques Roger has discussed the place of chemistry in this development in his "Chimie et biologie: Des 'molecules organiques' de Buffon à la 'physico-chime' de Lamarck," *History and Philosophy of the Life Sciences, 1* (1979), pp. 43–64.

30 Paul Jacques Malouin, Dr. Regent de la Faculté de Medecine de Paris, *Traité de Chimie, contenant La Maniere de préparer les Remedes qui sont les plus en usage dans la Pratique de la Medecine* (Paris: Guillaume Cavelier, 1734). Malouin's account of the early history of chemistry is found on pp. 1–9. The defense of his work will be found in the *Lettres d'un Medecin de Montpellier, a un Medecin de Paris Pour servir de réponse à la Critique du Traité de Chimie de M. Malouin,* 2nd ed. (Paris: Guillame Cavelier, 1735).

31 A. Fizes, *Leçons de Chymie de l'Université de Montpellier, Où l'on explique les Preparations avec la meillure Physique, & la usage de chaque Reméde, fondé sur la meillure Pratique de Médecine* (Paris: Guillaume Cavelier, 1750).

32 On Vieussens and early eighteenth-century iatrochemistry see José Maria López Piñero, 'La Iatroquimica de la Segunda Mitad del Siglo XVII,"in Pedro Laín Entralgo (Ed.), *Historia Universal de la Medicina* (Barcelona: Salvat Editores, 1973), 4, pp. 279–96 (292–3).

33 Antoine Deidier, Conseiller Medecin du Roy & Professeur Royal de Chimie, *Chimie raisonnée. Où l'on tâche de découvrir la nature & la maniere d'agir des Remedes Chimiques les plus en usage en Medecine & en Chirurgie. Conforment aux Leçons Latines de Chimie qui sont publiquement châque année dans le Laboratoire Montpellier* (Lyon: Marcellin Duplain, 1715), p. *xi^{r}.

34 Ibid., pp. 9–11.

35 Ibid., pp. 51–2.

36 Discussed by Lester S. King in *The Philosophy of Medicine: The Early Eighteenth Century* (Cambridge, Mass.: Harvard University Press, 1978), pp. 91–4.

37 On Chambon's life I have followed the account in the *Nouvelle Biographie Générale.* His earliest work (which I have not seen) is Cl. Guiron and Joseph Chambon, *E. sanitas a calidi, frigidi, humidi & sicci moderatione* (Paris, 1696).

38 Joseph Chambon, *Principes de Physique, Rapportez à la Medecine Pratique, & autres Traitez sur cet Art,* nouvelle ed. (Paris: dans la boutique de Claude Barbin, chez la Veuve Jombert, 1711), from the preface, pp. a v^{v}–a vi^{v}.

39 Ibid., p. a vii^{r}.

40 Joseph Chambon, *Traité des Metaux, et des Mineraux, Et des Remedes qu'on en peut tirer; avec des Dissertations sur le Sel & le Soulphre des Philosophes, & sur la Goute, la Gravelle, la petite Vérole, la Rougeole & autres Maladies: avec un grand nombre de Remedes choisis* (Paris: Claude Jombert, 1714), pp. á iv^{v}–á v^{v}.

41 Chambon, *Principes de Physique* (1711), p. ẽ iv^{v}.

42 Chambon, *Traité des Metaux,* p. 321.

43 Ibid., p. 459.

44 Ibid., pp. 459–60.

45 Chambon, *Principes de Physique,* p. ĩ ii^{r}.

46 Joseph Chambon, *Principes de Physique, Rapportés à la Médecine-Pratique,* nouvelle ed. (Paris: Charles-Antoine Jombert, 1750), p. 478.

47 Chambon, *Traité des Metaux,* p. 209.
48 Chambon, *Principes de Physique* (1711), p. 1.
49 Ibid., p. 2.
50 Ibid.
51 Ibid., pp. 4–6.
52 Chambon, *Traité des Metaux,* p. ẽ iv^{v}.
53 Chambon, *Principes de Physique* (1711), p. 23.
54 Chambon, *Traité des Metaux,* p. 390.
55 Chambon, *Principes de Physique* (1711), p. 170.
56 Chambon, *Principes de Physique* (1750), pp. 492ff.
57 Chambon, *Traité des Metaux,* pp. 210ff.
58 Ibid., pp. 376–70.
59 François Marie Pompée Colonne, *Abrégé de la Doctrine de la Paracelse, et de ses Archidoxes. Avec un explication de la nature des principes de Chymie. Pour servir d'éclairessement aux Traitez de cet Auteur & des autres Philosophes. Suivi d'un Traité Pratique de differentes manieres d'operer, soit par la voye Séche, ou par la voye Humide* (Paris: d'Houry fils, 1724), pp. iii^{v}–iiii^{r}.
60 Ibid., p. iiii^{r}.
61 Ibid., p. iiii^{v}.
62 Ibid., p. v^{r}.
63 Ibid., p. i.
64 Ibid., p. ii.
65 Ibid., p. iv.
66 Ibid., p. v.
67 Ibid., pp. vii–viii.
68 Crosset de la Haumerie (F. M. P. Colonne), *Les Secrets des plus cachés de la Philosophie des anciens découverts et expliqués, a la suite d'une Histoire des plus curieuses* (Paris: d'Houry, 1722).
69 A recent study of interest is Wallace Kirsop, "Alchemists and Antiquaries in Enlightenment France," *Australian Journal of French Studies, 12* (1975), pp 168–91. Kirsop analyzes the spirited midcentury discussion of Nicholas Flamel and reviews the works (among others) of Pernety and Lenglet du Fresnoy.
70 "J'y combat l'erreur de ceux qui cherchent la Pierre philosophale, en prouvant que la véritable ne consiste que dans l'agriculture." Pierre Brodin de la Jutais, *L'Abondance, ou Véritable Pierre Philosophale. Qui consiste seulement à la multiplication de toutes sortes de grains, de fruits, de fleurs, & généralement de tous les végetatifs* (Paris: Delaguette, 1752), p. iv. As a traditional chemical philosopher Brodin pictured the earth as a vast alembic and gave directions for the preparation of a menstruum useful for plant growth. He describes a trial of this fluid made at Vincennes in 1736 and assures the reader that the results were "magnificent" (p. 50).
71 Rigaud proceeds in scholastic fashion giving arguments against transmutation with their refutation. He also lists six reasons why transmutation is possible. His approach is very traditional. Etienne César Rigaud, *L'Existence de la Pierre Merveilleuse des Philosophes, Prouvée par des faits incontestables. Dédié aux Adeptes par un Amateur de la Sagesse* ("En France," 1765).
72 A survey of the secondary literature on Lenglet du Fresnoy is given by Kirsop, "Alchemists and Antiquaries," p. 188, n. 97.
73 Abbé Nicolas Lenglet du Fresnoy, *Histoire de la Philosophie Hermetique,* (Paris: Coustelier, 1742), 1, p. iii.

74 Ibid., 1, pp. xii–xiii.

75 Ibid.

76 Nothing is known of Richebourg; for the seventeenth-century edition see the account in John Ferguson, *Bibliotheca Chemica,* 2 vols. (London: Verschoyle Academic and Bibliographical Publications, 1954), 2, pp. 272–3.

77 Jean Maugin de Richebourg, *Bibliotheque des Philosophes Chimiques,* nouvelle ed., 4 vols. (Paris: André Cailleau, 1741, 1741, 1741, 1754), *1,* p. i.

78 Ibid., 1, p. vii.

79 Antoine-Joseph Pernety, *Dictionnaire Mytho-Hermétique, dans lequel on trouve les Allégories Fabuleuses des Poetes, les Métaphores, les Enigmes et les Termes barbares des Philosophes Hermétiques expliqués* (Paris: Bauche, 1758), p. iv. For a survey of Pernety's career and activities in Berlin, Avignon, and elsewhere see Viatte, *Les Sources occultes,* vol 1, pp. 92–103. He was deeply influenced by Swedenborg, whose eighteenth-century followers frequently expressed interest in alchemy.

80 Pernety, *Dictionnaire,* p. iv.

81 G. F. Venel, "Chymie," in M. Diderot and M. D'Alembert (eds.), *Encyclopédie, ou Dictionnaire Raisonné des Sciences, des Arts et des Métiers,* vol. 3 (Paris: Briasson, David, Le Breton and Durand, 1753), p. 408. Diderot's antimechanistic views have been noted by Charles Coulston Gillispie in his "The *Encyclopédie* and the Jacobin Philosophy of Science," in Marshall Clagett (ed.), *Critical Problems in the History of Science,* (Madison: University of Wisconsin Press, 1962), pp. 255–89 and in his *The Edge of Objectivity* (Princeton, N.J.: Princeton University Press, 1960), pp. 184–7.

82 Venel, "Chymie," p. 410.

83 Ibid., p. 431.

84 Ibid.

85 André Charles Cailleau, *Clef du Grand Oeuvre* (Corinte and Paris: Cailleau, 1777), sig. A ii^r.

86 Ibid., pp. 25–6.

87 Anon., *L'Art Hermetique à decouvert ou nouvelle Lumiere magique où sont contenus diverses Mystéres de Egyptiens, des Hebreux & des Chaldéens* (n.p., 1787). The chapters include topics such as "Du feu secrét de Philosophes," "De la Riviere des Perles," "Fontem perpetua Nature," and a letter of the Rosicrucians on the "Montagne invisible."

88 Anon., *Le Grand Livre de la Nature, ou L'Apocalypse Philosophique et Hermetique* (Au Midi: l'imprimerie de la vérité, 1790), pp. 12–13.

89 Pierre Joseph Buc'hoz, *Recueil de Secrets surs et Expérimentés, a l'usage des Artistes,* 2d ed., 2 vols. (Paris: Chez l'Auteur, 1783, 1785).

90 Robert Darnton, *Mesmerism,* p. 33.

91 Anon., *Cagliostro Démasque a Varsovie. Ou Relation Authentique de ses Opérations alchimiques & magiques faites dans cette Capitale en 1780* (n.p., 1786).

92 Etteila (or Alliette), *Les-Sept Nuances de l'Oeuvre Philosophique-Hermétique, suivies d'un Traité sur la Perfection des Métaux* (?Amsterdam: 1787), p. 16. A useful reprint of Saint Germain's most important work is the *La tres Sainte Trinosophie,* introduction by René Alleau (Paris: E. P. Denoel – Bibliotheca Hermetica, 1971). This includes a photographic reproduction of the manuscript.

93 Etteila, *Les-Sept Nuances,* p. 36.

94 Here see Karl Sudhoff's *Bibliographica Paracelsica: Besprechung der unter Hohenheims Namen 1527–1893 erscheinen Druckschriften* (1894; reprinted, Graz: Akademische Druck-U. Verlagsanstalt, 1958). Cited is *La Fausseté Des Miracles Des Deux Testamens, Prouvée par le parallele avec le semblables prodiges opérés dans diverses sects; Ouvrage traduit du manuscrit*

Latin intitulé: Theophrastus redivivus (London, 1775). There are also a number of midcentury editions of the *Europäischer Staats-Wahrsager* (from 1742), which seem to be connected with the *Assemblage de quelques Prophetie[s], qui paroissent cadrer aux Circonstances presentes des temps, tirees de Drabicius, Melanchthon & Theophraste Paracelse, traduit, en Francois sur la premiere traduction allemande* (Sudhoff, p. 652, states that he has not seen this translation of the 1741 tract, *Sammlung einiger Weissagungen . . .).*

Darnton states that "of the many systems for bringing the world into focus, mesmerism had most in common with the vitalistic theories that had multiplied since the time of Paracelsus. Indeed, Mesmer's opponents spotted his scientific ancestry almost immediately. They showed that, far from revealing any new discoveries or ideas, his system descended directly from those of Paracelsus, J. B. van Helmont, Robert Fludd, and William Maxwell, who presented health as a state of harmony between the individual microcosm and the celestial macrocosm, involving fluids, human magnets, and occult influences of all sorts." (Darnton, *Mesmerism,* p. 14.). A number of specific references to Paracelsus, Fludd, and other earlier chemical philosophers in the French literature on Mesmerism will be found in Viatte (*Les Sources occultes,* 1, pp. 223–31).

95 Darnton, *Mesmerism,* p. 64.

96 M. Joyand, *Dr. en Médecine de la Faculté de Besançon, Medecin de l'Hôpital militaire de Brest, Précis du Siècle de Paracelse* (Paris: de l'Imprimerie de Monsieur, 1787). An anonymous reply to this work will be found in *L'impulsion Triomphante, ou l'attraction foudroyée par le Dieu de la Lumiere* (Se trouve Dans la Planète de Mercure, chez les Impimeurs du Soleil, 1788).

97 M. Joyand, *Docteur en Médecine de la Faculté de Besançon, Médecin de l'Hôpital militaire de Brest, Lettre sur le siécle de Paracelse* (dated April 30, 1786) (Paris: de l'imprimerie de Monsieur, 1786), p. 3.

98 Ibid., pp. 3–4.

99 Ibid., p. 5.

100 Ibid., pp. 5–6.

101 Ibid., pp. 7–8.

102 Ibid., p. 9.

103 Ibid., p. 10.

104 Ibid., pp. 11–13.

105 Ibid., p. 13.

106 Ibid., pp. 14–16.

11

Inventing demography

Montyon on hygiene and the state

WILLIAM COLEMAN
University of Wisconsin, Madison

The power and wealth of the state rest on several foundations. Apart from sheer accumulation of specie, however, few appeared more important to sixteenth-century observers than the aggregate number of citizens. A century later statesmen had recognized that numbers alone offered an inadequate measure of national strength. The quality of the population also demanded close assessment. To the earlier, simplistic political calculus founded on mere number were now added economic considerations, the determination of those conditions of daily existence, principally employment and food supply, that tended to assure the health and happiness of the people and which thus stimulated its ever-greater procreation. Vauban in his *Projet d'une dîxme royale* (1707), realizing that only a well-fed, comfortably housed, and bodily sound people can and will increase rapidly in number, emphasized this indispensable economic element in effective populationist action. Vauban's views were widely shared in eighteenth-century France and not least by members of the royal administration.

Auget de Montyon, *intendant* in succession to the *généralités* of Riom, Provence, and La Rochelle and author of the *Recherches et considérations sur la population de la France* (1778), is now recognized as the "first French demographer really worthy of the name."[1] Montyon as demographer opens to view certain premises on which a populationist program must stand. These premises are largely biological in nature and necessarily direct attention to the basic conditions of human existence, to the air, food, water and other factors (designated by physicians the six things nonnatural) whose use or abuse human beings find unavoidable.

Political objectives will be attained by means of economic action. Economic action requires, however, an anthropology, that is, a sound and coherent knowledge of man, of his intrinsic being, and his relations to nature and society. "Whoever wishes to govern men," Montyon declared, "must seek to know them,

An abstract of this essay, entitled "L'hygiène et l'état selon Montyon" and prepared and translated by J. Guillerme, appeared in *Dix huitème siècle 9* (1977): 101–8. I thank Guillerme for his kind assistance as well as C. C. Hannaway and Dora B. Weiner, both friendly and incisive critics. To members of the Institute for Research in the Humanities, University of Wisconsin, and the staff at the Newberry Library, Chicago, I am grateful for hospitality and support. This study was begun with the aid of a Study Fellowship from the American Council of Learned Societies.

for nothing can be accomplished save by men and for men."[2] One's comprehension of man begins, therefore, in biology, for it is the basic conditions of human existence – air, food, water, and other unavoidable factors – that first require specification and whose direct influence presents an unrivaled opportunity for informed manipulation. Speculation on the biological factors conditioning population growth and limitation formed a common element in the views of both English and French population theorists of the later eighteenth century. Malthus and his adherents assumed the parsimony of nature and, above all, the impotence of man in rectifying or improving the basic conditions of his existence; they thereby reached notably dismal conclusions. Montyon, however, faced the problem of subsistence with strong optimism, an optimism founded on the conviction that man in society can become the master of his earthly condition and, given proper information and social powers, will do so without delay. Knowledge is power and the people must be improved. In these matters only the state and its responsible agents can act with wisdom and effect. The *Recherches* thus not only provides a masterful example of cogent reasoning on the raw data of population but also offers a carefully conceived and forcefully stated apology for intervention by public officials into the several controllable factors by which human welfare is determined.

The *Recherches* is not a manual of administrative practice but a multifaceted, programmatic statement of first principles for the populationist statesman. Drawing on his experience as well as his studies Montyon hoped to define and inspire appropriate action by the crown. Medicine and vital statistics, essential to the definition of man and his social condition, were drawn together in the *Recherches* in an often original and always striking manner, but neither constituted the heart of their proponent's objective. Montyon was, above all, an advocate of power, of royal power exercised through centrally controlled administrative channels. His rationale arose from the conviction that a large and healthy population is the unique source of national strength and that, in affairs of state, it is up to the royal administration to define goals and assure means for their attainment. Neither church nor town possessed the personnel appropriate to the task of building the population of France. Special agents were required, agents possessing expertise in a diversity of subjects, and only the crown could realistically hope to train and attract such persons. The collection and analysis of general statistics, begun formally if ineffectively, in the 1780s; inquiry into the incidence and causes of animal diseases throughout France; and investigation of institutions destined for the care and confinement of the sick, the aged, and the poor, were tasks whose nature demanded considerable scientific and medical competence and whose scope was well beyond the capacities of local charities and proved to exceed even the apparent reach of the church.[3] These and comparable tasks seemed not, however, beyond the grasp of the royal administrative elite, and of its medical and scientific officers.

In matters of population it is, of course, the procreative process that must be controlled. With regard to the exercise of political, social, or economic influence on procreation the statesman's justification for meddling is obvious: *Raison d'état* dictates that population is a public good and that all measures (rights of the illegitimate, regulation of marriage and divorce, family subsidies or whatever other measures might be imposed) that encourage its increase are desirable and proper. In the biological domain, however, the need for justification is as real as the terms of its expression are unfamiliar. By what right do we intervene in natural processes or, stated so as to be answered by Montyon's analysis, can we intervene at all in such processes and, if so, how is it to be done? The answer to these questions will not be given in terms of law, rights, or the demands of state. It is founded instead in natural knowledge. If the physical conditions of existence indeed influence human character and behavior, then effective action by both individual and state depends on awareness of the nature of man and of the relation of man to ambient conditions. Such knowledge immediately creates the potential for control and exploitation.

Man was born into nature, but it is by use of his reason and within and by means of society that he comes to create his own world. Montyon asserted unequivocally that "of all the causes which can influence population the most powerful are without contradiction physical causes."[4] Yet the primary claim of the *Recherches* was that it was society that can and must act. Obviously it can do so only through nature, by comprehending and ultimately manipulating the irreducible conditions of existence. Montyon made the highest virtue of this necessity, for it specified the scope, manner, and personnel involved in social control. What is man, what are the physical determinants of his bodily prosperity? Only medicine can answer these questions. How is society constituted? What are its numbers, how are they distributed among age groups, social orders, and geographic regions? Only the inquisitive and experienced public administrator can reply to these questions. But the medical and social facts are intimately related: bodily health and social structure exert a profound reciprocal influence on one another. The informed statesman, realizing this, dared ignore neither medicine nor society. He capitalized on his special knowledge in order to render service to the nation and also to demonstrate his singular and indispensable role in assuring the smooth functioning of the state.

Montyon in ambition was all too clearly the model bureaucrat, insisting that to him who possessed special knowledge of social affairs should fall the exclusive right to implement and enforce the laws of the state. Accepting the constitutional forms of contemporary France, he advocated an administrative service owing allegiance only to the crown and its representatives and guiding public affairs by means of an essentially apolitical science of society. Reality, however, was far different. Montyon's brief public career was marked by good intentions and uncertain accomplishment. Born (1733) into a bourgeois family aggressively seeking its fortune through royal service, A. J. B. R. Auget de Montyon ("de Montyon" was an assumed title, never legally accorded) abandoned a Parisian law practice to purchase (1760) the office of *maître des requêtes*, a crucial starting point in the royal service.[5] Administrative position was his obvious objective and it soon came with the intendancies of Riom and Auvergne (1767–71), Provence (1771–3) and La Rochelle (1773–5). Despite occasional successes in the provinces Montyon was clearly too often the wrong man in the wrong place. His splendid officiousness and severe bearing well disguised a genuine humanitarianism and he evidenced little tact in dealing with local interests. He was recalled from all three intendancies and after 1775 apparently became increasingly inactive in the royal administration. Fleeing France during the Revolution he later returned to Paris to superintend his enormous fortune, the basis of important pre- and postrevolutionary benefactions. He died there in 1820.

Montyon is known today, not for successful implementation of accepted or novel administrative principles, but for the clear-sighted collection and evaluation of statistical data. His reputation, however, has long faced another hurdle: Was it in truth he who conceived and prepared the celebrated *Recherches et considérations sur la population de la France,* a work actually signed by one "Moheau"? Here is a fine scholar's dispute and one to be resolved more probably by a commonsense approach to the likelihood of opportunity and capacity on the part of the author than by unimpeachable independent evidence of authorship.[6] Of real importance, however, is that the *Recherches,* whether prepared exclusively by Montyon or with the aid of the shadowy Moheau, gave forceful expression to the ideals of the newly ascendant royal administrative elite and thus provided direct focus on the problems of state and hygiene.

It was the joining of "researches" and "considerations" that constituted the methodological novelty and led to the doctrinal clarity of Montyon's contribution to social analysis. Whereas numerous others before had, with comparable methods and ambition, sought to determine the gross population of France, few among their number had begun to examine in the light of presumably applicable social and physical variables the data thus obtained and none had done so with anything like the care and comprehensiveness Montyon was to devote to their interpretation.[7] The necessary preliminary to such interpretation was, of course, the accumulation of reliable and ample population figures. Neither was truly available to Montyon, although he clearly believed the data at hand adequately supported the conclusions to be drawn. These data were derived from varied sources but principally from publications by previous authors (Buffon, De Parcieux, Messance, and others) and from a variety of local enumerations meticulously conducted by Montyon himself while in the provinces. A continuous history of the French census begins only in 1833. Previous enumerations, commencing in the 1780s and greatly expanded during the Revolution and Empire, were limited in geographical coverage and scope of inquiry and were prepared under variable central control and standardization. Their publication, moreover, was quite irregular. Montyon's effort to look beyond these disparate raw data was animated by a higher concern. My researches, he wrote, have been "so disposed that each assemblage of facts leads on to a consequence, [contributes] perhaps the basis of a conclusion and [serves as] the proof or indication of an important truth." Endorsing enthusiastically and uncritically the wondrous "method" of modern science Montyon desired above all else to rationalize the operations of government. "The [scientific] method consists in rising from the examination of facts to the establishment of principle; [it moves] from example to precept, from experience to theory. It is the most secure path leading to the knowledge of truth." Of this we have the assurance of the "father of modern philosophy," evidently Francis Bacon. The mind can reason correctly only on the sound foundations of experience. Hence, "experience, researches, calculations are the sounding line of every science. How many problems may thus be studied by the science of administration!" Even princes who lack genuine humanitarian feeling cannot fail to recognize that "man is at once the final expression and the source of all that is of value." Man may, in fact, be considered as having a price and consequently he is "the most precious treasure belonging to the sovereign." Anything of value, particularly where that value can be expressed in brute monetary terms, was not a matter for indifferent or ill-informed administrative practice. Montyon's "researches" thus offered a suitable foundation upon which "considerations" could be molded into a science, a body of fact and reasoned precept whose justification resided in its contribution, through augmentation of population, to the power and wealth of the state.[8]

Book I of the *Recherches* was devoted to examination of the size and condition of the population of France. Because it was the direct product of the population, the annual number of births received immediate consideration. Montyon's analysis was directed primarily against those claimants at midcentury (notably the Marquis de Mirabeau) who had spread the alarm that France was seriously depopulated and hence faced, in relation to the other European states, a desperate future unless strong measures were quickly instituted.[9] Lacking a nationwide census the major problem in answering this claim was to devise an index, derived from reasonably secure local enumerations, which might then serve as a basis for extrapolation to the overall population of France. Montyon accepted natality as this index, determining that in France two births occurred annually for every fifty-one inhabitants (hence, the celebrated ratio of 25.5:1, total population to annual number of births).[10] Multiplying this index by the sum of average annual natality (a critical figure unanalyzed in the *Recherches*) permitted direct computation of the population of the state (23.5–24 million). This figure, seemingly much in excess of that estimated for France in 1700 (circa 19–20 million), appeared decisively to refute depopulationist opinion and to confirm the view that conditions for nurturing a healthy and expanding populace in France were anything but hopeless.

Nonetheless, although natality seemed to offer a useful indication of population size and was subject to certain forms of social control, it was not and could not be the proper element upon which to launch a populationist program. By analyzing the annual number of births over a ten-year period, and in seven quite dissimilar localities, Montyon had discovered a striking regularity: The birth rate in each locality was virtually level.[11] Departures of up to 10 percent from the average of the decade were rare and were quickly obscured by the mass of population. The population was apparently reproducing at a rate near its biological limit. Natality was, therefore, a relatively fixed factor, being proportionate to the number of females of reproductive age in the population. Only by increasing the number of such females could society ensure greater productivity, and yet that increase could not easily be engendered by a population already approaching its natural limits of natality. The only solution was control of mortality.

The form of Montyon's argument is of central importance in understanding the widespread agitation after 1789 for positive action in favor of public health. To French populationists such as Montyon and to the later hygienic activists Jean Noel Hallé and Louis René Villermé, as well as to the more noted English sanitarians of the early nineteenth century led by Edwin Chadwick, the foremost concern was reduction of mortality. The excessive death rate in younger years (roughly one half, Montyon and numerous others had determined, of all children born in France were dead by age ten)[12] deprived the state of both productive new hands and mothers of additional children. If females who survived to reproductive age were the residuum of a seriously depleted population already procreating at a rate near its natural limit then the only hope for increasing overall biological productivity would be measures that effectively reduced mortality. Fewer deaths in the younger years (and, of course, also in the years of reproductive activity) directly increased the breeding population and thus indirectly contributed to general population growth. Death faced all mankind, yet populationist and sanitarian alike recognized that premature death was as needless as it was common. Montyon declared:

> Everyone carries within himself the seeds of his own destruction and the moment of birth is the first step towards death. But this progress towards the end proceeds more or less slowly in accord with the action of the different physical, moral, civil or political causes which influence our existence.[13]

Because death is subject to such manifold influences premature death is in reality preventable death (only thus does the adjective "premature" or "excessive" have meaning) and the nature of those influences is such that only comprehensive and stringent social action is efficacious. On these grounds Montyon, like Chadwick two generations later, both justified and demanded state hygienic action.

Montyon's perception of the factors influencing human mortality was conditioned by the singular data he had gathered and subjected to examination. Numbers now confirmed age-old generalizations and therein lay the force of the new vital statistics: "Men create theoretical systems from their reasoning and they deal with these either poorly or well; one does not, however, hold figures in doubt."[14] Vital statistics appeared to offer new and secure grounds for reasoning about society. The truths to be gained, although perhaps not notably different in kind, seemed distinctly more effective in their persuasive power than had been previous verbal generalizations. Properly marshaled and interpreted, vital statistics, Montyon obviously had concluded, rendered accustomed reasoning on the physical parameters of social causation both uncertain in foundation and inappropriate as a guide to needed social action.

Hippocratic medicine had first systematized thought regarding the role of these parameters in human affairs and had provided, in addition, initial consideration of the several factors that ultimately constituted orthodox climatic doctrine. This doctrine persisted in Western thought until the early nineteenth

Table 1. *Average expected duration of life in six parishes*

Age	High land or open level country (3 parishes)		Low and marshy areas (3 parishes)	
	Years	Months	Years	Months
1	26	11	21	3
3	41	6	32	10
5	44	5	35	6
10	44	7	36	3
15	41	7	34	—
20	37	7	30	8
40	25	—	20	10
60	12	11	11	8
80	3	7	2	7

century, enjoying during the Enlightenment a period of great prosperity.[15] While freely employing the verbal expression of the role of climatic factors (air, water, soil, and the like), Montyon also attempted to translate climatic doctrine into meaningful statistical language. Six parishes, whose record of mortality he had carefully verified, were selected. These (unidentified) parishes were then divided into two clearly defined groups, the one having a low and marshy setting and the other an elevated situation. Mortality data were tabulated to show average future duration of life at different ages in the two groups. The deleterious effect of marshland (other factors being deemed equal) on health was thus vividly exposed (Table 1).[16]

Residents of upland villages maintained throughout their lives an advantage over lowland residents. The discrepancy was particularly noticeable in the younger years, reaching a maximum of almost nine years at age five and not falling below five years until age thirty-one. Montyon attributed these striking differences directly to the influence of physical conditions.

Physicians had long recognized the seasonal incidence and age specificity of disease. Younger age groups tended to sicken and die during the early autumnal months, older persons were struck down in winter and early spring. Such conclusions were well-founded generalizations based on extended medical experience. Montyon now exhibited them anew and, possibly for the first time, in elementary statistical dress (Fig. 1). Mortality in the under-fifteen age group during September–November (25 percent of the year) represented 39 percent of total annual mortality for that group; for the fifteen to sixty age group the proportion dropped to 28 percent for for those over sixty it fell to 24 percent. For the period January–February (17 percent of the year) the order is reversed: 22 percent for those under fifteen, 25 percent for the fifteen–sixty group, and 29 percent for persons over sixty. Mortality obviously did not fall evenly over the year: a factor of almost three divided the months of maximum (September: 14

LIEUX.	Nombre d'enfans trouvés en dix ans.	Nombre de ces enfans décédés en dix ans.	Proportion par approximation.
Hôpital de Rouen. .	706	412	
Hôpital de Clermont. .	1110	641	
Hôpital de Tours. . .	1686	1167	
Total.	3502	2497	2/3
	Nombre d'enfans nés pendant 10 ans	Décès en dix ans.	
Mortalité ordinaire. .	45643	23637	21/40
	Enfans exposés.	Enfans décédés dans l'année du dépôt.	
Hôpital de la Rochelle. cinq ans.	517	286	11/20
Mortalité ordinaire. .	Enfans nés.	Enfans décédés dans l'année de leur naissance.	
	45643	12722	5/18

Fig. 1

percent) and minimum (July: 5 percent) mortality of children, and January killed persons over sixty at a rate double that of August.

To this evidence favoring situation and season as essential physical factors causing human mortality Montyon joined numerical confirmation of the role played by social causes. No social group under the ancien régime displayed a higher mortality rate than foundling children. Montyon's data (Fig 2) indicate that average mortality in the foundling hospitals at Rouen, Clermont, and Tours exceeded by at least one-fourth the already appalling death rate of children at large. Abandoned children at La Rochelle died before the age of one year at a rate double that of children in the population. Since children constituted a concrete promise for the future power and wealth of the state this extreme mortality was all the more deplorable and indicated a crucial point for administrative intervention. Economic status also influenced mortality, a fact forcefully confirmed by a comparative view of mortality among prosperous *rentiers viagers*

MOIS.	Juſques & compris 15 ans.		De 15 à 60.		Au deſſus de 60.	
	ISLE DE RÉ, 8 Par.	FRANCE 8 Par.	ISLE DE RÉ, 8 Par.	FRANCE, 8 Par.	ISLE DE RÉ, 8 Par.	FRANCE, 8 Par.
JANVIER. .	321	287	94	176	129	108
FÉVRIER. . .	293	280	92	142	90	83
MARS. . . .	278	325	103	174	92	103
AVRIL. . . .	239	290	99	212	104	106
MAI.	212	267	95	170	82	88
JUIN.	221	293	90	168	76	60
JUILLET. . .	194	306	70	98	65	49
AOUST. . .	363	341	89	120	59	44
SEPTEMBRE.	567	441	106	134	71	78
OCTOBRE. .	542	392	121	174	110	78
NOVEMBRE.	429	273	99	163	82	78
DÉCEMBRE.	307	253	97	149	118	87
TOTAL.	3966	3748	1155	1880	1078	962

Fig. 2

and the population at large.[17] Death among children under ten in the population exceeded 51 percent while that of the children of *rentiers* did not attain 13 percent. Montyon urged caution in using these figures (the *rentier* sample is quite small) but nonetheless vigorously affirmed the primary conclusion: Social condition, here defined by economic status, was determinative of physical well-being.

Statistical data of this nature clearly demanded a closer definition of the physical and social causes of population change and to this matter Book II of the *Recherches* was exclusively devoted. Montyon's assessment of the physical causes affecting population reflects throughout the abiding influence of the important ancient medical doctrine of the "six things nonnatural." Formulated by the Greco-Roman physician, Galen, in the second century A.D. and thereafter endlessly restated and revised by Byzantine, Arabic, and Latin commentators, the doctrine of the six nonnaturals provided Renaissance and Enlightenment physicians and medical publicists with a concise, carefully articulated, and seemingly well-founded language and conceptual structure for discourse regarding the influence of environmental factors on people.[18] The six nonnaturals included air, food and drink, sleep and rest, motion of the body and its parts, retention and evacuation of foreign materials, and the various passions. This list is not at all a miscellaneous one. It emphasizes those aspects of nature that exert an unavoidable influence on man. Thier influence is, indeed, beyond the direct control of man's will: Simply in order to live one must breathe, drink, eat, sleep, and evacuate bowel and bladder. Even the passions, always responsive and thus irrevocably joined to shifting external conditions, properly belonged in this broad category of the physical determinants of the human condition.

Climatic theorists of the eighteenth century agreed that air, the atmosphere, exerted the greatest influence on man's physical condition. To air and atmosphere were devoted two in a striking series of chapters in the *Recherches* (Book II, Part I, Chapters 1–8) that constitute a small treatise on the importance of awareness and control of the nonnaturals to any serious populationist program for France. Montyon here expounded theoretical foundations that justified and determined hygienic action by the state. Air served as a receptacle for a universal host of malign influences. Emanations from the earth were received by the air and then conveyed and insinuated into even the remotest corners of existence, including, of course, the human frame. The air was rarely if ever homogeneous. Corruption beset it at all times. Subterranean effluvia set free by volcanic action, mining, and other disturbances; marsh gases; the stench arising from improper interments; the effects of fire and of animals and men on air in an enclosed space – all of these to our eyes disparate factors were, in the opinion of Montyon and his contemporaries, contributory elements to the essential physical cause, air. They were also almost all harmful.[19] Humidity was of paramount concern, excessive dampness being a necessary concomitant of a low and marshy situation. High elevations with rocky soil and a brisk wind (but not a wind sweeping from off the sea or over swampy terrain) enjoyed singularly pure air. The latter presented a demonstrably higher life expectancy than the former, the critical *differentia* between the mountainous and the marshy setting being the quality of the air environing each. Forests, too, played a critical role in establishing atmospheric quality. Transpiration from heavy foliage vitiated the atmosphere with excessive humidity, rotting leaves gave rise to dangerous corruptions, stagnant pools formed and persisted. The eighteenth-century valuation of untamed wilderness was curiously ambivalent but Montyon, a shrewd and highly successful exploiter of his own domains, clearly believed that deforestation was both salubrious and profitable and therefore doubly necessary.[20]

Water, too, was a ready source of sickness and death. Not only did water also receive all manner of noxious contributions from the earth and from man's activities in field and city – "every place inhabited by a multitude of men," Montyon presciently and complacently observed, "creates a sewer which will discharge only into a flowing stream or into the sea, whose immensity can easily absorb all the filth" – but it was, of course, the source of atmospheric humidity. Standing water constituted by far the greatest threat to health.

> Stagnant waters are without contradiction the most unhealthy of all. Ponds and swamps produce an evaporation which gives an aquatic character to the air which we respire and thus alters its purity. Moreover, this water, not being renewed, and following the law governing all liquids, putrifies.[21]

Montyon illustrated his conclusion by pointing to the striking difference in the incidence of fever among the garrison troops at Lille and La Rochelle: at Lille fever patients numbered (on the average) forty-five throughout the twelve *bataillons* (circa 8,000 men) and never exceeded eighty; at La Rochelle the average number of cases reached eighteen to nineteen for each *bataillon* (circa 700 men) and rose in September–October to thirty-six to thirty-eight.[22] The fever rate at La Rochelle, lying low among salt marshes and open to winds off the ocean, thus exceeded that at inland Lille by a factor of at least five.

Montyon's remarks on food and general nutrition are brief but nonetheless disclose the state's responsibility in matters of diet. Human beings require not only pure water but vegetable and, above all, rich animal nutriments. A balance among the foodstuffs is in order. The relative nutritional value of the various foodstuffs must be established: Bread made from wheat sustains man better than do rye or barley breads but the exact relationship among them is unknown; it must be determined by scientific research.[23] Man is omnivorous and suffering during hard times can therefore be tempered, but only if one has obtained needed knowledge and possesses the means to act. The state alone can assure success in this enterprise.

The passions are dealt with by a set of well-worn platitudes and unpersuasive examples. The passions and the atmosphere, aliments, and bodily activities exert a reciprocal influence upon one another. Everyone requires some gaiety for physical well-being. Thus, slave traders prudently protect their property by providing the cargo with music and an opportunity for dancing. Passions vary with climate, being mobile and exaggerated in the south whereas northerners, slower perhaps to respond, are more profound in thought and circumspect in behavior. The passions also vary with social order: The *peuple* are dull in spirit, "their souls like their hands develop a callous." Reduced to mere physical existence they lack the spirituality and finesse of *gens du monde.* Meat eaters are more "ferocious" than vegetarians, wine drinkers are "more gay and lively but less steady." By no means are these facts a source of dismay for, as Montyon concluded, in bald expression of his deepest convictions:

> Happy are the people whose character does not lend itself to strong feelings, whose unique and constant condition is gaiety, who laugh at everything and sing away their cares! The regimen which serves so well the tranquility of the State is also that most favorable to the maintenance of health and the preservation of life.[24]

Montyon the statesman here as elsewhere easily intruded upon the inquiring philosopher, reaffirming his opinion that hygiene was to be encouraged by the state strictly for purposes of state, including internal control. The individual was largely lost in the mass; rather, the mass, the *peuple,* was simply not resolved into identifiable persons worthy of individual consideration. Montyon thus encouraged the transformation of the doctrine of the nonnaturals, radically individualistic in its original definition and traditional form, into an apologia for state-directed hygienic action. This was an essential step in the creation of an autonomous program and the delineation of suitable means for the improvement and protection of public health (discussed later).

When, Montyon declared, we contemplate the influence on health and mortality of climate and associated physical causes "it appears that these causes alone can determine the increase and loss of population." Nevertheless, he continued, "man, through prudence and industry, through malice and thoughtlessness, changes the physical order [of the world] and improves or degrades the original condition" of himself and his environment.[25] We do not stand passively before the forces of nature. We are inclined to action and history is but a record of our deeds. A crucial need in the assessment of a society and its institutions is determination of the scope and limits that dictate our capacity to master the basic conditions of existence, conditions that are both physical and social. To consideration, therefore, of physical causes must be added an appreciation of social agencies of change, for "social institutions, laws, *moeurs,* popular opinion, customs [and] taste, bring into being and destroy [a nation's] inhabitants and [thereby] create a new world."[26] It was the nature of man – an intelligent agent of his own worldly destiny – that effected the reconciliation between physical and social causes. Man exerted direct control over physical causes and was also able, by manipulation of social forces, to act in the religious, economic and political domains in a manner best calculated to encourage the beneficent aspect of those causes and their consequences.

Foremost among the "political, civil or moral causes" that affected health and population stood religion.[27] Superior to human law in its origin, religion (Roman Catholicism alone was meant, it being a faith in peculiar agreement with the "constitution of this realm," France) was subject to the administrative needs and desires of the civil power. Religion when guided by the wrong hands or directed to improper purposes was a dangerous instrument, and Montyon did not hesitate to castigate the "horrors of 1572 and 1685, monuments of delirium for which France still pays." But religion also commands pure moral behavior, charitable acts, rendering of justice, and respect for property. Even if the idea of an omnipotent supreme being were not an "eternal truth related [to us] by God himself" it would constitute the "greatest, most beautiful and most wise of human institutions." Religion to this political realist speaking Voltairean language stood firm as the critical foundation of crown and state – and also of population. Social disorder inevitably led to depopulation; strict obedience to the laws, together with a well-tutored sense of the sanctity of property, best maintained the harmony of the nation. The spiritual force of religion was here properly seconded by apposite and prevailing temporal forces: a French monarch tempered (at least in theory) by independent regional and judicial powers and the entire nation obedient to the laws of the land.[28]

But religion as a major influence on the increase of population most obviously derived authority from its moral force, from its encouragement of charity to the destitute and weak and, above all, from its jealously guarded sanction of marriage.[29] After hunger, sex makes the world turn; such was the Creator's will. Men and women do, of course, reproduce outside of marriage but this, in Montyon's eyes, merely represented practice "favorable to voluptuousness, not to propagation." Marriage alone could guarantee the "public health and means for raising children." The rampant spread of venereal disease Montyon confidently stated to be the inevitable consequence of promiscuity. It constituted a double threat to population, ruining the parents and destroying the children. Healthy children born out of wedlock fared little better. True propagation demanded not only sexual union but continued care of any children born. Marriage was an altogether admirable arrangement for assuring the protection of the young until they, too, reached breeding age.

Montyon's analysis of marriage was unsentimental and largely free from easy moralizing. The authoritarian bachelor acknowledged that his comments on "marriage and the happiness which it brings [should be viewed] from their human side and with regard to temporal affairs," marriage as a divine institution not being germane to the discussion. On this most important of social incentives to populationism his strictures were severe. Divorce he deplored. If divorce were tolerated women would follow their slightest desire, children would be abandoned, *moeurs* degraded. Only an unfruitful union provided "a just and necessary cause for divorce." The age of marriage was to be closely supervised. Early puberty was too dangerous: Men risked permanent impairment of their temperatment by excessive venery and women courted comparable peril through premature pregnancy and childbirth. The potential increase in human yield from early marriage being outweighed by the dangers of reproduction during adolescent years, the nation's overall harvest would be reduced. Parents, above all the father, were to enjoy absolute dominion over their children. Children were truly their most valuable *produit* and the state must join the parent in assuring the security of a mutually beneficial property. Man, of course, properly assumed total precedence over woman, he being the older, stronger, and more reasonable partner. Last, marriages between close relatives were to be forbidden. The intelligent farmer knew how to select his grain and the nations their sheep; men should express no less concern for their own well-being.

Celibacy Montyon condemned and also contraceptive practices. A powerful state can afford, indeed requires, a modest number of celibates. They serve the church and they can best serve the nation (a father makes his contribution other ways). But, hordes of idle churchmen and especially churchwomen were, like *rentiers viagers,* barren and therefore a complete loss to the state. Contraception moved Montyon to rare passionate outcry. Confessors and others, presumably rural surgeons or physicians, who have "a taste for physical researches important to the state," will inform you that rich women, for whom pleasure is the greatest interest and unique occupation, are not the only ones who regard propagation of the species as a deception of earlier times. Already these sad secrets, unknown to any animal other than man, have penetrated the countryside: they deceive nature even in the villages.[30]

These "homicidal tastes" resulted from an obvious corruption of *moeurs* and, if left unchecked, would devastate the nation as had the plague. Moral reform alone could mend France's decadent ways.

Montyon's strength lay not in the originality of his perception of the physical and social factors bearing on the increase of diminution of population. The designation of the six nonnaturals as the essential determinants of health and his inventory of social and political causes were utterly conventional. Much less so, however, was his union of this familiar causal approach with a statistical assessment of the condition of the population of France. Few of his correlations, perhaps, were as pointed or suggestive as those noted earlier, reporting on the incidence of fever at La Rochelle and Lille or evaluating the differential mortality of physiographically distinct parishes. Yet it is precisely these and comparable instances that render the *Recherches* so important a document in the development of a body of reasoned vital statistics, of the science of demography. Such instances also point directly to the ideal that motivated Montyon's entire endeavor and toward which the *Recherches* were throughout focused.

The administrative process must be rationalized. Power, of course, was the objective and Montyon was capable of expressing brutally its essential administrative conditions.

It is as difficult to render full justice to men as it is dangerous to tell them the truth. In every nation those who command have need of the greater or lesser ignorance and softness of those who obey . . . The French nation needs more breastworks in her streets, a larger guard in her cities, more laws governing the public order [and] more surveillance by her police . . . The wisest administration is that which is unperceived. In silence it foresees, disposes, prevents, directs, removes obstacles [and] attains its goal without its action having even been detected.[31]

Tyranny, surely, is espoused by these claims, yet it is a special and distinctively modern form of tyranny, that of an administrative elite wielding not only real social powers but guarding knowledge unavailable to others in the populace.

Must we concentrate the science of administration in the class destined to execute its functions or should it be open to all men?. . . . Without doubt administrative functions must be confined only to those who have made careful study of the science. And perhaps it would be wise to allow only those who have exercised administrative responsibility to write about these matters.[32]

It is knowledge, precise, comprehensive, and limited to a select corps, that stands foremost in establishing administrative control. Political power, social or religious compulsion, economic threat, and temptation were all, in Montyon's vision of the well-ordered state, indispensable to the ruler but none could be wisely directed or efficiently implemented without sound awareness of social composition and the influences that bear on it. The entire administrative edifice was thus ultimately reduced to the need for exhaustive knowledge of human nature and society.

Montyon's populationist program was one of action but, as may now be seen, was premised on knowing. By examining the social, political, and religious forms and practices of France he had gained the assurance that, despite occasional tempests and frequent confusion, the state survived and prospered by virtue of the rule of law. Man existed, however, not only in society but in nature. He had a biological being and it, too, proved subject to law. The administrator's task was to know the law, civil or natural, and assure its fullest exercise. With regard to health he no longer need bend impotently before the seeming chaos of natural causes. Recognizing the order behind confusion, he could perform his decisive task, that of assuring a disciplined utilization of the seemingly random influences of nature on man, expressed in the eighteenth century in the language of the nonnaturals, and doing so through state and municipal agencies and for the public good.

"The Sovereign," Montyon observed in opening the penultimate chapter of the *Recherches,* "can be portrayed in a light which renders his authority valued and worthy of respect only when he appears as the father of a family watching over his children, removing, by his beneficent care, all influences which might threaten their preservation." But, until the present, this aspect of "administration" simply had not received the consideration "a matter of such importance deserves." Demanding particular attention, therefore, were those "establishments and regulations of *police*" that might prove "useful to the population."[33] Montyon's usage of the term *police* followed appropriate contemporary meaning, the assurance by constituted authority of the comfort and tranquility of a circumscribed political entity, usually a municipality but embracing as well the nation at large. *Police* thus entailed determination and enactment of specific measures for the preservation of public health.

For such purposes traditional medicine seemed a blunt instrument. Montyon along with numerous others was still undecided whether the physician destroyed more people than he saved. Except for the efforts of a small number of urban practitioners, blessed as much by extensive experience as by formal education, medical care in France was, he believed, altogether wretched. His considerable experience in the provinces gave, no doubt, substantial grounds for this conclusion. In the small town or in the country one best trusted to nature, not to that omnipresent horde of charlatans who sold "error and death to the credulous and stupid belief" of the people. "Curative methods," fortunately, were but one aspect of medicine. A quite independent division was "preservative medicine, adhering to the general *police,* and essential to the conservation of public health."[34] The impotence of the medical art was a much-debated theme in the eighteenth century. Physicians of generous disposition tended to exaggerate their capacities and discerned in the not remote future a far brighter prospect for the alleviation and perhaps elimination of man's bodily ills. Their sanguine expectations have offered grounds for concluding that medical advances – already realized or merely anticipated – provided a strong conceptual underpinning for the hopeful philosophy of man of the Enlightenment.[35]

Montyon, in contrast, concluded that therapeutic measures, whether or not truly effective, were in any case unavailable to the vast bulk of the nation and that principal recourse in matters of health must therefore be made to the state.[36] Already certain measures, undertaken by various authorities and for diverse reasons, had had the welcome effect of rendering city air more salubrious. This was a capital event in the health of the nation and such measures were obviously to be extended. Thus, although some streets had been paved and much filth removed, the alignment of those streets had yet to be established so as to seize maximum benefit from incident sunlight and prevailing winds. Sewers still stood stagnant, their contents fouling the atmosphere; irreducible waste piles and mountainous dung heaps worked a comparable effect. The siting of houses and the materials used in their construction, especially stone and plaster, largely determined the dampness of the air within these habitations and thus contributed to the sickliness of their residents. Even more destructive of life was the poor quality of food and drink consumed, particularly that forming the diet of the poor who could afford no better. Urban interments had an altogether disastrous effect on the quality of the air: Here was the "most common, the most pernicious of all [urban] disorders."[37] Poisonous exhalations from the grave made church and churchyard a more than figurative meeting ground of the living and the dead and to the enormous debility of the former. If only France enjoyed different *moeurs,* if only increased taxation or private greed did not guide public affairs, this prime source of corruption might be vanquished. Hospitals were little better. They confounded the aged, the poor, and the orphan with the truly ill and had but death and misery as their harvest. Hospitals thus proved to be not benefactions but "monstrous products of a generous feeling."

These complaints, most of which refer to the vitiation of the quality of an important nonnatural (air), reiterate themes already familiar to the social observer, particularly the urban observer, during the ancien régime.[38] The city itself perhaps behaved as a nonnatural. It was man's own creation and the tangible manifestation of his insatiable desire to live contrary to his essential being.

> In the cities man produces less, he is more inclined to debauchery, he is more exposed to luxury, and consequently he fears having a large family. Experience has shown that the focus of these vices which destroy a population is located in the great cities and that from the cities they expand into the countryside.[39]

Under these conditions there could be but one realistic solution to the problem of assuring the growth of the population of France. The king must act. He must establish appropriate laws and provide means for their rigorous enforcement.

> It is not only by means of the regulations imposed by *police,* [by the creation of] useful institutions and by assuring advantages to the married condition that Kings can encourage population growth; the entire physical order seems to lie within their hands.[40]

In truth, the physical order did not, and cannot, lie within these or any other hands. At best one can so organize society that it will live as harmoniously as possible with the physical conditions of its existence. Montyon proposed no specific legislation in the *Recherches* and allowed one only to deduce public remedies from manifest social ills. Cleansing the city, preventing public abuses in the streets, regulating the building trade, overseeing the quality of foodstuffs offered for sale, improving the conduct of institutions for the relief of the poor and the sick – the feasibility and desperate need for measures to effect these changes are the implicit claim of the *Recherches.* Although the crown could not directly modify the climate, it could at least "direct the population" toward a more salubrious environment, that is, away from town and city. The power of centrally controlled institutions, heretofore acting only to bring people together in dense and unhealthy agglomerations, must henceforth be used to diffuse the population throughout the nation.

Yet it lies within the power of Kings to establish their subjects in places which the crown deems most healthy and, if it is possible that their habitations can be improved, it is all the easier to guide the people's choice in aliments and professions. But if from climate, regimen, custom and habitual tendency towards particular acts there results that unknown principle which forms character and mind, then one may conclude that Sovereigns, by means of wise laws and useful institutions, by the restraint and stimulus exercised by the imposition and suppression of taxation, exert dominion over the physical and moral being of their subjects.[41]

The crown, therefore, may fairly expect to determine at will "the *moeurs* and spirit of the nation" and to create a new and more populous France.

Here, then, were the means by which royal influence could be exerted on the population. Wise laws, suitable institutions, and judicious taxation constituted necessary preconditions to effective government. All presupposed in turn an enlightened monarch and above all, energetic and informed administrative officers. The latter, in regular contact either personally or through their numerous agents with events of both local and national concern occurring throughout France, were uniquely located to determine those conditions especially in need of legislative innovation and administrative action. The means for increasing population stood, as the author of the *Rechereches* had of course foreseen, in perfect accord with the structure and powers of the state administrative apparatus. Only the decision to act remained to be taken. Montyon's volume, as his peroration makes clear, was thus intended to illuminate the need for decision and perhaps to provoke that decision itself.

It is possible that, under Louis XVI, the Nation will assume a new character. Partisans of population, defenders of humanity, we proclaim our hope that the physical, moral and political order may be modified and directed in a manner most advantageous to the propagation and conservation of the human species.[42]

The hopeful expectations of reform expressed by publicists in the early years of the new reign of Louis XVI were not to be realized. Montyon's particular aspirations, certainly, received little or no recognition and led directly to no concrete innovations or improvements in the sanitary and hygienic practices of the regime. Indirectly, however, they are of great interest in defining the stages by which the broad concerns of "public health" became a self-conscious reality.

Montyon's populationist thought exhibited diverse tensions. He had, first of all, to confront and resolve the apparent opposition on the physical level of man to nature. Implicit in the resolution of this issue, which was achieved through an understanding of the influence on the nonnaturals, was the practical problem of defining means by which the clearly discerned goal, systematic intervention by the state on behalf of the bodily well-being of the nation at large, could be attained. But means and objective, as these were enunciated by Montyon, seemed grievously incompatible with one another. The nonnaturals, the doctrinal foundation of Montyon's public hygienic program, had been and remained an instrument tuned essentially to care of the individual. Greek hygienic doctrine, in which the nonnaturals came to play a major role and which, as already noted, continued to provide in the eighteenth century the basis for serious discourse regarding hygienic matters, emphasized the unique value of the individual.[43] Only persons enjoying sufficient means and leisure could pursue a hygienic program that emphasized exercise, diet and discipline of the passions, the elements of self-mastery shown by experience to best serve one's physical and mental constitution and needs. Hygiene in this, its accustomed, meaning was consequently highly idiosyncratic and its successful implementation was crucially dependent on that host of intrinsic and extrinsic factors that defined the individual.

The problem facing Montyon and his contemporaries was to adapt a hygienic doctrine designed for the self-preservation of individuals to the needs of the mass of citizens in the modern state. His solution was to preserve the vocabulary and conceptual structure of the doctrine of the nonnaturals and to apply both to the newly perceived social situation. Montyon himself failed to recognize the contradictions inherent in this radical shift in the application of the nonnaturals and it was a sign that a new era had begun in the definition of the proper domain of public health when for "choses non-naturelles" was substituted the expression "matière de l'hygiène."[44] Public health may encourage the individual to cultivate his personal well-being but it cannot halt at this modest beginning. "Public health" must by definition attend to those matters that affect the health of the majority or totality of the population: Its concern is truly public and no longer merely personal. The explanatory elements of the doctrine of the nonnaturals could, for this purpose, continue in force but, apart from an uninterrupted appearance in manuals of domestic medicine, their social reference would have to change. In so changing, this ancient medical doctrine provided initial orientation for approaching the health needs of a startlingly new urban and industrial society.

Montyon in 1778 was, of course, unaware of these impending changes. His outlook was agrarian and his easy solution to health problems in the town and city was, if his rhetoric may be accepted at face value, to abolish both. His demands, however, were so stated as to serve in reality all state interests, urban or rural. To Montyon as to others of his generation and social position humanitarianism melded nicely with defense of the royal interest and economic demands and all the better with the application of newly acquired technical expertise to social needs. His medical naturalism was not simply an invitation to further study but a call to action. Montyon's exposition of medical matters as well as his celebrated demographic investigations were, consequently, less exercises in scientific inquiry than indispensable elements in a greater objective, the articulation of a populationist program for France. These alternatives must not, however, be separated, for defending them all and integrating their distinct demands and contributions was that administrative corps whose superior wisdom and legitimate claim for power was to Montyon the very heart of the state's role in matters of health and hygiene.

NOTES

1 P. E. Vincent, "French demography in the eighteenth century," *Population studies 1* (1947): 57; also R. Gonnard, (ed.), Moheau [Montyon] *Recherches et considérations sur la population de la France* [1778], (Paris, 1912), introduction, xiii (this edition is cited hereafter); L. Chevalier, "Préface à Moheau" *Population 3* (1949): 211–32, esp. 219.

2 L. Guimbaud, *Un grand bourgeois au XVIIIe siècle. Auget de Montyon (1733–1820). D'après des documents inédits* (Paris, 1909), 65.

3 See C. C. Hannaway, "*The Société royale de médecine* and epidemics in the *ancien régime,*" *Bulletin of the history of medicine 46* (1972): 257–73; D. B. Weiner, "Le droit de l'homme à la santé: une belle idée devant l'Assemblée national constituante 1790–1791," *Clio medica 5* (1970): 209–23.

4 [Montyon] *Recherches,* 199.

5 See Guimbaud; *Un grand bourgeois* E. Zylberman, "Auget de Montyon et l'agriculture," in A. Rigaudière, E. Zylberman, and R. Mantel, *Etudes d'histoire économique rurale au XVIIIe siècle* (Paris, 1965), 105–50. On the central role of the *maitre des requêtes* see P. Goubert, *L'ancien régime. 2. Les pouvoirs* (Paris, 1973), 49–55. A fundamental contribution to understanding Montyon's mode of analysis is J. Lecuir, "Cruminalité et 'moralité': Montyon, statisticien du Parlement de Paris," *Revue d'histoire moderne et contemporaine 21* (1974): 445–93. The mercantilist context of Montyon's views is broadly stated by G. Rosen, "Mercantilism and health policy in eighteenth-century French thought," *Medical history 3* (1959).

6 Gonnard, ed. ("Notice," *Recherches* v–xx) reasserted the traditional ascription of the *Recherches* to its signatory (Moheau) and has subsequently been largely followed, notably by L. Chevalier, "Préface à Moheau." Reassignment of the work to Montyon has, however, more recently and persuasively been made; see Vincent, "French demography in the eighteenth century," and esp. E. Esmonin, "Montyon, véritable auteur des *Recherches et considérations sur la population,* de Moheau" *Population 13* (1958): 269–80. Moheau apparently served as Montyon's personal secretary in Paris between 1773 and 1776, thus rather late to have collaborated with his employer on a book planned in 1769 and whose dedication is dated 1774 (R. Le Mée "Jean-Baptiste Moheau," *Hommage à Marcel Reinhard. Sur la population française au XVIIe et XVIIIe siècles* [Paris, 1973], 417–25.)

7 J. J. Spengler, *French predecessors of Malthus. A study in eighteenth-century wage and*

population theory (Durham, 1942); L. Badley, "Buffon, precurseur de la science démographique," *Annales de géographie 38* (1929): 206–20; E. Esmonin, "L'abbé Expilly et ses travaux de statistique," *Revue d'histoire moderne et contemporaine 4* (1957): 241–80; A. des Cilleuls, "La Michodière et la statistique de la population," *Comptes rendus de l'Académie des sciences morales et politiques 141* (1894): 614–21; J. J. Spengler, "Messance: founder of French demography," *Human biology 12* (1940): 77–94. Previous analyses of Montyon's *Recherches* include J. J. Spengler, "Moheau: prophet of depopulation," *Journal of politcal economy 47* (1939): 648–77; M. Martin, *Moheau et les origines de la démographie en France au XVIIIe siècle* (Paris, 1944); L. Chavalier, "Préface à Moheau."

8 Montyon, *Recherches,* 5–7.

9 J. J. Spengler, *France faces depopulation* (Durham, 1938).

10 Montyon, *Recherches,* 25, 38. On the uncertainties of such estimates see Vincent, "French demography."

11 Montyon, *Recherches,* 90 and Table I, Question 5, Chapitre 10, 91.

12 Ibid., Table I, Question 1, Chapitre 11, 99.

13 Ibid., 96.

14 Ibid., 172.

15 G. Miller, "Airs, waters and places in history," *Journal of the history of medicine 17* (1962): 129–40; C. J. Glacken, *Traces on the Rodhian shore. Nature and culture in Western thought from ancient times to the end of the eighteenth century* (Berkeley, 1967); R. Shackleton, *Montesquieu. A critical biography* (London, 1961), 302–19; J.-P. Peter, "Disease and the sick at the end of the eighteenth century," in R. Forster and O. Ranum (eds.), *Biology of man in history,* trans. E. Forster and P. M. Ranum (Baltimore, 1975), 81–124.

16 Montyon, *Recherches,* 143–5; date in Fig. 1 selected from Table III, Question 3, Chapitre 11, wherein a record is given, by year, from ages one to ninety.

17 Ibid., Table II, Question 6, Chapitre 11, 220.

18 P. H. Niebyl, "The non-naturals," *Bulletin of the history of medicine 45* (1971): 486–92; O. Temkin, *Galenism. Rise and decline of a medical philosophy* (Ithaca, N.Y., 1973), 101ff; W. Coleman, "Health and hygiene in the *Encyclopédie:* a medical doctrine for the bourgeoisie," *Journal of the history of medicine 29* (1974): 399–421. A succinct contemporary statement of the doctrine is given by [Arnolfe d'Aumont], "Non-naturelles, choses," in D. Diderot and J. d'Alembert (eds.), *Encyclopédie,* original Paris edition (1765), XI, 217–24.

19 Montyon's etiological theory was thoroughly miasmatic, not contagionist. See G. Rosen, *A history of public health* (New York, 1958) 287–90; E. H. Ackerknecht, "Anticontagionism between 1821 and 1867," *Bulletin of the history of medicine 22* (1948): 562–93; A. H. Buck, *A treatise on hygiene and public health,* 2 vols. (london, 1879), I, 12–34.

20 See Zylberman, "Auget de Montyon," and G. Chinard, "Eighteenth-century theories on America as a human habitat," *Proceedings of the American Philosophical Society 91* (1947): 27–57. The rediscovery and aesthetic valuation of the wilderness during the eighteenth century are recorded by M. H. Nicolson, *Mountain gloom and mountain glory. The development of the aesthetics of the infinite* (Ithaca, N.Y., 1959).

21 Montyon, *Recherches,* 206–7.

22 Ibid., 207.

23 Ibid., 212–13.

24 Ibid., 224.

25 Ibid., 225.

26 Ibid., 226.

27 Ibid., 227–31.

28 Ibid., 232–5.

29 Ibid., 238–51.

30 Ibid., 258. See P. Ariès, "On the origins of contraception in France," and "An interpretation to be used for a history of mentalities," in O. Ranum and P. Ranum (eds.), *Popular attitudes towards birth control in pre-industrial France and England,* (New York, 1972), 10–20, 100–25.

31 Guimbaud, *Un grand bourgeois,* 230, 228.

32 Ibid., 228.

33 Montyon, *Recherches,* 285. See Boucher d'Argis, "Police," in Diderot and d'Alembert (eds.), *Encyclopédie,* XII, 904–12; L. Febvre, "*Civilisation:* evolution of a word and a group of ideas," in P. Burke (ed.), *A new kind of history,* trans. K. Folca (New York, 1973), 218–57, esp. 224–9; G. Rosen, "Cameralism and the concept of medical police." *Bulletin of the history of medicine 27* (1953): 21–42; "The fate of the concept of medical police," *Centaurus 5* (1957): 97–113.

34 Montyon, *Recherches,* 285. Only surgery, which is "not a conjectural art, whose method is certain and whose successes and failures are [always] evident," provided, in Montyon's estimation (ibid.), reasonably secure grounds for curative action. See T. Gelfand, "Empiricism and eighteenth-century French surgery," *Bulletin of the history of medicine 44* (1970): 40–53; O. Temkin, "The role of surgery in the rise of modern medical thought," *Bulletin of the history of medicine 25* (1951): 248–59.

35 P. Gay, *The Enlightenment, An interpretation,* 2 vols. (New York, 1966–9), II, 12–23.

36 Montyon, *Recherches,* 285–9.

37 O. Hannaway and C. Hannaway, "La femeture du cimetière des Innocents," *Dix-huitième siècle 9* (1977): 181–90.

38 See L. Bernard, *The emerging city. Paris in the age of Louis XIV* (Durham, 1970), 132–259; J.-P. Gutton, *La société et les pauvres. L'exemple de la généralité de Lyon, 1534–1789* (Paris, 1971).

39 Montyon, *Recherches,* 288.

40 Ibid., 290.

41 Ibid., 293.

42 Ibid.

43 L. Edelstein "The diatetics of antiquity," in *Ancient medicine,* ed. and trans. O. Temkin and C. L. Temkin (Baltimore, 1967), 303–16; Coleman, "Health and hygiene."

44 Niebyl "The non-naturals"; J. N. Hallé, "Hygiéne," *Dictionnaire des sciences médicales* (Paris, 1818), XXII, 508–610.

28. Ibid., [illegible]

29. Ibid.

30. [illegible] "[illegible] quantities," [illegible]

31. [illegible] 230–[illegible].

32. Ibid., [illegible].

33. Montesquieu [illegible]

[illegible] (New York, 1979), [illegible]

34. Montesquieu [illegible]

35. [illegible]

36. [illegible]

37. [illegible] (1971), [illegible].

38. See [illegible] (Cambridge, [illegible]) [illegible] (1973)

39. Montesquieu [illegible]

40. Ibid.

41. Ibid., 22[illegible].

42. Ibid.

43. [illegible] "[illegible]" [illegible]

44. [illegible] (Paris, 18[illegible]), XVI, [illegible].

12

Joseph Priestley, eighteenth-century British Neoplatonism, and S. T. Coleridge

ROBERT SCHOFIELD
Iowa State University

In interpretations of the transition from a sensationalist, materialist, empiricism of late eighteenth-century Britain to the transcendental idealism of the early nineteenth century, Joseph Priestley and Samuel Taylor Coleridge have played an anomalous role. Priestley is seen as representing a world view structured by Newton, Locke, and David Hartley that was overturned when Coleridge returned from Germany in 1799, bringing with him a "glowing enthusiasm" for Immanuel Kant, which he transferred to his contemporaries and followers.[1] There seems little reason to challenge the essential validity of the events described in this traditional interpretation, but an examination of their context reveals gaps and incongruities that mar the neat antitheses of literary ideology. Coleridge, it can be shown, became an idealist, prior to his journey to Germany, as an admirer of Jospeh Priestley and inheritor of a continuous strain of native idealism already present in Britain – a strain that served as a base from which his admirers moved to adopt Coleridge's views.

The traditional account, and most of its variants, has its origin in the writings, often incoherent and frequently contradictory, of Coleridge himself. In the *Biographia Literaria,* first published in 1817, Coleridge traces his intellectual itinerary and declares, "The writings of the illustrious sage of Königsberg, the founder of the Critical Philosophy, more than any other work, at once invigorated and disciplined my understanding." And when he moved beyond Kant into the more mystic reaches of Schelling's *Natur-Philosophie* and the *System des Transcendentalen Idealismus,* it was to their mutual possession of the "preparatory discipline" of Kant's philosophy that Coleridge credited the "genial coincidence with much that I had toiled out for myself."[2]

It matters little that Coleridge scarcely began a serious study of Kant before 1801, nor that he seems never entirely to have understood the details of what he studied. Enthusiasm does not depend on understanding nor, given Coleridge's gift for conversation, would the spreading of that enthusiasm. Certainly it is clear from the testimony of contemporaries as various as Humphry Davy, in

This essay was completed before the appearance of Trevor H. Levere's major study of Coleridge and science, *Poetry Realized in Nature: Samuel Taylor Coleridge and Early Nineteenth-Century Science* (Cambridge: Cambridge University Press, 1981). I have however, left the essay unchanged as it seems to me to supplement rather than contradict, or be contradicted by, Levere's work.

natural philosophy, and Frederick Denison Maurice, in theology, that he was a major influence. And it was this "Germano-Coleridgean doctrine" that made him, in the view of J. S. Mill, one of the two "great seminal minds in England" and, in that of Julius Hare, the "true sovereign of modern English thought."[3] One might, however, ask how Coleridge came so early to regard with such favor a philosopher whose works he had scarcely read.

There are two answers to this question, in addition to the usual invoking of innate sympathy of minds. First, Kant was by no means unknown in England when Coleridge left for Germany in September of 1798. By that time there had been published in London at least three books describing the Critical Philosophy.[4] And though Coleridge may not have known of them, he did know of Kant for he describes him as the "great German Metaphysician" in a letter to Thomas Poole, of May 1796. That same month, Coleridge's new Bristol friend, Dr. Thomas Beddoes, a man of broad learning and knowledgeable in metaphysics and German philosophy, had published a letter in the *Monthly Magazine* calling for an early translation of Kant's "celebrated code of metaphysics."[5]

Another, and more interesting reason for Coleridge's early acceptance of Kant can perhaps be found in his prior adoption of a British mode of idealism. The first support for this view comes again from Coleridge. By 1825 he was heartily tired of having "His" ideas linked always with the Germans and declared:

> I can not only honestly assert but I can satisfactorily prove, by reference to Writings (Letters, Marginal Notes and those in books that have never been in my possession since I first left England for Hamburg, &c.) that all the elements, the differentials as the Algebraists say, of my present Opinions existed for me before I had even seen a book of German metaphysics later than Wolff and Leibniz, or could have read it if I had.[6]

This statement does not necessarily contradict his previous encomium to Kant, for those "differentials" might well have needed disciplining, but it does encourage an investigation into possible sources of Coleridge's pre-German idealism.

Once sought, British sources of idealism are easily found. In 1792, for example, the Scottish geologist, James Hutton, published *Dissertations on Different Subjects in Natural Philosophy,* to be followed in 1794 with *An Investigation of the Principles of Knowledge and the Progress of Reason from Sense to Science and Philosophy* and, in 1795, with his famous *Theory of the Earth.* Together these works set forth a functional view of matter as force, or power, strikingly similar to the view identified as characteristic of Kant's natural philosophy. Note the declaration in the *Principles of Knowledge:*

> Having, in an accurate examination of natural bodies found that magnitude and figure, though commonly esteemed absolute qualities, were in their nature only conditional . . . I then found that there is nothing in those external things which strictly speaking, should be considered an absolute volume or real magnitude and figure; but there were only certain powers by which these conceived qualities may be produced in our mind.

Where Hutton obtained this idea of matter has not yet been discovered. When Dr. Beddoes reviewed the *Dissertations* in 1795, about the time of his meeting Coleridge, he remarked that the matter theory resembled that of Roger Joseph

Boscovich, whose *Theoria Philosophiae Naturalis* (Vienna, 1758), ostensibly based on Newton and Leibniz, had been praised by Joseph Priestley and was to be a persisting influence (on Humphry Davy, for example, and Michael Faraday) in British natural philosophy through the nineteenth century. Coleridge, who quoted Hutton's *Principles of Knowledge* in the "Gutch Memorandum Book," was to inscribe the title page of his own copy of the *Principles:* "The writer had made an important step beyond Locke, Berkeley and Hartley – and was clearly in the precincts of the Critical Philosophy." It is not known that Coleridge was introduced to Hutton's work by Beddoes, but that clearly is how the young Humphry Davy learned of it. And Davy, superintendent of research in Beddoes's Pneumatic Medical Institute at Bristol when Coleridge returned from Germany, adopted parts of Hutton's theories in early papers, as subsequently he adopted variations of Coleridge's Germanic idealism.[7]

Hutton's work shows that there was at least one strain of idealism in Britain known to Coleridge before he went to Germany. It seems unlikely however that this one was soon enough, or well enough, known to him to have induced the strong flavor of idealism in his poetry of that period. There must have been other sources. One of these, long discussed by Coleridge scholars, is perhaps to be found in the Neoplatonists that Coleridge is supposed to have read as a schoolboy at Christ's Hospital. How much of Plato, Plotinus, Proclus, Pletho, and the *Theologica Platonica* of Ficino, with which the boy of fifteen 'bewildered himself," carried over into and beyond Coleridge's days at Cambridge can only be guessed. Lucyle Werkmeister has provided a labored identification of Plotinian elements in Coleridge's writing between 1787 and 1791, but the stronger elements of Neoplatonism in the poems of 1794–6 can be traced to a different source.[8] In the "Eolian Harp," first published in 1796, we find:

> And what if all of animated nature
> Be but organic Harps diversely fram'd
> That tremble into thought, as o'er them sweeps
> Plastic and vast, one intellectual breeze
> At once the Soul of each and God of all?[9]

The same theme is to be found in "Religious Musing," begun in December 1794 and completed in March 1796.

> Contemplate Spirits! ye that hover o'er
> With untired gaze the immeasurable fount
> Ebullient with creative Deity!
> And ye of plastic power that interfused
> Roll through the grosser and material mass
> In organizing surge![10]

This view of matter, as inert but moved by a plastic spirit or power that constitutes the soul of each individual particle and yet is a single all-permeating soul, is to be found in a number of Neoplatonists (including Plotinus). It is, however, a particular focus of the work of two metaphysician-natural philosophers of the group known as the Cambridge Neoplatonists. Henry More writes of it in his

Enchiridion Metaphysicum (1671), and the specific term "plastic nature" is emphasized in Ralph Cudworth's *True Intellectual System of the World* (1678), which Coleridge borrowed (in the 1743 edition) from the Bristol Library in mid-May 1795 and again in November 1796.[11] Other references and allusions to Cudworth and More in Coleridge's notebooks and correspondence confirm his familiarity with the writings of these Neoplatonists of seventeenth-century Cambridge.[12]

Demonstration of Coleridge's knowledge of the Cambridge Neoplatonists goes a long way toward solving some, at least, of the ambiguities in his intellectual biography. Appreciation of their work would certainly have provided him with the basis of a sturdy British idealism to be encouraged and directed in Germany. John H. Muirhead writes of these Cambridge philosophers:

> Kant's merit, so far as England was concerned, consisted . . . in his having worked out with hitherto unequalled analytic power ideas that had already found expression and been made the foundation of a whole system in some of her own earlier writers.[13]

Nearly a quarter of a century earlier, Arthur O. Lovejoy said much the same thing, though less charitably, for Lovejoy compressed Kant's "unequalled analytic power" into his "peculiarly systematic and taxonomic mind." In comparison with the writings of the Cambridge Neoplatonists, Lovejoy declared:

> I also believe to be a precisely verifiable fact – that the Kantian doctrine was destitute of any radical originality; that none of the more general and fundamental contentions of the "Kritik der reinen Vernunft" were particularly novel or revolutionary at the time of their original promulgation.[14]

It would be a long, and not particularly fruitful, argument as to whether the frequently disjoint statements of that school of Cambridge theologians and metaphysicians, best represented in the work of Benjamin Whichcote (1609–83), John Smith (1618–52), Henry More (1614–87), and Ralph Cudworth (1617–88), could possibly be set equal to the architectonic systematizations of Kant.[15] Without entering into an invidious comparison of seventeenth-century British Neoplatonism and eighteenth-century German transcendental idealism, one can still inquire how the former might have retained sufficient intellectual vitality in Britain for Coleridge to have discovered it. The major representatives of the school were dead by 1700 and though editions of Cudworth and More were printed as late as 1743, even these would be half a century old when Coleridge began reading them. Is it necessary to rest content with some evasive mystification about "the deeper currents of his nature," or can there be found a more immediate explanation for Coleridge's apparently anachronistic reading?

It is well known that shortly after Coleridge entered Cambridge in 1791 he became a Unitarian, that he seriously contemplated becoming a Unitarian minister during the period 1796–8, and that he retained a Unitarian theological position as late as 1801. The intellectual champion of English Unitarianism had been Joseph Priestley until 1794, when he left England for the United States in self-imposed exile and Coleridge was, for a time, a warm admirer of Priestley. In the "Religious Musings" of 1794–6, he had written:

Lo! Priestley there, patriot and saint and sage,
Him, full of years, from his loved native land
Statesmen blood-stained and priests idolatrous
By dark lies maddening the blind multitude
Drove with vain hate. Calm, pitying he retired,
And mused expectant on these promised years.[16]

In January 1798, he wrote, "I regard every experiment that Priestley made in chemistry, as giving *wings* to his more sublime theological works."[17] Now Priestley's *Free Discussion of the Doctrines of Materialism and Philosophical Necessity* (1778) is referred to in the *Biographia Literaria.* It seems likely, therefore, that among Priestley's other "sublime theological works" Coleridge read would be the two that, with the *Free Discussion,* make up Priestley's metaphysical trilogy: *Disquisitions relating to Matter and Spirit* (1777) and *The Doctrine of Philosophical Necessity Illustrated; being an Appendix to the Disquisitions . . .* (1777). This is the more probable as the *Disquisitions,* in both the first and second editions, were among the books listed in the sales catalogues that included parts of Coleridge's library. Moreover, among the books Coleridge read during the period 1794–8 were Richard Baxter's *Immateriality of the Soul,* Cudworth's *Intellectual System,* David Hartley's *Observations on Man,* Andrew Ramsay's *Philosophical Principles,* Johann Mosheim's *Ecclesiastical History,* the *Works* of Berkeley, and possibly some of the writings of Giordano Bruno. This is an eclectic list, even for so omnivorous a reader as Coleridge, but it has one unifying feature – each book on it is referred to in the *Disquisitions.*[18] If Coleridge read Priestley's metaphysical trilogy (and he is known to have read at least one of the three), he would have been prepared for some of the idealist natural philosophy to be found in Germany. Note, for example, Coleridge's paraphrase of Priestley's argument:

For since impenetrability is intelligible only as a mode of resistance; its admission places the essence of *matter* in a mode of power which it possesses in common with spirit; and body and spirit are therefore no longer absolutely heterogeneous, but may without any absurdity be supposed to be different modes, or degrees of perfection, of a common substratum. . . . But as soon as materialism becomes intelligible, it ceases to be materialism. In order to explain *thinking,* as a material phenomenon, it is necessary to refine matter into a mere modification of intelligence . . . Even so did Priestley . . . He stript matter of all its material properties; substituted spiritual powers.[19]

Compare this with Kant:

We are acquainted with substances in space only through forces which are active in this and that space, either bringing other objects to it (attraction), or preventing them penetrating into it (repulsion and impenetrability). We are not acquainted with any other properties constituting the concept of the substance which appears in space and which we call matter.[20]

Coleridge soon criticizes Priestley for offering a "compendious philosophy, which talking of mind, but thinking of brick and mortar, or other images equally abstracted from body, contrives a theory of spirit by nicknaming matter."[21] But he found the cues for the rational elements, at least, of his early preparatory idealism in Joseph Priestley, for in Priestley there come together four of the five major streams in which Cambridge Neoplatonism flows into the eighteenth century.[22]

The first of the four streams is that of Sir Isaac Newton and Newtonian physicotheologians, such as Richard Bentley and Samuel Clarke. E. A., Burtt's *Metaphysical Foundations of Modern Physical Science* (1951) is a reminder of what many of Newton's contemporaries well knew, namely, that there were, in the *Principia,* ontological assumptions not derived from experience. There seems little doubt that those relating to the concepts of space and of force had been strongly influenced by the work of Henry More.[23]

Having been initially enthusiastic about Descartes, More reacted strongly against the Cartesian position that made matter, once created and put into motion, independent of the Creator. In his *Enchiridion Metaphysicum,* More proposed two answers to this problem. First, space, as infinite extension separate from extended matter, *is* the divine presence and becomes the connective device by which the divine and the natural (matter) can be related. Second, matter is essentially inert and material interaction occurs only as a consequence of a superadded motivating force, the "spirit of nature" that Cudworth was later to call "plastic nature," working according to natural law, which was the will of God.[24]

Each of these ideas, somewhat modified, is to be found in Newton's work. The General Scholium to Book III of the *Principia,* second edition, reads:

> He [God] is not eternity and infinity, but eternal and infinite . . . He endures for ever, and is everywhere present; and by existing always and everywhere, he constitutes duration and space . . . He is omnipotent, not *virtually* only but substantially, for virtue cannot subsist without substance.

Query 31 of the *Opticks* reads:

> Being in all Places, [God] is more able by his Will to move the Bodies within his boundless uniform Sensorium, and thereby form and reform the Parts of the Universe, than we are by our will to move the parts of our own bodies. And yet we are not to consider the World as the Body of God, or the several Parts thereof, as the Parts of God.[25]

And although some Newtonian commentators, particularly on the Continent and later in the eighteenth century, regarded Newtonian central forces in an empiricist, positivist manner, the early physicotheologians saw those forces as visible proofs of the continuing action of God in the universe. The point is cogently made in John Rowning's *Compendious System of Natural Philosophy* (1737–43), a text in frequent use into the 1790s. All matter (with respect to its substance) is ultimately homogeneous and its particles are solid, extended, mobile, and inert. To explain phenomena involving essentially inactive substance, principles of gravitation, cohesion, and repulsion must be added, principles that are so little the result of mechanical cause that "they are the very Reverse, and consequently can be no other than the continual acting of God upon Matter, either mediately or immediately."[26]

A second stream of Cambridge Neoplatonism into the eighteenth century is through the work of John Locke. To suggest that there was a streak of the Neoplatonic in the "father of British empiricism" might come as a shock to nineteenth-century Kantians, who thought Locke insufficiently idealist, or to Hume and latter-day logical positivists, who thought him insufficiently empiricist. It would not have surprised his immediate followers, as it has not surprised the few historians of ideas who have read all of Locke's works and judged what they read in context.[27] Locke commenced writing his most influential work, *An Essay concerning Human Understanding* (1690), in 1671, in consequence of some philosophical discussions in his circle of London friends. Who all of those friends were is not known, but among his frequent associates from 1667 were John Mapletoft, John Tillotson, and Simon Patrick; all disciples, in varying degrees, of John Smith and of Benjamin Whichcote, who commenced preaching in London in 1668 and became Locke's favorite preacher. Before he fled for Holland in 1683, Locke had, or at least commenced, a correspondence with Thomas, son of Ralph Cudworth. While in political exile, Locke's warmest friends were the Dutch Arminians (particularly Philip Limborch and Jean LeClerc), whose principal intellectual allies were the Cambridge Neoplatonists.[28] On his return to England Locke acquired a patroness in Lady Damaris Masham, daughter of Ralph Cudworth, and made the Masham house at Oates his permanent home after 1691. He continued his friendship with Limborch and LeClerc by correspondence until his death in 1704, the year after LeClerc began publishing extracts and analyses of Cudworth's *True Intellectual System,* continuing through nine volumes in his *Bibliotheque Choisie.*

That Locke had many, and close, associations with Cambridge Neoplatonism is clear, but one should not prove even Locke guilty by association. It is, indeed, not necessary to prove Locke a Neoplatonist (as he surely was not) so long as one can show that his work is congruent with some of the Neoplatonic ideas. For this it is essential to recognize that Locke's empiricism was an epistemological, not an ontological, concern, and to any seventeenth- or early eighteenth-century reader ontologies were quite as philosophically relevant as epistemologies. Examination of the *Essay* shows that Locke was a rationalist, not a sensationalist or a nominalist.[29] His denial of innate ideas *is* in opposition to a basic Platonic concept of knowledge held latent in the mind, but most of the Cambridge Neoplatonists opposed this as well. Like Cudworth, Locke believed that the mind had innate capacities for knowledge; through the concurrence of circumstances, the mind compounded sensations by intuition, reflection, and comparison into ideas. Henry More believed that the "soul possessed a sort of blank tablet which could reach out to the exterior world in perceptive understanding."[30] Ideas that for Locke, are complex, generated by the act of comparison, are, for example, the relation of cause and effect, of time and space, of identity and diversity, of equality and excess, and of right and wrong. These are all "creations and inventions of the Understanding," not contained in the existence of things, but added to the data of sensitive experience.[31] Compare this with Henry More:

There are a multitude of Relative Notions or Ideas in the Mind of Man as well Mathematical as Logical . . . that . . . are from the Soul . . . and are the natural Furniture of the Humane Understanding. Such are these, Cause, Effect, Whole and Part, Like and Unlike, Equality and Inequality . . . Proportion and Analogy, Symmetry and Assymmetry and such like, all which relative Ideas I shall easily prove to be no material impresses from without upon the Soul, but her own active Conceptions proceeding from herself whilst she takes Notice of External Objects.[32]

In the "science of morals," where Locke was most influenced by the Cambridge school, and especially by Cudworth, one sees him maintaining a doctrine of intuitive knowledge of morality, which he even spoke of in the Cambridge Platonist metaphor, as "a Candle of the Lord." In morality, the real essences of ideas are the same as the nominal essences and the mind can compare and judge as it does with mathematics.

Since our faculties . . . plainly discover to us the being of a God, and the knowledge of ourselves . . . our proper employment lies in those inquiries most suited to our natural capacities . . . morality is the proper science and business of mankind in general.[33]

Locke was not a Newtonian philosopher, for the greater part of the *Essay* was conceived and structured before he could have become significantly aware of Newton's work. The natural philosophy of the *Essay* is mechanical and corpuscular, derived out of Descartes and Gassendi, and probably strongly influenced by the work of Robert Boyle. But in the natural philosphy, as well, one can see traces of the Cambridge Platonists, in overtones strikingly idealist. To the traditional Cartesian primary qualities of matter (i.e., size, shape, and motion) Locke added solidity (or impenetrability) following a recommendation of Henry More, who had criticized the Cartesians for neglecting this essential distinction between physical and mathematical body. But Locke, although he declares that one knows of the existence of objects by experience, is well aware that one does not discover their essential qualities as a consequence of sensation. He is quite as explicit as Kant himself about the impossibility of ever achieving knowledge of objects in themselves (the *Ding an sich*):

The ideas of the unknown properties of bodies, based on rational and regular experiments, are still but judgement and opinion, not knowledge and certainty, and thus natural philosophy is not capable of being made a science.[34]

Another unexpected transmitter of certain aspects of Cambridge Neoplatonism through eighteenth-century Britain was David Hartley. Hartley is best known for his elaboration of the doctrine of associationism and for the mechanistic, physiological psychology of the first part of his *Observations on Man, his Frame, his Duty, and his Expectations* (1749). The whole of the *Observations*, however, had, as the author states in the preface, been written in the conviction that its "tenets . . . [were] highly conducive to the Promotion of Piety and Virtue amongst Mankind." The focus of the work lies, therefore, in the second part: on the being and attributes of God, of natural and revealed religion, and on the formation of a Christian rule of life. It is this second part of the *Observations* that

Robert Marsh has described as "a kind of unitarian Christian 'Platonism' with a pseudo-Calvinistic Flavor."[35]

As Hartley does not cite authorities (except Revelation) for his theology, evidence for its "Christian 'Platonism' " must be sought in congruences with familiar aspects of Plotinus – or, more immediately, such Cambridge Neoplatonists as John Smith and Ralph Cudworth. Such congruences permeate the *Observations.* Note, for example, Section VII, "Of the Regard due to the Pleasures and Pains of Theopathy in forming the Rule of Life." Here the notion that human miseries are due to moral imperfections – a want of benevolence – mirrors the Neoplatonist concept of vice, or evil, as a deficiency, a lack of good rather than positive being. "The existence of every thing else in only the Effect, Pledge, and Proof of his [God's] Existence and Glory." God is light, he is love, he is the only cause and reality, and the ultimate end of man is the love of God. Typical, also, is the argument, relating to the being of God, that matter is one of the works of the infinite power of God, that it is a "mere passive thing, whose very essence is to be endued with *Vis inertiae,*" and all of whose motions are entirely dependent on God.[36]

There is, however, or seems to be, a major incongruity in that aspect of Hartley's work alluded to in Marsh's "pseudo-Calvinistic flavor," namely, its determinism (or necessitarianism). It was Hartley's *Observations* that "established" Priestley "in the belief of the doctrine of necessity," and it was Hartley, by way of Joseph Priestley, who made Coleridge a necessitarian, from which he was supposedly "released" by the influence of Cudworth and other Cambridge Platonists or, these unaccountably failing, by the German idealist philosophers.[37] But the Cambridge Platonists or some of them, were determinists in just the way that Hartley was.

Hartley declared that omnipotence and omniscience, as natural attributes of God, make impossible "philosophical free-will" that is, the power to do either one thing or its contrary, when previous circumstances are identical. Human actions result "from previous circumstances of Body and Mind, in the same manner and with the same Certainty, as other Effects do from their Mechanical Causes." Philosophical free will would make us less able to comply with the will of God and not add, but take away from the requisites proposed by religion. Practical free will, the power of calling up ideas, resisting motives of sensuality, ambition, resentment; of deliberating, suspending, and choosing are all available to us, through mechanism and association.[38]

Now compare these statements with some of John Smith, of John Norris (1657–1777), one of the recognized heirs of the Cambridge tradition), or of Ralph Cudworth. The excellency and nobleness of true Religion, said Smith, lie in beholding all things, from beginning to end, as "acting that part which the Supreme Mind and Wisdom that governs all things hath appointed them." Freedom of the will, to Norris, lies not in the power of directly determining to overt action but in the power of attending or not attending more or less to an objective. And finally, from Cudworth, who is supposed to have liberated Coleridge from determinism: Unconditional liberty of choice is a spurious and adulterate notion, a vulgar view of freedom, not a power but a weakness, meaning the power to defy and abandon wisdom and goodness, and also the reason that dictates what is most agreeable to personal utility.[39]

The least-studied, but perhaps most important, mode for propagation of Cambridge Neoplatonism into eighteenth-century Britain was through the liberal dissenting academies, for they combined Locke, Newtonian physicotheology, and Hartley within a formal course of metaphysical instruction that also made explicit reference to members of the Cambridge school. The dissenting academies were founded by Nonconformist ministers, after the Act of Uniformity had closed Oxford and Cambridge for the training of dissenting clergy. Many of these academies (e.g., the strict Calvinist ones) had the constraints of sectarian religious dogmas to define their curricula, but for the more liberal schools, only the Cambridge Platonists, with insistence on reason and complete toleration of all Protestant Christians, provided a theological and metaphysical system consistent with their beliefs.[40]

The most influential of the liberal academies was that founded in 1729 at Northampton by Philip Doddridge, which supplied tutors for at least five other academies, including Daventry Academy, where Joseph Priestley studied between 1752 and 1755; and Warrington and Hackney New College, where he taught. Doddridge had studied at an academy whose curriculum developed (in three generations of schools) from that established by Richard Frankland, a graduate of Christ's College, Cambridge, where Henry More was fellow and tutor, and an M. A. the year following Cudworth's installation as master there. The framework of Doddridge's curriculum seems to have been that established by Frankland, as modified by his immediate tutor, John Jennings, but Doddridge was also influenced in details of his curriculum by the writings of Samuel Cradock (1621–1706), a graduate like most of the Cambridge Platonists of Emmanuel College, a relative of Benjamin Whichcote, executor to John Smith, and successor as rector to Ralph Cudworth. Still more influential in curriculum design was the personal advice of Doddridge's good friend, Isaac Watts.

Watts is now known primarily for his hymns, but in the eighteenth century, he was better known as the author of a series of texts, including one on grammar, another on logic (which Priestley read as a schoolboy), another on natural philosophy, and still another, on the *Improvement of the Mind,* which the young Michael Faraday read in 1809 to give system to his efforts at self-education. Watts had studied at Newington Green Academy, then headed by Thomas Rowe, an admirer of Descartes and of Henry More. Rowe, in turn, had studied under the founder of the academy, Theophilus Gale, whose major lifework, the *Court of the Gentiles,* was written to prove that the works of Plato, representing as they did the ultimate in truth and wisdom, must be dependent on a tradition of Hebrew learning, only fragmentarily represented in the Scriptures. The natural expectation that Watts, with this educational background, would encourage Doddridge in Neoplatonism is confirmed by reference to his *Philosophical Essays,* which, for example, refers admiringly to Cudworth, to Henry More's disciples,

Joseph Glanvill and Nathaniel Culverwell, and suggests that the human brain is designed by God to produce certain kinds of ideas from certain kinds of sensations – ideas, then, are not temporally innate, but are logically innate.[41]

The most detailed information about Doddridge's curriculum comes to us through the continuation of his academy at Daventry, under the direction of two of his students, Caleb Ashworth and Samuel Clark. Joseph Priestley enrolled at Daventry the first year of its existence and, through his *Memoirs* and a surviving fragment of the diary he kept while there, a list of the subjects studied and the texts used can be partially reconstructed. Priestley read, at Daventry, the *Essay* by Locke (which he had earlier studied on his own), Rowning's *Compendious System,* and Hartley's *Observations.* He also studied, in manuscript, the text in metaphysics prepared by Doddridge, which was subsequently edited and published by Samuel Clark, *A Course of Lectures on the Principal Subjects in Pneumatology, Ethics, and Divinity.*[42]

In its published form, Doddridge's *Course of Lectures* consists of the "Heads" of some two hundred and thirty lectures, curiously structured as Axioms, Definitions, Propositions, Corollaries, Scholia, and Lemmas. As the subject matter chiefly relates to the powers and faculties of the human mind, the existence and attributes of God, the nature of virtue and of civil government, the character of the soul, and evidences for revelation, the geometrical form might seem a clumsy mode of presentation. But Priestley wrote of it:

> If these definitions and axioms be laid down with due accuracy and circumspection, they not only introduce the easiest, the most natural, and cogent method of demonstrating any position, but lead to an easy method of examining the strength or weakness of the ensuing arguments.[43]

There is, or was supposed to be, a further unique characteristic of the *Course of Lectures.* It was designed to present, through a system of bibliographical references, conflicting opinions on every issue thought controversial, and students were expected to follow the controversies through the cited texts until they were thoroughly familiar with both sides of the arguments. This aspect of Doddridge's teaching was condemned by his more orthodox contemporaries and was praised by Priestley.[44] It is with considerable surprise, therefore, that one discovers the very narrow range of authorities cited in the *Course of Lectures.* John Locke, Isaac Watts, Henry More, Richard Baxter, Jean LeClerc, Descartes, and the Boyle lecturers, Richard Bentley, Samuel Clarke, and Thomas Burnet are the names most often cited. And for natural philosophy (not a major element in the *Lectures* and not, apparently, regarded as controversial) the range is even smaller: Locke, Watts, More, and Descartes again, to which are added Cudworth, John Ray, William Derham, and others of the physicotheologian cast. In short, next to Locke and Watts, the people most frequently cited in Doddridge's text are the Cambridge Neoplatonists or their allies, disciples, or sympathizers: the Dutch Arminians, Baxterian Presbyterians, Anglican latitudinarians, and physicotheologians.

These men did not always agree with one another, of course, but in their disagreement a general idealist position is maintained. Cudworth's "plastic nature" is opposed, for example, as is More's definition of space and time as attributes of God. Yet matter exists, moves, and acts on other matter only through the agency of a perpetual, omnipotent being, who proceeds by rules and methods called the laws of nature; both space and time are mere ideas and have "no real and positive existence without us."[45] Add to these the definitions for which there was no disagreement: "Natural Philosophy is that branch of learning which relates to *body*, giving an account of its various phaenomena, and the principles on which the solution of them depends." "We can have no conception of any substance distinct from all the properties of the being in which they inhere; for this would imply that the being itself inheres; and so on to infinity." "We have as clear an idea of spirit as we have of body; the essential properties of each being equally known, and the inward constitution equally unknown."[46]

As a result of his reading at Daventry Academy, of Locke and Watts, of Rowning, and Hartley, of Doddridge's *Course of Lectures*, and of the authors most cited there, Priestley graduated in 1755 with the foundations of a Neoplatonic, Christian idealism. Now Priestley was not a philosopher, in the sense that he self-consciously articulated a coherent system of thought. That he constructed, for himself, the elements of such a system can be inferred, however, from the general consistency of his metaphysical beliefs, as extracted from the work of some forty years on theology, science, and a bewildering multiplicity of other subjects. It cannot be argued that Priestley was uniformly a Neoplatonist or that his philosophical position remained unchanged. The natural philosophy of Priestley, which intrigued and inspired the young Coleridge, was put together from reading and experiments over the years. But Priestley began his career with a Platonist bias and themes from his education at Daventry recur throughout his writing.

From his earliest scientific publication, when he declared that electricity, optics, and chemistry were to be "inlets" into the inner properties of body, to his last, when be belittled "knowledge of the elements which enter into the composition of natural substances," in favor of continued investigation into the principles of their properties, Priestley remained a Natural Philospher, as the term had been defined at Daventry.[47] Yet the purpose of his study, in the final analysis, was always natural theology, for he knew the ultimate nature of matter. In the *Institutes of Natural and Revealed Religion*, published finally in 1772 but begun as a student at Daventry, he wrote:

> As the matter of which the world consists can only be moved and acted upon, and is altogether incapable of moving itself, or of acting, so all the powers of nature . . . can only be the effect of the divine energy perpetually acting upon them.[48]

From this position, wholly consistent with that of Henry More or John Rowning, he had certainly moved by 1778; in the *Free Discussion of the Doctrines of Materialism* he suggests that the Divine Being, in creating matter,

> only fixed *certain centers of various attractions and repulsions*, extending idefinitely in all directions . . . , these centers approaching to, or receding from each other, and consequently carrying their peculiar spheres of attraction and repulsion along with them . . .

matter is, by this means, resolved into nothing but the *divine agency*, exerted according to certain rules.[49]

But the new position, that read by Coleridge some twenty years later, is one of greater not lesser idealism, and Priestley's defense of the new view, offered for example, in his *Letters to the Philosophers . . . of France* of 1793, sounds like an echo of Kant, or of James Hutton, but far more likely is a reminder of the definition of substance from Doddridge's *Lectures:*

We know nothing of any substance, having no ideas of any thing but what we call properties, which, as we say, inhere in, or belong to, the several things, or substances, that we are acquainted with.[50]

And surely it is from the same combination of schooling in Neoplatonism, theological argument, and scientific experimentation that Priestley derived *his* confident a priori assertion of causality against the skepticism of David Hume. The plan of nature, from which the wisdom of God is inferred, requires intelligibility and intelligibility requires a system of natural laws. The constant temporal conjunction of some events is necessarily connected through those natural laws, and "we [therefore] conclude that the conjunction, if constant, is equally necessary, even when we are not able distinctly to perceive it."[51]

The rationalist idealism Priestley learned at Daventry was not commonplace in late eighteenth-century Britain, where the leaders of thought had selected, from Locke, Newton, and Hartley, just those elements of materialism, utilitarianism, and empiricism the Romantics of the nineteenth century were so vigorously to reject. And Coleridge, who might have championed that idealism, failed to do so. This was not simply because he was never conscientious in acknowledgment of the sources of his thought. As Fichte and Schelling had soon departed from Kant in their development of *Natur-Philosophie,* so Coleridge left the mathematical, mechanical idealism of Priestley for a spiritualized organicism.[52] It was an early development of his thought, and signs of it can be found in the Neoplatonic poetry of the pre-German years. An early draft of the lines, previously quoted, from "The Eolian Harp," had read:

Thus *God* would be the universal Soul,
Mechaniz'd matter as th' organic harps
And each one's Tunes be that, which each calls I.[53]

No doubt the final published form is better poetry, but it surely is worth noting, as well, that "Mechaniz'd matter" has been replaced by "animated nature" in the revision. Some ambiguity remains. Note the lines from "The Destiny of Nations: A Vision," first written in 1796:

. . . themselves they cheat
With emptiness of learned phrase,
Their subtle fluids, impacts, essences
Self-working tools, uncaused effects, and all
Those blind Omniscients, those Almightly Slaves,
Untenanting creation of its God.
But Properties are God: the naked mass
(If mass there be, fantastic guess or ghost)
Acts only by its inactivity.

> Here we pause humbly. Others boldlier think
> That as one body seems the aggregate
> Of atoms numberless, each organized;
> So by a strange and dim similitude
> Infinite myriads of self-conscious minds
> Are one all-conscious Spirit, which informs
> With absolute ubiquity of thought
> (His one eternal self-affirming act!)
> All his involved Monads, that yet seem
> With various province, and apt agency
> Each to pursue its own self-centering end.[54]

Here, in an attack on materialist natural philosophy, there is a balancing of the mechanical world view of Priestley against the "boldlier" vitalistic monadology of Leibniz. But bolder still, Coleridge soon departs even that link with the idealism of the seventeenth century as he goes on into his own version of mystic, organic idealism. Nonetheless, the path he traveled led through the idealism of the Cambridge Neoplatonists, Newton, Locke, and Hartley, as transmitted by Joseph Priestley. And if others, less bold then he, were inspired to idealism by Coleridge's example, surely their acceptance was made easier by the presence of a continuous, if scarcely recognized, native tradition of Neoplatonic idealism in eighteenth-century Britain.

NOTES

1 A convenient summary of the traditional Priestley-Coleridge antithesis can be found in Basil Willey, *The Eighteenth Century Background: Studies on the Idea of Nature in the Thought of the Period* (London: Chatto & Windus, 1950), pp. 168–204; *Nineteenth-Century Studies: Coleridge to Matthew Arnold* (London: Chatto & Windus, 1950), pp. 1–50. A full biography of Priestley does not yet exist; the most recent account is in the *Dictionary of Scientific Biography*, vol. 11, pp. 139–47. The phrase "glowing enthusiasm" of Coleridge is quoted from L. Pearce Williams, *Michael Faraday: A Biography* (London: Chapman & Hall, 1965), p. 62, who represents a moderate form of the traditional interpretation for the history of science. For a summary of Coleridge researches prior to 1950, see "Coleridge" and "Coleridge's Philosophy and Criticism" by T. M. Raysor and Rene Wellek, respectively, in Thomas M. Raysor (ed.), *The English Romantic Poets* (New York: Modern Language Association of America, 1950), pp. 66–117. Some more recent inquiries are cited later. See also E. S. Shaffer, "Coleridge and Natural Philosophy: A Review of Recent Literary and Historical Research," *History of Science* 12 (1974), pp. 284–95, which criticizes this traditional view.

2 Samuel Taylor Coleridge, *Biographia Literaria*, vol. 1 ed. J. Shawcross (London: Oxford University Press, 1907), pp. 102–3.

3 Quoted by Willey, *Nineteenth-century Studies*, pp. 1, 2.

4 F. A. Nitsche, *A General and Introductory View of Professor Kant's Principles* . . . (London: J. Downes, 1796); *The Principles of Critical Philosophy. Selected from the Works of Emmanuel Kant*, and expounded by James Sigismund Beck (London: J. Johnson, 1797); and A. F. M. Willich, *Elements of the Critical Philosophy* (London: T. N. Longman, 1798).

See also Rene Wellek, *Immanuel Kant in England* (Princeton, N.J.: Princeton University Press, 1931).

5 S. T. Coleridge letter to Thos. Poole, May 5, 1796, in Leslie Griggs (ed.), *Collected Letters of Samuel Taylor Coleridge,* vol. 1 (Oxford: Oxford University Press [Clarendon Press], 1971), p. 209. Thomas Beddoes, "Account of Kant's Philosophy," *Monthly Magazine* 1 (1796), pp. 265–7; the letter is dated March 28, 1796. See also E. S. Shaffer, *'Kubla Khan' and The Fall of Jerusalem: The Mythological School in Biblical Criticism and Secular Literature* (Cambridge: Cambridge University Press, 1975), pp. 24–7, 48; and Werner W. Beyer, "Coleridge's Early Knowledge of German," *Modern Philology* 52 (1954–5), pp. 192–200.

6 Letter of April 8, 1825, to John Taylor Coleridge, *Collected Letters,* vol. 5, pp. 421–1.

7 My treatment of Hutton is initially derived from Dr. Patsy A. Gerstner, "James Hutton's Theory of the Earth and his Theory of Matter," *Isis* 59 (1968), pp. 26–31. The quotation from the *Principles of Knowledge,* p. xlv, and Coleridge's inscription on its title page are taken from H. W. Piper, *The Active Universe: Pantheism and the Concept of Imagination in the English Romantic Poets* (London: Athlone, 1962), p. 111. Piper is the only person I know who has previously discussed Priestley's influence on Coleridge; see also his 'Pantheistic Sources of Coleridge's Early Poetry," *Journal of the History of Ideas* 20 (1959), pp. 47–59. Davy's admiration for Coleridge is treated by Williams, *Michael Faraday,* pp. 61–3, and his relation to Beddoes and to Hutton is discussed in Robert E. Schofield, *Mechanism and Materialism: British Natural Philosophy in An Age of Reason* (Princeton, N. J.: Princeton University Press, 1970), pp. 292–4, as is the *Theoria* of Boscovich, pp. 236–41.

8 Lucyle Werkmeister, "The Early Coleridge: His 'Rage for Metaphysics,' " *Harvard Theological Review* 54 (1961), pp. 99–123.

9 "Eolian Harp," lines 45–59, from Ernest Hartley Coleridge (ed.), *The Poems of Samuel Taylor Coleridge* (London: Oxford University Press, 1912), p. 102.

10 "Religious Musings," lines 402–7, *Poems of Samuel Taylor Coleridge,* p. 102.

11 George Whalley, "The Bristol Library Borrowings of Southey and Coleridge, 1793–8," *The Library* 4 (ser. 5, 1950), pp. 114–32. I owe this reference initially to W. Schrickx, "Coleridge and the Cambridge Platonists," *Review of English Literature* 7 (1966), pp. 71–91.

12 In addition to Schrickx's article, cited in n. 11, many books and articles relate Coleridge to the Neoplatonists. Claud Howard, *Coleridge's Idealism: A Study of its Relationship to Kant and to the Cambridge Platonists* (Boston: Richard G. Badger, 1924); John H. Muirhead, *Coleridge as Philosopher* (London, 1930, reprint, New York: Humanities Press, 1954). See also Wellek, cited in Raysor, *The English Romantic Poets.*

13 John H. Muirhead, *The Platonic Tradition in Anglo-Saxon Philosophy: Studies in the History of Idealism in England and America* (New York: Macmillan, 1931), pp. 13–14.

14 Arthur O. Lovejoy, "Kant and the English Platonists," in *Essays Philosophical and Psychological in Honor of William James* (New York: Longmans, Green, 1908), pp. 265–302.

15 There are several detailed studies of the Cambridge Neoplatonists. In addition to those of Muirhead and Lovejoy, one can cite Frederick J. Powicke, *The Cambridge Platonists: A Study* (1926; reprint, Hamden, Conn.: Archon Books, 1971); Paul Russell Anderson, *Science in Defense of Liberal Religion: A Study of Henry More's Attempt to Link Seventeenth-Century Religion with Science* (New York: Putnam, 1933); Ernst Cassirer, *The Platonic Renaissance in England,* trans. James P. Pettengrove, from German ed. of 1932 (Austin: University of Texas Press, 1953).

16 "Religious Musings," lines 371–6, *Poems of Samuel Taylor Coleridge,* p. 123.

17 Letter of January 16, 1798 to John Prior Estlin, *Collected Letters,* vol. 1, p. 372. In a letter of March 23, 1801, to Thomas Poole, vol. 2, p. 710, Coleridge was still talking of settling near Priestley in America.

18 The reference to the *Free Discussion* is in *Biographia Literaria,* vol. 1, p. 91. For other reading, in addition to Whalley's "Bristol Library Borrowings," see Piper's *Active Universe,* p. 33. The Bristol borrowings, incidentally, also included Priestley's *History of the Corruptions of Christianity* (1782).

19 *Biographia Literaria,* vol. 1, pp. 88, 91; quoted by Piper, *Active Universe,* p. 35.

20 Immanuel Kant, *Critique of Pure Reason,* trans. Norman Kemp Smith (London: Macmillan, 1953), p. 279; also quoted, in a slightly different translation, by Williams, *Michael Faraday,* p. 61. For a fuller discussion of Kant's position on matter and force, see Mary B. Hesse, *Forces and Fields: The Concept of Action at a Distance in the History of Physics* (London: Thomas Nelson & Sons, 1961), pp. 169–80.

21 *Biographia Literaria,* vol. 1, p. 163; this is probably a reference to Priestley's "Why, since I deny all solidity or impenetrability . . . should [I] choose to make use of so obnoxious a term as matter, when the less exceptionable one of spirit would answer my purpose full as well? . . . the cause of truth is best answered by calling every thing by its usual name, and I think it a mean subterfuge to impose upon mankind by the use of words." *Doctrine of Philosophical Necessity* &c., 2d ed. (London: J. Johnson, 1782), pp. 252–3.

22 The one that does not appear to concern Priestley is that of Christian Kabbalism, with a Cartesian flavor, which begins in Henry More's *Conjectura Cabbalistica . . . or a Conjectural Essay of Interpreting the Mind of Moses* (1653), continues in such works by John Hutchinson (1674–1737) as *Moses' Principia* (1724–7) and *Glory or Gravity* (1733), from these into the work of Hutchinsonians such as William Jones of Nayland, and from Jones, by way of the Glassite (Sandemanian) James Tytler, into the *Encyclopedia Britannica,* where the ideas were read by the young Michael Faraday.

23 E. A. Burtt, *Metaphysical Foundations . . .* (New York: Humanities Press, 1951), pp. 253–64; see also Richard S. Westfall, *Force in Newton's Physics: The Science of Dynamics in the Seventeenth Century* (London: Macdonald; New York: American Elsevier, 1971), chap. VII, pp. 323–423.

24 On More's natural philosophy, see Paul Russell Anderson, *Science in Defense of Liberal Religion,* esp. pp. 155–6.

25 General Scholium, Bk. III, Sir Isaac Newton's *Mathematical Principles of Natural Philosophy,* trans. Andrew Motte, from the 1727 ed., rev. Florian Cajori (Berkeley: University of California Press, 1947), p. 545; Sir Isaac Newton, *Opticks* (New York: Dover, 1952), based on 4th ed., 1730, Query 31, p. 403. Both quotations are to be found in Burtt, *Metaphysical Foundations,* pp. 257, 259, respectively.

26 John Rowning, *Compendious System* (London: Sam Harding, 1743), preface, p. xxxix. I have discussed Bentley, Clarke, and Rowning in my *Mechanism and Materialism,* pp. 34–39.

27 The best known of these is Richard I. Aaron, whose standard biography, *John Locke,* 2d ed. (Oxford: Oxford University Press, 1955), p. 127, declares, "Much of the fourth book of the *Essay* [*Concerning Human Understanding*] might have been written by one of the Cambridge School." See also Thomas E. Webb, *The Intellectualism of Locke: An Essay* (Dublin: William McGee, 1857); Georg von Hertling, *John Locke und die Schule von Cambridge* (Freiberg; Herder, 1892), esp. chap. III, pp. 159–243; and Mattoon Monroe Curtis, *An Outline of Locke's Ethical Philosophy* (Leipzig: Gustav Fock, 1890).

28 Most of the biographical information is taken from Aaron, *John Locke;* see also

Rosalie Colie, *Light and Enlightenment: A Study of the Cambridge Platonists and the Dutch Arminians* (Cambridge: Cambridge University Press, 1957).

29 I have used the twentieth edition of John Locke, *An Essay Concerning Human Understanding* (London, 1796), but compared the relevant passages to those printed in earlier editions to insure that no significant variations have occurred from what would have been read, e.g., by Joseph Priestley, in the 1740s and 1750s.

30 Anderson, *Science in Defense of Liberal Religion,* p. 159; also Webb, *Intellectualism of Locke,* pp. 41–2, 53.

31 Locke, paraphrased by Webb, *Intellectualism of Locke,* pp. 94–5; Locke's *Essay,* Bk. II, chap. 25, para. 11; chap. 26, para. 3, 5; chap. 27, para. 1; chap. 28, para. 1, 4.

32 Quoted from More, *Antidote against Atheism* (1712 ed.), p. 18, by Claud Howard, *Coleridge's Idealism,* p. 33.

33 John Locke, *Essay,* Bk. 4, chap. 12, para. 11; see also Aaron, *John Locke,* pp. 10–11.

34 Locke, *Essay,* Bk. 4, chap. 12, para. 8, 10; plus repeated passages, such as Bk. 4, chap 2, para. 11–13; Bk. 4, chap. 3, para. 12; Bk. 4, chap. 13, para. 16.

35 David Hartley, *Observations on Man . . .* (London, 1749; reprint, Hildesheim: Georg Olms, 1967), vol. 1, p. xviii; Robert Marsh, "The Second Part of Hartley's System," *Journal of the History of Ideas* 20 (1959), pp. 264–73, esp. 272.

36 Hartley, *Observations,* vol. 2, pp. 309–13; 31–3.

37 See Joseph Priestley, *Memoirs and Correspondence 1733–1787,* vol. 1, part 1, of *The Theological and Miscellaneous Works of Joseph Priestley,* ed. John Towill Rutt (London, 1831; reprint, New York; Kraus, 1972), p. 24, for the influence of Hartley; and W. Schrickx, "Coleridge and the Cambridge Platonists," pp. 79–80, 91, for the supposed incongruity of determinism with Neoplatonism.

38 Hartley, *Observations,* vol. 1, p. 500. vol. 2, pp. 53, 57, 66.

39 Muirhead, *Platonic Tradition,* pp. 30, 92 (quoting from Norris's *Theory and Regulation of Love,* 1688), and p. 68 respectively. See also Mattoon Monroe Curtis, "Kantean Elements in Jonathan Edwards," *Philosophische Abhandlung* (1906), pp. 34–63; where the great American necessaritarian is shown to have idealistic elements, derived from Cudworth, Newton, and Locke.

40 The best single and detailed work on the dissenting academies is that of J. W. Ashley Smith, *The Birth of Modern Education: The Contribution of the Dissenting Academies 1660–1800* (London: Independent Press, 1954), to whom I owe the initial suggestion of Neoplatonism in these academies.

41 Isaac Watts, *Philosophical Essays on Various Subjects . . . to which is subjoined a brief Scheme of Ontology,* &c., 2d ed. (London: Richard Ford and Richard Hett, 1734).

42 Philip Doddridge, *Lectures on . . . Pneumatology,* &c. (London, J. Buckland, J. Rivington, R. Baldwin et al., 1763). Priestley says, in his *Memoirs,* p. 23, that the printed version is substantially the same as the one he studied; the diary is extracted in the earliest edition of the *Memoirs of Dr. Joseph Priestley to the Year 1795 . . . with a continuation by his son,* vol. 1 (Northumberland, Pa.: John Binns, 1806), pp. 178–84.

43 Joseph Priestley, *A Course of Lectures on Oratory and Criticism* (London: J. Johnson, 1777), p. 46. One may note, here, the implicit agreement with Cudworth and Locke on the mathematical character of questions relating to morality.

44 See J. W. Ashley Smith, *Birth of Modern Education,* pp. 141–2; and Joseph Priestley, *Memoirs,* vol. 1, pp. 23–4.

45 Doddridge, *Course of Lectures,* Bk. II, Prop. XVIII, Dem. 1, 2, Cor. 1, 5 (pp. 72–4); Bk. II, Prop. XL (pp. 89–92).

46 Doddridge, *Course of Lectures,* Bk. I, Def VI (p. 3); Def. III, Cor. 2; and Def. V, Cor. 1 (p. 2), respectively. Note that Locke and Watts are the principal cited authorities for Cor. 2, and Locke one of those for Cor. 1.

47 See Preface, p. XIII and pp. 502–3 of *The History and Present State of Electricity* (London: J. Johnson et al., 1767), and Joseph Priestley, "Miscellaneous Observations relating to the Doctrine of Air," *New York Medical Repository* 5 (1802), pp. 264–7.

48 Joseph Priestley, *Institutes of Natural and Revealed Religion,* 2d ed., vol. 1 (London: J. Johnson, 1782), p. 35. See *Memoirs,* p. 27, for reference to the beginning of work on the *Institutes.*

49 *Free Discussion of the Doctrines of Materialism and Philosophical Necessity* (London: J. Johnson, 1778), p. 254.

50 Joseph Priestley, *Letters to the Philosophers and Politicians of France on the Subject of Religion* (London, J. Johnson, 1793), p. 8; and Doddridge, *Course of Lectures,* Pt. I, Def. III, Cor. 2.

51 Joseph Priestley, *Letters to a Philosophical Unbeliever,* 2d ed. (Birmingham: for J. Johnson, 1787), Pt. I, pp. 94–5, 208–9.

52 For early recognition of the differences between Kantian idealism and that of Fichte and Schelling, see J. B. Stallo, *General Principles of the Philosophy of Nature* (Boston, 1848), pp. xviii, 189, 210, 221–2, 338, and English summary, pp. 35–7, of Gunnar Eriksson, "Berzelius och Atomteorin," *Lychnos* (1965–6), pp. 1–34.

53 Quoted by W. Schrickx, "Coleridge and the Cambridge Platonists," p. 85.

54 "Destiny of Nations," lines 30–49, *Poems of Samuel Taylor Coleridge,* p. 132.

13
Enlightenment views on the genetic perfectibility of man

VICTOR HILTS
University of Wisconsin, Madison

The article on "perfection" in the seventh edition of the *Encyclopaedia Britannica,* published in 1842, contains the following definitions:

> *Physical,* or natural perfection is that by which a thing has all its powers and faculties, and those too in full vigor; and all its parts, both principal and secondary, and those in their due proportion, constitution, and adjustment. In this sense man is commonly said to be perfect when he has a sound mind in a sound body.
>
> *Moral* perfection is an eminent degree of virtue or moral goodness, to which men arrive by repeated acts of piety, beneficence, and self-restraint.
>
> *Metaphysical,* transcendental, or essential perfection is the possession of all the essential attributes, or of all the parts necessary to the integrity of a substance.[1]

These definitions had been stated in nearly identical form in earlier editions of the *Britannica* and were, in fact, translations of definitions that had been given in the *Lexicon rationale sive thesaurus philosophicus* edited in the seventeenth century by the French Protestant encyclopedist Étienne Chauvin.[2]

Recently John Passmore has surveyed differing views of human perfectibility in Western thought from antiquity to the present.[3] Of the three types of perfection mentioned, Passmore deals almost exclusively with the second two, but several writers are mentioned by him as placing their emphasis on the physical perfectibility of mankind. Passmore has appropriately called them believers in "genetic perfectibility," because they maintained, like later eugenicists, that children would be born with a "sound mind in a sound body" when their parents possessed these same qualities. One such writer was Tommaso Campanella, who published his *Civitas solis* in 1625.[4] Following the example set by Plato's *Republic,* Campanella suggested that children ought to be bred for the good of the race and that magistrates ought to arrange marriages according to philosophical rules. Also mentioned by Passmore was Francis Galton, the mid-nineteenth-century creator of the word "eugenics." Writing after the publication of Darwin's *Origin of Species,* Galton argued that human beings could take account of their own hereditary potential and direct the future evolution of the race.[5]

The writers mentioned by Passmore were far from alone in advocating the genetic prefectibility of mankind. In her history of utopian thought, Joyce Hertzler noted that several early utopians, including More and Bacon, as well as

Campanella, could be considered "prophets" of eugenics.[6] Other Renaissance names could be added to the list, including Juan Luis Vives, Juan Huarte, and especially Robert Burton. Burton, the English author of the *Anatomy of Melancholy*, believed that melancholy was a hereditary disease transmitted from generation to generation, and that because of such diseases, "it comes to pass that our generation is corrupt; we have many weak persons, both in body and mind, many feral diseases raging amongst us, crazed families, *parentes peremptores:* our fathers bad; and we likely to be worse."[7] More succinctly, said Burton, the world is full of "dizzards."[8] As a remedy Burton proposed severe restrictions on marriage, and he even went so far as to commend the ancient Scots for gelding all men "visited with the falling sickness, madness, gout, leprosie, or any such dangerous disease, which was likely to be propagated from father to son."[9]

Ideas concerning the genetic perfectibility of mankind in the writings of Campanella, More, Vives, Huarte, and Burton seem to establish a link between the protoeugenical thought of antiquity and that of the Renaissance.[10] This author believes that there are also reasons for thinking that ideas concerning genetic perfectibility never entirely disappeared from view after 1600. During the early nineteenth century these ideas were an integral part of the rhetoric of phrenological reformers.[11] Charles Rosenberg, who has examined nineteenth-century American attitudes toward hereditary disease, has shown that American physicians throughout the century were much concerned with the reverse side of genetic perfection – hereditary degeneration.[12] This chapter is an attempt to show that ideas of genetic perfectibility were present when they might have been least expected – during the Enlightenment.

HEREDITARY DISEASE

In a *Comparative View of the State and Faculties of Man with those of the Animal World*, published in 1765 from lectures given in Edinburgh in 1758, the physician John Gregory wrote as follows:

> We should likewise avail ourselves of the Observations made on tame Animals in those particulars where Art has improved upon Nature . . . Thus by a proper attention we can preserve and improve the breed of Horses, Dogs, Cattle, and indeed all other Animals. Yet it is amazing this Observation was never applied to the Human Species, where it would be equally applicable . . . A proper attention to this subject would enable us to improve not only the constitutions but the Character of our Posterity. Yet we everyday see very sensible people, who are anxiously attentive to preserve or improve the breed of their Horses, tainting the blood of their children and entailing on them not only the most loathsome diseases of Body, but Madness, Folly, and the most unworthy dispositions, and this too, when they cannot plead being stimulated or impelled by passion.[13]

In spite of Gregory's amazement, it was hardly surprising that Enlightenment philosophers did not recommend that human beings be improved as yet another breed of animals. For philosophers hereditary perfectibility was not a matter of great importance because the differences among mankind were less important than the similarities. This Enlightenment belief that "men must at *birth* be exactly equal" was far more than (in the words of John H. Randall) "a corollary that was drawn from Locke's sensationalism"; it was an integral part of the philosophical enterprise itself.[14] If philosophers did not assume that the reasoning process of other men was much like their own, how could they generalize from introspection? Rather than a corollary, the doctrine of equality was thus more like a postulate, one without which philosophy based on introspective experience could not succeed. Symptomatically, whereas Robert Burton was convinced that the world is filled with "dizzards," René Descartes took the position that "good sense is of all things in the world the most equally distributed."[15]

Although Enlightenment philosophers were not much concerned with either inheritance or the diversity among individuals, the case was very different with respect to Enlightenment physicians. When Gregory wrote about persons who tainted the blood of their children with loathsome diseases, madness, folly, and unworthy dispositions, he was reflecting a widespread medical concern. Perhaps unknown to Gregory, this concern had led the Academy of Science in Dijon, France, to offer a prize in 1748 for the best essay on the topic, "What causes the transmission of hereditary disease?" One competitor, the young French surgeon Antoine Louis, denied the existence of hereditary diseases – but few of his colleagues agreed.[16] No physician suggested that hereditary diseases did not exist when the Royal Medical Society of Paris offered another prize for the best essay on the subject in 1790.[17] By then hereditary diseases were universally recognized, although there was widespread debate concerning the cause of such diseases.

Two options were open to eighteenth-century physicians who wished to ensure that children were not born with hereditary afflictions. First, an attempt could be made to remove those factors leading to the original appearance to whatever diseases and defects were thought capable of becoming hereditary. Second, an attempt could be made to prevent the propagation of diseases and defects by the arrangement or even the prohibition of marriage. Usually Enlightenment physicians emphasized both approaches. The first approach led to traditional medical practice, whereas the second approach led to ideas concerning genetic perfectibility.

Shortly after the Paris competition, Erasmus Darwin included a long note in his *Temple of Nature* that illustrated both approaches to the problem of hereditary disease. The note was a commentary on the canto in Darwin's poem:

> The feeble births acquired diseases chase,
> Till Death extinguish the degenerate race.[18]

Because "hereditary diseases of this country have many of them been the consequence of drinking much fermented liquor; as the gout always, most kinds of dropsy, and I believe, epilepsy, and insanity," Darwin suggested that moderation could prevent many hereditary diseases.[19] He advocated a eugenical approach, however, when wondering why family names disappear and heiresses become the sole family survivors. "As many families become gradually extinct by hereditary diseases, as by scrofula, consumption, epilepsy, mania," wrote Darwin, "it is often hazardous to marry an heiress, as she is not unfrequently the last of a diseased family."[20] For this reason, he concluded that "the art to improve the sexual progeny of either vegetables or animals must consist in choosing the most perfect of both sexes, that is the most beautiful in respect to body, and the most ingenious in respect to mind."[21]

Erasmus Darwin restricted his advice concerning marriage to individuals, but others during the eighteenth century thought that the propagation of hereditary diseases should be regulated by the state. Especially on the Continent, writers on Cameralism and medical police took the stance that the health of the population was too important a matter to be left to individual discretion and should be regulated in the interest of the state as a whole.[22] In 1756 Johann Heinrich von Justi published his *Grundsätze der Policey–Wissenschaft* and advocated that persons with hereditary diseases as well as those unable to have children be prohibited from marrying.[23] More systematic was the advice contained in the first volume of Johann Peter Frank's famous *System einer völlstandigen medicinischen Policey*, published in 1779. According to Frank, "One cannot indiscriminately let people take part in a matter on which, in reality, the fate of society and of all mankind most intimately depend: first, because marriage under certain circumstances may be for the partner himself a disadvantageous, or even lethal, matter; second, because in such unhealthy marriages almost no children are born, or only such children who are only a burden to the community and who do not last long; third, because this sustains even more the transmission of hereditary diseases."[24]

Frank's advice resembled that of later eugenicists in several respects. Believing in the hereditary transmissibility of many diseases, including epilepsy, Frank wrote that "it is clearly the duty of the leaders of the community not to let those of their subjects who suffer from particularly grave and deleterious ills get married without thorough examination."[25] Laws should be passed "which would forever prohibit the marriage of all crippled, maimed, very stunted dwarfish persons, and leave the work of procreation to a healthier class of citizens."[26] More positively Frank advocated that "beautiful persons with well-built and healthy bodies, even if they have no means, be supported in concluding marriages with partners equally healthy and physically perfect, and in bringing up a large family of similar children."[27] He very greatly deplored the fact that the military generally took the best specimens of humanity, where they "are dead so far as the marital procreation of our race is concerned."[28]

Although Frank's *System einer völlstandigen medicinischen Policey* was authoritarian, it was not completely so. Motivating much of Frank's discussion was a vehement oppositon to the arrangement of children's marriage by their parents in the latter's own selfish interest. As Frank saw it, such arrangements, often made for the purpose of increased wealth or status, frequently produced matches for the aged and infirm – those most likely to be sterile or to transmit hereditary diseases to their offspring. Also mitigating the severity of Frank's authoritarianism was the fact that he placed great belief in the beneficial effects of the mixture of populations in marriage. Authors on hereditary disease since the sixteenth century had, in fact, frequently urged those with hereditary diseases to marry individuals as much unlike themselves as possible. Adopting this idea, Frank suggested that inbreeding be avoided by having persons from isolated villages seek partners from other such villages.[29] Frank also cited Buffon's conjecture that the divine law against incest might have been designed to prevent the degeneration of the race and thus have a sound biological basis in nature.[30]

NEW VARIETIES OF HUMANS

The thought of controlling marriages in order to prevent the transmission of hereditary diseases suggested to some eighteenth-century writers a more dramatic idea. Why could one not produce special varieties of humans as agriculturalists produced specialized varieties of other animals? This intriguing suggestion was raised by Maupertuis in 1746 in his *Venus Physique:*

> Children usually resemble their parents and the variations among them at birth are often the results of various resemblances. Could we follow these variations back, we might find their origin in a common but unknown ancestor.
>
> Nature holds the source of all these varieties but chance or art sets them going. So that people whose work is to satisfy the curiosity seekers become practically creators of new species.
>
> Why is this art restricted to animals? Why don't the bored Sultans in their seraglios, filled with women of all known races, have them bear new species? Were I reduced, as they are, to the only pleasure that form and features can give, I would soon have recourse to greater varieties.[31]

In explaining these thoughts, Maupertuis wrote, "Although we do not find among ourselves the creation of such new types of beauty, only too often do we see human beings who are of the same category for men of science, namely, the cross-eyed, the lame, the gouty, and the tubercular."[32] Maupertuis himself was not excessively worried about the propagation of the most undesirable human defects, since he believed that "nature . . . has not desired that they be continued."[33] However, he thought that beauty was more apt to be hereditary than deformity.[34]

Whereas Maupertuis only briefly hinted that mankind could create new varieties of its own type, the possibility was developed more extensively in 1756 in a little-known work entitled *Essai sur la manière de perfectionner l'espèce humaine* by the hygienist Charles Augustin Vandermonde. The son of a French physician who had practiced in Macao (China), Vandermonde (1727–62) attended classes in Paris and became a doctor of medicine in 1748. His reputation appears to have been rising in the 1750s, as he was appointed royal censor in 1757, taught at Paris, and became editor of the *Journal général de la médicine.*[35]

Vandermonde was only twenty-nine years old when he published his *Essai.* The work was accepted by contemporaries as having originality and it was given favorable reviews.[36] Though Vandermonde did not refer directly to Maupertuis's advice to bored sultans in their seraglios, he did refer to Maupertuis's analysis of that midcentury anthropological cause celebré: a white child supposedly born of Negro parents. Like Maupertuis, Vandermonde saw the phenomenon in hereditarian terms: "It is sufficient to be persuaded of the veracity of this case to think that these extraordinary children either have a white negro among their ancestors, or if they are the first white negroes of the race that they have had in their ancestry negroes who crossed with the white."[37] Vandermonde found a similar message in Reaumur's pigeon-breeding experiments and in that author's discussion of the polydactylous Maltese family in which a man possessing six digits on both hands and feet communicated the trait to his offspring.[38]

The case of the white Negro, polydactylous inheritance, and the experiments of animal breeders led to the same conclusion: that "nature contains all kinds of varieties" and that it is up to man to unravel them and to combine them with profit.[39] "Since one has been able to perfect the races of horses, cats, fowl, pigeons," he queried, "why not try some attempts with the human race?"[40] Vandermonde conjectured that one might even be able "to perfect the talents agreeable to society."[41] Imagining the consequences of marrying the best French male dancers with the best Italian or English female dancers, he speculated that such parents would be able "to raise their children to a yet higher degree of their art."[42] The same improvement could be done for the vocal arts, "if marriage was to unite the two best voices of the country, or if a talented French singer was matched with some Italian who had a distinguished voice."[43]

Despite this vision of new breeds of singers and dancers, however, Vandermonde's vision of perfectibility was not really very different from that of other medical writers. In the preliminaries to the *Essai,* Vandermonde took great pains to carefully state his chief goal: children of a beautiful and robust constitution.[44] His definitions of beauty and robustness were influenced on the one hand by Hippocratic medicine and on the other by eighteenth-century aesthetics: Beauty consists in the symmetry and proper proportions of the parts; robustness consists in that state of perfect health that results in an ability to endure fatigue, and injury – as well as too much drinking, eating, and debauchery. Having defined beauty and robustness. Vandermonde discussed the qualities necessary for a successful marriage, precautions to be observed during the marriage act, choice of marriage partners, crossing of races, and the biology of human generation. Then followed chapters on the effect of exercise, food, and sleep on the mother during pregnancy and a discussion of the nurture of infants. The *Essai* concluded with a discussion of custom.

Taken as a whole the *Essai sur la manière de perfectionner l'espèce humaine* was thus a manual for parents concerned about all aspects of their children's health. Many of the recommendations regarding choice of marriage partners were variants on advice familiar since antiquity. Too great a disparity between the ages of parents will lead either to infertility or weakness in the offspring. If a very large man marries a very small woman (or vice versa), the rules of nature will be broken and the offspring will suffer from hereditary defects that will subsequently perpetuate themselves through the race and become very difficult to eradicate. In avoiding the degeneration of the race, Vandermonde – strongly influenced by Buffon's theory of generation and Buffon's belief in the climatic modifiability of the race – placed great emphasis on the importance of choosing marriage partners from different climates.[45] Thus Vandermonde thought that successful horse breeders mated French horses with ones from either warmer or colder climates, whereas English sheep breeders sought varieties from abroad. The effect could also be observed among humans. Why are great men concentrated in Paris? The reason must be because of the mixture of people to be found in the French capital. The "grand genies, les hommes d'esprit, les gens a talent" to be found in Paris owed their existence to past marriages between provincials and Parisians.[46] "The more strangers there are in the city," he wrote, "the more celebrated it becomes."[47] Rather unexpectedly, Vandermonde used the same argument to explain what he took to be the natural vigor of kings; they owe their superiority to the royal custom of choosing marriage partners from different countries, and therefore from different climates.[48]

INHERITANCE OF MENTAL TRAITS

Though Vandermonde suggested the breeding of new races of dancers and singers, he did not make a parallel suggestion for new races of philosophers. But could one similarly perfect the powers of reason? Could the genetic perfectibility of mankind contribute as well to its moral perfectibility? Although Vandermonde failed to address himself directly to these questions, they were given affirmative answers by others.

Eighteenth-century medical thought provided the two postulates according to which a proposal for the genetic improvement of human mental and moral character might be deemed realistic, namely, that the mental differences among men are innate and also that they are hereditarily transmissible. Among medical men Robert Burton's belief that the world is filled with "dizzards" was more commonly accepted than René Descartes's contention that "good sense is of all things in the world the most equally distributed." Contrary to Descartes, in 1669 Walter Charleton wrote in his *A Brief Discourse Concerning the Different Wits of Men* that "although *Wit, or Natural Capacity of Understanding* seems to be the only thing wherein Nature hath been equally bountiful to Mankind; every one thinking he hath enough, and even those who in their Appetites and Desires of other things are insatiable, seldom wishing for more of that Excellent Endowment: Yet nothing is more evident than this, that some have more Wit than others, and that Men are thereby no less distinguishable each from other, than by their several Faces and Tempers."[49] Shortly afterward, Thomas Willis, the student of brain anatomy, wrote that "stupidity is sometimes original or born with one, and so it is either *hereditary*, as when Fools beget Fools, . . . or . . . is as it were accidental, to wit it frequently happens, that wise men and highly ingenious do beget Fools and Changelings."[50]

During the eighteenth century, medical belief in mental diversity was usually couched in terms of the doctrine of the four temperaments. As in classical writings, temperaments were considered to be modifiable by such influences as climate, diet, and age. For example, in 1705 Friedrich Hoffman's *De temperamento* ascribed the various mental peculiarities of nations to differences in temperament and thus to the physical influences of climate, which affected temperament.[51] The same plasticity was made use of by Montesquieu in his *L'Esprit des lois* of 1748. Adopting contemporary views that the temperaments are related to differences in the solid fibers of the body, Montesquieu argued that "cold air constringes the extremities of the external fibres, of the body," and that "warm air relaxes and lengthens the extremes of the fibres."[52] Because of these physical changes, ultimately northerners possess "more courage; a greater sense of superiority, that is, less desire for revenge; a greater opinion of security, that is more frankness, less suspicion, policy, and cunning" than their southern counterparts.[53]

Belief in the environmental modifiability of temperament might seem to lead diametrically away from a vision of genetic perfectibility. During the mid-eighteenth century, however, the dichotomy between environmental and genetic influences was not crucial. The important difference was rather between those who believed that the mind is an independent agent apart from nature and those who believed that the mind is affected by physical causes. Among the latter, a belief in temperamental diversity caused by climate was easily joined, as it had been in the Hippocratic corpus itself, with a belief in diversity perpetuated through inheritance. No one illustrated such a conjunction more clearly than did the materialistic De la Mettrie, who wrote as follows in his *L'Homme machine,* published the same year as Montesquieu's *L'Esprit des lois:*

> Look at the portraits of Locke, of Steele, of Boerhaave, of Maupertuis, and the rest, and you will not be surprised to find strong faces and eagle eyes. Look over a multitude of others, and you can distinguish the man of talent from the man of genius, and often even an honest man from a scoundrel. For example, it has been noticed that a celebrated poet combines (in his portrait) the look of a pickpocket with the fire of Prometheus . . .
>
> One nation is of heavy and stupid wit, and another quick, light and penetrating. Whence comes this difference, if not in part from the difference in foods and difference in inheritance, and in part from the mixture of the diverse elements which float around in the immensity of the void?[54]

In a footnote to the word inheritance, De La Mettrie wrote, "The history of animals and of men proves how the mind and the body of children are dominated by their inheritance from their fathers."

Those who shared De la Mettrie's belief in the influence of physical causes upon man's mental and moral dispositions likewise stressed the importance of inheritance. Thus in 1786 the American physician Benjamin Rush queried rhetorically in his *An Enquiry into the Influence of Physical Causes upon the Moral Faculty:* "Do we observe certain degrees of the intellectual faculties to be hereditary in certain families? . . . The same observation has frequently extended to moral qualities – Hence we often find certain virtues and vices as peculiar to families, through all their degrees of consanguinity, and duration, as a peculiarity of voice – complexion – or shape."[55] In his essay *On the Influence of Physical*

Causes in Promoting an Increase of the Strength and Activity of the Intellectual Faculties, delivered in Philadelphia in 1799, Rush took these observations yet a step further by suggesting that they could enable parents to predict "with certainty" the character of their future children:

> It is probable, the qualities of the body and mind in parents which produce genius in children may be fixed and regular, and it is possible, the time may come, when we shall be able to predict, with certainty, the intellectual character of children, by knowing the specific nature of the different intellectual faculties of their parents.[56]

Just as Vandermonde and Frank had sought physical perfectibility by bringing together marriage partners from different places and climates, Rush believed that the key to intellectual perfectibility was parental mixture: "As conjugal happiness, in its positive degree, is often the result of dissimilar tempers," wrote Rush, "so, strong intellects in children, may be the product of a difference in the mental faculties of the two sexes."[57]

CONDORCET AND CABANIS

Assuming that mental traits are inheritable, a new possibility was opened up to eighteenth-century thinkers. Can human hygienic, intellectual, and moral progress improve the organs of the mind, and can these improvements in mental organs be transmitted hereditarily? If so, then moral perfectibility and physical perfectibility would be two aspects of the same grand process. Indeed, this was the vision of none other than Condorcet, stated only briefly in 1793 in his *Esquisse d'un tableau historique des progrès de l'esprit humain,* but developed more extensively in *Fragment sur l'Atlantide,* written about the same time.

In *L'Atlantide,* Condorcet invented a scenario from mathematics to illustrate the acceleration of progress that might be possible if improvements in mental faculties were hereditarily transmissible.[58] Imagine, he said, the case of two mathematicians of the same outstanding mathematical ability, the first born in ancient Greece and the second born in the twentieth or twenty-first century. Because of the different degrees of mathematical knowledge in those two epochs, the first mathematical genius would not be able to solve problems considered routine by eighteenth-century mathematicians, whereas the second would be able to solve all these problems and many others. Imagine next, what would happen if the abilities of the mathematicians as well as the knowledge of mathematics were increased over time. "Is it possible to carry this progress to the point," asked Condorcet, "where a man, for example, who was a contemporary of Archimedes, after having mastered mathematics up to the point at which it has been left by the geometer of Syracuse, would be able to arrive in the short space of one lifetime to the point where it has been developed by Euler and Lagrange?"[59]

Although Condorcet wrote before Lamarck, his vision of mankind's genetic perfectibility assumed the inheritability of acquired characteristics. Admitting that little was known about such inheritability, he suggested in *L'Atlantide* that additional investigation should be made of the subject. Condorcet was less cautious, however, in his *Esquisse d'un tableau historique des progrès de l'esprit humain:*

But are not our physical faculties and the strength, dexterity and acuteness of our senses, to be numbered among the qualities whose perfection in the individual may be transmitted? Observation of the various breeds of domestic animals inclines us to believe that they are, and we can confirm this by direct observation of the human race.

Finally, may we not extend such hopes to the intellectual and moral faculties? May not our parents who transmit to us the benefits or disadvantages of their constitution, and from whom we receive our shape and features, as well as our tendencies to certain physical affects, hand on to us the intellect, the power of the brain, the ardour of the soul or the moral sensibility? Is it not probable that education in perfecting these qualities, will at the same time influence, modify and perfect the organization itself.[60]

Among those reading this passage was Thomas Malthus. Confronting the arguments of the *Esquisse* in 1798 in his *Essay on Population,* Malthus granted to Condorcet that "the capacity of improvement in plants and animals, to a certain degree, no person can doubt."[61] Granting also that "size, strength, beauty, complexion, and perhaps even longevity are in a degree transmissible," Malthus admitted "that by an attention to breed, a certain degree of improvement similar to that among animals might take place among men.[62] However, Malthus doubted that intellectual abilities were transmissible and doubted that the race could be indefinitely improved by Condorcet's methods. Overlooking the fact that Condorcet's scheme did not necessitate selective breeding but relied upon the inheritance of acquired characteristics, Malthus also countered that "as the human race . . . could not be improved in this way, without condemning all the bad specimens to celibacy, it is not probable, that an attention to breed should ever become general; indeed, I know of no well-directed attempts of this kind, except for the ancient family of Bickerstaff, who are said to have been very successful in whitening their skins, and increasing the height of their race by prudent marriages, particularly by that very judicious cross with Maud, the milk maid, by which some capital defects in the constitutions of the family were corrected."[63]

Who were the Bickerstaffs and who was Maud the milkmaid? Generations of social scientists ignorant of English literature may well have wondered. Isaac Bickerstaff was the pseudonym of the English satirist Richard Steele, who had provided the following apocryphal Bickerstaff genealogy in the *Tatler* of October 4, 1707:

We have, in the genealogy of our house, the descriptions and pictures of our ancestors from the time of King Arthur; in whose days there was one of my own name, a knight of his round table, and known by the name of Sir Isaac Bickerstaff. He was low of stature, and of a very swarthy complexion, not unlike a Portuguese Jew. But he was more prudent than men of that height usually are, and would often communicate to his friends his design of lengthening and whitening his posterity. His eldest son, Ralph, for that was his name, was for this reason married to a lady who had little else to recommend her, but that she was very tall and very fair. The issue of this match, with the help of high shoes, made a tolerable figure in the next age; though the complexion of the family was obscure until the fourth generation from that marriage. From which time, until the reign of William the Conqueror, the females of our house were famous for their needlework and fine skins. In the male line, there happened an unlucky accident in the reign of Richard III, the eldest son of Philip, then chief of the family, being born with a hump back and very high nose. This was the more astonishing, because none of his forefathers ever had such a blemish; nor indeed was there any in the neighborhood of that make, except the butler, who was noted for round shoulders, and a Roman nose: what made the nose the less excusable, was, the remarkable smallness of the eyes.

These several defects were mended by succeeding matches; the eyes were open in the next generation, and the hump fell in a century and a half: but the greatest difficulty was, how to reduce the nose; which I do not find was accomplished until about the middle of the reign of Henry VII or rather the beginning of that of Henry VIII.

But, while our ancestors were thus taken up in cultivating the eyes and nose, the face of the Bickerstaffs fell down insensibly into their chin; which was not taken notice of, their thoughts being so much employed upon the more noble features, until it became almost too long to be remedied.

But, length of time, and successive care in our alliances, have cured this also, and reduced our faces into that tolerable oval, which we enjoy at present. I would not be tedious in this discourse, but cannot but observe, that our race suffered very much about three hundred years ago, by the marriage of one of our heiresses with an eminent courtier, who gave us spindleshanks, and cramps in our bones and legs; until Sir Walter Bickerstaff married Maud the milk-maid, of whom the then garter king-at-arms, a facetious person, said pleasantly enough, "that she had spoiled our blood, but mended our constitution."[64]

Whereas Malthus was skeptical concerning Condorcet's vision of hereditary perfectibility, no such skepticism affected Condorcet's friend and disciple, Pierre Cabanis. Approaching the study of human nature from the standpoint of a physician, Cabanis exploited the various meanings of the word "temperament" to illustrate the close interaction between our innate endowment, the external influences, affecting that endowment, and our intellectual and moral life.[65] "In an extreme case," wrote Cabanis, "a temperament is an actual disease" and can be hereditarily transmitted like many other diseases.[66] But Cabanis also explained in those lectures delivered before the Institute Nationale in 1796–7 and later published as *Rapport du physique et morale de l'homme* how an innate temperament might be altered by climate, regimen, and other physical factors. Since differences of temperament affected man's mental world as well as his physical well-being, the physician could thus hope to improve the whole person through attention to regimen and physical eduction.

Though Cabanis argued that physical factors could do much to improve one's temperament, he recognized limits in the degree to which such factors could fundamentally alter the temperament with which one was born. Distinguishing in his *Rapport* between the *natural temperaments* with which one is born and the *acquired temperaments,* which are formed by the "long persistence of impressions," Cabanis admitted that the development of the latter seldom completely effaced the former.[67] "When the original imprint is strong and deep, it is rarely effaced," he wrote.[68] The accidental circumstances of life may modify that imprint by other more "superficial imprints," but ordinarily that is the "limit of their effect."[69] Moreover, "observation teaches us that the habits of the constitutions transmit themselves from parents to children and that they serve as an indelible mark in the midst of the most diverse circumstances of education, climate, occupation, and regimen.[70] By way of example, Cabanis noted that racial traits seemed to be propagated unchanged wherever the intermixing of races had not occurred.

Such observations did not lead Cabanis to a counsel of despair, however. Though there were limits with what could be done to alter the temperament of any particular individual, the race as a whole could be hereditarily perfected – the key lay in the fact that temperaments that are "acquired" in the individual can after a long sequence of generations become "natural" in the race. When the same means have been employed "for several successive generations," wrote Cabanis, "it is no longer, *caeteris paribus,* the same men, or the same race of men that exist."[71] Nor need this improvement be limited to man's physical organization. Referring to the opinions of Condorcet on the inheritability of mental and moral traits, Cabanis wrote that "several strong analogies, and the general laws of animal economy, lead us to believe that moral habits are . . . propagated by generation."[72] In a now familiar refrain, Cabanis contrasted what had been already done to perfect other animals with what should be done to perfect the human race:

> After we have occupied ourselves so minutely with the means of making better and stronger races of animals and useful and agreeable plants; after we have done a hundred times over the races of horses and dogs; after we have transplanted, grafted, worked over fruits and flowers in every manner, why is it not shameful to have totally neglected the human race! – as though it were more distant! – as if it were more essential to have large strong cattle, than healthy vigorous men; good smelling peaches and speckled tulips, than wise and good citizens![73]

Thus the physician and hygienist could transcend mere medicine and "aspire to perfect human nature in general."[74] But toward what goal should this striving be directed? Both Condorcet and Cabanis believed in the natural inequality of people at the same time that they believed in equal rights. Not surprisingly Cabanis's goal was not to yet further diversity of mankind, but to effect an ideal state of mental and moral health that would help restore natural equality. Thinking within the Hippocratic tradition, Cabanis envisaged a "well-tempered" individual whose bodily organs were in harmony and equilibrium.[75] Though Cabanis did not himself draw the parallel, the equality of organs within the body would thus mirror the equality of individuals within society. The goal itself, however, Cabanis believed would always remain elusive. Difference of occupation, regimen, and climate would always preserve diversity. Nonetheless, men might be made "equally fit for society, if not equally fit for all occupations within society."[76]

Cabanis's *Rapport du physique et morale de l'homme* contained a more profound and influential statement of the vision of genetic perfectibility than any other work published in the wake of Condorcet's *Esquisse.* Yet Cabanis's hereditarian ideas were not by any means unique in the medical community of postrevolutionary Paris. Thus in 1804 Henry-Alexandre Haren presented a dissertation before the faculty of the Paris School of Medicine entitled "Considérations sur l'Influence que peut avoir le tempérament des parents sur celui de leur enfants." Beginning his dissertation by quoting Cabanis, Haren continued by writing:

> All persons have been convinced that the son inherits the moral and physical qualities of his father. It is according to this theory that is founded the inheritance of kings among the policies of nations, and nobility, according to this, is only a perpetual caste established to honor genius and rulers. The order of chivalry diffused in Europe, and the legion of honor newly founded in France.[77]

Though most of Haren's dissertation was uncritical, he doubted more strongly than did Cabanis the possibility of modifying a well-established temperament. "The power of nature resists the strongest influences, and preserves the original type with which a man has been stamped."[78] Haren did not suggest that the race could be improved over several generations as the acquired temperaments became natural; instead he put his emphasis on the more traditional mixing of temperaments. Believing that the two sexes equally influence the temperament of their children, Haren found the best combination to be a mixture of sanguine and bilious temperaments.[79]

Another Parisian medical student published in 1801 what may be considered the culmination of eighteenth-century views on human hereditary perfectibility. Borrowing from the Greek the three words *megalos* (great), *anthropos* (man), and *genesis* (generation), Robert le Jeune entitled the book he published in 1801, *Essai sur la Mégalantropogénésia ou l'Art de faire des enfants d'esprit, qui deviennent de grands hommes.* The work gave rise to numerous responses in such places as *Le journal des Débats, La Gazette querrance, Le journal de Paris, Le Citoyen Français, Le Publiciste, La Feuille Économique,* and in a second edition of 1803 Robert included all of these replies plus much more material.[80] Meanwhile the argument of the *Essai* received a public hearing at the Medical School of Paris and Robert defended a dissertation entitled "Existe-t-il un art physico-medico pour augmenter l'intelligence de l'homme en perfectionnant ses organes, ou le mégalantropogénésie n'est-elle qu'une erreur?" Robert believed that the key to producing great men lies in educating women, so that both parents would be persons of ability.[81]

CONCLUSION

In spite of the philosophic emphasis on reason as the primary agent of man's improvement, the vision of what Passmore has called "genetic perfectibility" never disappeared during the eighteenth century. During the middle of the eighteenth century the concept of medical police and the development of hygiene served to revive the idea of Renaissance utopians that the state should regulate marriages in order to avoid the propagation of hereditary diseases and defects among its citizens. By the end of the eighteenth century this idea was joined by a more positive vision of genetic perfectibility that aimed not only at the prevention of physical degeneration but also at the improvement of human intellectual and moral traits.

As had been done since Plato's *Republic,* those who seized upon the possible hereditary perfectibility of mankind often contrasted the care with which man bred his animals with the unconcern with which he treated his own propagation. The vision of genetic perfectibility during the Enlightenment was not, however, primarily motivated by the success of animal breeders. It was medicine that provided the context of eighteenth-century hereditarian thought – contributing observations upon the inheritability of disease, a recognition of human diversity, and a belief in the influence of physical causes on the human mind. Unlike during the Renaissance, medicine did not contribute greatly to philosophical views concerning human nature during the eighteenth century. Enlightenment medicine, however, provided the link between Renaissance ideas on genetic perfectibility and similar ideas that became espoused during the early nineteenth century by phrenologists and anthropologists.

NOTES

1 *Encyclopaedia Britannica,* 7th ed., 1842, vol. XVII, p. 228.

2 Stephanus Chauvin, *Lexicon philosophicum,* with an intro. by Lutz Geldsetzer (Dusseldorf: Stern-Verlag Janssen, 1967), pp. 480–1.

3 John Passmore, *The Perfectibility of Man* (London: Duckworth, 1970), pp. 186–9.

4 Campanella, "City of the Sun," in *Famous Utopias,* with an intro. by Charles M. Andrews (New York: Tudor, n.d.), pp. 275–317.

5 Francis Galton, *Hereditary Genius: An Inquiry into its Laws and Consequences* (New York, 1870).

6 Joyce Oramel Hertzler, *The History of Utopian Thought* (New York: Cooper Square, 1965), p. 288.

7 Robert Burton, *The Anatomy of Melancholy,* Mem. I. Subs. 6 (Philadelphia, 1836), vol. I, p. 93.

8 Ibid.

9 Ibid.

10 Gabriel A. Pérouse, *L'Examen des esprit du docteur Juan Huarte de San Juan. Sa diffusion et son influence en France auz XVI[3] et XVIIe siècles. Bibliothèque de la faculté des lettres de Lyon,* XIX (Paris: Societé d'édition "les belles lettres," 1970). Alfredo M. Saavedra, "La Eugenesia de Juan Juis Vives," *Medicina, Revista Mexicana,* t. XLI, no. 861 (Feb. 10, 1961), supp. pp. 17–19. Also Gustavo Tanfani, "Principi di Eugenetica e di Profilassi morale negli scritti de Girolamo Cardano," *Congresso (VIII) Storia della Medicina* (1930), pp. 427–32. Also Allen G. Roper, *Ancient Eugenics* (Oxford: Blackwell, 1913). Although Renaissance eugenical thought is well-known, there has been no adequate or critical study of its importance.

11 Victor L. Hilts, "Obeying the Laws of Hereditary Descent: Phrenological Views on Inheritance and Eugenics," *Jour. Hist. Behav. Sciences* (in press).

12 Charles Rosenberg, "The Bitter Fruit; Heredity, Disease, and Social Thought in Nineteenth-Century America," *Perspectives in American History* VIII (1974), pp. 189–235.

13 [John Gregory], *A Comparative View of the State and Faculties of Man with those of the Animal World* (London, 1765), pp. 15–18.

14 John H. Randall, *Making of the Modern Mind* (Cambridge, Mass: Houghton Mifflin, 1940), p. 315.

15 René Descartes, "Discourse on Method," *The Philosophical Works of Descartes,* trans. Elizabeth S. Haldane and G. R. T. Ross (Cambridge: Cambridge University Press, 1931), vol. I., p. 81.

16 [Antoine] Louis, "Dissertation sur la question . . . *Comment se fait le transmission des Maladies héréditaires?*" (Paris, 1749). A biography of Louis is Pierre Huard, "Antoine Louis," *Biographies Médicales et Scientifiques, XVIIe* siècle (Paris, 1972), pp. 35–117. Louis's arguments in his prize essay were called "more ingenious than well founded" by Antoine Portal, "Considerations on the Nature and Treatment of some hereditary or Family Diseases" (1808), trans. *Medical and Physical Journal,* conducted by T. Bradley, M.D., and J. Adams, M.D., XXI (1808–9), p. 332.

17 *Histoire de la Société Royale de Médecine.* Année 1786 (Paris, 1790), pp. 17–18. The prize question was originally announced in 1787 but the date for entries was extended to May 1, 1790, after the first submissions were judged inadequate. The winning entry was Joseph Claudius Rougemont, *Abandlung über die erblichen Krankheiten* (Frankfurt, 1794). Two other published entries were Alexis Pujol, "Essai sur les Maladies Héréditaires," *Oeuvres de Médecine Pratique d'Alexis Pujul* (Paris, 1823), t. ii. pp. 211–420, and the article "Hereditaires (maladies)" in the *Encyclopédie Methodique. Médecine.* (Paris, 1798), t. 7, pp. 161–76) by Pages d'Alaix. Rougemont cited (p. 3) thirteen different authors whose works on hereditary disease had been consulted although he said that he was distant from a large library.

18 Erasmus Darwin, "The Temple of Nature," in *The Poetical Works of Erasmus Darwin* (London, 1806), vol. III, p. 250.

19 Ibid., p. 252.

20 Ibid., p. 253. Francis Galton, who was Erasmus Darwin's grandson, was also intrigued by the heiress problem.

21 Ibid., p. 253.

22 George Rosen, "Cameralism and the Concept of Medical Police," *Bull. Hist. Med.* XXVII (1953), pp. 21–42.

23 Johann Heinrich Gottlobs von Justi, *Grundsätze der Policey-Wissenschaft* (Göttingen, 1756), pp. 63–70.

24 Johann Peter Frank, *A System of Complete Medical Police,* ed. with an intro. by Erna Lesky (Baltimore: Johns Hopkins University Press, 1976), p. 46.

25 Ibid., p. 48.

26 Ibid., p. 50.

27 Ibid., p. 49.

28 Ibid., p. 50

29 Ibid., p. 63. The translators of Frank's *A System of Complete Medical Police* have not included most of Frank's discussion concerning hereditary diseases and the consequences of inbreeding by animals and humans. Frank's most important discussion of the latter topic is Johann Peter Frank, *System einer medicinischen Policey,* zweite, verbesserte Auflage (Mannheim, 1784), Bd. I, pp. 452–8. For a sixteenth-century treatise on hereditary disease see David F. Musto. "The Theory of Hereditary Disease of Luis Mercado," *Bull. Hist. Med.* XXV (1961), pp. 346–73.

30 Johann Peter Frank, *System einer medicinischen Policey*, Bd. I, pp. 452–3.

31 Pierre Maupertuis, *Earthly Venus*, trans. with intro. by George Boas (New York: Johnson Reprint, 1969), pp. 77–8.

32 Ibid., pp. 78–9.

33 Ibid.

34 Ibid.

35 "Charles-Augustin Vandermonde," *Biographie Universelle, Ancienne et Moderne* (Paris, 1827), t. 47, pp. 433–4.

36 "Essai sur la manière de perfectionner l'espèce humaine," *L'Anée Littéraire.* Année 1756 (Amsterdam), t. iii, pp. 13–22.

37 C.A. Vandermonde, *Essai sur la manière de perfectionner l'espèce humaine* (Paris, 1756), t. i, p. 88.

38 Ibid., p. 98.

39 Ibid., pp. 90–1.

40 Ibid., p. 94.

41 Ibid., p. 95.

42 Ibid., p. 95.

43 Ibid., p. 95.

44 Ibid., "Préliminaries," pp. 1–64.

45 Ibid., "Du Croisement des Races," chap. iv., pp. 99–118.

46 Ibid., p. 108.

47 Ibid., p. 108.

48 Ibid., pp. 110–11. Subsequent writers were to take exactly the opposite view of the health of royalty and to attribute their defects to inbreeding.

49 [Walter Charleton], *A Brief Discourse Concerning the Different Wits of Men* (London, 1669), pp. 4–5.

50 Paul E. Cranefield, "A Seventeenth-century view of mental deficiency and schizophrenia: Thomas Willis on 'Stupidity' or foolishness," *Bull. Hist. Med.* XXXV (1961), p. 297.

51 Friderich Hofmann, "De temperamento fundamentorum et morborum in gentibus," *Dissertationes physico-medicae,* curiosae selectiones, 1708, ii, pp. 288–9.

52 Montesquieu, *The Spirit of the Laws,* trans. Thomas Nugent with an intro. by Franz Neumen (New York: Hafner, 1949), vol. I, p. 22

53 Ibid.

54 Julien De la Mettrie, *Man a Machine,* philos. and hist. notes by Gertrude C. Bussey (Chicago: Open Court, 1953), pp. 96–7.

55 Benjamin Rush, "An Enquiry into the Influence of Physical Causes upon the Moral Faculty," in Benjamin Rush, *Two Essays on Mind,* intro. by Eric T. Carlson (New York: Brunner/Mazel, 1972), p. 5.

56 Benjamin Rush, "On the Influence of Physical Causes in Promoting an Increase of the Strength and Activity of the Intellectual Faculties of Man," in Rush, *Two Essays on Mind,* p. 119.

57 Ibid.

58 Condorcet, *Oeuvres,* nouvelle impression en facsimilé de l'édition Paris 1847–1849 (Stuttgart: Friedrich Frommann Verlag, 1968), t. vi., pp. 618–31. A portion of *L'Atlantide* is translated in Keith Baker, *Condorcet: Selected Writings* (Indianapolis: Bobbs-Merrill, 1976), pp. 283–300, but Baker's translation omits Condorcet's discussion of genetic perfectibility.

59 Ibid., pp. 626–8.

60 Condorcet, *Sketch for a Historical Picture of the Progress of the Human Mind,* trans. June Barraclough, with an intro. by Stuart Hampshire (New York: Noonday Press, 1955), p. 20.

61 Thomas Malthus, *Population: The First Essay* (Ann Arbor: University of Michigan Press, 1959), p. 60 (chap. 9). This was first published anonymously in 1798 as *An Essay on the Principle of Population, as It Affects the Future Improvement of Society, with Remarks on the Speculations of Mr. Godwin, M. Condorcet, and Other Writers.*

62 Ibid., p. 60.

63 Ibid., p. 60.

64 *The Tatler,* no. 75 (Saturday, October 1, 1709). What may have motivated Steele to make up his story is unknown. Possibly he was influenced by a knowledge of eugenical ideas in antiquity and the Renaissance. In the *Spectator* no. 307 (February 21, 1711–12), Steele made reference to Juan Huarte's sixteenth-century *Examen de Ingenios,* in which the main thesis was that parents could control the particular "genius" of their offspring by a careful choice of marriage partners.

65 Pierre-Jean-Georges Cabanis, "Rapports du Physique et du Moral de L'homme," in *Oeuvres Philosophiques de Cabanis,* texte établie et présenté par Claude Lehec et Jean Cazeneuve (Paris: Presses Universitaires de France, 1956), t. i. pp, 105–631. For Cabanis see especially Martin S. Staum, *Cabanis: Enlightenment and Medical Philosophy in the French Revolution* (Princeton, N.J.: Princeton University Press, 1980).

66 Cabanis, *Oeuvres Philosophiques,* t. i, p. 356.

67 Ibid., pp. 618–19.

68 Ibid., p. 631.

69 Ibid., p. 631.

70 Ibid., p. 355.

71 P. J. G. Cabanis, *Sketch of the Revolutions of Medical Science and Views Relating to Its Reform,* trans. with notes by A. Henderson (London, 1806), pp. 22–3. Also Cabanis, *Oeuvres Philosophiques,* t. ii, p. 78.

72 Cabanis, *Revolutions of Medical Science,* p. 338.

73 Cabanis, *Oeuvres Philosophiques,* t. i, pp. 356–7.

74 Ibid., p. 356.

75 Ibid., pp. 353–5.

76 Ibid., p. 358.

77 Henry-Alexander Haren, "Considérations sur l'Influence que peut avoir le tempérament des parents sur celui de leurs enfants, suivies de quelques réflexions relatives a cet object." Présentées et soutenues à l'Ecole de Médecine de Paris, le 11 fructidor an XII (Paris, 1804), p. 13.

78 Ibid., p. 6.

79 Ibid., p. 11.

80 Leon Élaut, "Note sur la *Mégalantropogénésie ou l'art de faire des enfants d'esprit* de Robert Le Jeune, Docteur en Médecine." *Monspeliensis Hippocrates,* 1964 (autumne), pp. 3–8.

81 Robert Le Jeune, "Existe-t-il un Art Physico-Médical Pour augmenter l'intelligence de l'homme en perfectionnant ses organes?" (Paris, 1803), pp. 30–2.

14

Anatomia animata

The Newtonian physiology of Albrecht von Haller

SHIRLEY A. ROE
The Wellcome Institute for the History of Medicine

Albrecht von Haller (1708–77) made significant contributions to several fields of study in the eighteenth century. The "last universal scholar" according to one historian,[1] Haller published works on anatomy, physiology, medicine, and botany, among the sciences, and on political and theological topics. In his youth, he wrote poetry, and, in later years, he composed three political novels.[2] More remarkable still, Haller is reputed to have published some ten to twelve thousand book reviews, covering not just scientific topics but literary, political, and philosophical ones as well.[3] He was active during his lifetime as a physician, university professor, academician, civil servant, and politician.

The details of Haller's life and his contributions to these different areas of knowledge have been documented by a number of historians.[4] With regard to the sciences, Haller's work on muscle physiology, especially on irritability and sensibility, is most frequently discussed.[5] Yet one major aspect of Haller's scientific work has received far less attention, namely, Haller's own philosophical views on the nature of science.[6] As a physiologist, Haller contributed not just factual knowledge concerning anatomical and physiological subjects, but, even more importantly, he contributed a physiological method – a way of looking at physiological phenomena, of asking questions and defining answers within clear philosophical limits. It is this aspect of Haller's physiological work that I shall concentrate on here, for his philosophical views on the nature of biological explanation form a significant chapter in the history of Enlightenment thought.

Three principal topics will form the focus for my analysis. First, I shall discuss Haller's views on observation and experiment, and his proempiricist and antirationalist attitudes. Second, I shall deal with Haller's attempt to create an animal mechanics, founded, I will argue, on a Newtonian model. Haller's views on matter and forces will be treated here, as well as his debates with the animist Robert Whytt and the materialist La Mettrie. Third, I shall turn to Haller's religious beliefs and to his desire to promote science within clearly defined limits. I shall argue that Haller's religious convictions and his belief in mechanical and empirical explanation form a consistent philosophy of biology, one that reflects a number of key concerns of the eighteenth century.

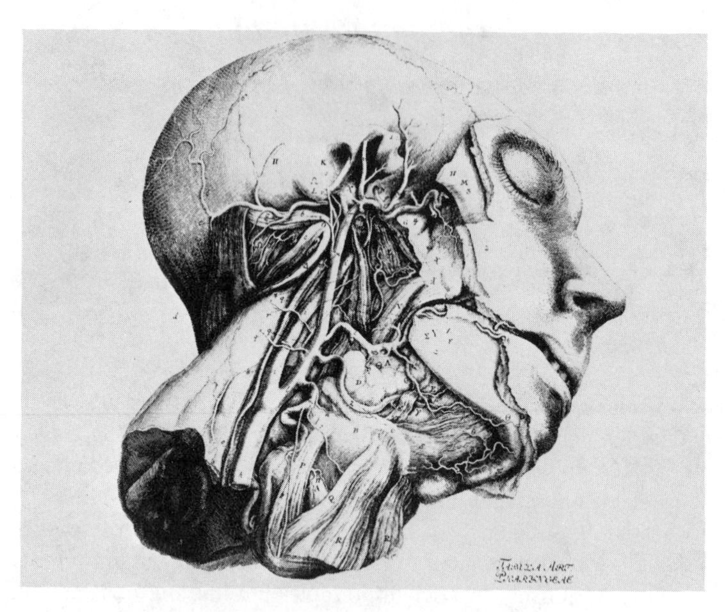

Fig. 1. The arteries of the pharynx, as illustrated in Haller's *Icones anatomicae.* (Courtesy of the Francis A. Countway Library of Medicine, Harvard Medical School.)

EXPERIMENTAL EMPIRICISM

"I am persuaded," Haller declares in his famous paper on irritability and sensibility (1752), "that the greatest cause of error [in physiology] has been that most physicians have made use of few experiments, or even none at all, but have substituted analogy instead of experiments."[7] Under the term experiment, Haller included both observation and experimentation, and he recommended anatomical dissection, comparative anatomy, and vivisection as proper experimental procedures. In his own work Haller relied heavily on anatomical observation, as can be seen in his *Icones anatomicae* (1743–54), where Haller utilized the technique of injecting colored wax into blood vessel networks, developed by Frederik Ruysch, to produce his masterful illustrative plates (see Fig. 1). Haller claimed in the opening pages of his paper on irritability and sensibility that his theory was based on six years of observations, many with his student Johann Zimmermann, culminating in a series of 190 experiments on live animals in the preceding year alone. To cite a third example, Haller's first major embryological work, *Sur la formation du coeur dans le poulet,* published in 1758, contains an entire volume of hour-by-hour reports of observations on hundreds of incubated chicken eggs.

Haller cautions against experimenting with prejudices in mind; one must dissect, "not with the intention of seeing what a classical author has described, but rather with the desire to see that which nature herself has brought forth."[8] He recommends repeated experiments to sift out extraneous factors, leaving only "the pure things . . . that constantly happen the same way because they flow out of the nature of the thing itself,"[9] and to prevent hasty generalization from isolated facts. "Why do we err?" Haller asks. "We have seen many cases [of something], and we conclude [that it must be so] for all without having seen all."[10]

Haller had definite opinions about the relationship of his views on observation and experiment to the traditions of rationalism and empiricism. "Better telescopes, rounder glass drops [lenses for simple microscopes], more precise divisions of measurement, syringes and scalpels did more for the enlargement of the realm of science," Haller claims, "than the imaginative mind of Descartes, than the father of classification Aristotle, and than the erudite Gassendi. With each step one took nearer to nature, one found the picture unlike that which the philosophers had made of it."[11] Championing the cause of empiricism, Haller frequently praises Bacon, who "had shown the way to come to know nature through experiment," in contrast to Descartes, who "was too hasty for the experiment, the street was too long for him."[12]

Haller's criticisms of Descartes rest principally on the latter's rationalist method. In a preface to the German translation of volume I of Buffon's *Histoire naturelle,* Haller defends at length the use of hypotheses in science as the most successful route to true conclusions. But one must be cautious in using hypotheses; they must always be tied to experiments or they will lead us astray. This was the downfall of Descartes, who "applied a mechanical method to the building and construction of the world, and took the freedom to give to the smallest parts of matter such figures and to impart to it such manners of movement as was necessary for his explanation."[13] But, Haller notes, "this convenient custom did not last as long as the idle natural philosophers had wished. The inventions of the imagination are like an artificial metal; it can have the color but never the density and the indestructible solidity that nature gives to gold."[14] Arbitrary hypotheses lead to false systems of knowledge; only hypotheses used in conjunction with the proper experimental method are allowable in science.

Haller's rejection of Descartes's method is reflected in his opinions of his German rationalist contemporaries. The reign of scholasticism and metaphysics, Haller writes in a review of a work on logic by fellow Göttingen professor Samuel Christian Hollmann, has fortunately come to an end in almost all countries. Germany, however, is an exception, where, led by Christian Wolff, philosophers have taken a step backward, introducing terms "that had become barbaric" into science, "that domain that Bacon and Galileo had torn away from the schools."[15] Scholastic rationalism has little to say about the real world, in Haller's opinion, for "God has created individuals, bodies, movements, and one amuses oneself by contriving classes, modes, and qualities."[16]

Why did Haller, a Swiss physician and professor at a major German university, reject rationalism, in both its Cartesian and Wolffian forms? In an amusing story, Haller relates an early encounter with Cartesianism, when he was studying with his stepuncle, Johann Rudolf Neuhaus, in 1722–3. "The old man was a determined Cartesian," Haller reports. "He began by making me study the principles of Descartes, and every page revolted me. 'From where do you know that the particles of the second element are round and that those of magnetic material are like a screw?' These questions came again at all moments and drew from me poor response."[17] This attitude toward Descartes is indicative of Haller's opinion of all rationalists, including his German contemporaries. Part of the reason for this lies in Haller's education, which, except for one year at Tübingen, took place in his native Switzerland and, more significantly, in Holland. In a comparative study of Haller and Leibniz, Richard Toellner notes that Haller's intellectual home was in fact not Germany but the early Enlightenment of the Netherlands.[18] Here, under the influence of Boerhaave, Haller's philosophical views were forged. Having missed a classical German education, Toellner claims, Haller did not come to grips with rationalism until he began teaching at Göttingen at the age of twenty-eight, after his own opinions had been developed. In my opinion Toellner's argument is essentially sound, for Haller, I shall argue, was influenced by his Leiden education in far more than his attitude toward rationalism. To show what the full extent of this influence was, let me turn now to Haller's views on physiological explanation.

ANIMAL MECHANICS

"Whoever writes a physiology," Haller declares in the preface to the first volume of his *Elementa physiologiae,* "must explain the inner movements of the animal body, the functions of the organs, the changes of the fluids, and the forces through which life is sustained." Haller believed that the science of physiology is the science of movement in living bodies, movement based on mechanical forces. The physiologist, he continues, must explain the forces "through which the forms of things received by the senses are presented to the soul; through which the muscles, which are governed by the commands of the mind, in turn have strength; the forces through which food is changed into such different kinds of juices; and through which, finally, from these liquids both our bodies are preserved and the loss of human generations is replaced by new offspring."[19] Sensation, motion, digestion, assimilation, growth, reproduction – these are the functions of the living organism the physiologist must explain. And his explanation must provide, through force mechanics, "a description of those movements by which the animated machine is activated."[20] Physiology, Haller proclaims, is "animated anatomy."[21]

Haller endorsed, like his teacher Boerhaave before him, the application of mechanical laws derived from physics to living processes. Yet this must be done cautiously, for, Haller remarks, "in the animal machine there are many things that are very different from the common mechanical laws."[22] Water flowing through a pipe, for example, is not totally analogous to liquids flowing through living vessels, which act in various ways to speed up or slow down the movement of fluids. Simple hydraulics will not explain the motion of the blood through the organism. However, Haller concludes, "I would not for this reason believe in discarding the laws by which moving forces outside of the animal body are governed; I propose that they never be transferred to our animated body machines unless experiments agree."[23]

Haller was a mechanist in his physiological outlook, yet he was not a total reductionist. Rather than reducing vital phenomena to the known laws that govern inorganic bodies, Haller proposed to create a distinct "animal mechanics." Here the basis for explanation was to be laws that operate in the same manner as physical laws, but which are not necessarily the *same* laws. Living organisms thus may possess forces that are not found in nonliving matter; yet these forces operate mechanically in exactly the same way as physical forces do.

The clearest example of Haller's nonreductionist mechanism is his concept of irritability. The foundation of his physiological system, Haller's theory of irritability, and its distinction from sensibility, was his most significant contribution to eighteenth-century physiology. As such, it forms the most important example of Haller's physiological method and his philosophy of science. Through his definitions of irritability and sensibility, and his defense of these notions against criticism, the philosophical basis for Haller's animal mechanics becomes manifest.

Haller referred to a concept of irritability first in 1739, in his edition of Boerhaave's teachings, the *Praelectiones academicae in proprias institutiones rei medicae.* Here in a footnote to Boerhaave's description of the systolic motion of the heart, Haller comments that since movement persists in hearts in animals that have recently died, the heart must beat from some "unknown cause" that "lies hidden in the fabric of the heart itself."[24] Although Haller made similar remarks in other publications,[25] his first major exposition of the theory of irritability and sensibility was his famous paper, "De partibus corporis humani sensilibus et irritabilibus," which was presented in 1752 to the Royal Society of Sciences of Göttingen and published in its journal, *Commentarii Societatis Regiae Scientiarum Gottingensis,* in 1753. Appearing within three years in French, German, English, Italian, and Swedish translations, Haller's paper became the focal point for discussion and debate throughout the medical world.[26]

Consistent with his experimental method, Haller based his concepts of irritability and sensibility, which were to form, he claimed, a new basis for classifying the parts of the body, on clear laboratory definitions: "I call that part of the human body irritable, which becomes shorter upon being touched . . . I call that a sensible part of the human body, which upon being touched transmits the impression of it to the soul; and in brutes, in whom the existence of a soul is not so clear, I call those parts sensible, the Irritation of which occasions evident signs of pain and disquiet in the animal"[27] (Fig. 2). Haller's procedure was to expose various parts of a live animal's body, and to irritate each one by the touch of a needle or scalpel, or by using air, heat, salt, alcohol, or various caustic chemicals.[28] From these experiments, Haller concluded that irritability is a property of the muscle fiber alone, whereas sensibility depends entirely on the nerves. By completely separating irritability from sensibility, Haller was able to

Fig. 2. Researching irritability and sensibility, as illustrated in the frontispiece to Haller's *Mémoires sur la nature sensible et irritable des parties du corps animal,* vol. 1. (Courtesy of the Francis A. Countway Library of Medicine, Harvard Medical School.)

demonstrate that muscular contraction is independent of the soul and is a product of an inherent property of muscles.

With his concept of irritability, Haller stepped into a debate that had occupied physiologists for more than a century. The nature of muscular contraction had been the subject of widely varying explanations, from Descartes's mechanical theory based on animal spirits, to the fermentation-combustion models of Borelli and Willis, to Stahl's notion of an immaterial *anima* governing all life functions.[29] Haller added a short historical section to his paper, in which he credits Glisson with the first use of the term irritability, and Bellini and Baglivi with having attributed an independent power of contraction to muscles.[30] Haller distinguishes his own theory from Glisson's, noting that Glisson had believed irritability to be present in all fibers and had based its action on a faculty of "natural perception."[31] Haller was most concerned to combat the Stahlian explanation of muscular contraction, especially as reformulated by Haller's contemporary Robert Whytt. By claiming that muscular contraction is independent of the nerves and the soul, Haller opposed those who thought that all contraction is based on some kind of sensation, be it conscious or unconscious. For Haller, the nerves act just like any other irritant in causing muscular motion; when I will my arm to move, for example, whatever is transmitted from my brain to the muscles of my arm acts simply as a stimulus to activate the inherent irritability of the muscles themselves.[32] In involuntary motion, such as the movement of the heart, other stimulants are involved to cause contraction (in this case the irritation of the entering blood).

"What therefore should hinder us," Haller proclaims, "from granting irritability to be that property of the animal gluten in the muscular fiber,[33] such that upon being touched and provoked it contracts, to which moreover it is unnecessary to assign any cause, just as no probable cause of attraction or gravity is assigned to matter [in general]. It is a physical cause, hidden in the intimate fabric, and discovered through experiments, which are evidence enough for demonstrating its existence, [but] which are too coarse to investigate further its cause in the fabric."[34] As a physical property of muscle fibers, irritability acts automatically when stimulated and is totally separate from the will. Although ascribing it to the "fabric" of muscular tissue, Haller refuses to speculate on irritability's ultimate cause.

Beginning with the fourth volume of the *Elementa physiologiae* (1762), Haller refers to irritability as the *vis contractilis musculis insita,* the contractile force innate to muscles, a phrase shortened to *vis insita* in the *Primae lineae physiologiae* (1765). Haller distinguishes the *vis insita* from both a *vis mortua,* a general contractile power resident in almost all animal tissue, and a *vis nervosa,* Haller's new name for sensibility. The *vis insita,* Haller explains, "seems to constitute an altogether peculiar power, unique to animal [muscle] fiber . . . It is a peculiar force distinct from every other power, which should be classed among the springs for generating motion whose further cause is unknown. This force is innate in the fiber itself and does not come from some other source."[35] In the *Primae lineae physiologiae,* Haller reiterates his refusal to speculate on the ultimate cause of irritability, about which "we do not indeed inquire," explaining only that the *vis insita* "seems to be a more brisk attraction of the elementary parts of the fibre by which they mutually approach each other, and produce as it were little knots in the middle of the fibre. A stimulus excites and augments this attractive force, which is placed in the very nature of the moving fibre."[36]

The two classes of evidence that Haller pointed to in asserting that irritability is a power innate in muscles were the involuntary motions in the live animal that are entirely independent of the will, and the contractions that occur in muscle tissue in a dead animal or in muscles separated from the body. This latter issue in particular formed the focal point for Haller's debate with the Scottish physician Robert Whytt. The controversy, which ensued for fifteen years, took place at times on a rather acrimonious level, leading one Edinburgh student to remark that "neither Baron Haller nor Dr. Whytt were deficient in irritability."[37]

Whytt published his *Essay on the Vital and Other Involuntary Motions of Animals* in 1751, one year before Haller presented his paper on irritability and sensibility in Göttingen. Whytt proposed that all muscular movement, both voluntary and involuntary, is produced by a "sentient principle." A capacity of the immaterial soul, the sentient principle resides in the nerves and is thus coextensive with the body, animating the muscles in response to stimuli. When an organ is irritated – when air, for example, remains too long in the lungs or when food enters the intestines – an "uneasy perception" arises. The sentient principle then initiates motion designed to remove the irritating cause. An automatic process, involuntary motion is nevertheless dependent on the soul, which governs all vital functions, and not on any property possessed by the muscles themselves. Consequently, Whytt argues, we must conclude "that the all-wise Author of nature hath endued the muscular fibres of animals with certain active powers, far superior to those of common matter, and that to these the motions of irritated muscles are owing."[38]

Whytt's theory of muscular motion was based not on a separation of irritability from sensibility, as Haller's was, but on a dependence of one upon the other. Sensibility is widespread throughout the animal's body, and only through sensation does irritability operate. Thus, for Whytt, all motion depends ultimately on sensation, either conscious or unconscious.[39] For Haller, on the other hand, motion itself is completely independent of sensation, since the latter, in Haller's terms, must always be conscious.

Part of the controversy between Haller and Whytt lay in their differing views on the location and operation of the soul. Both agreed that sensation depends on the nerves and their communication with the soul. Yet Haller would not allow the soul any location other than the brain,[40] nor would he allow any unconscious sensation to occur in the body. Sensibility exists only because a part of the body is connected to the soul through the nerves. Whytt, on the other hand, could accept the idea of an immaterial principle residing in the nerves, which operates independently of consciousness or rational thought.

Haller believed that he had experimental proof that sensibility and irritability are entirely distinct capacities of the organism. The most irritable parts of the body, the muscles, are not by themselves sensible, and the agents of sensibility, the nerves, are not irritable. "What is insensitive moves and what is unmoved senses," Haller remarks; "therefore, one should remove perception from the contractile force."[41] More importantly, Haller pointed to the overwhelming evidence that muscles whose nerves have been tied or severed, and those that have been separated from the body, still retain the power to contract. How could this movement occur, if it is dependent on the soul, since the soul is no longer connected to the muscle? The soul is the source of consciousness and feeling. "But a finger cut off from my hand," Haller claims, "or a bit of flesh from my leg, has no connexion with me, I am not sensible of any of its changes, they can neither communicate to me idea nor sensation; wherefore it is not inhabited by my soul nor by any part of it; if it was, I should certainly be sensible of its changes . . . Irritability therefore is independent of the soul and the will."[42] How can one attribute motion to the nerves, when contraction occurs in dead animals and in muscles no longer connected to nerves? "Certainly it is necessary to rescind all the laws of reasoning," Haller contends, "if it is permitted to call something the cause of an effect when the removal, destruction, or absence of it neither eliminates nor changes the effect."[43] Irritability is thus entirely independent from the nerves, from the soul, and from sensibility.

That Whytt was able to accept the idea of an immaterial agent operating in the body and that Haller could not, is symptomatic of a deeper and, for our purposes, more important disagreement between the two. Haller and Whytt held widely divergent views on the nature of motion in general and on its relationship to material versus immaterial forces. Revealing of this difference is the use each made of the analogy of gravity as a model for how motion is produced in general, Whytt likening gravity to his sentient principle, Haller to his *vis insita.* And both pointed to the fact that gravity itself had not been fully explained to sanction their own incomplete explanations.

Whytt argued that since we use gravity to explain phenomena on the basis of its observed effects, why should we not utilize the observed power and energy of the mind to explain animal motion? Furthermore, Whytt maintains, partly in response to Haller's own comparison of irritability with gravity, both the sentient principle and gravity are *immaterial* in nature. Haller cannot liken irritability, an alleged property of matter, to gravity, Whytt claims, since gravity is an immaterial agent, just as is the sentient principle in organisms.[44]

To Whytt's attribution of an immaterial status to gravity, Haller strenuously objected. "Will therefore effervescence, putrefaction, fermentation, gravitation, elasticity . . . be works of some soul that falls in a stone, that turns unfermented wine into bubbles, that uncoils the wound blade of a clock?"[45] This, to Haller, was unacceptable. "I piously acknowledge," he proclaims, "that God is the mover of all nature. For this reason, neither the elasticity of expanding air, nor the weight of a stone, nor the effervescence of acids mixed with alkalis, nor the contraction of a dissected muscle ought to be attributed to incorporeal forces." Immaterial agents can in no way be responsible for such physical phenomena. "God gave to bodies," Haller continues, "an attractive force and other forces, which once received are exercised and which they owe to no other soul or spirit than God. I have doubts about this [Whytt's] spirit being the cause of motion: I derive all from God."[46]

Haller believed that matter is essentially passive and that to it forces are added by God. The motion that results is thus not produced by matter itself. "Indeed," Haller comments, "the great world bodies move themselves, and thus all parts of the earth and sun with them. But this movement is foreign to them; it is imparted to them . . . Consequently, movement of matter is not grounded in its being."[47] And although motion, and the forces responsible for motion, are identifiable through observation, we can never, Haller claims, fully understand motion itself. "The measure of forces," he explains, "consists in their effects, for the nature of motion itself, which is a most familiar thing, no one in philosophy has yet comprehended."[48] Speaking specifically about irritability, Haller remarks, "One will never know the mechanical source, from which the movements that follow irritation arise, but one will approach [this], one will perhaps succeed in measuring exactly the effect."[49] All forces operate on a mechanical basis; all can be known through observing and measuring their effects – but their ultimate source is not matter but God. With regard to muscular movement Haller writes, "We create no movement: our soul wills that the arm lift itself. But it gives to it neither the force nor the movement; God has put the force in the muscle."[50]

Haller thus distinguished his views from those of the animists by claiming that both physical and vital phenomena are based upon mechanical forces. Furthermore these forces are not innate to matter but are given to it by God. Yet this did not prevent his theory of irritability from being identified with materialism. As Whytt charges, "If irritability be a property of the muscular glue, why may not sensibility and intelligence be properties of the medullary substance of the brain?"[51] If properties of matter can be responsible for some vital phenomena, why can they not be the cause of consciousness and thought as well?

Haller was further identified with materialism through his relationship with La Mettrie. In 1748, the first edition of La Mettrie's *L'Homme machine* appeared with a dedication devoted, in ironical terms, to extolling Haller's virtues, explicitly crediting Haller as the spiritual father of La Mettrie's materialist views.[52] Haller had previously charged La Mettrie with plagiarism, on the grounds of his having issued a French version of Haller's edition of Boerhaave's *Praelectiones academicae*, making wholesale use of Haller's notes but giving no credit to Haller as their author.[53] There was no friendship to be lost between Haller and La Mettrie, and La Mettrie's intimation in his dedication of *L'Homme machine* that he was a pupil and friend of Haller's, though false, only added fuel to the flames. To this Haller responded in a public letter printed in the *Journal des sçavans* in November 1749: "I disavow this book as entirely opposed to my views. I regard its dedication as an affront more cruel than all of those that the anonymous author has made to so many honest men, and I beg the Public to be assured that I have never had any connection, acquaintance, correspondence, nor friendship with the author of *L'Homme machine,* and that I regard any conformity with his opinions as the greatest of misfortunes."[54]

Haller's wish to dissociate himself from La Mettrie's views was prompted by more than La Mettrie's dedication (which incidentally appeared in only the first printing of *L'Homme machine*). In his book, La Mettrie used irritability phenomena to bolster his argument for materialism. Never using the term irritability, La Mettrie nevertheless argues that "given the least principle of motion, animated bodies will have all that is necessary for moving, feeling, thinking, repenting, or in a word for conducting themselves in the physical realm, and in the moral realm which depends on it."[55] As evidence for the existence of this "least principle of motion" La Mettrie cites ten physiological observations, including the ability of muscles to contract when removed from the body and the fact that a decapitated chicken is able to run and flap its wings. "Here we have many more facts than are needed to prove, in an incontestable way," La Mettrie contends, "that each tiny fibre or part of an organized body moves by a principle which belongs to it. Its activity, unlike voluntary motions, does not depend in any way on the nerves."[56] If animal organisms possess the property of movement, then this can form a basis for denying the existence of a separate spiritual soul. "Let us then conclude boldly," La Mettrie proclaims, "that man is a machine, and that in the whole universe there is but a single substance differently modified."[57]

Haller responded to La Mettrie's use (or misuse) of irritability notions in his paper on irritability and sensibility. "The deceased M. De La Mettrie has made Irritability the basis of the system which he advances against the spirituality of the soul; and after saying that Stahl and Boerhaave knew nothing of it, he has the modesty to assume the invention to himself, without ever having made the least experiment about it."[58] Claiming that La Mettrie had actually learned of irritability indirectly from himself, Haller argues that his own experiments totally refute La Mettrie's conclusions. "For if irritability subsists in parts separate from the body, and not subject to the command of the soul, if it resides every where in the muscular fibres, and is independent of the nerves, which are the *satellites* of the soul, it is evident, that it has nothing in common with the soul, and it is absolutely different from it; in a word, that neither Irritability depends upon the soul, nor is the soul what we call Irritability in the body."[59] Haller claims that he has proved that irritability is totally separate from the soul, since the former is dependent on the muscles and the latter acts through the nerves. There is no basis for La Mettrie's assertion that irritability can be used to explain away the existence of the soul. Consequently, Haller's theory is not materialism.

In the eighteenth century, materialism was perceived to involve two key assumptions. First, the materialist position rested on a denial of the existence of a spiritual soul. Second, it implied that matter possesses active powers, powers that were attributed to no source outside of matter. Together, these notions added up, for many, to atheism. This identification of materialism with atheism, with the view that science if carried far enough could lead to irreligion, was responsible in large measure for the vehement reaction materialism prompted among so many eighteenth-century intellectuals.

Haller's response to La Mettrie in his paper on irritability and sensibility answered the first threat of materialism, the denial of the soul. Haller simply claimed that the soul exists and that his research on irritability and sensibility did nothing to challenge this. Haller was extremely concerned about the second aspect of materialism as well, for, as we have seen, he made it very clear that he did not believe that matter possesses forces on its own. Rather, the author of all movement and of all material forces is God. Consequently, materialism and atheism are not the inevitable result of a mechanistic approach to physiological phenomena.

SCIENCE WITHIN THE LIMITS OF RELIGION

Haller was deeply concerned with religious questions and in particular with the relationship of religious belief to scientific knowledge. In his later years, he devoted two works to theological issues, *Briefe über die wichtigsten Wahrheiten der Offenbarung,* published in 1772 (which also appeared in 1780 as *Letters from Baron Haller to his Daughter on the Truths of the Christian Religion*), and *Briefe über einige Einwürfe nochlebender Freygeister wieder die Offenbarung,* a three-volume work published in 1775–7, which was directed against the atheistic views of Voltaire. Haller's personal commitment to religion can also be seen in his poetry, most of which was written when he was a young man in his twenties.[60]

A number of scholars, especially those that have dealt with Haller as a literary figure, have pointed to a fundamental disharmony in his thought.[61] Principally, this is seen as a conflict between science and religion, in particular between knowledge based on science and truths based on revelation. Yet, although there is evidence, especially in Haller's early poetry, that this issue was of deep concern to him, when one looks at his scientific work, this conflict seems to disappear. What one finds instead is that Haller carefully defined for himself the boundaries of scientific explanation so that a clash between scientific knowledge and revelation could not occur. A major factor in Haller's many controversies with other scholars of his day – with Buffon, La Mettrie, Voltaire, C. F. Wolff – was indeed just this issue: Science must be carried on within the limits of religion. Where there is a danger of scientific theories forming a basis for materialism and atheism they must be rejected. As Erich Hintzsche has expressed it, "Haller saw the task of scientific investigation not to arouse doubt but rather to produce confirmation for belief in a Creator."[62]

Haller believed that God had created the world in such a manner that only human beings can understand its harmonious design. "Man alone," Haller writes, "knows the beauty, the splendor of nature. The wonderful construction of animals, of plants, can raise us through its accordance with their revealed design to knowledge of a Creator: so much we know, it is man alone who knows and enjoys the variety, the order, and the relationships among the parts of the world."[63] Through human ingenuity and proper methods of study, nature's secrets can be revealed, at least to the extent intended by God. The world is a Divine Creation, and it is through studying its laws that man can come to appreciate the wisdom of the Creator.

Perhaps the most significant example of the relationship between Haller's scientific work and his religious beliefs is his work on embryology. I have commented elsewhere on the changes that Haller's embryological theories underwent during his lifetime, and I will not discuss them in detail here.[64] But I would like to underscore the tie between what Haller saw as an acceptable explanation of embryological development and his belief in a divinely created world.

The two prevalent embryological theories during the eighteenth century were preformation and epigenesis. The former position argued that the embryo develops from a preformed structure that preexists in either the mother or the father before conception. Although significant differences existed among preformationists, most in the eighteenth century also believed in *emboîtement,* the idea that all organisms were created at one time by God and encased within one another, so that each original member of a species contained all future generations. The alternative explanation was epigenesis, which was based on the belief that the embryo forms gradually out of unorganized material. The greatest difficulty with this position in the eighteenth century was how to account for the source of the embryo's complex organization if it is not preformed in the generative matter.[65] It was partly this problem that accounted for the popularity of preformationism during this period.

Haller's views on embryological development underwent several changes during his lifetime. The most significant one for our purposes here is his conversion from epigenesis to preformation in the early 1750s. As an epigenesist Haller had argued in the 1740s that the formation of the embryo proceeds through attractive forces, which operate in a manner analogous to crystal formation. Yet Haller was soon to abandon this position, principally because he found it difficult to explain how a mere force could be responsible for the complexity of the resulting organism. How could an attractive force know what to do?

Haller's thinking on this issue was articulated especially in his comments on Buffon's theory of generation, which had appeared in 1749 in volume II of Buffon's *Histoire naturelle.* Wrestling with Buffon's account of formation through an "internal mold" that operates in part on the basis of a "penetrating force," Haller came face to face with the problem of how much forces could or could not do on their own. In response, he concluded, "If matter has forces that allow it to build things, it does not possess them blindly. They are surrounded by eternal limits, and build always perfectly not mechanical equals but something similar, something that is prescribed in an inviolable plan: but with a diversity that excludes the constraint of blindly working matter."[66] Blind forces of matter could not on their own be responsible for the developmental process. Rather they operate within "eternal limits" prescribed by God.

Haller's concern over the role forces play in embryological development played a significant part in his conversion to preformationism. For if the embryo is preformed, one can utilize forces in explaining its development without worrying about how they know what to do. The structure preexists, and forces act to produce its unfolding. In Haller's preformationist theory, announced in 1758, it is primarily the force of irritability that governs the developmental process.[67] The tiny preformed heart of the embryo is stimulated during fertilization, and, because of its inherent irritability, begins to beat. By pumping fluids through the folded, preexistent vessels of the embryo, the heart directs the gradual development of the organism. Solidification of structures, increasing opacity, differential rates of expansion – all of these are involved in the transformation of the embryo to the mature organism. All are based on forces of nature that operate in a mechanical fashion. Yet none is responsible for development itself, since this was preordained by God at the Creation.

Haller's explanation of development through preformation enabled him to account for the formation of the embryo in a manner consistent with both his scientific and his religious beliefs. He expressed just this sentiment in 1766 in his *Elementa physiologiae:*

> If the first rudiment of the fetus is in the mother, if it has been built in the egg, and has been completed to such a point that it needs only to receive nourishment to grow from this, the greatest difficulty in building this most artistic structure from brute matter is solved. In this hypothesis, the Creator himself, for whom nothing is difficult, has built this structure: He has arranged at one time, or at least before the male force [of fecundation] approaches, the brute matter according to foreseen ends and according to a model preformed by his Wisdom.[68]

Each particular generation of a new life is the result of causes established by God at the Creation. The material of future generations was organized in such a way that, through natural forces, also added to matter by God, development would proceed automatically at each instance of generation.

"Enough, there is a God," Haller proclaims in an early poem; "nature shouts it out,/ The whole construction of the world shows signs of his hand."[69] The stars in the heavens never lose their course; animal organisms function according to a Divine plan, with their blood circulating in a perfectly adjusted manner. And finally there is the highest creation: "Man, whose word commands the earth,/ Is a composite of pure masterpieces;/ In him are united the art and splendor of bodies,/ No part of him exists, that does not show him to be the master of Creation."[70] Through science Haller saw man's route to knowledge of God's universe. Science can never lead to atheism and materialism but rather to a deeper appreciation of Divine wisdom and power.

HALLER'S NEWTONIANISM

Haller's ideas on the experimental method, on the nature of forces, and on the role of God in the universe did not develop within a vacuum. Rather they emerged in the context of Haller's education and were influenced, in particular, by the two figures Haller held in highest esteem, Herman Boerhaave and Isaac Newton.

After a brief period at the University of Tübingen, where he began his medical studies, Haller moved to Leiden in 1725 to study under the famous Boerhaave. Here Haller was inculcated with a markedly empirical approach to anatomical and medical studies, with a negative view of Cartesian rationalism and its nonobservational method, and with a strong emphasis on the usefulness of mechanical reasoning in physiological explanation. Also at Leiden was Bernhard Siegfried Albinus, professor of anatomy and surgery, whose own work on bones and on muscles was reflected in Haller's later choice of research topics.[71]

Prior to receiving medical training, Boerhaave had studied for a divinity degree, intending to become a clergyman. In later years, his religious beliefs remained an integral part of his life, and carried over, by his own example, into his teaching. "Dr. Boerhaave was a religious and modest man," the *Gentleman's Magazine* reported the year of his death; " . . . he never made mention of the Supreme Being but to admire and exalt him in his Works."[72] Boerhaave was a "truly Christian" man, Haller relates; " . . . to him I owe eternal affection and everlasting gratitude . . . Perhaps future centuries will produce his equal in genius and learning, but I despair of their producing his equal in character."[73]

Boerhaave was also an ardent supporter of the Newtonian philosophy (according to Lindeboom one of the first on the Continent).[74] In his "Vie de M. Herman Boerhaave," La Mettrie, who studied at Leiden in the early 1730s, recounts that "Boerhaave was a Newtonian, convinced and convincing . . . he regarded . . . Newton as the favorite of Nature, as the organ which it used to illuminate the universe, and to reveal to him its mysteries." La Mettrie contrasts Boerhaave's attitude toward Newton with his opinion of Descartes, who built a system "without consulting nature." Newton, however, "set out only after the experiment, and established nothing that was not on the most solid foundation."[75] Boerhaave, in his own strongly empirical approach, emulated the Newtonian method.

During Haller's years of attendance, another Newtonian proponent taught at Leiden, the professor of mathematics, Willem Jacob 'sGravesande, whose *Physices elementa mathematica,* subtitled "an introduction to Newtonian philosophy," was published in 1720–1. Haller notes in his Leiden diary in 1726 that 'sGravesande "through his acquaintance with Newton comprehends his system so well, that he makes a perfect frame of it. His best are the experiments, which, because of his beautiful instruments, he does very exactly and very frequently. One can see all of this in his new edition of the *Physique.*"[76] That Haller was familiar with 'sGravesande's work is clear; it is even possible that he studied with him, given Haller's own interest in mathematics. Haller thus received through Boerhaave and 'sGravesande, if not his introduction to Newton, a strongly positive grounding in Newtonian physics and philosophy.

After finishing his medical studies, Haller traveled in 1727 to England, where he was enormously impressed with English intellectual life. "In the sciences," Haller writes in his diary, "it appears that no land is superior to England . . . in the investigation of nature, excellent experiments and all that concerns surveying and the nature of substance, they surpass all former times and present lands."[77] The reasons for this, Haller suggests, are the wealth of the land, the prizes and rewards that are given to scholars, and the "meditative and ambitious temper of these people."[78] Even the court values science; Haller notes that the queen herself is interested in the controversy between Newton, Clarke, and Leibniz.[79] Haller visited Newton's recent grave in Westminster Abbey (Newton died the preceding March), which impressed him in being as magnificent and as costly as that of the English royalty. "The extraordinary veneration of the whole people for this great spirit," Haller remarks, "testifies that unusual scholarship is held here in as much regard as elsewhere are nobility and war service."[80] Haller lists as exemplary of the high state of English science, Wallis, Hauksbee, Keill, Boyle, Desaguliers, Raphes, Pemberton, and Clarke.[81]

Although Haller could not read or speak English at the time of his first trip to London, he came away with a veneration for English culture that was never to leave him. After returning to Switzerland in 1728, Haller studied mathematics in Basel with Johann Bernoulli. Holding Bernoulli's abilities in highest esteem, Haller ranked him among the mathematical giants of his time. "Leibniz was a Columbus," Haller wrote to a friend, "who caught sight of several islands of his new world, but a Newton, a Bernoulli are born, to be the conquerors of the same."[82]

During Haller's stay in Basel he also met Benedikt Stähelin, who furthered Haller's interest in English intellectual life. Stähelin, himself fluent in the English language, was an ardent admirer of English literature, to which he introduced Haller. Learning English himself, Haller became acquainted with the works of Shaftesbury, Pope, and others. Haller wrote his own poetry during these years, and in his "Gedanken über Vernunft, Aberglauben, und Unglauben" (1729), dedicated to Stähelin, Haller proclaims that he is following the model of English philosophical poetry to show that such a style can be successfully used in the German language as well.[83]

In this same poem, which extols human reason and reverence for God, and combats superstition and atheism, Haller praises Huygens and Newton for their scientific achievements. "A Newton exceeds the limits of created minds, / Finds nature at work and appears as master of the universe; / He weighs the inner force, that is active in bodies, / That makes one fall and moves another in a circle, / And he breaks open the tables of the eternal laws, / Once made by God and never broken."[84] Newton figures also in "Die Falschheit menschlicher Tugenden" (1730), where Haller lists Newton's accomplishments – the infinitesimal calculus, gravitation, color theory, and Newton's explanation of the tides. "He fills the world with clarity," Haller proclaims, "He is a continual source of unrecognized truth."[85]

That Newton should appear in such a laudatory manner in eighteenth-century poetry does not seem surprising. Yet its significance for our understanding of Haller's philosophical views should not be underestimated. True, we find numerous references to Newton in English poetry. Yet, as Karl Richter has pointed out, this is not the case in the German-speaking world. Haller, Richter claims, is one of the first, if not *the* first, to refer to Newton in his poetry.[86]

As one would expect, references to Newton occur in Haller's scientific work as well. In dedicating his *Enumeratio methodica stirpium Helvetiae indigenarum*

(1742), one of Haller's major botanical works, to the Prince of Wales, Haller praises English science, especially the works of Newton.[87] Newton figures again as the champion of the hypothetical method in Haller's preface to the German translation of volume I of Buffon's *Histoire naturelle.*[88] Finally, Newton's own speculations concerning the aether's role in muscular movement and sensation are referred to in the *Bibliotheca anatomica* and discussed more fully in the *Elementa physiologiae.*[89] In the latter, Haller also cites the "first law of Newton," that causes should not be multiplied unnecessarily in explaining phenomena, and Newton's views on the nature of motion.[90]

Evidence thus exists to document Haller's familiarity with, and admiration for, the work of Newton. We know that he owned the *Opticks,* that he read and admired the *Principia,*[91] that he was acquainted with the Leibniz-Clarke correspondence, and that he read both 'sGravesande's book on the Newtonian system and Henry Pemberton's *View of Sir Isaac Newton's Philosophy* (1728). Haller characterizes Pemberton's work, in a letter to Johannes Gessner, as "a splendid book in which the mysteries of the Newtonian philosophy are revealed in an easy style and without algebra."[92] There also exists an unpublished review of Pemberton's book, written by Haller in English, in which Haller seems particularly impressed by Pemberton's critique of the common errors of philosophy, such as concluding too hastily from too few observations, and by his defense of the Newtonian system against the charge that it reinstated occult qualities into natural philosophy. Concerning the latter, Haller notes, "He shews farther the injustice of those that accuse the N. P. [Newtonian philosophy] of occult quality's and says with a great deal of Truth, that Philosophers are to be content, if they attain one of the intermediate marshes of the scale of cause's, and noway oblig'd to come to the first, hidden in an impenetrable obscurity."[93]

It is not surprising, given the scope and volume of Haller's reading, that he should have been acquainted with the major Newtonian publications of his day. Yet their effect on him was far deeper than mere familiarity. In each of the principal aspects of Haller's philosophy of science discussed earlier, one can trace a clear, and fundamental, Newtonian influence. In my opinion, Haller consciously sought to emulate the Newtonian program in his scientific work and to construct, in particular, a new physiology based upon the canons of the new philosophy.

Concerning the experimental method, Haller was in close agreement with Newton. Newton's famous declaration that he would "feign no hypotheses" concerning the cause of gravity, since hypotheses not based on phenomena "have no place in experimental philosophy"[94] finds its echo in Haller's silence concerning the causes of irritability and sensibility. "But the theory," Haller cautions, "why one or the other of these properties either is not in these parts, or is in other parts of the human body, such a theory, I say, I certainly do not hope to give. For I am persuaded that the origin of both abilities lies hidden in the intimate fabric, and is placed far beyond the power of the scalpel or the microscope: beyond the scalpel or microscope I do not make many conjectures."[95] And even though we do not know the cause of irritability, we can postulate its existence from its observed effects "to which moreover it is unnecessary to assign any cause, just as no probable cause of attraction or gravity is assigned to matter [in general]. It is a physical cause, . . . discovered through experiments, which are evidence enough for demonstrating its existence."[96] Irritability is itself, Haller postulates, an attractive force, which operates in the inner structure of muscle tissue.

That gravity, and especially the fact that its origins are unknown, should have been used by Haller as an analogy for his own unexplained force is not particularly unusual. In the eighteenth century, gravity was frequently called upon by physiologists to sanction a myriad of unknown properties and powers that could not be explained but must, it was argued, be postulated to exist. As we saw earlier in discussing Haller's debate with Robert Whytt, the example of gravity was often used by both sides of the same controversy.[97] Yet in Haller's case, we find a utilization of Newtonian elements that far exceeds the simple analogy of irritability with gravity. Haller's entire approach to physiology as a force mechanics was Newtonian in outlook. One can almost hear Haller's definition of physiology in Newton's proclamation that "the whole burden of philosophy seems to consist in this – from the phenomena of motions to investigate the forces of nature."[98] Furthermore, Haller's views on the relationship between matter and forces – that matter is passive and must be activated by forces – were similar to Newton's. As Newton argued in Query 31 of the *Opticks,* matter alone possesses only the *vis inertiae,* a passive principle through which bodies tend to remain at rest or in motion. Yet "by this Principle alone there never could have been any Motion in the World."[99] Forces are required, Newton contends, to account for gravity, fermentation, and other kinds of motion; for "we meet with very little Motion in the World, besides what is owing to these active Principles."[100] That the motion of animal muscles might also be an active force added to passive matter seems but a simple extension of Newton's own views.

Haller champions Newton's ideas on the nature of motion against those of Leibniz and Descartes, who argued that motion can never be created or destroyed and that the total amount of motion is constant in the universe. Haller ties this view to that of the Stahlians, who attempted to show that it is therefore only the immaterial soul that can cause motion in the human body. "Newton, however," Haller asserts, "determined that new motion is generated and old destroyed, and all nature agrees with him."[101] Consequently, "the motion that . . . lifts the arm does not arise from the soul, but that motion is generated in the body according to a law established long before."[102] Irritability, the physical cause of the muscle's contraction, produces motion in the body in the same way that motion is created or destroyed in "all nature."

Yet if forces are responsible, through motion, for the phenomena we observe in the universe, where do they come from? The answer lies, for both Newton and Haller, in the religious context of their scientific views. Haller, as we have seen, saw God as governing the operations of the world through the forces that He imparted to matter at the Creation. As P. M. Heimann has aptly expressed it, this was a salient feature of Newtonianism in the eighteenth century: "Newton's ideas were originally presented and disseminated (by the Boyle lectures) in a form which stressed the theological dimension to Newton's philosophy of nature. For early eighteenth century thinkers, Newtonian doctrines of the passivity of matter, of the primacy of forces in nature, and of gravity as a power not essential to but imposed upon matter expressed a theology of nature."[103] For Newton, as for other English voluntarists, nature and the laws of nature were seen as dependent on Divine Providence. The laws of motion are an expression of the will of God; forces of nature are the result of Divine power. An intelligent agent created the world, not blind fate.[104] "Such a wonderful Uniformity in the Planetary System," Newton declares in the *Opticks,* "must be allowed the Effect of Choice. And so must the Uniformity in the Bodies of Animals."[105] Or, as Newton phrases it in the General Scholium to the *Principia,* "Blind metaphysical necessity, which is certainly the same always and everywhere, could produce no variety of things. All that diversity of natural things which we find suited to different times and places could arise from nothing but the ideas and will of a Being necessarily existing."[106]

Haller's views on the relationship of God to His creation were, as we have seen, similar in many respects to Newton's. Discussing matter and the forces that it possesses, Haller remarks that "it could be without gravity, without elasticity, without irritability . . . These qualities do not take part in its essence; they are foreign to it; they are not common to all parts of matter . . . A first cause has thus allotted to different classes of matter abilities and forces calculated according to a general plan, and it is there that we recognize the hand of the Creator."[107] The constancy we observe in the world testifies to God's will, not chance, as the cause of phenomena. And the diversity is evidence for God's intelligence, not necessity, as the guiding factor. Speaking of the similarities and varieties one finds among flowers, Haller observes, "All is not chance; otherwise the carnation would become a tulip; all is not necessity, otherwise the carnation would remain always such as was the first carnation."[108] An intelligent God, through His own free choice, created the universe and the laws that govern it. It is the task of science, for both Newton and Haller, to uncover these Divine laws and to show how the phenomena we observe are produced by them.

Haller saw science as leading toward a deeper appreciation of and reverence for God, and away from the dangers of atheism and materialism. Newtonian philosophy was also seen as a bulwark against irreligion both by Newton and by his contemporaries. Richard Bentley's Boyle lectures, the final three of which were titled *A Confutation of Atheism from the Origin and Frame of the World,* were designed to make just this point, as were those delivered by other Boyle Lecturers. Through Newton's system of the world, Roger Cotes declares in his preface to the second edition of the *Principia,* "We may now more nearly behold the beauties of Nature, and entertain ourselves with the delightful contemplation; and, which is the best and most valuable fruit of philosophy, be thence incited the more profoundly to reverence and adore the great Maker and Lord of all."[109] With the Newtonian philosophy, the dangers of a world based on necessity have been overcome. "Newton's distinguished work," Cotes concludes, "will be the safest protection against the attacks of atheists, and nowhere more surely than from this quiver can one draw forth missiles against the band of godless men."[110] Haller saw his own work as adding another arrow to the quiver and as furthering the Newtonian cause of science in support of religion. Of skeptics and atheists Haller wrote, "None has known nature well enough to be able to discover for himself the traces of the Diety that beam forth so abundantly and so luminously in the ends and in the order of created things. Where a Hobbes doubted, a Newton believed; where an Offray [La Mettrie] scoffed, a Boerhaave worshipped."[111]

In the three major areas of Haller's philosophy of science – his experimental method, his creation of an animal mechanics, and his belief that science exists to serve religion – he was fundamentally inspired by Newton and the Newtonian philosophy. Much of this influence came not from Newton directly but through other Newtonian proponents, especially Herman Boerhaave, whose own example as a scientist and as a religious man Haller sought to emulate. The views of Haller and Newton were certainly not identical in all respects, nor was Haller's the only variety of "Newtonianism" extant in the eighteenth century. Nevertheless, the influence is too marked to overlook, for the biological viewpoint that resulted carried with it the imprint of these shared beliefs.

CONCLUSION

The impact of mechanism on physiology forms an important episode in the history of seventeenth- and eighteenth-century biology. With Descartes's strict separation of soul from body and his attempt to explain the operations of the latter through matter and motion alone, iatromechanism received one of its most influential statements. Championed by Italian physicians Borelli and Malpighi, by Friedrich Hoffmann, and by Herman Boerhaave, among others, the mechanical method became a dominant influence in early eighteenth-century anatomy and physiology. And although differences existed among these various proponents of mechanical physiology, all looked to the categories of matter and motion, and to mechanical analogies, as the foundation of physiological explanation.

With Haller we reach a new level of mechanical physiology, for Haller's animal mechanics was significantly different from the theories of his iatrophysical predecessors. Based as it was on the concept of force, Haller's physiology added a new dimension to the matter and motion dyad. In this aspect, Haller's work is based on Newtonian rather than Cartesian mechanism, for it was Newton who added force to matter and motion as the primary elements of physical explanation. Haller's animal mechanics was a force mechanics, and it was this that gave it its versatility and its extraordinary appeal.

But Haller's animal mechanics was not reductionistic in outlook. Physiological explanation was to remain parallel and equal to physical explanation. Both should be based on mechanically acting forces, yet physiology, for Haller, was to be grounded on its own forces. Irritability, the key force of living organisms, operates in exactly the same manner as physical forces do: It causes movement when the proper conditions arise. Yet irritability is its own special force, which cannot be reduced to other known forces. Haller's animal mechanics is thus based on a Newtonian model; it is an extension into the biological realm of Newtonian categories of explanation. But animal mechanics retains its independent status.

Haller's opposition to the Stahlians and to La Mettrie, and his attempt to steer a middle course between animism and materialism, is closely related to his nonreductionist view of physiology. Haller opposed animism because it undercut the mechanical approach to physiology, because it extended too far the sphere of the soul in the animal body. The soul is immaterial; it is not scientifically, that is, mechanistically, treatable. The soul definitely exists – this Haller is quick to argue – yet its role is not to govern vital functions but to create consciousness and thought. Haller's animal mechanics allows no room for the soul in physiological explanation, for the forces governing bodily processes are not immaterial.

Yet these forces are not entirely material products either. Matter itself is passive; to it must be added activating forces by God. Thus Haller's animal mechanics ruled out materialism also. Not only are physiological forces not reducible to physical forces, but, more importantly, no forces are the product of matter alone. All are dependent on the will of God, whose wise intelligence created a world that operates on the basis of these mechanical forces. Thus animal mechanics leads not to atheism but to a deeper understanding of and reverence for the Almighty Creator.

Haller's contemporaries and followers saw his physiological work as a momentous step, as a break with past erroneous traditions. Condorcet, in his eulogy of Haller, declared that irritability had created "a revolution in Anatomy." At last, Condorcet continues, "Physiology, for too long resting on metaphysical and uncertain ideas, can finally be based on a general fact, proven by experience."[112] Tissot, in the introduction to his French translation of Haller's paper on irritability and sensibility, hails irritability as one of the "keys of nature," ranking it with the previous discoveries of the properties of air, of the circulation of the blood, and of electricity, discoveries that occurred, Tissot notes, within less than a century. "It is today the turn of Irritability," Tissot proclaims.[113] "The whole of animal mechanics revolving around this principle, it is easy to imagine what changes this discovery will produce in the explanations of facts: to England we owe natural philosophy, to Switzerland we shall owe physiology, and the Memoir on Irritability will form its immovable base."[114] Haller's concepts of irritability and sensibility figure, in various forms, in numerous physiological and medical treatises of the late eighteenth and early nineteenth centuries. Looking back a century later, Pierre Flourens was to note, "In 1752 appeared his beautiful discoveries on irritability and sensibility; and, from this moment, a new horizon was opened."[115]

It was not just Haller's physiological discoveries that had an impact on successive generations. More importantly, Haller's physiological method, based as it was on experimentalist empiricism and on a nonreductionist mechanism, made its influence felt as well. Haller became the example of the first modern physiologist, the model to emulate in philosophical attitude. Haller thus transcended his eighteenth-century surroundings and, like his predecessor Newton, entered the realm of historical legend.

NOTES

1 Richard Toellner, *Albrecht von Haller: Über die Einheit im Denken des letzten Universalgelehrten,* Sudhoffs Archiv Beihefte, no. 10 (Wiesbaden: Steiner, 1971).

2 For a nearly complete listing of Haller's published works, including their editions and translations, see Susanna Lundsgaard-Hansen-von Fischer, *Verzeichnis der gedruckten Schriften Albrecht von Hallers,* Berner Beiträge zur Geschichte der Medizin und der Naturwissenschaften, no. 18 (Bern: Haupt, 1959).

3 Pioneering work on identifying and classifying Haller's anonymous reviews has been done by Karl S. Guthke. See his *Haller und die Literatur* (Göttingen: Vandenhoeck and Ruprecht, 1962); "Haller als Kritiker: Neue Funde," in *Literarisches Leben im achtzehnten Jahrhundert in Deutschland und in der Schweiz* (Bern and Munich: Francke, 1975), pp. 333–53; and *Hallers Literaturkritik,* ed. Karl S. Guthke (Tübingen: Max Niemeyer, 1970). See also Otto Sonntag, "The Idea of Natural Science in the Thought of Albrecht von Haller" (Ph.D. diss., New York University, 1971).

4 Some of the major sources include, in addition to those listed in the preceding notes, Rüdiger Robert Beer, *Der grosse Haller* (Säkingen: Hermann Stratz, 1947); Stephen d'Irsay, *Albrecht von Haller: Eine Studie zur Geistesgeschichte der Aufklärung,* Arbeiten des Instituts für Geschichte der Medizin an der Universität Leipzig, no. 1 (Leipzig: Georg Thieme, 1930); Heinrich Ernst Jenny, *Haller als Philosoph, ein Versuch* (Basel: Basler Druck- und Verlags-Anstalt, 1902); Christoph Siegrist, *Albrecht von Haller* (Stuttgart: Metzler, 1967); and Otto Sonntag, "Albrecht von Haller on Academies and the Advancement of Science: The Case of Göttingen," *Annals of Science, 32* (1975), pp. 379–91, "Albrecht von Haller on the Future of Science," *J. Hist. Ideas, 35* (1974), pp. 313–22, and "The Motivations of the Scientist: The Self-Image of Albrecht von Haller," *Isis, 65* (1974), pp. 336–51. See also Erich Hintzsche, "Einige kritische Bemerkungen zur Bio- und Ergographie Albrecht von Hallers," *Gesnerus, 16* (1959), pp. 1–15.

5 See especially Richard Toellner, "Anima et Irritabilitas, Hallers Abwehr von Animismus und Materialismus," *Sudhoffs Archiv, 51* (1967), pp. 130–44; Gerhard Rudolph, "De partibus sensilibus," *Sudhoffs Archiv, 49* (1965), pp. 423–30, and "Hallers Lehre von der Irritabilität und Sensibilität," in *Von Boerhaave bis Berger,* ed. K. E. Rothschuh, *Medizin in Geschichte und Kultur,* vol. 5 (Stuttgart: Gustav Fischer, 1964), pp. 14–34; and Erna Lesky, "Albrecht von Haller und Anton de Haen im Streit um die Lehre von der Sensibilität," *Gesnerus, 16* (1959), pp. 16–46.

6 The most original, though controversial, contribution on this subject to date is Toellner's *Albrecht von Haller.* See also Sonntag's dissertation (n. 3, this chapter).

7 "De partibus corporis humani sensilibus et irritabilibus," *Commentarii Societatis Regiae Scientiarum Gottingensis, 2* (1752), p. 115. See also "A Dissertation on the Sensible and Irritable Parts of Animals," ed. and with an intro. by Owsei Temkin, *Bull. Hist. Med., 4* (1936), pp. 651–99.

8 *Elementa physiologiae corporis humani,* 8 vols. (Lausanne: M. M. Bousquet; Bern: Societas Typographica, 1757–66), 1 (1757), p. iv.

9 Ibid., p. v.

10 *Briefe über die wichtigsten Wahrheiten der Offenbarung* (Bern: Neue Buchhandlung, 1772), p. 45.

11 Haller's preface to *Allgemeine Historie der Natur,* by Georges Louis Leclerc, Comte de Buffon, 8 vols. (Hamburg & Leipzig: G. C. Grund and A. H. Holle, 1750–72), 1 (1750), p. x.

12 Review of *Oeuvres de M. Thomas,* vol. 4 (containing Thomas's "Éloge de Descartes"), in *Zugabe zu den Göttingischen Anzeigen von gelehrten Sachen* (1773), p. 371. Haller's authorship is established by the presence of an "H" in the margin of Haller's own copy of the *Göttingische Anzeigen von gelehrten Sachen,* now in the possession of the Stadtbibliothek in Bern, Switzerland. (See Guthke references in n. 3, this chapter).

13 Preface to *Allgemeine Historie der Natur,* 1 (1750), p. ix.

14 Ibid., p. x.

15 Review of *Philosophia rationalis, Pars I,* by Samuel Christian Hollmann, in *Bibliothèque raisonnée, 37* (1746), pp. 356–7. See Guthke, "Haller als Kritiker," pp. 348–53, for a list of Haller's reviews in this journal.

16 Review of *Philosophia rationalis,* p. 356.

17 *Von und über Albrecht von Haller: Ungedruckte Briefe und Gedichte Hallers sowie ungedruckte Briefe und Notizen über denselben,* ed. Eduard Bodemann (Hannover: Carl Meyer, 1885), p. 89.

18 "Haller und Leibniz, zwei Universalgelehrte der Aufklärung," *Studia Leibnitiana Supplementa, 12* (1973), pp. 249–60.

19 *Elementa physiologiae,* 1 (1757), p. i.

20 Ibid., p. v.

21 Preface to *Primae lineae physiologiae in usum praelectionum academicarum* (Göttingen: A. Vandenhoeck, 1747), p. 5. "Erunt, qui objiciant, meram me scripsisse anatomen. Sed physiologia est animata anatome."

22 *Elementa physiologiae,* 1 (1757), pp. v–vi.

23 Ibid., p. vi.

24 Herman Boerhaave, *Praelectiones academicae in proprias institutiones rei medicae,* ed. and with notes added by Albrecht von Haller, 6 vols. in 7 (Göttingen: A.Vandenhoeck, 1739–44), 2 (1739), sec. 187, n. *i,* p. 129. Based on Boerhaave's *Institutiones medicae,* Haller's edition includes two sets of notes, one explaining Boerhaave's doctrines and one containing Haller's own amplifying and critical remarks.

25 *Praelectiones academicae,* 4 (1743), sec. 600, n. *a,* p. 586; and *Primae lineae physiologiae* (1747), sec. 113, p. 51; 2d ed. (1751), sec. 408, p. 252.

26 For a list of these translations and editions, see Susanna Lundsgaard-Hansen-von Fischer, *Verzeichnis der gedruckten Schriften Albrecht von Hallers,* pp. 30–2.

27 "De partibus corporis," p. 116.

28 Haller first refers to electricity as a stimulus for irritability in the *Elementa physiologiae.* See for example vol. 4 (1762), pp. 448, 458, 535, 553, and 556. See also Roderick W. Home, "Electricity and the Nervous Fluid," *J. Hist. Biol., 3* (1970), pp. 235–51, where Haller's rejection of electricity as the agent of nerve action is discussed.

29 See E. Bastholm, *The History of Muscle Physiology from the Natural Philosophers to Albrecht von Haller* (Copenhagen: E. Munksgaard, 1950).

30 "De partibus corporis," pp. 154–8.

31 See Owsei Temkin, "The Classical Roots of Glisson's Doctrine of Irritation," *Bull. Hist. Med., 38* (1964), pp. 297–328; and Walter Pagel, "Harvey and Glisson on Irritability with a Note on van Helmont," *Bull. Hist. Med., 41* (1967), pp. 497–514.

32 "De partibus corporis," pp. 138–9; *Elementa physiologiae,* 4 (1762), pp. 516–7.

33 Haller thought that muscle fibers consist of earthy particles and a glutinous mucus, and that the property of irritability most likely resided in the gluten. See "De partibus corporis," p. 152.

34 Ibid., p. 154.

35 *Elementa physiologiae,* 4 (1762), pp. 460–1.

36 *First Lines of Physiology* (New York: Johnson Reprint, 1966), sec. 407, p. 236. A reprint of William Cullen's 1786 English translation of Haller's *Primae lineae physiologiae* (3d ed., 1765), this work contains significant additions to Haller's chapter on muscular motion that did not appear in the editions of 1747 and 1751.

37 From a marginal note in the Edinburgh University copy of *The Works of Robert Whytt,* cited in R. K. French, *Robert Whytt, the Soul, and Medicine* (London: Wellcome Institute of the History of Medicine, 1969), p. 11. See also chap. 6, pp. 63–76, on Whytt's debate with Haller, where the major publications and issues of the controversy are discussed.

38 *An Essay on the Vital and Other Involuntary Motions of Animals* (Edinburgh: Hamilton, Balfour, & Neill, 1751), p. 241.

39 See French, *Robert Whytt,* chap. 7, pp. 77–92, for Whytt's views on involuntary and reflex motion.

40 *First Lines of Physiology,* sec. 370–2, pp. 216–18.

41 *Elementa physiologiae,* 4 (1762), pp. 460.

42 "De partibus corporis," p. 138.

43 *Elementa physiologiae,* 7 (1765), p. iv. Haller's preface to this volume also appeared separately as *Ad Roberti Whyttii nuperum scriptarum apologia* in 1764.

44 See Whytt's *Essay,* p. 275, and *Physiological Essays* (Edinburgh: Hamilton, Balfour, & Neill, 1755), pp. 185–8.

45 *Elementa physiologiae,* 4 (1762), p. 531.

46 *Elementa physiologiae,* 7 (1765), p. xii.

47 *Briefe über einige Einwürfe nochlebender Freygeister wieder die Offenbarung,* 3 vols. (Bern: Typographische Gesellschaft, 1775–7), 1 (1775), p. 223.

48 *Elementa physiologiae,* 1 (1757), p. 426.

49 "Oeconomie animale," *Supplement à l'Encyclopédie,* 4 vols. (Amsterdam: M.M. Rey, 1776–7), 4 (1777), p. 105. For a list of Haller's contributions to the supplementary volumes of the *Encyclopédie,* see Erich Hintzsche, "Albrecht von Hallers Tätigkeit als Enzyklopädist," *Clio Medica, 1* (1966), pp. 235–54.

50 *Briefe über einige Einwürfe nochlebender Freygeister,* 3 (1777), p. 148.

51 *Physiological Essays,* p. 185.

52 La Mettrie's dedication is reprinted in Aram Vartanian's *La Mettrie's "L'homme machine": A Study in the Origins of an Idea* (Princeton, N.J.: Princeton University Press, 1960). For an English version, see Ernst Bergmann, "The Significance of La Mettrie and Pertinent Materials," *Open Court, 27* (1913), pp. 411–32. This and "The Satires of Mr. Machine," *Open Court, 28* (1914), pp. 412–23, are translated portions from Bergmann's *Die Satiren des Herrn Maschine: Ein Beitrag zur Philosophie- und Kulturgeschichte des 18. Jahrhunderts* (Leipzig: E. Wiegandt, 1913). Considerable controversy has arisen over whether Haller was the author of the anonymously published *L'Homme plus que machine* (1748). For definitive establishment of E. Luzac as the author, see Karl S. Guthke, "Haller, La Mettrie und die anonyme Schrift *L'Homme plus que machine,*" *Études Germaniques, 17* (1962), pp. 137–43; and Aram Vartanian, "Elie Luzac's Refutation of La Mettrie," *Mod. Lang. Notes, 64* (1949), pp. 159–61. See also Erich Hintzsche, "Neue

Funde zum Thema: L'homme machine und Albrecht Haller," *Gesnerus, 25* (1968), pp. 135–66.

53 Herman Boerhaave, *Institutions de médecine,* trans. and ed. Julien Offray de La Mettrie, 8 vols. (Paris: Huart, 1743–50). La Mettrie mentions Haller in his preface but does not indicate that most of his commentary is a direct translation of Haller's. See Haller's review of La Mettrie's edition in the *Göttingische Zeitungen von gelehrten Sachen* (1745), pp. 377–8. (For Haller's authorship, see Hintzsche, "Neue Funde," p. 138.)

54 *Journal des sçavans, 149* (1749), pp. 332–3.

55 *Man a Machine,* French-English edition, ed. Gertrude Carman Bussey (La Salle, Ill.: Open Court, 1961), p. 128.

56 Ibid., p. 130.

57 Ibid., p. 148.

58 "De partibus corporis," p. 158.

59 Ibid.

60 These were published by Haller as *Versuch Schweizerischer Gedichten* (Bern: E. Haller, 1732 and many later editions). The standard modern collection is *Albrecht von Hallers Gedichte,* ed. and with an introductory monograph by Ludwig Hirzel (Frauenfeld: J. Huber, 1882).

61 See Margarete Hochdoerfer, *The Conflict between the Religious and the Scientific Views of Albrecht von Haller (1708–1777),* University of Nebraska Studies in Language, Literature, and Criticism, no. 12 (Lincoln, 1932); Karl S. Guthke, *Haller und die Literatur,* pp. 17–19, and "Glaube und Zweifel: Hallers Rezeption des Christlichen Erbes," in *Literarisches Leben im achtzehnten Jahrhundert,* pp. 174–92; and Toellner, *Albrecht von Haller,* pp. 3–6, 21–7.

62 Cited in Toellner, *Albrecht von Haller,* p. 5.

63 *Briefe über einige Einwürfe nochlebender Freygeister,* 3 (1777), pp. 56–7.

64 Shirley A. Roe, "The Development of Albrecht von Haller's Views on Embryology," *J. Hist Biol, 8* (1975), pp. 167–90, and *Matter, Life, and Generation: Eighteenth-Century Embryology and the Haller-Wolff Debate* (Cambridge: Cambridge University Press, 1981).

65 See Shirley A. Roe, "Rationalism and Embryology: Caspar Friedrich Wolff's Theory of Epigenesis," *J. Hist. Biol, 12* (1979), pp. 1–43; and Jacques Roger, *Les Sciences de la vie dans la pensée française du XVIIIe siècle: La génération des animaux de Descartes à l'Encyclopédie* (Paris: Armand Colin, 1963; 2d ed., 1971).

66 Preface to *Allgemeine Historie der Natur,* by Buffon, 2 (1752), p. xv. This preface was also published as *Réflexions sur le système de la génération de M. de Buffon* (Geneva: Barriollot, 1751).

67 See *Sur la formation du coeur dans le poulet,* 2 vols. (Lausanne: M. M. Bosquet, 1758), and *Elementa physiologiae,* 8 (1766).

68 *Elementa physiologiae,* 8 (1766), part 1, p. 143.

69 "Gedanken über Vernunft, Aberglauben, und Unglauben," *Albrecht von Hallers Gedichte,* p. 57, lines 325–6. The original reads: "Genug, es ist ein Gott; es ruft es die Natur, / Der ganze Bau der Welt zeigt seiner Hände Spur."

70 Ibid., p. 58, lines 347–50. Original text: "Der Mensch, vor dessen Wort sich soll die Erde bücken, / Ist ein Zusammenhang von eitel Meister-Stücken; / In ihm vereinigt sich der Körper Kunst und Pracht, / Kein Glied ist, das ihn nicht zum Herrn der Schöpfung macht."

71 For Haller's pioneering work on the formation of bone, see his *Deux mémoires sur la formation des os, fondés sur des expériences* (Lausanne: M. M. Bousquet, 1758).

72 *Gentleman's Magazine, 8* (1738), p. 491.

73 *Bibliotheca anatomica qua scripta ad anatomen et physiologiam,* 2 vols. (Zürich: Orell, Gessner, Fuessli and Socc., 1774–7), 1 (1774), p. 757.

74 G. A. Lindeboom, *Herman Boerhaave: The Man and his Work* (London: Methuen, 1968), p. 268.

75 Cited in ibid., p. 270.

76 *Albrecht Hallers Tagebücher seiner Reisen nach Deutschland, Holland und England 1723–1727,* ed. Erich Hintzsche and Heinz Balmer. Berner Beiträge zur Geschichte der Medizin und der Naturwissenschaften, n.s., vol. 4 (Bern: Huber, 1971), p. 80. 'sGravesande met Newton in 1715.

77 Ibid., p. 93. Haller's diary on his trip to London was apparently compiled from notes after he returned to Basel in 1728. See Hintzsche's introduction to the *Tagebücher,* p. 15.

78 Ibid., p. 93.

79 Haller refers to the "Newton, Clarke und Leibniz Streitschrifften," undoubtedly a reference to the published Leibniz-Clarke correspondence, which appeared in English in 1717 and in French in 1720. See *Tagebücher,* pp. 91, 93.

80 Ibid., p. 94. See also p. 98 on Haller's visit to Newton's grave.

81 Ibid., p. 94.

82 Cited in Johann Georg Zimmermann, *Das Leben des Herrn von Haller* (Zürich: Heidegger, 1755), p. 49.

83 For discussions of various English influences on Haller's poetry, see Howard Mumford Jones, "Albrecht von Haller and English Philosophy," *Pub. Mod. Lang. Assoc., 40* (1925), pp. 103–27; Lawrence Marsden Price, "Albrecht von Haller and English Theology," *Pub. Mod. Lang. Assoc., 41* (1926), pp. 942–54; Lura May Teeter, "Albrecht von Haller and Samuel Clarke," *J. Eng. Germ. Philol., 27* (1928), pp. 520–3; and Karl Richter, *Literatur und Naturwissenschaft: Eine Studie zur Lyrik der Aufklärung* (Munich: Wilhelm Fink, 1972), pp. 57–111.

84 *Albrecht von Hallers Gedichte,* p. 46, lines 51–6. Original text: "Ein Newton übersteigt das Ziel erschaffner Geister,/ Findt die Natur im Werk und scheint des Weltbaus Meister;/ Er wiegt die innre Kraft, die sich im Körper regt,/ Den einen sinken macht und den im Kreis bewegt,/ Und schlägt die Tafeln auf der ewigen Gesetze,/ Die Gott einmal gemacht, das er sie nie verletze."

85 Ibid., p. 73, lines 267–8. "Er füllt die Welt mit Klärheit,/ Er ist ein stäter Quell von unerkannter Wahrheit."

86 Richter, *Literatur und Naturwissenschaft,* p. 66.

87 *Enumeratio methodica stirpium Helvetiae indigenarum* (Göttingen: A. Vandenhoeck, 1742), p. lvi.

88 Preface to *Allgemeine Historie der Natur,* 1 (1750), pp. xiii–xiv, xvii, xviii–xix.

89 *Bibliotheca anatomica,* 1 (1774), pp. 621–2; *Elementa physiologiae,* 4 (1762), pp. 536, 552. Haller refers to Newton's *Opticks* (which he indicates that he owned by an asterisk) and to Thomas Birch's *The History of the Royal Society of London* (London: A. Millar, 1756–7). In the bibliography to the *Elementa physiologiae,* Haller gives the *Opticks* a double asterisk, a designation reserved only for the "best" books.

90 *Elementa physiologiae,* 4 (1762), pp. 533, 557.

91 Haller reviewed a Swiss edition of the *Principia* (Geneva, 1739–42) in the *Bibliothèque raisonnée, 37* (1746), pp. 54–61. (For Haller's authorship, see Guthke, "Haller als Kritiker.") Haller had clearly read the *Principia* in its original version since he compares the additions and commentary (by Thomas Le Seur and François Jaquier) with Newton's edition.

92 *Albrecht von Hallers Briefe an Johannes Gessner (1728–77)*, ed. Henry E. Sigerist. Abhandlungen der Königlichen Gesellschaft der Wissenschaften zu Göttingen, Math.-Physik. Klasse, n.s., vol. 11, no. 2 (Berlin: Weidmann, 1923), pp. 22–3. Pemberton himself studied under Boerhaave, and it was at Leiden that he was introduced to the work of Newton.

93 Review reprinted in Karl S. Guthke, "Zur Religionsphilosophie des jungen Albrecht von Haller," *Colloquia Germanica, 1* (1967), p. 150. Haller's English never attained a level of proficiency.

94 *Mathematical Principles of Natural Philosophy,* trans. Andrew Motte, rev. Florian Cajori (Berkeley: University of California Press, 1960), p. 547. I have changed the word "frame" to "feign" according to the translation suggested by I. Bernard Cohen in his *Introduction to Newton's "Principia"* (Cambridge, Mass.: Harvard University Press, 1971), p. 241, esp. n. 9.

95 "De partibus corporis," p. 115. I am reading "partibus" for "particulis" in the phrase "Theoriam autem, cur utravis proprietas aut in his partibus nulla sit, aut in aliis corporis humani particulis aliqua."

96 Ibid., p. 154.

97 This sanctioning of unexplained properties is seen by Thomas Hall as characteristic of Newton's impact on biology. Yet one must be careful in using this as a criterion for Newtonianism, for one cannot assume that every physiologist who called upon the analogy of gravity was in fact a Newtonian in philosophical outlook. See Hall's "On Biological Analogs of Newtonian Paradigms," *Phil. Sci., 35* (1968), pp. 6–27.

98 *Mathematical Principles of Natural Philosophy,* p. xvii.

99 *Opticks* (New York: Dover, 1952), p. 397.

100 Ibid., p. 399.

101 *Elementa physiologiae,* 4 (1762), p. 557. Although Haller gives no reference, he undoubtedly had in mind Newton's discussion in Query 31 (pp. 397–400) of the creation and decay of motion in the universe.

102 *Elementa physiologiae,* 4 (1762), pp. 557–8.

103 "Newtonian Natural Philosophy and the Scientific Revolution," *Hist. Sci., 11* (1973), p. 1. See also J. E. McGuire, "Force, Active Principles, and Newton's Invisible Realm" *Ambix, 15* (1968), pp. 154–208; and P. M. Heimann and J. E. McGuire, "Newtonian Forces and Lockean Powers: Concepts of Matter in Eighteenth-Century Thought," *Hist. Stud. Phys. Sci., 3* (1971), pp. 233–306.

104 See Francis Oakley, "Christian Theology and the Newtonian Science: The Rise of the Concept of the Laws of Nature," *Church History, 30* (1961), pp. 433–57; David Kubrin, "Newton and the Cyclical Cosmos: Providence and the Mechanical Philosophy," *J. Hist. Ideas, 28* (1967), pp. 325–46; and Martin Tamney, "Newton, Creation, and Perception," *Isis, 70* (1979), pp. 48–58. See also I. Bernard Cohen, "Galileo, Newton, and the Divine Order of the Solar System," in *Galileo, Man of Science,* ed. Ernan McMullin (New York: Basic Books, 1967), pp. 207–31, and "Isaac Newton's *Principia,* the Scriptures and the Divine Providence," in *Philosophy, Science, and Method,* ed. Sidney Morgenbesser et al. (New York: St. Martin's Press, 1969), pp. 523–48.

105 *Opticks,* p. 402.

106 *Mathematical Principles of Natural Philosophy,* p. 546.

107 Review of *Histoire naturelle,* vol. 2, by Georges Louis Leclerc, Comte de Buffon, in *Bibliothèque raisonnée, 46* (1751), p. 86. (For Haller's authorship, see Guthke, "Haller als Kritiker.")

108 Ibid., p. 87.

109 *Mathematical Principles of Natural Philosophy,* p. xxxii.

110 Ibid., p. xxxiii.

111 *Briefe über die wichtigsten Wahrheiten der Offenbarung,* p. 6.

112 "Éloge de M. de Haller," *Histoire de l'Académie Royale des Sciences, année 1777* (Paris: L'Imprimerie Royale, 1780), p. 140. See also I. Bernard Cohen, "The Eighteenth-Century Concept of Scientific Revolution," *J. Hist. Ideas, 37* (1976), pp. 257–88.

113 Haller, *Mémoires sur la nature sensible et irritable des parties du corps animal* (Lausanne: M. M. Bousquet, 1756), p. iii.

114 Ibid., p. xiv.

115 *De la vie et de l'intelligence* (Paris: Garnier Fréres, 1858), pt. 2, p. 64.

PART III

Science in America

15

Creating form out of mass

The development of the medical record

STANLEY JOEL REISER
University of Texas, Houston

Physicians have seemed always mired in data. Medieval doctors complained of the torrent of details necessary to master pulse feeling in diagnosis and bleeding in therapy, eighteenth-century healers felt stifled by the many disease syndromes created by contemporary nosologists, and twentieth-century physicians (more pained than predecessors from believing their situation unique) have retreated into ever smaller medical specializations as a shield against "the knowledge explosion."

The growth of medical science and technology and the insights into the process of medical care gained from humanities and social sciences are written of in books and essays – and also in clinical records. In records this knowledge is revealed in a form transmuted by passage through the mind of the clinician, a pragmatic essence stripped of the embellishments of book writing or the crafted phrases of essay prose that reflect the temporal pressures and hasty decisions of practice. The form and content of these records are important evidence of how knowledge is applied and practice organized in medicine. This essay focuses on the development of the clinical record in general medicine in the United States from the early nineteenth to the early twentieth century, a time of change and ultimately reform for such records.

During the nineteenth century most American physicians and hospitals failed to keep systematic or adequate records. There was little pressure to do so from professional organizations, the law, or the prodding of scholars; in fact, few comments on the subject are found in the medical literature of the period. The medical record, generally, was considered a personal affair that each doctor handled as he saw fit. There were, of course, exceptions, of which the Massachusetts General Hospital is noteworthy. The records of this hospital not only display a keen interest on the part of its physicians and administrators in retaining data about the course of an illness and the therapy accorded it, but also reveal important events and innovations that would create a significant change of attitude toward medical records by the twentieth century.

From its beginning in 1821 the Massachusetts General sought to maintain good records on patients. Nevertheless the results in early years were not entirely satisfactory to its staff. This produced in 1837 an amendment to hospital by-laws stating that it was a formal requirement of patient care for each practitioner and house officer to "keep a daily record of every important fact in the history of the patient and as soon as possible to enter it in a handsome manner in the case book of the department."[1] It was recognized by some at that time, and clearly stated by the Boston physician James Jackson, that exact records were not only important for the medical care of the patient but also could be helpful in promoting the advance of medical science.[2] This view was also reflected in the 1851 minutes of the hospital's board of trustees, discussing the appointment of a person as a chemist and microscopist. It declares that one of his responsibilities are "Records – The appearances and peculiar phenomena met with in any of the above examinations shall be inserted in the Records of the cases to which they belong, and when by the above course judiciously followed out, an accumulation of valuable knowledge shall have been made, the officer may under the patronage of the Hospital and at the discretion of the Surgeons and Physicians publish it to the world as the result of his labors in this department."[3]

By the mid-nineteenth century indexes by diseases were being compiled in the hospital for all patients who had entered since its founding. Dr. William Thayer, who directed the project, classified under the head of 260 disease categories the patients treated, the results of the therapy, and the book and page of the clinical records where the details of each case might be found. The records were placed in the charge of the hospital librarian in 1857, who was to bind and letter them.[4] In 1897 Grace Whiting Meyers was appointed as custodian of records, an unusual step for the time.

Until the beginning of the twentieth century record keeping at the Massachusetts General reflected a relative neglect of outpatients, whose files were not bound in volumes or kept as systematically as those of inpatients. Starting in 1904, however, records of outpatients were put in one room, with an elaborate card index enabling a case to be readily found. A note in the Report of the Trustees reads: "The spirit of the Hospital has always been: first, to do everything that it can for its patients; second, to use the patients, so far as it is right and consistent with their best care, for educational purposes; third, to have its data so accurately recorded that it furnishes the maximum aid to the advancement of medical and surgical science. This has always been true of the In-Door Patients, and during the last year the Out-Patient Department has been able to make wonderful progress in the same direction."[5]

As for the content of the records themselves, during the 1820s and 1830s little was written about a patient's past medical history, social habits, or family. Generally the record was begun by brief references to age and occupation, followed by an elaboration of the circumstances of the present illness, and the findings of the physical examination. That concluded the record at admission. Then came a brief, almost daily report of progress focusing on pulse rate, the patient's sensations, and therapy.[6]

By the 1840s we find considerably more attention given to the past history of the patient, before inquiry is made into present illness and a physical examination conducted. A good example is the case of

Ebenezer Soule at 53.[7] Married. Stone-cutter [in the city of] Plympton. Entered House this P.M. Jan. 20, 1840. Has resided in Boston the last 5 yrs. Parents died at an advanced age. Mother had a severe cough a few months before her death, which was consequent on copious "bleeding from the Lungs." 2 sisters – at 66 – died of "Consumption." Also, one brother died of "Consumption" – had a Paralytic stroke a year or two prior to his death. Has one brother living – much troubled when young with "Scrofula."

Patient reports – previous health not good. Never very robust. When young – subject to swellings and "kernels" in neck, about lower jaw – which never ulcerated. Has always been troubled with spells of shortness of breath.

18 yrs. ago – while travelling in the State of Maine – took cold – and in the night came on "Bleeding from the Lungs" to the amount of a quart or more. Blood, bright-florid-frothy. Was a long time after, under Medical treatment – "was pronounced to be consumptive – and was reduced to a skeleton." About a year after – Engaged in Mackerel Fishing – 4 or 5 months – regained flesh and strength – felt perfectly well. Had no bleeding from lungs for 5 yrs. afterwards – when it returned after taking cold as before – and has had the same every year since – sometimes 3 or 4 times a year.

8 yrs. ago had – "Influenza" – was confined from Jan. 10 till middle of March – suffered much dyspnea and wheezing.

3 yrs. ago – met with an accident by which several ribs of L. side were fractured and dislocated. Had a slight cough previously – which became more troublesome after this accident – and has continued ever since.

A fortnight ago – after exposure – raised in the night by coughing . . . clear, florid blood – and continued to raise for 10 or 12 days after.

Now – 7 P.M. – up and dressed. Aspect pretty good. Flesh sufficient. Is much troubled with spells of dyspnea and wheezing – which come on after exercise – and sometimes while sitting still – after going to bed – and particularly on rising on morning – accompanied with a dry cough. Has frequent chills in night – with heats or sweats. Appetite and strength not good. Bowels regular. P. 64. T. natural. [On the next day the admitting physician completed the examination.]

[Jan.] 21. In both upper front chests, Resp. noisy, rather puerile than bronchial below Clavicles, but with expiration exceeding inspiration in length. Above L. Clavicle, same thing occurs – but above r. resp. is decidedly bronchial, voice more resonant, and Percuss. dull. Also dull upon Clavicle. Resp. bronchial, with strong blowing Expiration and great resonance of voice over r. scapula, and duller percuss. with some thrill to the hand. A less degree of same signs over L. Full vesicular Resp. in lower half of L. back – obscure ditto in r. Percuss. very resonant throughout L. – very dull in whole lower third of r. as far forward as axilla. A considerable prominence in L. front chest, about an inch from sternum – and extending vertically more than five inches. Signs of heart appear normal. [The patient was treated for two months and discharged feeling better. Although the working diagnosis of the staff was presumably "consumption," this is never directly stated in the record].

In 1870 a feature begins to appear that indicates a new approach to the type and organization of medical data in the record, namely, a chart for the twice-daily entry of the respiratory rate, pulse rate, and the temperature (numerical changes in which are depicted graphically).[8] Concern with such data reflects a view emerging at this time that quantitative was superior to qualitative evidence.

During the 1880s and 1890s a number of other charts are added to display the outcome of chemical analyses of urine, and to pictorially show the location of physical signs through outlines of the chest[9] and abdomen.[10] The record also reflects a growing attention by physicians to other aspects of body chemistry such as the state of the blood, and to the use of bacteriological techniques. The case of a clerk, Gilbert Fales,[11] is typical.

Nov. 11, '95 *F.H.* [family history] father died "liver disease"; mother, 1 brother, 1 sister well.

Habits 5¢ tobacco chewed and smoked in a week, occasional beer.

Prev. History

Measles, scarlet fever, whooping cough, in childhood. Health excellent save for these illnesses til 15 yrs. of age, then noticed little bunch forming on rt. side. (about 5th interspace in axillary line). This was lanced and cupful of pus taken out. No illness of any sort, for several years previous to formation of abscess, no cough, no injury to chest. General health not specially affected, no fever, kept on at school as usual. Considerable discharge of pus from this sinus ever since, less now than formerly. (Dr. Porter saw case 20 years ago and advised letting it alone). 15 years ago in N.Y. hospital, doctors called it "fistula of liver." 2 other openings made which discharged a little pus, one closed up 10 years ago, and one 1 year ago. Health always good, no loss flesh or strength till recently, bowels regular, digestion good, no pain.

Pres. Illness.

Last July, bad cold, cough, with considerable yellowish sputa, never red, feeling weak but kept on with work. Feverish at times 2–3 chills in last 3 mos. Dyspnea, no pain, appetite poor. wt July 225 lbs. wt at present 191 lbs. Night sweats, sleeps well. Kept on working through July feeling miserable, gave up work Sept 1st since then keeping quiet in house, not in bed.

Phys. Exam.

large, well developed and nourished, fair color.

tongue pale, whitish coat, no glands.

pulse regular, good volume and tension.

respiration a trifle rapid, lies partly on rt side.

Blood. Reds 4 000 000 Whites. 11,300 4.30 pus. Haemoglobin. 66%

Urine	Color pale-turbid	Reac. acid	S.G. 1020 *No Diazo.*
	Alb. 0	Sugar 0	Sed.

Heart sounds normal, not displaced.

Liver not pushed down.

Abdomen normal. *Spleen and liver* normal size

Knee jerks normal. *Sputum.* No Tubercle Bacilli.

Liquids and Soft Solids [prescribed by the physician].

[Nov.] 13 *Chest tapped* below angle of scapula, also in anterior axillary line. Needles firmly held, no fluid.

[Nov.] 16 Tuberculin 1 milligram and 24° later 10 mg injected in back. After 1st dose, temperature rose to 101.2°, after 2d dose to 102.2°, duration of fever after 2d dose, 24 hours.

[The reader is now directed to the continuation of the case in a different part of the record volume.]

Marked reaction, dyspnea, languor, pain in back and limbs, chilly, headache.

[Nov.] 18 Surgeon, (Dr. Porter) saw case, advised operation on rib and possibly further operation if necessary to relieve condition of chest. Patient refuses further treatment, and against advice, insists on going . . . [Diagnosis] tuberculosis.
Discharged Unrelieved

As the twentieth century dawned the medical record was becoming overspread and oversized, with an increasing profusion of specialized laboratory reports, X-ray findings, and consultant notes often pasted on the edges of pages and thereby obscuring other clinical reports and observations. The confusion thus introduced into the process of integrating the findings was further exaggerated by physicians beginning new cases in the record volumes before old ones were finished. Thus (as illustrated in the previous case) long histories were continued on scattered pages throughout the volumes. Although compared to other American hospitals records at the Massachusetts General were well kept and superior, the net result of these actions was a confusing mass of data spread in layers on pages and often discontinuously inscribed in the record books. And those who were at different times both outpatients and inpatients had two different files in the hospital.

Directed attention to the record, which would lead ultimately to its improvement in the United States, began at this time and was centered around the issue of education. In 1900 Walter Cannon, while still a student at Harvard Medical School, suggested the idea of using printed records of cases in the teaching of medicine.

Cannon was concerned about the dissemination of knowledge through the lecture system: It caused students to become passive listeners and prevented their engagement with the real-life situations they would encounter as physicians. Cannon asserted that students could not gain the ability to "reason clearly in medical matters, weigh conflicting evidence or draw just conclusions, when their chief practice is taking lecture notes."[12] He believed experience at the bedside was the best way to study medicine. Yet he recognized the impossibility for novice, or even advanced, students to have a series of real patients to follow who could demonstrate, in an orderly sequence, the facts necessary for medical learning. He found a satisfactory substitute for such experience, one that could teach both the logic of medical thinking and its necessary facts, in a systematic study of actual cases revealed in clinical records. This "case method" had been introduced into American legal education thirty years earlier at Harvard Law School by C. C. Langdell and diminished the influence of textbook learning there and elsewhere in the legal world. Cannon acknowledged this contribution and recommended its application to medicine: "Cases of all the types, variations and complications of almost every disease are to be found in hospital records or in records of the private practice of instructors. These records include a history of family tendencies, notes of previous illness, an account of the onset of the attack, the results of physical examination at the hospital, the story of the ups and down in the course of the disease, the treatment with its modifications as the symptoms changed, and, in case of death, possibly the findings at autopsy."[13]

Data abstracted from them would be analyzed by students, who then would be required to provide a diagnosis, prognosis, and therapeutic plan. It could be compared with the clinical outcome and autopsy findings, known only to the instructor.

A commentator on Cannon's exposition of the case method at a meeting in 1900 of the Boston Society for Medical Improvement was Harvard president Charles Eliot. He, like Cannon, recognized this technique for what it was – the method of scientific induction – applied first to law, now medicine. "Professor Langdell got his facts from recorded cases, and then studied in those recorded cases the induced generalization or principle, which is precisely what a chemist does when he experiments in the laboratory . . . Now Mr. Cannon advocates the use of cases in medicine in a strictly analogous manner." Eliot recognized an important advantage of this method in medicine: "It enables a vast number of cases to be utilized for the instruction of students – not only those cases that are to-day visible in the Massachusetts [General] Hospital and the [Boston] City Hospital, but all recorded cases of value for teaching."[14] Medical educators became as enthusiastic about the use of cases as had those in the law.

A parallel use of the medical record occurred to Richard C. Cabot at the Massachusetts General Hospital. In 1910 he began to discuss clinical histories of medical cases as weekly exercises for house officers and medical visitors to the hospital, with pathologist James Homer Wright commenting on autopsies he performed in the cases. In this proceeding the clinician discussed the evidence and offered a diagnosis, which then was confirmed or denied by the pathologist's findings. The event had the atmosphere of a heroic contest. Wrote an observer: "In the hands of masters of diagnosis the exercise becomes as absorbing as a game. It is skill and scholarship pitted against the unknown. For the moment the dingy pathological ampitheater [sic] becomes a jousting field. At the end of the discussion there is a moment of tense suspense while the pathologist, not without a sense of the dramatic value of the situation, comes forward to herald the outcome of the encounter. Defeated or in laurels, the clinician goes his way – in either case a better diagnostician by virtue of as rigorous a bit of mental athletics as a man can well undertake."[15]

Yet for many the correct diagnosis became the most important element of the exercise. Cabot had to stress that its principal value lay in the weighing of evidence, the balance of probabilities, the tracing of relationships between clinical and autopsy findings. As he told his classes: "If I make the diagnosis, I win. If I am wrong, I learn something. Either way I win."[16]

In April 1915 the hospital began publishing these discussions as "Case Records of the Massachusetts General Hospital" and sent them to doctors by subscription. They achieved rapid and worldwide notoriety. "Never have I, in the medical literature that I know seen anything more original, suggestive, and practical than this teaching," wrote one physician.[17] "To a great many of us, these cases are the only post-graduate work we have at the present time," wrote another.[18] In 1923 they began to be published as a regular feature in *The Boston*

Medical and Surgical Journal, renamed in 1928 *The New England Journal of Medicine.* Their fame continued to grow.

Cannon's and Cabot's introduction of the clinical case as a critical medium of instruction drew attention to the value of accurate and complete medical records. Their work served as backdrop for the searching review of records that occurred in the second decade of the twentieth century, when the medical record became a central object in a series of reforms in the United States directed at enhancing the clinical and scientific work of the hospital. These reforms led to great improvement in hospital functioning and in the keeping of complete and accurate records. The key mover in this effort was the American College of Surgeons, a professional association dedicated to improving surgical practice. It initiated what became known as the hospital standardization movement.

The College, founded in 1913, was concerned from its inception with upgrading hospital records. This was a result of a decision to ask candidates for admission to the College to furnish 100 records of operations in lieu of a formal examination. These records were so incomplete, fragmentary, and different in form that the College became convinced of the need to initiate a campaign to improve record keeping generally. Although the First World War slowed the process, the College still moved forward on the problem, motivated by the necessity to develop minimum standards of care for military hospitals. As the war neared its end the College refocused its attention on the conditions of civilian hospitals.

By this time nearly 7,000 hospitals and sanatoriums existed in the United States, containing almost 650,000 beds. Physicians and patients now viewed hospital care not as a luxury but a necessity, even a right.[19] Hospitals assumed a role as public service institutions with the privilege of requesting public confidence, good will, and fiscal support. In return, they recognized the need for public accountability: "Such an accounting," the College of Surgeons declared, "is inevitable. If the initiative is not taken by the medical profession, it will be taken by the lay public; and this entire accounting is what we mean by hospital standardization."[20]

In 1917 at its annual meeting the College appointed a committee to recommend action on standardization. It drafted a questionnaire to collect data from general hospitals in the United States and Canada to use in the development of a minimum standard of efficiency. From these data, and with the consultation of leading authorities in medicine and hospital administrators, a standard was devised as a test for competent treatment of patients in hospitals. It was confined to fundamentals and was broad enough to be applicable to all general hospitals. The standard dealt with staff organization and conduct, clinical and x-ray laboratories, and case records. It called for restricting staff privileges to those ethically and technically competent and asked hospitals to develop specific rules and regulations to guide the conduct of its members. The growth of scientific medicine had made the use of clinical and x-ray laboratories urgent, and the standard required hospitals to provide adequate facilities of this kind. It had become evident that there was a general deficiency in the number and type of laboratory tests offered in hospitals.

Finally great emphasis was given to the need for dependably documented case records. Recognizing that most doctors maintained relatively poor office files, the College singled the hospital out as the logical repository for the medical records of the community. It alluded to the economic loss that resulted from the duplication of results not entered. The standard asked hospitals to keep systematic records of all patients, together with a convenient summary of each case, and to use the records to analyze medical and surgical results and efficiency.[21] Personal visits to hospitals of representatives of the American College of Surgeons would encourage attention to its suggested reforms; perhaps even more potent would be annual publication of lists of hospitals meeting the standard.

At the time the standardization meetings were beginning, the state of medical records throughout the United States seemed dismal. John Hornsby, editor of *The Modern Hospital,* wrote in 1917 that in most "hospitals in this country, large and small, general and special, the record as it is kept today is practically valueless."[22] In many instances it lacked data on the history of the illness, or evidence of physical examination or laboratory study. Records of many patients about to undergo treatment, even surgical treatment, often lacked even a tentative diagnosis. This reflected not only a failure by doctors to transcribe what was done but, in some cases, a failure to do anything at all. The one part of the medical record that seemed reasonably maintained in a majority of the nation's hospitals was the nursing chart, with its entries of temperature, pulse, respiration, and medication. Only rarely did one find in the record a running medical story of the case written from day to day.[23] The first survey of the American College, for the year 1918, bore out Hornsby's judgments. Of the 617 American hospitals of 100 beds or more examined, less than half were found to have adequate records.[24]

Along with the negligent behavior of incompetent physicans, part of the problem was the continuing habit of physicians to depend on memory as the storage bank for clinical facts. Many still viewed records principally as notes to jog memory. This habit of practice, though never good, was less troublesome in the nineteenth century when medicine was basically an individual pursuit. Its persistence into the early twentieth century, when medicine became a more organized effort among growing cadres of specialized technicians and physcians, created significant problems. Many doctors did not recognize the need to give these others explicit insight into their thoughts or findings: "Individuals here and there have kept histories and used them but most of these have not learned the value of coöperation . . . History taking and record keeping as a coöperative enterprise is new."[25] Further, physicians who kept office notes on their patients failed to see the need to have another file on them at the hospital. As the hospital became more central to the health care of people such idiosyncratic behavior became intolerable, and rules and standards to gain cooperation among the staff became necessary to impose: "The mistake that we made in former times was to leave the recording of cases to the pleasure and convenience of the surgeon.

Usually he never found it convenient . . . Under the old régime we looked to the individual and not to the organization."[26]

The issue of confidentiality also stood in the way of record keeping. Although physicians did not object to maintaining hospital records on charity patients, they hesitated with private ones. They feared that with the data out of their personal control in the hospital record room unwarranted breaches of confidentiality might occur. But the necessities to practice medicine as a group required physicians to adopt a new view toward confidentiality. As surgeon William J. Mayo noted: "We hear objection sometimes to complete records for private patients . . . The poor have had records because, let us say, they could not help it; the rich and the middle class are just as deserving as the poor and they haven't the records."[27] This imposed, however, a responsibility on the hospital to protect records against unauthorized intruders such as insurance company agents, who too often had success in gaining access to them.

Basically, for reasons of cost, and an inadequate understanding of the value of records, hospitals themselves interfered with good record keeping by not supplying an adequate administrative apparatus to deal with the task. Dictation machines and stenographers could help, as could an office with a trained staff to collect and oversee the storage and retrieval of records. Even a small hospital was now thought to need a full-time record historian, whose job it was to encourage prompt entry of medical findings, determine if all requisite data were entered such as a history, physical examination, and working diagnosis before operation, and to assure the keeping of progress notes. In one Philadelphia hospital many of the 285 records it lost in a year were attributed to poor management.[28] If hospitals were going to insist that certain standards be maintained on what went into the record, they seemed obliged to supervise how it was kept.

Although problems such as these had retarded the writing of a useful medical record, the effort toward standardization was bolstered by an emerging recognition that accurate, comprehensive, and honestly kept records served important medical purposes. Doctors in the early twentieth century began increasingly to turn to them in scientific work. The Massachusetts General Hospital, for example, reported a greater use of clinical data for investigation: "Much more research is being done than ever before. It makes it necessary that all work upon records should be promptly and intelligently carried out."[29] Careful maintenance of records was seen as a commitment to scientific methodology, a key dimension of which was careful notation of evidence. The record was also appreciated as a valuable tool to oversee the work of medical students and house officers, and gain insight into their progress as learners. Case records stood too as tangible evidence to patients and communities that the hospital conscientiously carried out tasks and accepted scrutiny. As its reputation was changing from a dangerous boarding house for the sick to a scientific institution increasingly successful in curing and helping, the need for such proof was particularly strong. An analogy was made between hospitals and banks. If one could not imagine banks being run without good books or records, could hospitals, entrusted with a far greater human responsibility, be allowed to function without them?[30] The hospital began to enunciate the principle that since physicians practiced in its facilities, it was entitled to learn what they did. Doctors too began to speak about the benefits of keeping track of their work through records: "Records are facts that you find, that you filter, that you focus, and then face fearlessly."[31] Some declared that physicians were the greatest beneficiaries of good records. Wrote Harvey Cushing: "Without records your work is really for naught; however much good those records may be for others, they are most vital to yourselves."[32] Thus, the medical record was singled out as a vehicle through which doctors and hospitals could review whether they were caring only for people whose illnesses they were qualified to treat.

Still, convincing the medical world at large of the need for such review was no easy task, a fact no one knew better than Boston physician E. A. Codman. He had presented his "End Result System" to the College of Surgeons as it began its deliberations on hospital standardization, and with some effect. Codman argued that in earlier times the public and hospital trustees had been content to know patients had been treated and cared for, without examining the efficiency of the therapy given. Physicians viewed each practitioner as an island unto himself and were relatively unconcerned about the competence of peers. Who were good and who were poor physicians was judged by the vagaries of reputation and personality. Codman urged hospitals to attempt systematically to specify causes for the success or failure of each case treated by recording truthfully the following sort of data: What was wrong with the patient? What was done to him? What was the result? If poor, why? Was it the fault of the physician, patient, disease, hospital organization, or equipment?[33]

The record system thus would become an organ for measuring success and failure and for fixing responsibility. It would be a source of information to patients who wanted to know their chances of cure: "Let us suppose that a workman enters a hospital and learns that he needs an operation for hernia. It is not reasonable that he should ask: 'Based on your figures for other cases like mine during the past year, what chances have I to be at work again after the operation?' . . . Can the hospital answer it? Can the staff and officers claim with easy conscience that they protect the welfare of this man by every safeguard known to medical science? If so, how can they prove the claim?"[34]

The implementation of such a review required specific institutional action by hospitals, central to which were the documentation, collection, and analysis of clinical results by trained workers; a follow-up system that brought the patient back some time after discharge from the hospital to examine the outcome of the treatment; regular meetings of the hospital staff in which the results of medical action could be discussed; and publication of the findings. Significant strides were made by hospitals under the impetus of the College of Surgeons toward these ends, most particularly in its emphasis on monthly staff meetings to evaluate results. A hospital superintendent in 1923 wrote that now physicians "know that in order to keep the hospital up to the standard, they must do a definite work at their monthly meetings, must go over the records of the hospital, and that [a] man who has had a mortality among his patients must explain why."[35]

Progress was also made in the documentation of findings and in their analysis by improving the way in which data were recorded and stored. Codman and others, including the College of Surgeons, had recommended that careful abstracts of records be made on summary cards to facilitate review of data, and many hospitals implemented this suggestion. Attention was also being given to developing a standardized format for the writing of the medical record, which included use of a consistent terminology to describe disease. The American College of Surgeons, and institutions like the Presbyterian Hospital of Columbia University published clinical record forms, whose use would assure a more standard way of seeking and stating evidence and eliminate omission of important medical data.[36] To those physicians who took umbrage at being asked to relinquish their independence of action and follow a uniform procedure in recording clinical data, the Johns Hopkins statistician Raymond Pearl responded: "This argument is perfectly valid. It will inordinately cramp such portion of their individuality as finds its expression in carelessness, inaccuracy, forgetfulness, and inattentive observation. In so far as it is desirable to foster and preserve these intellectual qualities, and embalm their results in the permanent archives of a hospital, clinicians and surgeons should be encouraged to go on writing histories in the old way."[37] This idea encountered then, and encounters now, a good amount of opposition. Although noting clinical findings on forms can produce uniformity in records, it can also introduce stereotypic responses to complicated and diverse medical observations. However, a second reform directed at ordering the record has received widespread acceptance – the concept of the unit record.

In 1916 at the Presbyterian Hospital, a system of compiling records was inaugurated in which data from all the encounters of patients in any of the divisions of the hospital were put in a single file.[38] It replaced a system operative throughout hospitals in the United States in which a patient might have several independent records of care in the same hospital spread throughout its several divisions. This had occurred at the Massachusetts General Hospital, as noted previously. Although the unit system required more personnel and thus was more costly, advocates claimed the expense was justified. Histories were more complete and accurate under it because the staff felt their observations would be used more effectively. Colleagues also had more opportunity to inspect the judgments of each other and offer friendly criticism. Patients, their medical findings in one place, received better care. The systematic recording and filing of results under a unit record were also crucial for the success of evaluating the results of therapy: "Only those who have used the antiquated and uncorrelated forms of in-patient and out-patient histories and later studied the unit histories in follow-up work can appreciate the advantages of this system."[39]

The Presbyterian Hospital was notable for the concerted action it took to establish accurate records. And as a whole, the work of the College of Surgeons had resulted in a general up-grading of record keeping and use.[40] With all the gains made by the mid-1920s, however, many problems remained. Boards of trustees at many hospitals remained ignorant of the value of records and failed to adequately support the work of record rooms. Few well-trained record librarians existed. Many physicians remained indifferent to the quality of their records. They had to be prodded to keep them complete, to keep them coherent by singling out and relating disparate observations in the chart, and to use a standard terminology that enabled comparison of results. Incomplete charts and diverse terminology made compiling reliable statistics for monitoring the level of hospital care and for research purposes quite difficult. Even when accurate statistical tabulation as well as accurate and complete charts were available, many doctors did not use them to the fullest extent possible to advance medical knowledge and to evaluate medical care. And the burden of maintaining record standards seemed to rest too heavily on laymen, such as hospital superintendents and record librarians, rather than on the community of physicians at large.[41]

The first twenty-five years of this century was an era of significant reform of the medical record. It became recognized as a crucial basis for evaluating the performance of doctors and hospitals, legitimizing their call for community support, and integrating the human and technological resources of medicine for the benefit of patient care, teaching, and knowledge seeking. Subsequent physicians generally agreed that changing the medical record to serve these purposes was a worthy goal. Yet they have been unable to mobilize the will or devise methods of recording and organizing clinical data to satisfactorily achieve it. The reforms left the medical record a far better instrument of information than it had been, but they were only a beginning.

NOTES

1 *Massachusetts General Hospital Acts, Resolves, By-Laws and Rules and Regulations* (Boston: James Loring Press, 1837), Chap. 5, Art. 7.

2 James Jackson, in his letter of resignation to the MGH in September 1837, urged the record be sustained for such purposes. Archives of the Massachusetts General Hospital.

3 *Records of the Trustees of the Massachusetts General Hospital*, vol. 4 (November 1851), p. 452.

4 *Records of the Medical Board of the Massachusetts General Hospital*, vol. 1 (February 27, 1857).

5 *Ninety-first Annual Report of the Trustees of the Massachusetts General Hospital* (1904), pp. 44–5.

6 Case record of William Steward, *Medical Records, Massachusetts General Hospital*, February 23, 1823, p. 9, is a typical example.

7 Case record of Ebenezer Soule, *Medical Records, Massachusetts General Hospital*, vol. 100, January 20, 1840, pp. 1–2.

8 Case record of Leonora C. Johnson, *Medical Records, Massachusetts General Hospital*, vol. 292, August 11, 1870, provides an example.

9 Case record of Jester J. Crocker, *East Medical Records, Massachusetts General Hospital*, vol. 381, July 30, 1885, p. 229, provides an example.

10 Case record of Albert G. Freeman, *East Medical Records, Massachusetts General Hospital*, vol. 447, March 14, 1893, p. 251, is a good example.

11 Case record of Gilbert Fales, *East Medical Records, Massachusetts General Hospital,* vol. 473, November 11, 1895.

12 W. B. Cannon, "The Case Method of Teaching Systematic Medicine," *Boston Medical and Surgical Journal,* vol. 142 (1900), p. 32.

13 Ibid., p. 33.

14 C. W. Eliot, "The Inductive Method Applied to Medicine," *Boston Medical and Surgical Journal,* 142 (1900), p. 557.

15 F. M. Painter, "Extending the Influence of a Hospital," *The Modern Hospital,* vol. 2 (1918), p. 356. For a discussion by Cabot of the origin of this exercise, see Frederic A. Washburn, *The Massachusetts General Hospital: Its Development, 1900–1935* (Boston: Houghton Mifflin, 1939), pp. 115–17.

16 Painter, "Influence of a Hospital," p. 356.

17 Ibid., p. 355.

18 Ibid., p. 356.

19 "Standard of Efficiency, First Survey of the College," *Bulletin of the American College of Surgeons,* vol. 3, no. 3 (1918), p. 1.

20 Ibid.

21 Ibid., pp. 2–5.

22 John Hornsby, "The Hospital Problem of Today–What Is It?" *Bulletin of the American College of Surgeons,* vol. 3, no. 1 (1917), p. 7.

23 Ibid., p. 8.

24 "Hospital Standardization Series: General Hospitals of 100 or More Beds," *Bulletin of the American College of Surgeons,* vol. 4, no. 4 (1919), p. 5.

25 Carl E. Black, "Securing, Supervising and Filing of Records," *Bulletin of the American College of Surgeons,* vol. 8, no. 1 (1924), p. 65.

26 John Wesley Long, "Case Records in Hospitals," *Bulletin of the American College of Surgeons,* vol. 8, no. 1 (1924), p. 65.

27 William J. Mayo, Introductory Remarks to American College of Surgeons Meeting, October 24, 1919, *Bulletin of the American College of Surgeons,* vol. 4, no. 3 (1919), p. 4.

28 J. Lawrence Evans, "Access to Patients' Records," *Bulletin of the American College of Surgeons,* vol. 8, no. 1 (1924), p. 55.

29 *Ninety-ninth Annual Report of the Trustees of the Massachusetts General Hospital* (1912), p. 90.

30 Allan Craig, "The Approved Hospital as a Factor in Advancing Scientific Medicine," *Bulletin of the American College of Surgeons,* vol. 8, no. 1 (1924), p. 11.

31 Charles Moulinier, cited in George A. Ramsey and Roy C. Kingswood, "Case Records," *Bulletin of the American College of Surgeons,* vol. 7, no. 2 (1923), p. 23.

32 Harvey Cushing, Moderator's Comments, *Bulletin of the American College of Surgeons,* vol. 8, no. 1 (1924), pp. 11–12.

33 E. A. Codman, "Case-Records and Their Value," *Bulletin of the American College of Surgeons,* vol. 3, no. 1 (1917), pp. 24–7.

34 John G. Bowman, Introduction to "Case Records and Their Use," *Bulletin of the American College of Surgeons,* vol. 4, no. 1 (1919), pp. 2–3.

35 Robert Jolly, "Sidelights of Hospital Standardization," *Bulletin of the American College of Surgeons,* vol. 7, no. 2 (1923), p. 14.

36 "Newsletter Concerning Hospital Standardization," *Bulletin of the American College of Surgeons,* vol. 4, no. 1 (1919), pp. 1–4.

37 Raymond Pearl, "Modern Methods in Handling Hospital Statistics," *Bulletin of the Johns Hopkins Hospital,* vol. 32 (1921), pp. 187–8.

38 H. Auchincloss, "Unit History System," *Medical and Surgical Report of the Presbyterian Hospital in the City of New York,* vol. 10 (1918), pp. 30–72.

39 Allen O. Whipple, "The Interval Result System," ibid., p. 17.

40 Malcolm T. MacEachern, "The 1923 Survey – Problems and Conclusions from a Study Thereof and Plans for 1924," *Bulletin of the American College of Surgeons,* vol. 8, no. 1 (1924), p. 7.

41 Zula Morris, "Difficulties Encountered in Case Records and How to Overcome Them," *Bulletin of the American College of Surgeons,* vol. 8, no. 1 (1924), pp. 67–70.

16

"Frankenstein at Harvard"

The public politics of recombinant DNA research

EVERETT MENDELSOHN
Harvard University

On February 7, 1977, the Cambridge (Mass.) City Council passed an ordinance embodying the recommendations of the Cambridge Experimentation Review Board (CERB).[1] It established conditions for the conduct of research, involving recombinant DNA, within the city. It then added a series of recommendations to other agencies (primarily federal) concerning their role in monitoring and regulating research and safeguarding the public and research staffs who might be affected by this form of DNA research.

This was probably the first time that a local governmental authority examined a research effort and then imposed restrictions directly on the processes of research itself.[2] This final legislative outcome followed more than half a year of lively, and at times bitter, public debate on the nature of recombinant DNA research, the potential health hazards of the research, and some conjectures about longer-range potentials. As early as June 1976 the *Washington Star* gave a particularly dramatic gloss of the topic in a front-page headline: "Is Harvard the Proper Place for Frankenstein Tinkering?"[3] By mid-July 1976, the City Council in Cambridge had held a set of public hearings and taken its first step of intervention: It banned all recombinant DNA research in the city for three months. A member of the City Council commented to me at the time that had a public vote been held among the lay citizens, it would have favored the ban 99 to 1.[4] For the scientific commentators in Cambridge there was the specter of a poorly educated lay public interfering with the conduct of basic research. The confrontation between what the scientists considered their prerogatives to design and conduct research, and what worried lay citizens considered their rights for protection against potential health hazards, seemed starkly drawn.

This paper had its genesis in a study prepared jointly with colleagues for a U.S. Congressional Commission. See Everett Mendelsohn, James Sorenson, and Judith Swazey, Appendix 3, "Recombinant DNA–Science as a Social Problem," Special Study, Scholarly Adjunct, *Report to the National Commission for the Protection of Human Subjects of Biomedical and Behavioral Reasearch,* December 1976. An earlier and shorter version appeared in E. Mendelsohn, D. Nelkin, and P. Weingart (eds.), *The Social Assessment of Science* (Bielefeld, 1978).

BACKGROUND ISSUES

The background scientific issues involved have been quite fully reviewed elsewhere and need not be rehearsed here.[5] It should be noted, however, that the recombinant DNA research methods provided for scientists working in the field just the kind of strong new experimental techniques that are often dreamed of, but seldom realized.

There are several broader background isues that should be brought to the fore so that the remainder of this case study can be examined against the matrix of conflicting and overlapping issues, interests, and concerns that they suggest. An often hidden assumption in debates about science and society is that applied science and technology should be open to public scrutiny and even control, whereas basic (or pure) science should be the sole preserve of scientists. Even if this arrangement were to be accepted, it remains unclear as to who assigns the categories and what happens when the middle ground is unclear or a basic field rapidly develops utilitarian potentials.[6]

A second issue worth noting, even though it does not stand at the focus of our discussion, is the dynamics of the research process within science. In the middle of one of the meetings at Harvard, one senior scientist queried his DNA researcher colleague, "Why the rush with this research?" The answer was immediate and revealing. "The field is 'hot,' and if we don't get this research done, we will be 'scooped' by another group and we will stand the chance of losing our younger researchers to other laboratories." This issue of scientific competition adds another dimension to the series of factors that may separate scientific and public interests in the design and conduct of research. Numerous recent sociological studies have explored the dynamics of new fields or research fronts and have identified several key factors: the existence of a limited group with a fairly high level of internal communication; a period of rapid growth shown by increased publication and migration to the field from neighboring areas; a self-identification of being involved with a "hot" topic and a keen awareness of the forms of recognition and reward that come with breakthroughs.[7]

In spite of the fact that scientists dealing with radioactive materials, and others working in applied fields such as pharmacological research and bacterial research, have had to deal with problems of potential harm to sectors of the public as well as to researchers themselves, no well-formulated policy of the interests and rights of populations put at risk or hazard has been developed.[8] The individual subject of research has gained some protection under the guidelines and rules for research using human subjects, but the broader issues of those potentially threatened who lie wholly outside the research population need further attention.[9] It is this population that often finds itself perceiving its interests as being at odds with those of the working scientists. It is this rather ill-defined population that has recently opted into the assessment of science (and technology and medicine) through public interest groups, citizens' initiatives, and so on.[10]

The final background issue that I want to raise concerns the differentiation between responsibility and accountability. Through the years there has been a fairly consistent pattern of scientists emerging within the research community who focus attention on the need for the community explicitly to take responsibility for examining the potential consequences of scientific and technical activity.[11] By and large, the focus of accountability was back to their peers. This closed circle of experts has not been accepted as adequate in recent years. The past decade and a half has seen numerous attempts to shift accountability to more public bodies, both legislative and popular.

DNA AND CAMBRIDGE

The city of Cambridge inherited DNA as a political issue from other locations. A potential danger to health arising from recombinant DNA experimentation was initially recognized by a group of truly "insider" scientists during the summer of 1973 while they were gathered at a Gordon Research Conference.[12] These get-togethers of relatively small groups of biological scientists working in a specific field have long served the purpose of increasing the informal flow of information and keeping those in the field apprised of work at the research frontiers. That they were conducted in the relaxed atmosphere of New England summer resorts, or colleges and schools emptied for the holidays, may help account for the fact that more than just technical issues were discussed. In this instance, the group gathered to discuss DNA and drafted a letter to the National Academy of Sciences (NAS), urging that it undertake an exploration of the issues raised by recent research breakthroughs in recombining genetic materials. A copy of the letter was published in *Science* and brought directly to the attention of the broader scientific community the potential dangers from certain forms of recombinant DNA experimentation using the bacterium *E. coli.*[13] This group of working scientists accepted a limited responsibility for the consequences of their work and saw the organized scientific community in the United States, through the agency of the National Academy of Sciences, as the venue of their accountability.

It is worth digressing for a moment to note the political context of this step. The United States was at the tail end of a long and controversial involvement in war in Indochina. During the course of this conflict academic and intellectual communities had been the focus of intense political activity and moral soul searching. Military-sponsored research had been challenged and by implication many other areas of scientific and scholarly activity had been subjected to social and political scrutiny.[14] Self-criticism was at a high level and responsible and responsive behavior strongly encouraged. The consequences of research, as well as the relevance of research, was a much talked of issue. In this context it seemed natural for a group of scientists concerned about potential biohazards arising from their research to bring forward a request for a broader "look." It also seemed quite natural for the NAS rapidly to appoint a committee to undertake an examination.

What seems, in retrospect, somewhat out of the ordinary was the response of the NAS committee brought together under the chairmanship of Paul Berg, chairman of the Department of Biochemistry at Stanford University and himself involved in recombinant DNA experimentation. In a letter published simultaneously in *Science* and *Nature* (July 1974) they urged scientists around the world to join the committee members in voluntarily deferring certain classes of experiments.[15] They called for an examination of potential hazards arising from recombinant DNA molecules, and for development of containment methods for safe conduct of certain types of experimentation. They called on the director of the National Institutes of Health to establish an advisory committee to deal with experimentation in this area and to establish a series of guidelines for further research. Finally they planned an international meeting for early 1975 to discuss the proper means for dealing with potential biohazards from this research.

Although the report of the NAS committee explicitly noted "that adherence to our major recommendations will entail postponement, or possibly abandonment, of certain types of scientifically worthwhile experiments," the tone was generally cautious and the focus limited. They directly examined potential biohazards for experimentation, but they did not deal with broader questions of possible applications of the research, should it be successful, in areas such as biological warfare or genetic engineering. The focus of responsibility and accountability, as they developed it, remained wholly within the scientific community. The main thrust was the establishment of research guidelines.

In March 1975 the international meeting was convened at the Asilomar Conference Center, Pacific Grove, California.[16] More than one hundred conferees from seventeen countries (including a small group of science reporters) heard David Baltimore, a member of the NAS committee and himself a vigorous and often socially concerned scientist, call for a sharp segregation of the scientific issues from ethical and moral issues. The meeting was not held to take up "peripheral" issues such as genetic engineering or the potential for chemical and biological warfare. Instead, the aim was to develop a strategy for continued research that would maximize benefits and minimize hazards.

From the conference came a report that suggested guidelines for the conduct of research, including recommendations for the construction of facilities with varying degrees of physical and biological containment.[17] The temporary halt on experimentation was to be lifted and the potentially revolutionary effects of recombinant DNA research were saluted.

These scientific contributions to the monitoring and regulating of research were augmented by the meetings and hearings called by the National Institutes of Health (the major financial supporter of biomedical research in the U.S.). The first was held in the summer of 1975 at Woods Hole, where experimental guidelines, based on the Asilomar report, were developed,[18] followed by NIH hearings in February, designed to "beef-up" the earlier guidelines and for the first time to open the discussion to public input, limited though it might be. The guidelines were discussed, revised, and ultimately promulgated June 23, 1976.[19] One additional step instituted by the scientific bureaucracy was the filing, by the NIH director, of an Environmental Impact Statement required under the new environmental legislation.[20]

Thus far the discussions were almost wholly contained within the scientific community, and indeed, almost wholly limited to that group of geneticists interested in conducting research in the new area. Scientists are seen exhibiting "responsibility" for the possible consequences of their work; they established links of accountability to the scientific community, especially their knowledgeable peers. The guidelines they suggested and developed, however, were limited in applicability to those researchers working on government-funded projects. Industry (particularly pharmaceuticals) that had evinced an immediate interest lay outside the authority of the new guidelines.[21] The discussion of issues by the scientific community had been sharply and consciously limited: immediate, potential biohazards. Longer-term issues such as "genetic engineering" were consciously put aside. Ethical and moral concerns were eschewed in favor of direct "technical issues." The interests served seemed twofold: Scientists had found a means to continue their research on a challenging set of problems and, second, steps had been taken to protect the public health, though so far no move had been made to include the public itself in the discussion at any level. It should be pointed out that there was no unanimity within the community of DNA researchers. No less a figure than Erwin Chargaff (one of the key figures in early research on DNA) challenged the experiments, the experimenters, and the promulgators of the guidelines, referring to the latter as the "Bishops of Asilomar."[22]

This point of dissonance within the circle of knowledgeable biologists is important, for it seems almost certain that the public would not have had much to say at all in the early stages of the biohazards discussion even if they had been invited to join the decision-making process. It was the breakdown in the consensus among the biologists that seems to have triggered the first public concerns and to have given the various public interest groups their point of entry into what became a large-scale public debate.

But what kinds of steps occurred that allowed public participation in what could otherwise be classed as a very esoteric and technically specialized field? First, there was the individual scientist of great enough distinction to command attention, Erwin Chargaff, who alerted others to what appeared to him to be a problem. (Leave aside for now the questions of what motivated one of the founders of DNA chemistry to challenge a large segment of the new generation of researchers.)[23] Second, a group of politically active scientists, the inheritors of the political attitudes of 1960s – Scientists and Engineers for Social and Political Action, more commonly known as Science for the People – formed a "Genetics in Society" group that had already cut its teeth on the heredity and intelligence debate and now turned to recombinant DNA experimentation.[24] How were they different from the "Bishops of Asilomar"? They were younger; they were antiestablishment; they were conscious believers in participatory politics.

These were the forces that opened the semipublic debates on university campuses, for example, the University of Michigan and Harvard. But, they were largely still within the boundaries of the scientific community, even if some would have rather disowned them. If the break into the public domain came in the hearings called by Senator Edward M. Kennedy before the Senate Health Subcommittee, April 1975, there can be little doubt, however, that the most striking shift from internally contained discussions occurred when the venue for debate became the Cambridge City Council chambers, in the summer and autumn of 1976.[25]

THE DEBATE IN CAMBRIDGE

Cambridge became involved in the recombinant DNA controversy through the research interests of scientists at Harvard and the Massachusetts Institute of Technology, although it was the former institution that became the focal point of argument. The molecular biology group at Harvard applied to the university for clearance to build a laboratory for DNA experiments under the newly established guidelines of the NIH. The P-3 facility (i.e., fairly high protection) was to be constructed in the Biology Laboratories and the standard steps toward approval were set in motion. First the biology department itself held two meetings devoted to the topic and it rapidly became apparent that there would not be smooth sailing. Several senior department members, reflecting the vigorous questions raised at the University of Michigan and by Chargaff and others, challenged the safety precautions. They particularly doubted the ability of the new lab to contain potentially harmful microorganisms since the biological laboratories had been unable to eradicate a community of red ants that had roamed unhindered since they escaped from W. M. Wheeler's lab in the 1930s. Although a sort of grudging approval was given by the biology department, it was clear that not all its members would give up their opposition, particularly since several additional semipublic steps had to be taken.[26]

The construction proposal made its way through the Biohazards Committee (an insiders' group with one hastily recruited student member), but since controversy was in the air, it was brought to public session of the facultywide Committee on Research Policy. The committee, primarily made up of natural scientists, with a scattering of social scientists, adopted the policy of having an open hearing in a large auditorium following suitably broad publicity. Clearly they were going public. Why? A product of the late 1960s, when students and faculty were questioning such issues as Defense Department funding of university research, the Committee on Research Policy had adopted a somewhat participatory mode of operation. Their expectation was that experts and laypersons alike should have the opportunity to be heard. One of the biologist proponents of the research was able to express in private his confidence that the committee itself was conservative enough that he fully expected approval of the proposed experimental facility, but the mode of decision making, in form at least, had been partially opened to the public (albeit primarily from within the university community). And the public came and participated! The discussions were lively and at times acrimonious.[27] Science for the People brought out some of its ablest spokesmen, themselves molecular biologists. The proponents came in number and were equally well prepared. The discussion itself ranged from technical issues of biological containment to public policy issues of the decision-making process. The dramatic high point came as two Nobel laureates, both on the Harvard faculty and members of the Department of Biochemistry and Molecular Biology, squared off against each other and argued, not technical biology, but rather risk and benefit calculus and politics of research. Those two, who had both been considered members of the progressive faculty caucus and had shared sides in many previous political-scientific debates (nuclear weapons testing, bombshelters, Vietnam War, social responsibility of scientists) now found themselves on opposite sides of the controversy. Indeed, their split was symptomatic of the break that was much more widespread among previous political allies. In short, simple political allegiance was by no means a good predictor of where a person would stand in the debate. Nor was it a simple age or generation phenomenon, although the Science for the People members were more nearly newcomers to the ranks of established scientists. Neither the weight of scientific evidence nor the strength of sociopolitical argument would have provided any firm indication of how a decision would be made. The only theme that could be discerned in the committee members' questions seemed to be that "research" should not be interfered with unless the necessity to do so seemed overwhelming. The implicit autonomy of the scientific community was reinforced and the attempts to "politicize" research, that is, bring nontechnical issues into the discussion, were rejected.

The issue seemed as though it might remain wholly within the university community (as it had in Michigan, albeit it went right up to the Board of Regents), had it not been for one small intervention toward the end of the open hearings.[28] Barbara Ackermann, one-time Cambridge mayor and currently a member of the City Council, announced her presence and indicated that the Council was keeping its eye on the discussions.

Just how the City Council became actively involved in the dispute remains a point of controversy. Some claimed that the opponents of the research prevailed on members of the Council (often considered prone to fight the university – the feisty mayor often proposed paving Harvard Yard for a parking lot) to take up their cause since it seemed lost in the university's own decision-making processes. For others it seemed a natural outcome of the growing controversy, which would have gone public somewhere, if not Cambridge.

The legal authority under which the Council operated was simple – the need for a building occupancy permit – and relied on the same legislation that governed the placement of automobile service stations. Council members saw it as their responsibility to deal with any threats to the health and safety of citizens. Council meetings were always open affairs and it was not at all unusual for the public to turn out in large numbers for any controversial issue. On the other hand, there was a fair amount of open hostility within the university to the idea of an issue like this being brought to public discussion. It seemed something of an offense to academic mores, at least, and perhaps also a threat to more traditional intellectual freedoms.

After an initial skirmish, a pair of Council meetings were set aside for open hearings. They were packed sessions in which stellar figures from the scientific community were pitted against each other with Council members alternately probing, attacking, and defending. Some of the scientific "brass" from Washington were on hand and several quickly found themselves embroiled in argument. Laypersons entered the discussion at a number of points, but it was clearly the scientists who carried the bulk of both sides of the controversy.

Mayor Alfred Vellucci, who in addition to his feisty nature always had a flair for the dramatic, invited the experts on both sides to bring their cases to the public. Two booths were set up at the Saturday Market in Kendall Square (on the "other side" of town) and for several summer weekends Cantabrigians could find scientists in shirtsleeves with molecular models, illustrations, and texts hard at work explaining the "basics" of genetics and DNA experimentation to an amused, but often quite attentive citizenry.[29] Science was literally taken to the citizen.

But if outdoor public scientific education was one beneficial outcome of the City Council's intervention, another and potentially equally important one was the enactment of two three-month bans on recombinant DNA research during which time the newly established Cambridge Experimentation Review Board was to meet and formulate recommendations for Council action. The board called into being by the Council was conceived of as a lay body, and in actuality came very close to meeting the ideal. No Harvard or MIT people were involved and the chair was given to a former city councillor who operated a small home heating oil business in North Cambridge. Other members included a Catholic nun, an engineer, a former councillor with a long interest in health and hospital matters, a physician, an urban sociologist from Tufts University (a Cambridge resident who taught in a nearby town), two long-time "concerned citizens" from the nonuniversity quarter, and the city health commissioner as an ex officio member. Staffing was provided by a technical assistant on loan from the local management consulting firm, Arthur D. Little, Inc.

THE CAMBRIDGE EXPERIMENTATION REVIEW BOARD

As people returned to Cambridge from their summer holidays in the autumn of 1976, the big question was whether a local citizens' group could handle problems of the sort posed to them by the Council. Could they successfully evaluate an esoteric form of biological experimentation and reflect the public's interest?

In the first instance the board decided to interpret its mandate narrowly and to deal solely with the problem of immediate biohazards posed by the conduct of research in the city. They decided that they would recommend actions and procedures suitable for translation into city ordinances. They eschewed any attempts to deal with longer-term issues of public concern such as genetic engineering. From the beginning they were conscious of setting precedents and were determined not to stumble unnecessarily. They turned out to be remarkably successful, beyond even their own expectations. They involved themselves in what amounted to a crash course in biology, including laboratory visits, and quickly demonstrated to the experts they called in for testimony that they would not be spoken down to or easily misled. Their sessions, though open to the public, were generally orderly and actually fairly quiet. During seventy-five hours of testimony involving thirty-five witnesses, they heard from all parties with a desire to testify. Midway through the sessions they conducted a mock trial to see what type of information and views an adversary procedure might produce. They concluded their several months of activities with some twenty-five hours of planning, deliberation, and report writing.

The report they delivered to the City Council was unanimous and surprisingly did not represent a compromise to a least common denominator. Equally surprising, both sides in the conflict accepted the report, despite the fact that it did not produce the ban on research requested by the most vigorous opponents, or leave the researchers to proceed unhindered.[30] In fact, the board recommended an increase in restraints on research beyond those included in the NIH guidelines. They also addressed several issues not dealt with by the guidelines: proposals to monitor the health of those working in the labs and those living in proximity to them; and plans to monitor leaks from the P-3 facilities on a continuing basis. Two other recommendations for City Council action reflected confidence in the board's own experience in lay participation in decision making in areas involving society–science interactions: They asked that lay community members be included as members of the university biohazards committees and that these bodies be cross-disciplinary in nature and include laboratory technicians as well as the traditional professional participants. Further, they proposed that their own experience not be considered a one-time affair, but that the city appoint a Municipal Biohazards Committee to continually review the conduct of research with potential for public harm. Such a committee could also oversee *all* research and monitor compliance with the guidelines and the local additions to them. These are the elements that became part of the city ordinance and, in theory at least, remain in force. (In practice, the newly established city committee did not meet for the better part of a year and the scientists went about their business with everyone tacitly assuming good faith.) The implications of these municipal actions were clear: Local government *was* an appropriate place for the social assessment and social regulation of traditional scientific activity.

The second set of CERB recommendations implied, however, that local authority was not adequate for the whole task. The board suggested that the City Council call for uniform federal legislation to govern all recombinant DNA research (responding here to the threat on the part of some scientists to take their research to other locales if Cambridge became too restrictive), that health monitoring become a federal requirement, and that a national registry of all recombinant DNA researchers be compiled (similar to the registry of those working with radioactive materials); and finally, recognizing the limited financial resources of the municipality, they called for federal funding of "preventive" research.

Neither the CERB nor the City Council directly addressed the question of whether federal regulations and guidelines ought to supersede local laws in this area, although they implied as much by strengthening the guidelines for local use. The mayor, however, was strongly in favor of local supremacy and continued to support it in the course of Congressional testimony.[31]

EPILOGUE

Although the first phase of the recombinant DNA debate is over it is still too early to draw final conclusions about the efficacy of full public involvement in the setting of research guides and restraints. In this instance it appears that the initial fears of biohazards emerging from the research were overdrawn, reflecting in part, at least, distrust of science and scientists. But it also reflected concern for the unexplored long-term issues of genetic manipulation and bioengineering.

The U.S. Congress, though still considering some sort of uniform national legislation covering both research laboratories and industrial producers, has backed away from its early, intense interest in the problem. Several municipalities have taken action on biohazard protection, and recombinant DNA experimentation has become an item on the agenda of several major environmental groups, the current focus being on large-scale agricultural field testing. But there also was a backlash, with many key experimenters in the field, anxious to get on with research they considered exciting and important, organizing their own political action to block, or at least soften, any legislative proposals that came before the Congress.[32]

Several important points have emerged in the course of the debate and the years immediately following. First, there has been the success of the public in inserting itself into an area of discussion and decision from which it not only had been previously excluded, but in which it had never felt it had any reason to enter. Whether one applauds or decries this involvement seems to rest on political considerations concerning the nature of the scientific enterprise. It is far from clear whether the decisions made from the scientific or the public view are of improved quality, but the process itself brought a limited change in the relationship between the scientific community and the wider society.

More than one of the initiators of the early warning and call for a moratorium have retreated from those earlier efforts to assume social responsibility in the scientific realm.[33] J. D. Watson, for example, is not alone in believing that his stand was foolish and had he known what the public reaction would be, he would have muted his fears. He might well be correct that in retrospect there was nothing to worry about, but the attitude he and others now exhibit brings focus directly to the point of where accountability lies. This issue is unresolved and will almost certainly be the focus of conflict between those in the scientific community who believe that they and the public are best served by maintaining accountability within the body of knowledgeable, scientific peers, versus those in the public who opt for participation in decision making on issues with potential consequences. The focus for social analysts should be on the manner in which this dispute is ultimately resolved and on the modes of participation and interaction that can serve the needs of both communities.

There is another element, however, that stands out in this survey and involves the set of questions not directly addressed within the scientific community and initially muted among the public critics of DNA research, namely, the issues of longer-term consequences of genetic manipulation or genetic engineering. For the biologists, as such figures as David Baltimore and J. D. Watson made very clear, the issue seemed like a diversion – either too distant in the future to be real, or too trivial to be relevant. Yet I believe it is accurate to claim that "fears" of genetic manipulation were high on the "hidden agenda" of many of the public critics. Neither the Asilomar gathering nor the city of Cambridge deliberations nor the Congressional efforts provided any entrance into a set of problems that almost certainly stands between science and the public. Modes for internally and externally airing such issues, outside of science fiction, still seem largely out of reach.[34]

Several other facets of the controversy, only slightly visible, or, not seen at all in the early stages of debate, have only now become clear. The threats made by some scientists to take their research elsewhere if restrictions became too cumbersome in the United States have actually been carried out. The fact that one of the "breakthroughs" leading to an important industrial interest in recombinant DNA research – the production of human interferon – involved in an international team based in Switzerland, with significant U.S. participation, indicates the limits of local or even national research guidelines and restrictions.[35] Although the European Science Foundation and the European Molecular Biology Organization have adopted, in general fashion, the NIH guidelines from the United States, enforcement procedures are very limited and compliance relies instead on the good will of the experimenters. The ESF and EMBO committees responsible for evaluating recombinant DNA research are composed almost entirely of scientists working in the field with little room for direct public involvement. Only Great Britain, through its Genetic Manipulation Advisory Group (GMAG), has explicitly included public representatives in the policy and decision process. If public risk were indeed an important issue, it is clear that some processes of internationalizing public safety would need to be developed to match the very active current internationalization of the research process.

But the controversy now is marked by two very different and striking trends, that is, reduction in official regulation of research and an increase in industrial and commercial involvement in the research process and the production aspects of recombinant DNA. The move away from the elaborate guidelines and downgrading of potential dangers is taken by the scientists as a vindication of their early reluctance to have the government involved in restricting the research process in an area they considered "basic" research. After all, no significant accidents have occurred and no health hazards have developed. "What," they seem to ask, "was all the fuss about?" The almost unspoken additional claim is that scientists *do* know best and public busybodies were doing little more than interfering in the fundamental processes of research.

It is the second trend – industrial involvement – that has kept interest in "regulating the gene-swappers" alive.[36] An editorial in the *Boston Sunday Globe* restated the claims of public interest, especially now that "DNA research has begun to pay off (and) industrial applications have become more apparent." Private industry, they noted, was exempted from the NIH guidelines and compliance was completely voluntary. It was feared that laboratory precautions would be neglected and dangerous experiments undertaken "as the possibility of profits appears." As a direct follow-up to the widespread commercial interest in using recombinant procedures, the Occupational Safety and Health Administration has moved into the situation with proposals for greater precaution in the workplace. Paralleling these administrative efforts, and perhaps serving to prod them, was a move by Senator Adlai Stevenson, Jr., to introduce legislation to require private industry to tighten its requirements through procedures for the registration of all appropriate recombinant DNA research.[37] The Congressional moves languished and there is little expectation that they will be revived.

Although the scientific breakthrough involved in the creation of bacteria capable of producing human insulin potentially assured supply of this needed drug at reasonable cost, the discovery that a bacterium could be genetically coded to produce human interferon created even more excitement. Interferon, a protein normally produced in the human body, is utilized by the body to fend off viral infection and perhaps some cancers. The commercial potential was recognized early and indeed the research was financed by a firm created for the purpose and involved a significant number of key recombinant DNA researchers. The experimentation itself was carried out in Switzerland by Biogen, with large segments owned by International Nickel Company of Toronto and the pharmaceutical company, Schering-Plough.[38] As befit the public and commercial stake in the research, the new discovery was announced at a press conference and was featured on the front pages of major newspapers.[39] The size of the economic interest in these new finds of gene splicing is significant enough to cause large fluctuations in the stock prices of those corporations with major investments in the work and to promise continued commercial interest in, and exploitation of, the new genetic techniques.[40]

In addition to the commercial euphoria, however, some rather basic issues of science–society relations have been raised by the rapid transformation of research into economically valuable techniques. Only recently commented on in the scientific press have been the widespread industrial ties entered into by the same scientists working in university research labs. Success promises financial gain for both the science and the scientists, but has also crucially altered the position of the scientists. The claim of "disinterest" is gone and is replaced instead by an allegiance to industrial enterprises and economic interests. This is not to make a normative judgment of such ties, but rather to indicate a marked shift in structural and institutional relationships. A strong economic and industrial factor is added to decision making and problem choice among those scientists working in the field and by implication to those in similar or related fields.

Another newspaper headline, also featuring Harvard University (just four short years after the Frankenstein reference), clearly indicated the strains that commercial success was bringing to academe: "Harvard Shuns the Apple but Doesn't Step on the Serpent" (*New York Times,* November 23, 1980). The story chronicled the deliberations, by Harvard's administration and governing boards, about whether the university should establish a business partnership with several of its molecular biologists to commercially exploit the new discoveries emerging from the recombinant DNA laboratories. The story's subhead told the moral: "Potential Profits in Bioengineering raised Questions . . . that Universities Cannot Comfortably Answer." The problem was parodied by an irreverent Art Buchwald quoting the answer of a mythical university gene splicer to the obvious question:

"But won't it compromise your academic ideals if you start doing research just for profit?"
"Academic ideals my foot. We're making money, and that's what a university is for . . . "We're going to become another Xerox."[41]

The strains put on the traditional university relationship with its own scientists and with industry have become the focus of phase two of the recombinant DNA story. A series of reports in *Science* under the general title, "The Academic-Industrial Complex," published during 1982, gives a good indication of the range of problematic situations scientists, universities, and industrial firms encountered.[42] Perhaps the most striking evidence for the existence of real conflicts of interest between the academic and industrial demands on a university scientist was the resignation of Walter Gilbert, Nobel Prize winner, from his professorship at Harvard in order to pursue his other role as chief executive officer of Biogen, the Swiss-based firm.[43]

One consequence of the strong commercial involvement is that the claim for the scientific community itself serving as the focal point of responsibility and accountability becomes vitiated. Instead, corporate interest through pharmaceutical, chemical, and other industrial firms will exercise significant control. The record of sensitivity to public interest on the part of such enterprises is mixed and suggests that the direct role of the public in assessing the social consequences of an area of scientific research will be weakened. Will the scientific community feel less threatened and more at home with corporate scrutiny and influence in the place of public assessment? The history of such relationships in other areas of science suggests that transformations in the research process will be no less profound.

As scientists retreated from their early efforts at accepting responsibility for potential research consequences and judged their concerns as "overdrawn" and fears as "too catastrophic," one consequence they viewed as more promising than threatening was the rapid utility of their discoveries. As we review the discussion, the focus was almost wholly limited to potential biohazards with little attention paid to the deeper implications of commercial research. There was little or no discussion of the shift in the balance of power in decisions and the possible commanding influence of new industrial priorities. The scientists were probably correct in their growing judgment that immediate health hazards were slight, but in that context, as in the new commercial context, the implications for genetic engineering are still being largely ignored – ignored even while all the techniques for selectivity in gene splicing are being improved.

The early approach on the part of scientists to taking responsibility can be judged as well intentioned, but amateurish; likewise, the initial public interest.[44] But the real issues of accountability remain obscured, specifically, who and to whom.

The current shift in the legitimating sources for recombinant DNA research from the university research community to the industrial research group raises a new set of questions. In the shift to the industrial framework, and thereby to a new and broader set of interest structures, the special autonomy claims of the university research laboratory are lost and the activity clearly becomes opened to the same public scrutiny and criticism as all other economic enterprises. The readiness with which the geneticists entered into commercial relationships brings into question the very claims they made just a short while ago that public assessment would endanger the autonomy of scientific research and severely weakens claims to a special separateness for science. There are, in addition, implications for other research areas where a claim for autonomy on the basis of the fundamental nature of the research can also be weakened by potential, rapid utility. What emerges once again in the example of recombinant DNA research is that the relationship between research and utilization is not accidental but structural, and therefore leaves no area of scientific activity immune from implied and explicit social relations.

The academe–industry link has been especially troubling since the fear of "spillover" to other university claims for the autonomy of scientific research and the potential for new forms of external regulation has taken on new life. Congressional critics of science might have given up efforts to regulate against potential biohazards but they have begun to assert the need for guidelines that would protect the interests of both universities and companies.[45] Perhaps it was the prodding of Congressman Albert Gore, or just the internal pressures building up within universities; in any case it led to a major university–industry conference convened in March 1982 by the presidents of five key research universities and including some thirty corporate and academic leaders. This was followed fairly rapidly in December 1982 by an expanded meeting at the University of Pennsylvania, on "university-corporate relations in science and technology."[46] In the months that followed several of the major universities, Yale and Harvard among them, established new rules for faculty–corporate relationships.[47]

As the new focus of concern develops around problems engendered by the industrialization of recombinant DNA research, it becomes apparent in retrospect that the critics themselves also touched only lightly on this basic structural-social problem. Instead, they had followed the scientists' own lead and dwelt primarily on the fears of biohazards. In its own peculiar way, the Cambridge City Council came as close to placing the issue in its broad perspective when, much to the anger of working scientists, they applied the same public safety and health rules involved in gasoline station siting to their intervention in recombinant DNA research. They were adopting the legal and procedural safeguards available for protecting the interests of the public against any commercial or other special interests.

As we step back from the details of recombinant DNA research and its travails in the Harvard context, it is worthwhile to attempt to place the specific events in an explicit interpretive framework. What has been witnessed has been the rapid deinstitutionalization of recombinant genetic research in the university laboratory setting. First there was the breakdown of the traditional internal structures of responsibility and control of research and in their place the development of public *roles* and publicly sponsored *rules.* These, in turn, were quickly joined and probably superseded by the reinstitutionalization of the research within the industrial context and the adoption by the researchers of the practices and norms of the commercial world. Thus, there occurred not merely a politicization of recombinant DNA research (largely rejected by the scientists involved), but also, and probably more importantly, an "economization" of research activities (warmly embraced by the research community).

This deep interpenetration of science and the economy is not a new phenomenon, but in many ways it resembles the long-term relationships developed by chemistry and chemists as well as by other areas of applied biology dating back to the nineteenth century and the days of Louis Pasteur. Neither the explicit public involvement nor the commercial ties of these earlier examples meant an end to science, but rather the emergence of new institutional structures and the involvement of strong elements (political and economic) from outside the scientific research community. The boundaries of the science–society relationship were altered and new expectations were created for such processes as responsibility and accountability.[48]

The early and strong defense of freedom of scientific research in the area of genetics gave way too rapidly to the industrialization and utilization of the research for any strong model of "free research" and "autonomous science" to be a useful analytical tool. After all, even the notoriously short "public memory" can see the direct links drawn from the university laboratory to industrial production. And for those whose memory is too short, the business pages of *Newsweek, Time,* the *Wall Street Journal* and the *New York Times* provide florid reports. What attempts to interpret the recombinant DNA phenomenon clearly cry out for is a new model of the sciences and their social relations. The model must be capable of accounting not only for those persisting institutionalized structures of science, but also for the shifting boundaries of the institution and the implied deinstitutionalization that occurs both when scientists "go political" and when they "go industrial." The model must, in addition, provide understanding and explanation for the perceived new interests of the public in the research process as well as in research utilization. Indeed, as the recombinant DNA case shows so clearly, the feedback loop is too short to see these parts of research as separately institutionalized activities. The long-cherished mythology of a sharp bifurcation between the production of knowledge and its utilization is no longer tenable.

The next problem area for recombinant DNA, genetic engineering (the hidden agenda of much of the early controversy), surely calls for prior thought and planning rather than post hoc arguing and adjusting. In this area of knowledge utilization, certainly closer now than researchers had previously calculated, the demands for dealing with explicit social and political elements will strain to the breaking point any old-fashioned model of the separation of research and its uses. The history of the present controversy and the changing relationships and structures of science and the public provide both fertile fields and strong imperatives for a reconstruction of the modes of interaction between the public and its sciences.

NOTES

1 Report of the Cambridge Experimentation Review Board, "Guidelines for the Use of Recombinant DNA Molecule Technology in the City of Cambridge," submitted to city manager, Jan. 5, 1977.

2 Two issues should be noted: first, the entrance of local governmental authorities in contrast to those at the national level; and second, the action directed at the research process in science. Therapeutic experimentation in medicine was more often the site of direct interference and explicit limitation. Several recent studies by Judith Swazey have examined limits of research in medicine, with special concern focused on human subjects used in research. See Judith P. Swazey, "Protecting the 'Animal of Necessity': Limits to Inquiry in Clinical Investigation," *Daedalus* (Spring 1978), pp. 129–45; also the considerably earlier paper, Judith P. Swazey and Renée C. Fox, "The Clinical Moratorium: A Case Study of Mitral Valve Surgery," in Paul Freund (ed.), *Experimentation with Human Subjects* (New York: Braziller, 1970), pp. 315–57.

3 The sensational headline covers a rather mild story referring to the decision of the dean, at Harvard University, to recommend that construction begin on a containment laboratory for recombinant DNA research. *Washington Star,* June 16, 1976, carrying a UPI dateline.

4 At this same time, a staff member of the undergraduate paper, the *Harvard Crimson,* guessed that the student body would have divided about evenly, with a small majority *supporting* continued research.

5 The issues are summarized at a semipopular level in two articles: Stanley N. Cohen, "The Manipulation of Genes," *Scientific American,* July 1975, pp. 24–33; and Clifford Grobstein, "The Recombinant DNA Debate," *Scientific American,* July 1977. A whole issue of *Science,* April 8, 1977, was devoted to reports of recent experimental work using recombinant DNA techniques.

6 Some sense of the confusion that exists can be seen in the almost simultaneous claims for r-DNA as "basic research" and for the significant fruits that society should expect to gain in such fields as medicine, agriculture, and pharmaceuticals. See, e.g., the following by supporters of the research: Roy Curtiss III, "Genetic Manipulation of Micro-organisms: Potential Benefits and Biohazards," *Ann. Rev. Microbiol.,* 30 (1976), pp.

507–33; Harrison Schmitt (letter), "A Scientist-Senator on Recombinant DNA Research," *Science*, July 14, 1978, pp. 106–8.

7 See, e.g., Gerard Lemaine, R. M. MacLeod, Michael Mulkay, and Peter Weingart (eds.), *Perspectives on the Emergence of Scientific Disciplines* (The Hague: Mouton, 1976); also studies of the transformation of special fields, e.g., David O. Edge and Michael J. Mulkay, *Astronomy Transformed, The Emergence of Radio Astronomy in Britain* (New York: Wiley, 1976).

8 Indeed, the issue of "population at risk" has become one of the most vigorously debated areas in the relations between science and the public. Tests of vaccines and birth control pills, the establishment of nuclear power generators, and even the less specific issues of atmospheric pollution have seen a new public awareness of issues of risk. The advent of the "Nader movement" and other paralegal public interest groups focusing on medicine and the sciences has raised the visibility of various science-related issues and has provided investigatory, and often legal, frameworks for public intervention. One systematic bibliography of the law–science interface gives access to the field: Morris L. Cohen and Jan Stepan, "Literature of the Law-Science Confrontation, 1965–1975," *Newsletter* of the Program on Public Conceptions of Science, June 1975, January 1976, and April 1976.

9 See Paul Freund (ed.), *Experimentation with Human Subjects;* also the National Academy of Sciences Forum, *Experiments and Research with Humans: Values in Conflict* (Washington, D.C., 1975).

10 See, e.g., Dorothy Nelkin, *Nuclear Power and Its Critics: The Cayuga Lake Controversy* (Ithaca, N.Y.: Cornell University Press, 1971) for an early discussion of public involvement in the United States. She has recently participated in a joint study of nuclear energy and public involvement in Europe. Dorothy Nelkin and Michael Pollak, *The Atom Besieged: Antinuclear Movements in France and Germany* (Cambridge, Mass.: MIT Press, 1980).

11 The pages of the *Bulletin of Atomic Scientists* and the Federation of American Scientists' *F.A.S. Public Interest Report* regularly reflect the range of controversies. Both devoted significant space to the recombinant DNA issue; see, e.g., *F.A.S. Pub. Int. Rep.*, 29 (April 1976); *Bull. Atom. Sci.*, 33 (May 1977).

12 A vivid account of the discussions is given in William Bennett and Joel Gurin, "Science That Frightens Scientists," *Atlantic*, Feb. 1977. See also the brief but valuable book by *Science* news reporter, Nicholas Wade, *The Ultimate Experiment. Man-Made Evolution* (New York: Walker, 1977).

13 *Science*, September 21, 1973, pp. 1, 114.

14 Although the intensity of feeling varied at different university centers, few will forget the vigorous debates and active student–faculty demonstrations that surrounded military linked research efforts. The Mansfield Amendment requiring the Department of Defense to give specifics on the military utility of department-supported research at the federal level and local efforts to remove universities from military ties brought careful scrutiny of previously ignored research grants. At Harvard University, for example, a new practice was introduced that required publication on a regular basis of all grants, indicating source and amount of funds and summary of research goals.

15 P. Berg et al. (letter) "Potential Biohazards of Recombinant DNA Molecules," *Science*, July 26, 1974, p. 303.

16 See the summary statement of the report submitted to the National Academy of Science, Paul Berg et al., "Asilomar Conference on Recombinant DNA Molecules," *Science*, June 6, 1975, pp. 991–4. Nicholas Wade reports on the conference in Chap. 5 of his book, *The Ultimate Experiment*, pp. 41–53.

17 Paul Berg et al., "Asilomar Conference"

18 See the reports of the meeting and the controversy it stirred in Nicholas Wade, "Recombinant DNA: NIH Group Stirs Storm by Drafting Laxer Rule," *Science,* November 21, 1975, pp. 707–69; also Wade, "Recombinant DNA: NIH Sets Stricter Rules to Launch New Technology," *Science,* December 19, 1975, pp. 1,175–9.

19 National Institutes of Health, "Recombinant DNA Research: Guidelines," *Federal Register,* July 7, 1976, pp. 27,902–43.

20 National Institutes of Health, "Recombinant DNA Guidelines: Draft Environmental Impact Statement," *Federal Register,* September 9, 1976, pp. 38,426–83.

21 Part of the effort that developed to attain federal legislation was designed to bring industry into compulsory adherence to the "Guidelines." Several states moved in the interim to compel such adherence. See N. Wade, *The Ultimate Experiment,* pp. 136ff.

22 Erwin Chargaff, "Profitable Wonders, A Few Thoughts on Nucleic Acid Research," *The Sciences,* August/September 1975, pp. 21–6; (letter) "On the Dangers of Genetic Meddling," *Science,* June 4, 1976, pp. 938–40.

23 Some autobiographical remarks are found in Erwin Chargaff, "A Fever of Reason. The Early Way," *Ann. Rev. Biochem.,* 44 (1975), pp. 1–18.

24 The group included several prominent younger genetic researchers such as Jonathan King (MIT), Jonathan Beckwith (Harvard), and Richard Goldstein (Harvard), as well as a variety of other activists from various scientific and medical fields.

25 The pages of the student newspaper, the *Harvard Crimson,* and the weekly town paper, the *Cambridge Chronicle,* are filled with lively details. The Cambridge hearings brought forth a sympathetic editorial in the *Boston Globe,* July 10, 1976. See also the report by Nicholas Wade, *The Ultimate Experiment,* chap. 11, pp. 127–44.

26 The department itself never voted, but instead left the decision to be made by the Faculty Committee on Research Policy.

27 See the reports cited in n. 25, this chapter.

28 See the *Harvard Crimson* and the official *Harvard University Gazette.*

29 There was an attractive cover story in the *Harvard Magazine* and good coverage in the weekly *Cambridge Chronicle.*

30 See the *Crimson* reports for February 8 and 9, 1977. The mayor, however, continued his opposition, but was not successful in blocking passage.

31 Mayor Vellucci went to Washington to testify in favor of giving local government the option to strengthen research restrictions.

32 See Barbara Culliton, "Recombinant DNA Bills Derailed: Congress Still Trying to Pass a Law," *Science,* January 20, 1978, pp. 274–7, for a report of the activities of the scientists. See also David Dickson, "Friends of DNA Fight Back," *Nature,* April 20, 1978, pp. 664–5.

33 James D. Watson repeated his arguments in favor of DNA research and lamented his support for the initial moratorium in several publications ranging from the *New Republic* (June 25, 1977) to the editorial page of the *International Herald Tribune* (May, 15 1978).

34 One excellent collection of papers has been brought together by the *Southern California Law Review,* 51 (September 1978): "Biotechnology and the Law: Recombinant DNA and the Control of Scientific Research."

35 See interferon story, *New York Times,* January 17, 1980.

36 *Boston Sunday Globe* editorial, September 16, 1979.

37 *Nature,* February 7, 1980, p. 512; see also *Science,* Feb. 15, 1980, p. 745.

38 *Science,* February 1, 1980, p. 507.

39 *New York Times,* January 17, 1980.

40 See some of the discussions in *New York Times,* September 11, 1979, February 3, 1980; *Science,* March 21, 1980, p. 1,326; *Newsweek,* March 17, 1980, pp. 62–71.

41 *Washington Post,* November 18, 1980.

42 See, e.g., Barbara Culliton, "The Academic-Industrial Complex," *Science,* May 28, 1982, pp. 960–2. The issue is also explored in W. J. Whelan and S. Black (eds.), *From Genetic Experimentation to Biotechnology: The Critical Transition* (New York: Wiley, 1982).

43 *Harvard Crimson,* March 25, 1982.

44 The fullest discussion to date of the first phase of the recombinant DNA controversy is to be found in the excellent study by Sheldon Krimsky, himself a member of the Cambridge Experimentation Review Board, *Genetic Alchemy: The Social History of the Recombinant DNA Controversy* (Cambridge, Mass.: MIT Press, 1982).

45 *Science,* January 14, 1983.

46 *Science,* December 24, 1982, and Jan. 14, 1983.

47 For Yale see *New York Times,* December 20, 1982, p. 40; Harvard published on May 2, 1983, "Proposed Guidelines for Research Projects Undertaken in Cooperation with Industry."

48 The external guidance of science was extensively investigated by the "Starnberg Group"; see the recent English edition, Wolf Schäfer (ed.), *Finalization in Science: the Social Orientation of Scientific Progress* (Dordrecht: Reidel, 1983). The concept of deinstitutionalization in science has been the focus of recent studies by Peter Weingart. Peter Weingart, "Verwissenschaftlichung der Gesellschaft – Politisierung der Wissenschaft," *Zeitschrift für Soziologie,* 12/3 (July 1983), pp. 225–41.

17

William Ferrel and American science in the centennial years

HAROLD L. BURSTYN
U.S. Geological Survey

"If I have seen farther," wrote Isaac Newton to Robert Hooke on February 5, 1675–6, "it is by standing on the shoulders of giants."[1] This well-known observation comes instantly to the mind of a person asked to acknowledge the debt that he owes to his teachers. I first met Bernard Cohen in early 1948, my freshman year at Harvard College. Uncertain of a future in chemistry, I was unwilling to spend the springtime in the smelly laboratory of qualitative analysis. The reexamination of the curriculum going on around me, in response to *General Education in a Free Society* (1945), suggested that tutorials were the most distinctive feature of a Harvard education. My search through the handbook of fields of concentration turned up History and Science, which offered both tutorial instruction and a chance to study science. Choosing a major thus led me first to Cohen's study in Eliot House and then into a career. Of the original four graduates in History and Science, I alone have continued with the history of science. The others entered traditional professions: Two chose medicine; one, law.

These reminiscences may seem irrelevant to the high scholarly purpose of this volume; they are not. For my subject arises directly from those undergraduate years when I sat at Bernard Cohen's feet and from topics he was working on as he extended his studies of Franklin, the Newton of eighteenth-century America, back to Sir Isaac himself. I proposed to meet the history requirements for my bachelor's degree by taking all the intellectual history that was offered: medieval, modern European, American, and so on. My plan was short-lived; the history department insisted that I take courses in one region or period only. Thus I became a student of American history at the time that Cohen, having edited Franklin's experiments in electricity, was about to publish his study of Harvard's scientific apparatus of the eightenth century.

To the historian of colonial American science, the American scientific scene in the middle of the nineteenth century must appear profoundly disappointing. A society that had produced in Benjamin Franklin an outstanding contributor to the newest field, electricity, seemed a century later to be lagging behind the

Portions of this study were supported by a grant from the National Science Foundation to the Woods Hole Oceanographic Institution. I am indebted to Nathan Reingold for comments on an earlier version and to my colleagues at the Geological Survey for their support.

science of western Europe. Franklin had proved himself both the answer and the answerer to the contempt of sophisticated Europe for the newfound land, with its difficult climate and its different, perhaps degenerate, fauna and flora. Yet a century later after independence there had not arisen in America a culture, a scientific culture, to equal Europe's. Cohen put the matter succinctly in 1961: "A historian of science contemplating the role of America in the advancement of sciences during the nineteenth century cannot avoid a profound sense of disappointment . . . The bleakness of the picture presented to us by an examination of American science during most of the nineteenth century is enhanced by a comparison with the impressive list of distinguished contributions made by Americans to science prior to national independence."[2]

To his general judgment of nineteenth-century American science, Cohen made an important caveat. "In one science a first-class contribution may have come from America,"[3] he says in an aside to a series of comparisons otherwise unfavorable to Americans. The science he had in mind was oceanography; the scientist, Matthew Fontaine Maury. Here, in my opinion, Cohen unerringly put his finger on the most important achievement of American science in the middle years of the nineteenth century. Here is a "single commanding idea" that arose in America to "dominate a whole segment of future scientific activity."[4] But oceanography is too general a name for the science that resulted, and Maury, though he furnished the inspiration, did not make the achievement. The science I speak of had no name during much of its history. Now it is called geophysical fluid dynamics: "fluid dynamics" because it treats the immensely complicated motions of the ocean and atmosphere; "geophysical" because these motions are further complicated by the rotation of the earth on its axis. And the scientist who gave this science its major impetus in the nineteenth century was an obscure American named William Ferrel.

In order to explain the nature of Ferrel's achievement, I must first discuss the science he helped found and show that it meets Cohen's criterion that it be "central to our understanding of matter or of life."[5]

What is the nature of our physical environment? This is the central question posed by those pre-Socratic thinkers whose speculations we call the beginnings of science. In their answers we identify the first attempts at human understanding of matter. The ideas of the pre-Socratics were reworked by Aristotle into the theory of the four elements, and this theory, with the elements renamed the geospheres, is the basic principle of physical geology. I shall speak here only of the hydrosphere and the atmosphere, the two fluid spheres that surround the earth or lithosphere. They are held to the solid earth by its gravity, and the motions under gravity of their fluids, air, and water (and of solids) make up the natural motions of Aristotelian physics.[6]

When in the early modern period people began to search for more precise descriptions of matter, they transformed the four elements into principles of chemical composition. There began a research tradition that led to the periodic table of modern chemistry and the fundamental particles of modern physics. But the equally fundamental problems of motion in the earth's fluid envelopes could not be treated in the same way. The treatment of fluids in motion did originate in the research tradition of particle mechanics that began with the motion of planets and projectiles and led also to fundamental particles. However, a full treatment of fluid motion demanded continuum mechanics, which branched off from particle mechanics. Begun by Newton in Book Two of his *Principia*, continuum mechanics received its principal formulation from Euler in the eighteenth century.[7] In the nineteenth century, Laplace, Airy, Kelvin, Stokes, Helmholtz, and Poincaré were among those scientists of distinction who contributed fundamental advances in theory. To apply their theoretical advances to the fluids of the earth required careful observations. Because the data were hard to obtain in long time series, progress in explaining the motions of the atmosphere and ocean was sporadic. Ferrel's central contribution was a demonstration that the rotation of the earth caused the observed pattern of winds and ocean currents. This mid-nineteenth-century result was the culmination of three hundred years of observation.

When in the sixteenth century people first voyaged across the open ocean, their framework for observation was wholly Aristotelian. The writers on winds and ocean currents discerned two cycles. The first was an east–west current of air (the trade winds) and of water (the equatorial current, which was conflated with the tides). The second was a north–south current, which moved to the poles as air returned as water. The north–south (or meridional) cycle was driven by solar heat; the east–west (or zonal) cycle, by either the rotation of the earth, according to followers of Copernicus, or the rotation of the heavens, according to his opponents.[8] In the seventeenth century a new science based on particle mechanics began to replace the old. Aristotle's idea that winds are meteors – exhalations of the earth – gave way before the idea that winds are air in motion from areas of high to areas of low pressure. In other words, the invention of the barometer in the seventeenth century allowed the problem of atmospheric circulation to be separated from the problem of evaporation and precipitation. Within this new scientific context Edmund Halley developed a theory of the trade winds. In Halley's theory the relative motion of the sun around the earth, that is, the earth's rotation on its axis, set up a zonal current of air.[9]

In addition to the first non-Aristotelian theories of winds, the seventeenth century saw the first experiments to prove the rotation of the earth. Galileo made an equivocal suggestion that the earth's rotation would deflect bodies falling from a great height, and Newton proposed to Robert Hooke in 1679 that this experiment should be performed to prove the earth's motion. Hooke complied; the experiment was repeated at intervals for more than two centuries. The knowledge thus provided of the deflecting force of the earth's rotation led to the first modern theory of the trade winds, announced by the English lawyer George Hadley to the Royal Society in 1735. Hadley revived the notion of a meridional convection cell, this time confined to atmosphere, and he showed that it must be deflected zonally by the earth's rotation. Newton's disciple, the Scottish mathematician Colin Maclaurin, applied the same idea to ocean currents in 1740.[10]

Maclaurin expressed this idea in a paper on the tides, the problem of planetary hydrodynamics in which in the eighteenth century the greatest theoretical advances were made. These advances were expressed mathematically, and they culminated in Laplace's theory of the tides. In this theory there appeared for the first time the equations of fluid motion in a rotating coordinate system such as the earth.[11]

These equations expressed the principles from which the general circulation of the atmosphere and ocean might be deduced. Yet it took two generations of rigorous effort to extract from them the general result they implied. The scientists of the nineteenth century lacked both the data against which to test their ideas and the confidence in mathematical deduction that characterizes twentieth-century science. Hence half a century separated Laplace's theory from Ferrel's, the first general statement of how the winds and currents move around the earth. What was the sequence of events that culminated in this pathbreaking paper of 1856?

George Hadley's rough explanation for the trade winds, namely, that the earth's rotation deflects zonally air that is moving meridionally, was by no means universally accepted in the century after he announced it in 1735. Though Benjamin Franklin, Immanuel Kant, and John Dalton accepted Hadley convection, J. L. R. D'Alembert, the Comte de Buffon, Richard Kirwan, and Elisha Mitchell denied it.[12] Finally in the 1830s the most prominent meteorologist of his day, Heinrich Wilhelm Dove, erected Hadley convection into a dogma. In the form in which Dove stated it, that only meridional motions are possible and they are deflected only zonally, the principle is false. The earth's rotation deflects bodies moving along its surface in any direction. The deflection is to the right in the northern hemisphere, to the left in the southern. By making explicit what Hadley had only implied, Dove led meteorology away from rather than toward understanding.[13]

He did so at a time when the observational evidence was increasing rapidly. Benjamin Franklin began, on some of his many sea voyages to Europe, to take the surface temperature of the sea in order to show where the currents were. By 1830 the Englishman James Rennell had delineated many of the surface currents of the Atlantic, and in the 1840s the Frenchman Georges Aimé developed techniques to measure ocean temperature beneath the surface.[14] These are only the highlights of a growing series of investigations. By the 1840s also, Matthew Fontaine Maury of the U.S. Navy provided global charts of surface winds and currents. As the electromagnetic telegraph spread over Europe and North America, there developed the first synoptic charts of the atmosphere.

By the 1850s, as the winds and currents were plotted, they showed that Dove's ideas, even though they were shared by Sir John Herschel, August Colding, and others, could not be correct. The work in the United States in the 1830s and 1840s on cyclonic storms had suggested that these storms acquired their pronounced rotation from the motion of the earth. This work had been bitterly disputed; in the 1850s it began to be accepted. In 1853 an American, James Coffin, announced to the American Association for the Advancement of Science that winds blow along, rather than across, the isobars, with pressure higher to the right of the moving wind. Coffin's paper appeared in 1856, and in the following year the same result was announced independently by the Dutch meteorologist Christoph H. D. Buys Ballot, after whom it is usually called Buys Ballot's Law. Also in 1857, F. Vettin began in Germany the first of what are known as dishpan experiments, the modeling of atmospheric and oceanic circulation in rotating tanks of liquid.[15]

All these were steps along the way to William Ferrel's first general treatment of natural motions on the surface of the earth. His "Essay on the winds and currents of the ocean," published in Nashville, Tennessee, in 1856, gave the first modern account of the general circulation of atmosphere and ocean. In it he showed that the observed motion of air and water along the isobars resulted from the rotation of the earth on its axis. That is, he offered the first explanation from theory of the observations of Coffin, Buys Ballot, and others. Ferrel followed up this qualitative paper with a comprehensive mathematical treatment published in 1858–61.[16]

How had it fallen to an unknown school teacher in a small frontier city to solve a major problem in earth science? What in William Ferrel's background thrust him from obscurity to the center of the world scientific community? If ever a scientist lived the life of a folk hero, Ferrel did. Like Abraham Lincoln, hero not only to Americans but to people all over the world, Ferrel was born into poverty in a log cabin on the frontier. Like Lincoln, Ferrel moved with his family from place to place in search of better times, catching an education by firelight after a day's heavy labor. Like Lincoln, Ferrel had no formal schooling in his early years. Yet he rose to be a scientist of world renown in one of those theoretical, highly mathematical fields that have traditionally claimed our homage as the bases of scientific progress.

Ferrel was born in 1817 in south-central Pennsylvania, a mountainous area just north of the Mason-Dixon line. He was Scots-Irish on his father's side, German on his mother's, the eldest of their eight children. In 1829, when William was 12, the Ferrels moved about fifteen miles south across Maryland to Virginia. Here William had the usual schooling of the time, English and arithmetic taught in a one-room log schoolhouse over a couple of winters, the only time an able-bodied boy could spare from the unrelenting toil of farm labor. Reading matter was scarce on his farming frontier, and the glimpse of an arithmetic book fired the imagination of the introverted Ferrel. On a trip to the nearest town he bought a copy with fifty cents earned by helping a neighbor with his harvest. Ferrel mastered this simple book, and in 1832, when he was fifteen, a solar eclipse inspired him to attempt to work out the motions of the sun and moon. With only farmers' almanacs and a geography book to guide him, he spent two years developing crude methods for predicting eclipses. By 1834 he had taught himself to predict those for the following year, and the agreement of his predictions with those found in calendars spurred him on. He then borrowed a book on surveying and two on geometry, mastering each in turn. In 1837, when he was twenty, Farrel found an elementary work on natural philosophy, from which he "learned the law of gravitation, and that the moon and planets move in elliptical orbits."[17]

By this time Ferrel, like many young men on the frontier who had both intelligence and book learning, had begun to teach school. In 1839 he took his savings from school teaching and entered Marshall College in Mercersburg, Pennsylvania, for a summer course in mathematics. At age twenty-two he thus started his formal studies with algebra (for the first time), geometry, and trigonometry. The work required so little effort from one used to so much that Ferrel began, in his spare time, to learn Latin and Greek. After another fall and winter of teaching, he returned to college for a second summer with financial help from his father. Two more years' teaching, and Ferrel resolved to continue formal education for at least one full year. Rather than return to Mercersburg, he went a greater distance to the newly opened Bethany College in Virginia's northern panhandle, not far from Pittsburgh, founded by the Reverend Alexander Campbell for his Disciples of Christ, the church to which the Ferrel family belonged. Admitted to the senior class though he had spent only two semesters at Marshall College, Ferrel was graduated in 1844 as a member of Bethany's first class. He was twenty-seven years old.

Ferrel now headed west. For two years he taught at Liberty, Missouri, where, in a bookstore in this frontier town, he found a copy of Newton's *Principia,* ordered by some itinerant scholar but never claimed. Returning home to Virginia in 1846 after a bout of ill health, Ferrel was diverted via Tennessee to Todd County, Kentucky, a wealthy agricultural area just over the Tennessee border. During his seven years here, in the time he could spare from his teaching, he mastered the *Principia,* becoming especially interested in the tides. Learning of Laplace's monumental summary of the century of astronomy after Newton in his *Mécanique céleste,* Ferrel asked a merchant going on the long journey to Philadelphia to fetch a copy for him. With Laplace's great storehouse before him, translated and annotated by the American Nathaniel Bowditch, Ferrel could work out an idea that had occurred to him in reading Newton: "The action of the moon and sun upon the tides must have a tendency to retard the earth's rotation on its axis."[18] Ferrel's paper, written in the summer of 1853 when he was thirty-six, was his first contribution to the scientific literature. It appeared in Benjamin Gould's *Astronomical Journal,*[19] published in Cambridge, Massachusetts, the first American scientific journal devoted to a single subject. In the spring of 1854 Ferrel moved to Nashville, Tennessee, the only city in which he had ever lived.

The first half of his life establishes Ferrel's place in the Lincoln tradition. Early poverty, a life completely rural, an education obtained largely outside institutions and entirely outside major universities or urban cultural centers: Ferrel's life fits the pattern of the American myth. Against all odds of birth and upbringing, against our firmly held notions of the time of life for mathematical creativity, Ferrel had by age thirty-seven made himself a scientist, committed to expressing in mathematical language his understanding of the workings of the universe.

Yet one published paper does not make a scientific career, even when the contribution is as original and important as Ferrel's on tidal retardation. Ferrel's claim to our attention rests on what came after his move to Nashville. Though his occupation of school teacher did not change, Ferrel found in the city, a bustling state capital of nearly 13,000 people, what none of his earlier residences had had: bookstores and people interested in science. Nashville had a university with a medical faculty; one of its founders, Dr. William Bowling, published the *Nashville Journal of Medicine and Surgery.* Bowling's interests were wider than medicine, and he encouraged Ferrel to contribute on physical subjects.[20]

The talk of the world of science, and even of the general world in the early 1850s, was Léon Foucault's pendulum and gyroscope. It is hard for us, with our confidence in the power of mathematical deduction, to understand how electrifying were these simple demonstrations of the rotation of the earth, a physical fact that had not been in doubt for over a century, the confirmation of a mathematical result on the deflecting force of the earth's rotation that had been impeccably derived decades before by the French masters of analytical mechanics. But Foucault's two experiments, the pendulum of 1851 and the gyroscope of 1852, were repeated around the world in an astonishingly short time and their elucidation filled the pages of the journals of science.

Ferrel, too, became interested in the gyroscope, drew some conclusions from his theoretical study that he verified by experiment, and published the results in Bowling's journal late in 1856.[21] Meanwhile his browsing for scientific books yielded a copy of the enormously popular *Physical Geography of the Sea,* by Matthew Fontaine Maury. Pressed by Bowling to develop his own views, which were contrary to Maury's, Ferrel wrote for the *Nashville Journal* his epoch-making "Essay on the winds and currents of the ocean."[22] In this qualitative paper Ferrel outlined the first satisfactory theory of the general circulation of ocean and atmosphere, thus solving a problem with which scientists had been grappling since the age of the discoveries. Ferrel showed that the pattern of winds and ocean currents over the globe resulted from their deflection by the rotation of the earth. Half of this great truth had been glimpsed a century before by Hadley and Maclaurin, and by Ferrel's day the half-truth had become Dove's dogma. Foucault's pendulum and gyroscope made it possible to see the whole truth. William Ferrel, the self-taught school teacher in the small western town, was the first to grasp its meaning for our understanding of the world.

Meanwhile Ferrel published a second paper on the tides in Gould's *Astronomical Journal.*[23] His work thus became known to the leaders of the American scientific community, and his abilities in the more mathematical branches of science made him someone whose career had to be furthered. School teaching might be a suitable occupation for a minor talent, but someone of Ferrel's achievements ought to be working full-time in science.

As in any developing country, in mid-nineteenth-century America there were few institutions in which one might pursue advanced science. Only the more enlightened of the colleges encouraged their faculty members to carry out research. Various state geological surveys provided some professional employment, but not to workers in the exact sciences. A community in which those sciences that are mathematical would have pride of place as they did in Europe required professional employment for the American heirs of Newton and Laplace. But, as the French observer Alexis de Tocqueville pointed out, in the egalitarian society of the United States the willingness to apply knowledge was sure to outrun the willingness to increase it. Establishing a viable scientific community meant finding some interest that science might serve. The leader in developing native American scientific institutions in the 1840s was Benjamin Franklin's great-grandson Alexander Dallas Bache, distinguished graduate of West Point and a leading scientist in Philadelphia. Bache and his friends, self-consciously building scientific institutions for America, found in seaborne trade, the lifeblood of the Republic, a commercial interest that could be tapped to support science.

While William Ferrel was growing up on the frontier, two agencies of the federal government began in the late 1830s to provide what we now call research and development services to the shipping industry. One was the Coast Survey, which came into Bache's hands in the 1840s. The other was the Hydrographic Office and National Observatory. By the late 1840s, so far from furthering the aims of the leaders of American science, the observatory had been captured by Matthew Fontaine Maury, first and foremost a naval officer, ill-equipped to promote the spirit of pure science, the disinterested search for truth about nature.[24]

The jobs available in the Coast Survey were limited both in number and scope, and the Observatory was in hostile hands. Bache and his friends therefore organized a third agency, the Nautical Almanac Office. The Almanac was founded in 1849 by Lieutenant Charles Henry Davis, brother-in-law of Benjamin Peirce, then Perkins Professor of Astronomy and Mathematics at Harvard College and the most distinguished mathematician in America. In order to steer clear of Maury, Davis set up the Almanac Office in Cambridge, using Harvard's Observatory rather than the Navy's. In a period when Boston was the Athens of America and Washington a small provincial town, there were advantages for the progress of science in the strange location of this arm of the United States Navy. With Benjamin Gould's *Journal* to provide opportunities for publication, the Harvard telescope for observing, and the Nautical Almanac for jobs, Cambridge grew rapidly into a leading center of astronomical research. Benjamin Pierce lured Joseph Winlock from Kentucky to join the Almanac staff of computers who laboriously calculated the numbers that went into the tables of the positions of the sun, moon, and stars. When Winlock arrived in Cambridge in 1852, he found himself in distinguished company. Already working for the Almanac were Simon Newcomb, who became America's most distinguished astronomer; Maria Mitchell, the prodigy from Nantucket, later a professor at Vassar College; J. D. Runkle, later president of Massachusetts Institute of Technology; and Chauncey Wright, the founder of the distinctly American philosophy known as pragmatism. Unknown to any of them, William Ferrel was in Kentucky teaching school and studying Newton's *Principia.*

After Charles Henry Davis's return to sea duty in 1856, Winlock – though a civilian – became superintendent of the Almanac. At Benjamin Gould's suggestion, Winlock invited William Ferrel to leave Nashville for Cambridge. Now forty years old, Ferrel had his first scientific post. In the time he could spare from the endless calculations for the Almanac, Ferrel developed in quantitative form his theory of the general circulation, publishing it briefly in Gould's *Journal* in 1858, extensively in Runkle's *American Mathematical Monthly* in 1859 and 1860, and in popular form in the *American Journal of Science* in 1861.[25] When the National Academy of Sciences was established in 1863, Ferrel was among the fifty incorporators. He continued his work on tidal retardation in a paper published by the American Academy of Arts and Sciences in 1864.[26]

In 1867, when Benjamin Peirce took over the Coast Survey on Bache's death, he was followed from Cambridge to Washington by Ferrel. There he made a number of contributions to the theory of the tides and built one of the earliest tide-predicting machines, the first piece of large-scale computing machinery actually put into use. In 1870 the Signal Office of the United States Army began a national weather service, and Cleveland Abbe came to Washington the following year to run its scientific side. Long a disciple of Ferrel's, Abbe brought his master into the Signal Service in 1882 (until his retirement four years later at age seventy), and Abbe was responsible for the energetic propagation of Ferrel's ideas throughout the world.

"The whole life-work of Ferrel," said Simon Newcomb in 1891, "we might describe . . . as the theory of cosmical fluid motion."[27] Our current name for this subject is geophysical fluid dynamics, and it is one of the exacting branches of theoretical physics. Like any other predominantly mathematical field, it taxes its practitioners to the utmost; one physicist has spoken of the "sense of frustration that is the lot of all theoretical physicists most of the time."[28] The majority of our heroes among scientists seem to come from theoretical physics. How does William Ferrel measure up?

In his 1859 paper, "The motions of fluids and solids relative to the earth's surface," Ferrel began the modern study of geophysical fluid dynamics. At his death thirty-two years later he was still one of its foremost students. Though he spent the first half of his seventy-four years on the culturally backward frontier, Ferrel became one of the most distinguished theoretical scientists America has ever produced. Yet, in spite of an early life that lends itself to myth almost as well as Lincoln's, Ferrel remains to this day virtually unknown even among historians dedicated to reconstructing the past of American science. Can we learn anything from our neglect of him?

First, I think historians of American science have been looking in the wrong places for the distinguished scientists of the mid-nineteenth century. The exact sciences in America, though they may not have bloomed as they did in Europe, certainly did not languish. In the use of mathematics to understand the physical environment Americans were predominant from the 1840s for at least three decades, and foremost among them was William Ferrel. Rather than lamenting the backwardness of American science in the nineteenth century, I want to ask why we have accepted so distorted a view of the past.

In my opinion historians have been looking at the exact sciences in the nineteenth century from the standpoint of mid-twentieth-century physics. The march of science since 1900 has focused our attention on the very small. From the quantum revolution to the present, the progress of physics has lain in the atomic and subatomic realm, and our prejudice for mathematical science makes physics the touchstone of scientific progress. The narrowing of physics into microphysics in our century has become our guide to the past. Yet, though physicists may proclaim it as an article of faith, it is by no means certain that the behavior of gross matter can be understood solely in terms of elementary particles, any more than living organisms can be understood solely in terms of biological molecules. A historiography rooted in contemporary physics cannot, in my judgment, enlighten us about science in the nineteenth century. Such a historiography bears about the same relation to what happened in history as the unilinear Great Chain of Being bears to Darwin's notion of an evolutionary tree with myriad branches. What we need is a broader viewpoint, one that will do justice to the complexity and size of the scientific enterprise.

There were other major achievements of the exact sciences in the nineteenth century besides those that led to quantum physics. Mathematical science a century ago concerned itself more with macroscopic than with microscopic problems. C.F. Gauss with terrestrial magnetism and astronomy; Ferrel, Kelvin, and Helmholtz with fluid dynamics, both ordinary and geophysical; Maxwell and Kelvin with electrodynamics – these were among the chief developments. Until we revise our historiography to take account of what scientists were actually doing, our interpretation of what Americans contributed to world science in the nineteenth century is likely to be incomplete.

If we do change our perspective, we shall, I feel sure, come to see American science in the nineteenth century in a new light, no longer solely the stepchild of Europe, but preeminent in those studies of the environment that deserve the high title of theoretical physics. Among the workers in this difficult field, William Ferrel stands out. The story of his life in science, coming into full flower in the years between the end of the Civil War and the celebration of the centennial in 1876, can inspire us a century later as does the life of Ferrel's contemporary Abraham Lincoln. William Ferrel is the person with the strongest claim to be the folk hero of nineteenth-century American science.

Ferrel may indeed be, as I have claimed, a genuine hero among American scientists, a person whose stature in the nineteenth century is comparable to Franklin's in the eighteenth. But is a quest for heroes appropriate to the history of science? Does not a sophisticated historiography make finding a hero beside the point? For the very idea of heroes, those whose contributions to science were of exceptional significance, is an internalist idea. It is integral to a view of science as a branch of thought with its own inherent logic. Scientific ideas, the internalist argument runs, follow their own dialectic; they wax and wane in response to no currents of thought outside themselves. There is no control worth mentioning exerted by the wider culture over the development of scientific ideas.

This internalist perspective has recently fallen under attack. A search for great people and great ideas belongs to an outmoded view of science and progress, as "systematized positive knowledge," in the words of George Sarton.[29] The newer historiography enjoins us to study the group of scientists, its membership, social origins, and aspirations, rather than to hunt great ideas through the great minds that contributed to their growth. We are told to pay attention to the politics, the economics, the literature, the art, and the material circumstances surrounding those whose ideas we wish to unravel.

This newer, externalist view of the history of science arises, I believe, from a fundamental difference between science and the other forms of human creativity. Unlike art, literature, or music, science is a communal enterprise. Its results are not embodied in artifacts – paintings, poems, performances. Rather the achievements of science are intangible, the common possession of a number of individual minds. Though ideas grow in each mind independently, scientific research is done in groups whose members find it hard to distinguish individual contributions. Not only is the research communal, its aim is to reach consensus about what the outside world is like. This notion of consensus, of a shared paradigm, implies a group among whom one has a consensus.[30] It is these groups who ought to be the subject of our historical analysis.

One such group is the American scientific community, and the increasing attention paid to it is an index of the growth of external history of science. As an example of science in one country, the history of American science falls by definition into the externalist camp. Yet the question most often asked, what did those who were Americans contribute to world science? is not an externalist question. It demands for its answer some standards of judgment to evaluate the achievements of American scientists, and these standards are unequivocally internal. To answer questions about American achievements in science requires that we be well versed in the details of the principal paths of scientific development. Otherwise we may overlook significant American scientists or misunderstand their contributions. (I have suggested earlier that we have heretofore overlooked a significant American scientist of the nineteenth century, William Ferrel.)

To ask what Americans contributed to world science is to confuse internal and external history. For the external question, the one that leads us to study science in America in the first place, is a different one. It is: What was science in America like? To answer this question one begins by enumerating scientists, their institutions, and their papers, classifying them respectively by region, by social standing (for individuals) or type (for institutions), and by field. These statistics of people, their publications, and their workplaces, show in outline what the community was like. To discern the shape of a community and its concerns, a study of heroes and their feats is out of place.

Nevertheless, the external point of view does not suffice by itself to make clear the history of science in America. For science is not a collection of individual contributions. Unlike other forms of culture, science produces no artifact that can be understood in isolation. The historian of art can describe the nature of an individual painting; the historian of literature, a novel or poem; the historian of architecture, a building. These objects stand by themselves, palpable in their reality. To them the critic applies the principles of judgment. From this assessment of the merit of individual objects of creation, the critic is led to identify those persons who have made the most distinguished contributions to the field under critical inquiry and to order them. An inventory of the objects created, each of them measured by some canon of critical judgment, then becomes an index to the creative achievement of the community the critic studies. Hence in the assessment of American creativity in art, music, literature, painting, sculpture, or architecture in a particular period, enumeration – number and origin of practitioners, number and location of works – becomes a useful tool of historical analysis.

I do not believe that the the same thing is true for the history of science. Science comes closer than do other forms of human creativity to being the common possession of a number of disparate minds. Rather than a collection of discrete contributions, science is a synthesis of them. To be sure, this synthesis is not identical for each contributor. It is shared, since a scientific truth is what a group of scientists believes has been tested against reality and confirmed. Before the group agrees on a scientific result, it must be repeatable. Repeatability is thus the guarantee that the scientific knowledge is objective.

Because the result must be repeated to be accepted and accepted by the community to be part of science, it does not have a single locus in the literature. Hence the individual truths that make up the body of scientific knowledge cannot be precisely located, nor are the various kinds of literature of the same value in assessing scientific creativity. Hence counting scientific papers, as one might count sonatas or symphonies, cannot make science understandable. Since scientific papers are not additive, one cannot learn the nature of science in a given period by enumeration, even of papers ordered in a hierarchy. The nature of science will escape historians if they do not adopt the internalist concern for the leaders and their ideas.

Without knowing where Americans stood in the world scientific community, the question of what science was like in America becomes purely parochial, like what were American railroad stations like, or American post offices. The communal nature of science requires that the standards by which the community operates by made explicit before the historian can grasp its nature. Hence to understand American science in the centennial years, the years between the Civil War and 1876, we must know which Americans had standing among the world's scientists and why. So a search for heroes, though it may appear old fashioned, motivated by internal criteria that are irrelevant to the American scene, is in fact necessary to our grasp of what science was like in America a century ago.

Necessary, but not sufficient. For the search for heroes can never be more than the first step. Though the contributions of the leading scientists require detailed comprehension by historians, these contributions do not exist apart from the community of scientists – followers as well as leaders – that shares them.[31]

My conclusion, then, is a paradox. In order to do better external history of science, to discern the nature of the various scientific communities in which we are interested and their relations to each other and to the wider culture, we need, I believe, more internal history of science, more exploration of other areas than the traditional ones of mathematics, physics, and chemistry. We need to trace back to their roots the concerns of the leading scientists of whatever period we choose to study.[32] I suggest these roots will lie as often in questions about the environment as in questions about the basic structure of matter and energy. But is this tracing of things back to their origins not precisely the example that Bernard Cohen has set for us in all his three decades of meticulous scholarship? I for one am a better scholar for his teaching and example.

NOTES

1 Quoted by Alexandre Koyré, *Newtonian Studies* (Cambridge, Mass.: Harvard University Press, 1965), p. 227. This well-known aphorism is traced to its origins in Robert K. Merton, *On the Shoulders of Giants* (New York: Free Press, 1965).

2 I. Bernard Cohen, *Science and American Society in the First Century of the Republic* (Columbus: Ohio State University, 1961), pp. 1, 3.

3 Ibid., p. 2.

4 Ibid.

5 Ibid. Responding to criticism (esp. Nathan Reingold, "American Indifference to Basic Research: A Reappraisal," in George H. Daniels, ed., *Nineteenth-Century American Science. A Reappraisal* [Evanston, Ill.: Northwestern University Press, 1972], pp. 38–62), Cohen has recently softened the language of his judgment and omitted the reference to Maury ("Science and the Growth of the American Republic," *Review of Politics, 38* [1976], pp. 359–98).

6 Harold L. Burstyn, "The Empirical Basis of the Four Elements," *Actes du XIIième Congrès Internationale d'Histoire des Sciences* (Paris: Hermann, 1971), *3A*, pp. 19–24.

7 C. Truesdell, "Rational Fluid Mechanics, 1687–1765," *L. Euleri Opera Omnia*, ser. 2, vol. 12 (Zürich: Orell Füssli, 1954), pp. IX–CXXV.

8 Harold L. Burstyn, "Theories of Winds and Ocean Currents from the Discoveries to the End of the Seventeenth Century," *Terrae Incognitae, 3* (1971), pp. 7–31.

9 Harold L. Burstyn, "Early Explanations of the Role of the Earth's Rotation in the Circulation of the Atmosphere and the Ocean," *Isis, 57* (1966), pp. 167–87.

10 Ibid.

11 P. S. de Laplace, "Recherches sur plusieurs points du système du monde," *Mémoirs de l'Academie Royale des Sciences,* 1775, pp. 75–182; 1776, pp. 177–267.

12 Benjamin Franklin, "Physical and Meteorological Observations, Conjectures, and Suppositions," in Leonard W. Labaree, ed., *The Papers of Benjamin Franklin*, vol. 4 (New Haven, Conn.: Yale University Press, 1961), pp. 235–43. Immanuel Kant, *Neue Anmerkungen zur Erlauterung der Theorie der Winde* (Königsberg, 1756), in *Kants Werke* (Berlin: de Gruyter, 1968), vol. 1, pp. 489–503. John Dalton, *Meteorological Observations and Essays* (London, 1793), Essay 2. J. D'Alembert, *Réflexions sur la cause générale des vents* (Paris, 1747), p. iv. G. de Buffon, *Théorie de la terre* (Paris, 1749), articles 12–14. Richard Kirwan, "On Winds," *Philosophical Magazine, 15* (1803), pp. 311–19. Elisha Mitchell, "On the Proximate Causes of Certain Winds and Storms," *American Journal of Science, 19* (1831), pp. 248–92.

13 H. W. Dove, "Über den Einfluss der Drehung der Erde auf die Strömungen ihrer Atmosphäre," *Annalen der Physik, 36* (1835), pp. 321–51; *Meteorologische Untersuchungen* (Berlin, 1837) and *Das Gesetz der Stürme* (Berlin, 1857; 2d ed., 1861), the latter translated as *The Law of Storms* (London, 1862).

14 Margaret Deacon, *Scientists and the Sea 1650–1900* (London: Academic Press, 1971), pp. 202–3, 220–4, 288–90.

15 James H. D. Coffin, "An Investigation of the Storm Curve Deduced from the Relation Existing Between the Direction of the Wind, and the Rise and Fall of the Barometer," *Proceedings of the American Association for the Advancement of Science, 7* (1856), pp. 83–101, on p. 88. C. H. D. Buys Ballot, "Note sur le rapport de l'intensité et de la direction du vent avec les éscarts simultanés du baromètre," *Comptes Rendus de l'Académie des Sciences, 45* (1857), pp. 765–8. F. Vettin, "Meteorologische Untersuchungen," *Annalen der Physik, 100* (1857), pp. 99–110.

16 William Ferrel, "An Essay on the Winds and Currents of the Ocean," *Nashville Journal of Medicine and Surgery, 11* (1856), pp. 287–301, 375–389; "The Motions of Fluids and Solids Relative to the Earth's Surface," *The Mathematical Monthly, 1* (1858–9), pp. 140–8, 210–16, 300–7, 366–73, 397–406; *2* (1859–60), pp. 89–97, 339–46, 374–90, republished as a pamphlet with the same title (New York: Ivison, Phinney, and London: Trübner, 1860).

17 Quoted from p. 291 of William Ferrel, "Autobiographical Sketch," *Biographical Memoirs of the National Academy of Sciences, 3* (1895), pp. 287–99. This printed version, which follows Cleveland Abbe, "Memoir of William Ferrel. 1817–1891," ibid., pp. 265–86, differs only in orthography from the holograph original in the Harvard College Library.

18 Ferrel, "Autobiographical Sketch," p. 294.

19 Ferrel, "On the Effect of the Sun and Moon upon the Rotatory Motion of the Earth," *Astronomical Journal, 3* (1853), pp. 138–42.

20 F. Garvin Davenport, *Cultural Life in Nashville on the Eve of the Civil War* (Chapel Hill: University of North Carolina Press, 1941).

21 Ferrel, "The Rotascope," *Nashville Journal of Medicine and Surgery, 11* (1856), pp. 209–11.

22 Ibid., pp. 287–301, 375–89 (see n. 16, this chapter). Ferrel followed this paper with a longer one on the gyroscope: "A Complete Solution to the Problem of the Rotascope, Based upon Well-known Mechanical Principles," ibid., pp. 463–73.

23 Ferrel, "The Problem of the Tides with Regard to Oscillations of the Second Kind," *Astronomical Journal, 4* (1856), pp. 173–6.

24 Harold L. Burstyn, "Seafaring and the Emergence of American Science," in Benjamin W. Labaree, ed., *The Atlantic World of Robert G. Albion* (Middletown, Conn.: Wesleyan University Press, 1975), chap. 5.

25 Ferrel, "The Influence of the Earth's Rotation upon the Relative Motion of Bodies Near its Surface," *Astronomical Journal, 5* (1858), p. 97; "The Influence of the Earth's Rotation upon Rotating Bodies At its Surface," ibid., pp. 113–14; "The Motions of Fluids and Solids Relative to the Earth's Surface," n. 16, this chapter; "The Motions of Fluid and Solids Relative to the Earth's Surface," *American Journal of Science, 31* (1861), pp. 27–61.

26 Ferrel, "Note on the Influence of the Tides in Causing an Apparent Secular Acceleration of the Moon's Mean Motion," *Proceedings of the American Academy of Arts and Sciences, 6* (1864), pp. 379–83; supp. comm., ibid., pp. 390–3.

27 Simon Newcomb, "Ferrel's Early Astronomical Work," *The American Meteorological Journal, 8* (1891), pp. 337–9.

28 Lincoln Wolfenstein, "The Tragedy of J. Robert Oppenheimer," *Dissent, 15* (January–February 1968), pp. 81–5.

29 *The Study of the History of Science* (Cambridge, Mass.: Harvard University Press, 1936), p. 5. See Arnold Thackray and Robert K. Merton, "On Discipline Building: The Paradoxes of George Sarton," *Isis, 63* (1972), pp. 473–95.

30 These conclusions come from my reading of Thomas S. Kuhn, *The Structure of Scientific Revolutions,* 2d ed. (Chicago: University of Chicago Press, 1970); John M. Ziman, *Public Knowledge* (Cambridge: Cambridge University Press, 1968); and Jerome R. Ravetz, *Scientific Knowledge and Its Social Problems* (Oxford: Oxford University Press [Clarendon Press], 1971), parts 2 and 3. Ravetz offers one of the clearest statements: "Scientific knowledge, or facts, can exist only when the overlap between the private understandings of the objects among the members of the relevant community is sufficiently great for arguments to be communciated and univocally assessed" (ibid., p. 240). A similar point is made by Patrick A. Heelan: "Learning science is . . . an apprenticeship to a tradition possessed by a community of (more or less) expert practitioners. Science is not developed in isolation from a community and a tradition" ("Horizon, Objectivity and Reality in the Physical Sciences," *International Philosophical Quarterly, 7* [1967], pp. 375–412, 378. I am indebted to David Hemmendinger for this reference.

31 For an impassioned plea against the popular concept of scientific achievement as individual rather than communal, see Philip Siekevitz, "On Prizes," *Science, 202* (1978), p. 574.

32 For a trenchant critique that comes to similar conclusions, see Paul Forman, "Geneses of Scientific Ideas as Historiographic Goal," unpublished paper delivered to the History of Science Society, October 1978.

18

The American occupation and the Science Council of Japan

NAKAYAMA SHIGERU
Tokyo University

The creation of the Science Council of Japan was an epoch-making event in the history of science. The council is a representative organ for scientific researchers. Its members are elected by the entire body of professional, working scientists – not only by the established groups but by rank and file researchers as well. Council members, like those in the American National Academy of Science, are recognized and given official status by the government. The council was undeniably, however, a product of America's occupation of Japan following World War II. Had it been created independently, separate from America's occupation policy, there is no doubt that the council would have taken on a substantially different character. This essay is an attempt to examine the influence of occupation policy on the birth of the Science Council of Japan. The analysis is from the vantage point of the occupation forces and is based on the documents of the Economic and Scientific Section, Scientific and Technical Division (ESS ST) of General Headquarters (GHQ).[1]

HARRY C. KELLY AND THE JAPAN ASSOCIATION OF SCIENCE LIAISON

Harry C. Kelly (1908–78) was a former professor at Lehigh University, Lehigh, Pennsylvania. As a physicist, during the war Kelly worked on radar research at the Massachusetts Institute of Technology. Shortly after the war he joined the American occupation forces in Japan as a civilian chief of the Fundamental Research Branch of ESS ST and later became the associate director of the division. Kelly can be regarded as one of the New Dealers active in the early years of the occupation. His aspiration was to contribute to Japan's social and economic development by helping to organize its scientific research. His first act was to create the private Japan Association of Science Liaison as a conduit to Japan's scientific community. This organization later evolved into the Preparatory Committee (Sewaninkai) of the Scientific Research Organization Renewal Committee, finally emerging as the Science Council of Japan. Kelly was, therefore, the true parent of the Science Council, the driving force creating Japan's postwar scientific organization system. Consequently, of the GHQ documents related to the recreation of Japanese science, those concerning Dr. Kelly are among the most significant.

The first task undertaken by GHQ in academic and scientific matters was the dissolution of arms research. Beginning with atomic energy, aviation, and naval research, the occupation authorities ultimately banned research on television, radar, and other radio-related activities as well. A well-known story has it that GHQ went so far as to have Japanese cyclotrons destroyed.

Japanese researchers had no idea whatever about what to expect for the dismemberment of wartime scientific mobilization or about the extent to which scientific activity would be authorized under the occupation. It was widely believed that GHQ authorization had to first be obtained to carry out research plans, so researchers flooded ESS ST with letters outlining research projects and requesting permission to pursue them. The authorities responded to each individual, pointing out that there was no restriction on normal, routine research. GHQ in time came to feel the necessity of employing measures to correct misunderstandings about occupation policy.

The question of permission was not the only problem. Researchers also faced at that time the problem of research materials. In the early years of the occupation even tickets for railway travel were best obtained by having the authorization of the occupation forces. As a result, Japan's scientists harbored the vague hope that given the absence of even daily necessities, if they could only get the signature of the occupation authorities by petitioning them directly they could avail themselves of the usual supply routes, or even receive materal from GHQ itself and not have to rely on the black market. Although Japan's scientists were interested in material assistance, as scientists their first desire was to bridge the information gap that had cut them off from foreign developments during the war. Researchers, consequently, flocked to the Scientific and Technical Division seeking, most frequently, information on the American scientific world. Their primary objective then was to pursue the latest developments that had taken place outside Japan. To do so they would like to have traveled to the United States themselves but as this was very likely not possible they hoped to forge scientific ties by inviting American researchers to Japan.

Japanese scientists had many and varied problems for which they sought solutions. The occupation authorities – particularly Kelly and Dr. G. W. Fox – recognized that to respond to these problems was a matter of some urgency. They therefore had scientists create a liaison group of Japanese researchers that later evolved into an organization known as the Japan Association of Science Liaison (JASL). GHQ for its part intended to use Japan's scientific organization to further the objectives of the occupation. To that end it requested that a report be drafted on wartime research activities and ordered a monthly submission of reports on research-related activities. A joint Japanese–American community was also established for atmospheric and oceanic surveys, oil exploration, and similar projects.

Projects of this kind, however, were the fruits of individual, personal contacts between GHQ scientists and their Japanese counterparts and not the products of institutional initiative. As a consequence, those researchers who sought relations with the Scientific and Technical Division represented that part of Japan's scientific community that was competent in conversational English. The concern over whether this group was representative of the scientific world as a whole was a question that dogged Kelly. In order to build a more reliable and meaningful linkage with Japan's academic world Kelly conceived of and attempted to put together JASL as a more rational means of fostering contact. This idea is expressed in GHQ documents as early as April 1946.

From that time until the birth of the Science Council in 1949 Kelly remained the occupation's point man for issues related to the rebuilding of Japanese science. Kelly's superiors, including Brigadier General O'Brien, chief of ST, allowed him a great deal of freedom, generally supporting Kelly's positions and protecting him from outside interference while to a certain extent controlling his activities. In the early years of the occupation, other branches of GHQ harbored the misplaced notion that if Japan's research institutions were strengthened, the wartime mobilization system would be resuscitated and disarmament dead. To deal with important problems of this kind Kelly convened a meeting of various GHQ branches to decide on an initial general policy. More important, however, Kelly used this meeting to win the approval of his proposals from other GHQ divisions and was thus able to proceed immediately after his negotiations with the Japanese.

In his travels around Japan, Kelly developed friendly relations with a number of Japanese scientists. One of these men, Horiuchi Juro* of Kokkaido Imperial University, introduced him to professor Tamiya Hiroshi of the botany department, science faculty, Tokyo Imperial University.[2] With Tamiya, Kelly began discussions on the organization of Japanese science. The living standards of an Imperial University professor had fallen to such an extent that Tamiya later recalled that he had been drawn to Kelly and his plans and was induced to accept responsibility for the reorganization task Kelly envisaged by the mere offer of American cigarettes.[3]

Through Tamiya's efforts a nucleus of mainly Tokyo University scientists, beginning with Kaya Seiji and Sagane Ryokichi and several others, began to be assembled. Quickly Kameyama Naoto and Yukawa Hideki were included as the circle of participating researchers expanded.

Kelly may have originally intended to create only a liaison comittee of Japanese scientists to aid in the execution of the occupation's policies, but in the preparation process it became clear that this modest objective was not sufficient. In his meetings with important Japanese scientists Kelly saw that the existing academic associations had become an impediment to scientific development and concluded that a wholly new organization had to be devised. GHQ, particularly Kelly, must therefore be identified as the catalyst in the ultimate fulfillment of this task: the creation of the Science Council of Japan.

Three academic organizations in postwar Japan were directly affected by this

* Japanese personal names here follow the Japanese style of family name first.

discussion about reorganization: The Japan Academy, the National Research Council of Japan, and the Japanese Society for the Promotion of Science. Since its founding in 1906 the Japan Academy, despite being frequently referred to as an old folks' home, had been an eminent institution. The Research Council, which had been created following World War I to foster international technical exchange, during World War II guided the mobilization of scientific activity. The Society for the Promotion of Science was a government-endowed nonprofit foundation whose purpose was to disperse research funds. Along with eliminating the wartime mobilization system and disbanding the Research Council, questions arose about reorganizing the Japan Academy and the Society for the Promotion of Science and dividing between them the functions of the Research Council. The reorganization activities were moved forward by the Japanese themselves under the direction of Nagaoka Hantaro, the director of the Japan Academy.

On the fifth and sixth of June 1946, Kelly held the first meeting of the Japan Association of Science Liaison. Kelly seems to have used this organization and this opportunity to embark on his major objective of reorganizing Japan's technical and research structure. Kelly, in a letter of June 22, indicated the extent of his interest in this reorganization. He suggested to Tamiya that he should

> write a letter from the Japan Association of Science Liaison to the National Research Council in America telling what [the] functions of the organization are, what the problems are confronting the organization and what contact you would like to have with the National Research Association and what kind of aid you need from the American scientists. Also if you would like to have a member from NRC come over here not for just a few months but so that there will always be one member in Japan.
>
> I don't think we in America appreciate the problems here. If we could get some indication of the eagerness of the Japanese scientists to help your country and cooperate. I think American scientists would probably be in a much better mood to help out in this situation if they have this information. They know nothing except what is read in the papers. I will go back and see the NRC and ask them if they will help.[4]

Meeting the same day, Kelly and Tamiya discussed the proposal that had been outlined in the letter. The following is a portion of the transcript of their discussion.

K: Do you think it is a good idea?
T: Yes. But there is the National Research Council of Japan and also the Imperial Academy.
K: I would not worry about that. There is too much confusion now and I will tell them that they need help to get out of the confusion.
T: You see, we have stated our interests as messengers and our duty is only to transport the message.
K: I disagree with you.
T: A letter from our organization would not be representative of the Japanese scientists as the Japan NRC and Imperial Academy of Science are representative of Japanese science.
K: They are not doing the job right. I don't think they are doing very much good.

T: Then I will try to write a letter as you said.
K: How about writing it from the Japan Association of Scientific Liaison and telling them the group is representative of Japanese scientists?
T: We cannot do that, we are not intended to represent Japanese scientists.
K: I think the time has come when we have to come out in the open – we are trying to be that representative. That is really what we are driving to do.
T: Well, this group, the NRC and Imperial Academy has accepted our association with limitation: that it should not intend to reform Japanese science.
K: Those organizations are not good, they need reorganization. Therefore, it is up to your group to help reorganize this group.
T: Our present group is far from it. We have the hope, confidence and worth to be called active scientists but among such a political situation, it is not yet time to express this thought. It is too idealistic.
K: If anything is good, it is not too idealistic. Do you think it is a good idea?
T: Yes, but it takes time, but the name is a poor one if it is this idea.
K: It is necessary today that this be done gradually – not too quickly because if done quickly, you will do a poor job.
T: It will be easier if you yourself grasped the Japanese scientists' leg and organized them. I do not think this group is representative of Japanese nor does it have history – only 40 members.
K: Even if there were five members it could represent a nation.
T: Yes, many of the members of this new association do represent Japanese science – for example, Yukawa, Horiuchi, but then others are not.
K: Why do you think there is no history? This organization has history and is going to make history.
T: We started only one month ago. Frankly speaking, you Americans always judge only a cross section of these people and culture.

Kelly the New Dealer is clearly in evidence in this conversation. Secure in his authority Kelly moved in impatient pursuit of radical, rational renovation without regard for existing institutions.

Tamiya felt obliged to produce the letter to the American NRC as Kelly requested. By the following day a draft was completed. During a discussion at the JASL meeting of July 9, however, the question of the organization's scope and significance was, once again, a point of contention between Kelly and Tamiya. Tamiya's concern was that by overcomitting himself to Kelly and his proposals he would lose the support of Japanese researchers. He feared that if he courted GHQ privately and strengthened JASL's position without its having obtained the authority or right to represent Japanese scientists as a whole its actions would be criticized and its negotiations repudiated.

Kaya Seiji's view of the issue was quite different. In contrast to Tamiya's cautious attitude at the July 9 meeting, Kaya aligned himself with Kelly and expressed his agreement with Kelly's demands. Kaya, as one observer has noted, possessed no unified, coherent body of thought. He was a pragmatist. He acted on the basis of a cogent analysis of existing conditions and possessed the flexibility necessary to achieve compromise.[5] Among those scientists and engineers who had been a part of the wartime research mobilization program, Kaya became a representative postwar technocrat attempting to find a voice for Japanese scientists at the policy-making level of government and he ultimately achieved the greatest prominence in this effort.

The year 1946 was the most severe period of postwar Japan's economic and food crises. Like David Lilienthal, the first head of the Atomic Energy Commission and the planner of TVA, who after the depression urged economic rebirth through science and technology, Kelly suggested at least as early as April 1946 that GHQ "encourage only those research programs which are directed toward improvement of Japan's economy." In an internal policy memo Kelly stated that

> one of the immediate objectives of the Occupation is to insure that the Japanese secure for themselves the necessities of food, clothing, shelter and education. The scarcity of raw materials requires that the maximum use be made of such materials as are available and that substitutes be developed. Research should be directed toward these ends.

Fundamental research under such circumstances as Japan was then experiencing could only be considered, in Kelly's view, a "luxury." Kelly did not propose that basic research be forbidden but rather that Japanese scientists be encouraged "to accept their obligations in this emergency." Later, when these views were put before the JASL the members were generally apathetic. One reason for this response was that Kelly's initial contact with the Japanese scientific community was limited to the elite group of pure scientists at Tokyo University who were incapable of such an approach. A second reason was that Japanese researchers in general had lost their vitality and self-confidence in the postwar period, making such a course impossible. Among them, however, the views of Kaya and Kameyama Naoto, who later became the first president of the Science Council, were closest to the New Deal line laid down by Kelly, and they and their group later became the founding nucleus of the Science Council of Japan.

THE DEPARTURE OF MEIJI ACADEMISM

Kelly returned to Japan from the United States in late September 1946. On September 28 he convened a meeting of JASL members representing the existing academic organizations along with the relevant GHQ sections and from each group demanded the presentation of a proposal for the reform of Japan's scientific and technical infrastructure. Kelly, however, from early on, had decided to push for the evolutionary development of JASL itself.

It quickly became apparent, however that a movement had emerged that sought to defend and indeed broaden the functions of the existing research organizations. Members of JASL led by Tamiya, Sagane, and Kaya met with Kelly on November 15 to alert him to this problem. Kelly, determined to check such sentiments, met with the head of the Imperial Academy, Nagaoka Hantaro, on November 18. Kelly attempted to impress upon Nagaoka the urgency of the problems facing Japan and Japanese science and the need to quickly and effectively reorganize scientific and research institutions. The following discussion then ensued:

N: The problem is how to organize the center for directing research and we hope that the Academy would be the best place and we are now going to have new young members – 150 in number.
K: But that may not be the complete answer.
N: We are going to hold conventions and discuss the matter. We had a conference with NRC and also with the Society for Promotion of Science. The NRC wishes that 300 new members should be elected to the Academy but that was too much. We discussed in the Academy that the number should be limited to 150 and not for whole life term but [a member's] term would be four years.
K: That doesn't mean anything to me. Have you thought this way – What is the function of organization of science? Have you talked about it? Now is the time to discuss that. No where else in the world do they have the opportunity to do something about it that you people have.
N: Since the establishment of the Society for the Promotion of Scientific Research, the number of research items is about 300; that involves all the researches that can be undertaken at that time which was most useful to the country.
K: That still doesn't mean anything to me. That is still not worrying about the basic problems which are facing Japan and the world – not the number of research items or papers – the question is what the function of a research organization is and what is the responsibility to other countries, to your people, and to the world.
N: That is national and international.
K: I have seen no discussion of that; you are just shuffling things around and nothing will happen. Science today has a frightful responsibility in the world and we can't just talk about it in this way.
N: It is a rather difficult problem and everybody will have his own opinion and they must be connected together and restudied.
K: But you're doing nothing about it.
N: Well, we are always considering that problem.
K: Your proposed changes have no consideration of these problems.

The difference between the two men stemmed largely from the fact that to Nagaoka the conduct of research was not controllable and was best left to the spontaneous creativity of individual scientists. Kelly, on the other hand, opposed the individualistic attitude of Japanese scientists and criticized them for their inability to cooperate to solve larger social problems. He was therefore clearly warning Nagaoka of the reorganization of Japan's scientific community by GHQ – or in Kelly's words, with GHQ as a "catalyst" – as a means of dealing with the problems that confronted Japan.

At this time Nagaoka, who was more than eighty years old, and Kelly, who was not yet forty, were separated in large measure by a difference in generations. Nagaoka had studied in Germany and was the representative in Japan of the highest tradition of German academic scholarship. He was also one of the first Japanese scientists to be involved in and compete in international research front activities. Japan's Nobel laureate physicist Yukawa Hideki is reported to have said that "without Nagaoka Hantaro we would not have amounted to anything."

It is for his role as a pioneer of Japanese theoretical physics that Nagaoka is today remembered.

Nagaoka, however, also typified the Japanese legacy of nineteenth-century German academism. It is erroneously believed that the Japanese national universities were modeled after the German university system. German universities' philosophy faculties, however, possessed in general institutional freedoms and organizational flexibility absent in the Japanese system. What Japan's universities did resemble most closely was the Prussian bureaucratism rooted in law faculty graduates. From around the 1880s the upper reaches of the Japanese governing bureaucracy came to be dominated by graduates from the Faculty of Law at Tokyo University. As a result Japan's national universities and its scientific community resembled the national government's organization and were subordinate to it.

Nagaoka disliked bureaucratism and formalism and as a result was continually uneasy about the strong bureaucratic strain in the academism of Japan's universities. He had participated in the highly elite late nineteenth-century German academic world, which at that time was on the eve of the quantum theory and which was an academic structure he felt to be ideal. To Nagaoka Japanese academism, while differing in time and place, was not very far from this ideal type.[6] During the war Nagaoka criticized the National Research Council of Japan, which conducted academic research under the auspices of the scientific mobilization program. At that time he emphasized that what is called research is less a function of organization and faculties than of human beings, the originality of individual creativity. Such views were the very essence of the scientific ideals of German academism. Given Nagaoka's ideals it is quite natural that he held opinions at variance with Kelly's New Deal approach to scholarship and science. Even such approaches as those taken by J. D. Bernal's *The Social Function of Science* to Nagaoka represented a strategy identical to that of the wartime mobilization system.

FROM THE PREPARATORY COMMITTEE TO THE RENEWAL COMMITTEE

The model for Japanese scientific organization considered by Kelly was the American National Academy of Science (NAS). Like Japan's Imperial Academy, NAS was an honorary institute. But unlike its Japanese counterpart NAS was more a working institution. Responding to Woodrow Wilson's call NAS organized the National Research Council for the purpose of mobilizing wartime research activities during World War I. NAS's members were limited to natural scientists, excluding engineering, the social sciences, and humanities.

Following the war Japan's established scientific organizations – the Imperial Academy and other established scientific groups – had lost much of their strength and as a result their ability to resist overhauling. Kelly was convinced he could build an organization of unprecedented strength and effectiveness without interference or intrusion from existing power bases. It was felt, therefore, that if a complete organizational renovation were to take place this was the best time. Given the occupation's role as a "godfather," Kelly developed the ambition to leave a lasting legacy in the history of Japan's scientific policy. For Kelly and for the other New Dealers of the occupation it was an opportunity to grapple with and bring to a successful conclusion the great problems – such as agrarian reform – that had become impossible in their own country. It is this sense in which Kelly's activities too must be seen.

Kelly's first task was to try to eliminate the existing scientific organizations. The organizations for their part, recognizing the circumstances that prevailed under the occupation, attempted to come up with their own plans for internal reform. These proposals were essentially hollow and were put forth in an attempt to forestall GHQ intervention. Kelly, however, took no notice of the organizations' considerations. He wanted to give official sanction to JASL and make it over into a more public forum. JASL began, however, as a private group centered on Kelly. Nishina Yoshio and others were very concerned by the complaints that had arisen in the scientific community about JASL's symbiotic relationship with GHQ and its reliance on the prestige and the power of the occupation forces. On October 16, 1946 they visited GHQ and warned the occupation authorities once again about this matter.

The Imperial Academy, despite its geriatric character, did possess the machinery for electing new members and it operated under formal rules. JASL had been provided with no such authority. JASL was criticized as being an undemocratic organ. To offset these complaints the more publicly visible Preparatory Committee (Sewaninkai) was created. This group acted as a supervisory board overseeing the election of members to the Scientific Research Organization Renewal Committee, insuring that the election of members was conducted democratically and that the elections were representative of the Japanese scientific community.

The task of scientific reorganization moved step by step from JASL to the Preparatory Committee to the Renewal Committee, and in so doing became more public and formal. Kelly, who had originally conceived of JASL as a private intermediary organ, was unable to control the more public Renewal Committee. Japanese scientists had pushed for this opening of the reorganization process and Kelly could do little but leave its details in their hands. At the time of JASL, for example, it was Kelly's intention to broaden the organizational base from basic research to applied sciences such as engineering, medicine, and agriculture. A JASL expansion of this kind Kelly could control.

With the rejection of the reorganization proposals drafted by the Imperial Academy and the Society for the Advancement of Science, however, organizational reform plans were devised in which the humanities and social sciences would be included. Such a conception was quite contrary to Kelly's American Academy model and represented an institutional expansion he was unable to control. At the same time the Ministry of Education's Scientific Education Division Chief, Shimizu Kinji, called for intermediaries representing the seven faculties of Tokyo University: law, letters, economics, science, engineering, agriculture, and medicine. On January 17, 1947 the Preparatory Committee was launched with forty-four members selected in accordance with the Tokyo University seven-faculties model. The development of the Preparatory Committee based on Shimizu's "Scientific Organizational Reform Plan" was a situation with which Kelly was not wholly content. He expressed his dissatisfaction in a handwritten memo of August 18, 1947 (presumably to be addressed to the visiting Scientific Advisory Group).

1. Sewaninkai [Preparatory Committee] created contrary to our ideas and members appointed by Mombusho [Ministry of Education].
2. [Sewaninkai] never submitted election procedure until too late to alter.
3. Procedure sounds good on paper but wrong in practice for following reasons: The members of the Executive of the societies are mainly old school, nominated old school people as electors, continuing process of old school ties until Renewal Committee [is] comprised mainly of them.
4. Mombusho, having such a hold, made representation of social sciences strong.
5. The Renewal Committee being comprised of a group Tokyo University men, it is inevitable they will fight to get an organization in which they retain their privileges and power.
6. The younger scientists and those from the universities other than Tokyo, not having respect for the Renewal Committee, will not abide by its findings (no matter how good they are) although they will pay lip service.
7. If we have to apply pressure to ensure findings of Renewal Committee are in accordance with our ideas we lose all the force of our policy to let the Japs decide for themselves: better to ensure that Renewal Committee is truly representative by assisting with the procedure for election – give the Renewal Committee some basic principles and let them ride, merely keeping a watch briefly to see they keep within the scope of the enunciated principles.

The Preparatory Committee began the process of electing members of the Renewal Committee in early April 1947. The election was completed by August 10 when 108 members had been selected.[7] On the American side the Scientific Advisory Group, which was sent to Japan by NAS to "advise General MacArthur's staff on the decentralization of Japanese scientific research and to evaluate the proposals considered by Japanese scientific organizations," arrived in Japan on July 19. The arrival of this group represented the fruit of Kelly's visit to the United States the previous year.

After its arrival the advisory group visited universities and research groups in various parts of the country. The group submitted its report on August 28 and on the following day returned to America. Kelly, who acted as the group's guide during its stay in Japan, explained to it the activities of the Preparatory Committee. Inevitably, Kelly's criticisms of the committee – which were expressed so clearly in his memo – found their way into the advisory group's report.

There was also criticism from a number of Japanese about the absence of democracy in the Preparatory Committee's activities. Comments on the preservation of the influence of the Tokyo University clique on the Renewal Committee appeared in various newspaper editorials of the time, which shows that Kelly's opinions were not unique to him.[8] It might be pointed out that since the adoption of the seven-faculty organization by the Science Council of Japan it has been very difficult to win support for newly developed areas of scholarship. The possibility of an organizational flexibility that would allow for the creation of new divisions to accommodate new knowledge was absent from the Preparatory Committee's thinking on the issue.

Both Kelly and the American Advisory Group reluctantly authorized the inclusion of social science in the incipient scientific organization. Kelly's reorganization efforts began with the sciences and broadened to include the applied sciences. This in large measure explains the organization and structure of JASL. The representatives of the humanities and social sciences on the Preparatory and Renewal Committees were not as interested in the work of the committees as were the natural scientists. As a result, representatives from the science area attended committee meetings in the greatest number followed by engineering, agriculture, medicine, humanities, law, and economics. Moreover, later at the first meeting of the Science Council the percentage of humanistic area representatives who voted was low compared to the natural sciences. Several factors may account for this difference in interest. For example, unlike the natural sciences, there was no national learned society in the humanities or social sciences. Also unlike the physical sciences, which moved toward reorganization at the time of JASL, discussion of organizational reform of the humanities was largely absent. Finally, the natural science division, which was the birthplace of JASL, was most reform minded and the average age of its representatives was the lowest.

GHQ RESPONSE TO THE RESEARCH RESTORATION COUNCIL

Of the various organizations of left-wing scientists the most well known is the Association of Democratic Scientists, an umbrella group for leftist scientists and science organizations. The association was launched as early as January 1946 but does not appear in ESS ST documents on scientific organization until much later.

On June 10, 1947, members of the Preparation Committee (Junbi Iinkai) for the Research Restoration Council met with Kelly and General O'Brien, head of the Scientific and Technical Division. The chairman of the Imperial Inventions Association, Ono Shunichi, acted as the group's spokesman. Ono was a skilled engineer – he had graduated in electrical engineering from Waseda University – and was fluent in Russian. According to friends of that time, his pleasant personality enabled him to deal with people of all ideological stripes.

The creation of the Research Restoration Council was sought as an organ designed to complement the Economic Restoration Council, made up of labor and popular groups. It was to comprise scientists' and technicians' unions in private, governmental, university, and industrial research laboratories. Its intention, announced in a petition, was to put pressure on Kelly and O'Brien, under whose authority the reorganization of science, it was believed, was being carried out. Shortly before the meeting between ESS ST leaders and the representatives of the Preparatory Committee the new socialist government of Katayama Tetsu had been formed. To Ono, the opinions of the Economic Restoration Council were not listened to during the former Yoshida Shigeru cabinet. With the change of government, however, he believed that the council's views would be given their proper weight. Being a Socialist Party member himself, and having a number of friends in the new cabinet, Ono felt inclined to apprise Kelly and O'Brien of his potential influence. Thus for Ono the inauguration of the Katayama cabinet was a welcome event and represented a propitious time for his group to force itself on GHQ.

GHQ for its part was greatly concerned with the activities of the Economic Restoration Council and in the preparation process of the Scientific Restoration Council. It was, however, bogged down with conferences on the Science Council organization issue. So it was Ono who first approached Kelly. He pointed out the Preparatory Committee (Sewaninkai) was merely a collection of Tokyo University professors. It did not, he argued, reflect the opinions of researchers in other universities, government and private laboratories nor did it represent the views of younger scientists. In response Kelly stated:

> One of our wishes is that the younger scientists and technologists be represented and if they are not, there are two things to do, find out how they can be better represented in this group, Sewaninkai, than they are now. We would like to hear your suggestions. But I should also like to point out that if you want to set up a group like this you would have identical problems. You too would be criticized for not being representative, and I think you can help most by finding the best way of amalgamating and finding the best way to see that these younger scientists are represented. I am surprised that they do not have a vote in the Renewal Committee.

Ono then proposed a union of the preparation committees of the Renewal Committee and the Research Restoration Council. Giving O'Brien and Kelly a list of organizations that participated in the Restoration Council preparation process he introduced the members of the Preparation Committee who represented those bodies. The following discussion then took place:

K: We don't know why you need this group as there is already the Sewaninkai and the Renewal Committee.

Ono: One point is clear that when the ultimate organization of science will be established, the Council for Research Restoration will amalgamate so you can think that this Council is of temporary nature, but instead of establishing the final organization of science in Japan, it is clear we must prepare and endeavor in such manner that Sewaninkai will not eliminate the necessary younger strata and minor elements who are very essential for the final.

K: Your job is to see that Sewaninkai and the Renewal Committee do a good job.

From Kelly's point of view consultations with other organizations would simply cause confusion. As a result, despite their limitations, he believed there to be little choice but to work through the Preparatory Committee and the Renewal

Committee. Although he had complaints about the direction of the Preparatory Committee and its ability to fulfill its functions, Kelly bureaucratically rejected the appeal of the Research Restoration Council. He then left the meeting and the conversation was carried on by O'Brien and Technical Division chief Colonel Allen. To Ono's group the Preparatory Committee's expressed methods of carrying out the election were not democratic but oligarchic and underlined the committee's lack of responsiveness to younger scientists and those in industry in particular. Beyond criticizing the Preparatory Committee, however, Ono and his group derived no satisfactory results from the meeting.

The quoted conversation between Kelly and Ono perhaps gives the impression that GHQ regarded the Research Restoration faction lightly and that it felt it could deal with the group as it wished. In fact, however, if occupation documents are examined carefully we see that Ono and his group made a very strong impression on Kelly, the result being that from this time research was started and a Research Restoration Council file began to be assembled.

Even within the Scientific and Technical Division, assessments of the Research Restoration Council differed. A memo written by Allen and circulated in mid-June suggested that if the Preparatory Committee (Sewaninkai) were likened to an employers' group GHQ should then provide appropriate proportional representation on the Renewal Committee to the Ono faction. If this were done the Research Restoration Council, for which Ono spoke, would renounce its idea of trying to control the Preparatory Committee. Moreover, the memo argued, such a course could serve to create a mediating organ for the two sides within the Economic Restoration Council.

A memo authored by Kelly countered, saying there was no reason for confusing matters by comparing academic reorganization with a labor–management contract. He stated that GHQ's goal was the creation of an organization of scientists to facilitate research, not the establishment of a political organization. To deal with the issue as a political question, he argued, would not be very useful in prodding Japanese scientists into action.

Both Allen and Kelly were invited to attend a meeting of the Research Restoration Council held on June 30 in the Science Museum in Tokyo. Kelly's speech to the assembly emphasized the necessity of conducting scientific reorganization through the Preparatory Committee. Pursuing other approaches at that point, he claimed, carried the danger of disrupting the reorganization process. Allen in his speech was more sympathetic to the Restoration Council.

Ono himself recognized that it was too late to reorganize the Preparatory Committee. Indeed, the new Renewal Committee, chosen in accordance with the original Preparatory Committee formula, met for the first time on August 25, 1947. On the following day a departmental memo penned by Kelly stated, "We have diplomatically and tactfully announced to the science Restoration Council that we support the Renewal Committee alone." On September 5, General O'Brien finally spelled out the division's policy concerning the problem of Ono and the Restoration Council and in doing so affirmed the role of the Renewal Committee. He said that the questions brought by Mr. Ono and the Research Restoration Council are ones that must be decided by the Japanese themselves. The inaugural message of the Renewal Committee was endorsed by both the committee and Prime Minister Katayama. It should, therefore, henceforth begin its work. Opinions on the scientific reorganization issue which are presented to the Renewal Committee are not the province of GHQ. GHQ will involve itself only in situations from which the Renewal Committee abstains.

In October 1947 the Association of Democratic Scientists submitted a plan for the renewal of scientific organization to the Renewal Committee. The plan demanded the creation of a high-level conference of scientists whose advice and opinions the government would be obliged to follow. According to ESS ST documents (particularly a February 23, 1948 report from Paul Henshaw to Kelly) this conference plan was Ono's way of responding to the Renewal Committee. But, the report said, the plan "has been completely ignored and has in no way influenced reorganization activities so it need not concern us."

The Renewal Committee, which had been formed in August 1947, held a number of general conferences. GHQ, however, received only intermittent progress reports on its reorganization deliberations and had no opportunity to influence them. In late January 1948 Prime Minister Katayama sent a note to Renewal Committee chairman Kaneshige Kankuro proposing that the reports be terminated until the end of the fiscal year on March 31, when preliminary investigations of the issue had been completed.[9] The Renewal Committee sought GHQ's approval for such a plan. At this final stage GHQ held a series of staff section meetings beginning from March 1 to study and finalize GHQ's position. In a memo of March 19 General O'Brien informed ESS chief General Marquat that agreement on the broad principles of scientific reorganization had been obtained from all GHQ sections. These principles were formally approved by the various sections on March 23.

Concerning this period, Tetu Hiroshige has argued that "between the seventh (February 23–5) and eighth (March 25–7) conferences of the Renewal Committee there must have been powerful pressure placed on the Renewal Committee by GHQ."[10] This assertion is borne out in GHQ records of the staff section meetings of that time. One of the clearest expressions of this problem can be seen in the meeting of March 10. The main subject of this particular meeting was the relationship between the government and the Science Council.

The first to speak was the government section representative, Mr. Porter. He complained that the structure of the scientific advisory body as proposed by ESS ST was becoming unnecessarily complex and that what it seemed to desire was an organization that would coordinate Japanese scientific activity. Porter suggested that an advisory organ appointed by the prime minister alone might be a simpler and better alternative. Kelly and General O'Brien of the Scientific and Technical Division, however, felt it necessary to preserve the organization's independence from government and to maintain the dignity of scholars. To achieve these ends they expressed their wish to build an advisory body similar to the British Royal Society or the American NAS. Thus, it was necessary to avoid government appointment of members and government-paid salaries for them. If not, the members would be unable to criticize the government or resist its actions. Moreover, they suggested, the organization's dependence on government funding might be unconstitutional.

Colonel Johnson from the public health and welfare section brought up another question. He wondered what the objection was to having the prime minister appoint members from a list prepared by the scientific electorate as Porter had proposed. Rather than reducing, he asked if such a procedure might not in fact enhance the organization's prestige. On this question the following discussion took place.

O: No, you have 500 members elected. You have 50,000 people electing the members. You then present them . . . qualified people should pick the best 210, not the Prime Minister. [Cites Ono and his group of political ideologists. If PM is to retain party support he would have to elect Ono's group, if they were in power] . . . let them be elected by the scientists themselves and not have 210 members culled out for political reasons.

K: I understood you to say we would like this to have scientific prestige and not political prestige.

P: You say represent Japan in international science . . . it has both the weight of the scientists and the [government] back of it. Would it not have more prestige?

K & O: Political pressure groups like Ono's would influence the Prime Minister. They are interested only in political ideologies and trade union activities which do not fall into this category. I think you would lose [a] tremendous amount by taking [the] Prime Minister into this purely for the securing of funds, when it may be unconstitutional.

On March 10, when discussions on the nature of the scientific organization were taking place, the new Ashida Hitoshi cabinet was formed. On February 10 the Katayama cabinet fell, creating a political vacuum and a long interregnum imbued with considerable confusion. It was under such circumstances that the GHQ debate on government–science council relations was taking place. Kelly's and O'Brien's greatest concern was that Ono, who had earlier flaunted his relationship with the Katayama cabinet, might influence the government, pressuring it into accepting a left-wing takeover of the Science Council. To prevent this they sought to restrict the government's ability to intervene in the Science Council's activities. The draft law for the Science Council of Japan, presented to GHQ by the Renewal Committee following its seventh meeting, included the provision that the "government should refer the following matters to the Japan Science Council." On the ESS ST copy of the draft law the words "should refer" were underlined in pencil and in the margin a question mark was written. In a revised copy of the draft, this passage was changed to read, "The government shall seek the opinions of the Science Council of Japan on the following. . . ." The word "shall," however, was later crossed out and the word "may" written above it. This correction appeared in the final draft and was later adopted in the approved Science Council Act.

The results of this process are well known to Science Council members. Council recommendations, having no binding authority, have steadily come to be ignored by the government, a development that defines the course of the Science Council of Japan.

CONCLUSION

In the early GHQ files on academic reorganization a name that frequently appears is Watanabe Satoshi. Watanabe bitterly denounced the old Japanese establishment, giving forth views highly critical of Tokyo University, the Ministry of Education, and the existing scientific organizations. Watanabe's opinions can be considered to have influenced Kelly's views of the Japanese scholarly establishment. Thus Kelly paid little heed to the desires of the established scientific groups. Instead he believed it necessary to eliminate them. Some other Japanese group must have assumed a central role in the process of reorganization. At the outset that role was played by the Japan Association of Science Liaison (JASL), Kelly's private, scientific advisory group. Later, the principal position was expanded to include the Preparatory Committee (Sewaninkai) and the Renewal Committee, and was finally assumed by the Science Council of Japan. The most important activists in the Science Council were among others Kaya Seiji, Kameyama Naoto, and Kaneshige Kankuro. It was these men who would become the spokesmen and technocrats for the later rapid growth of Japanese science and technology.

Kelly and other members of ESS ST had a keen interest in the activities of Japan's scientific community. If we are to judge from the data in GHQ files, however, their sources of information were surprisingly limited. Of the very large and influential Association of Democratic Scientists they were largely ignorant and they were unprepared to accept recommendations from Ono's group, which they perceived as threatening to the reorganization process they sanctioned.

Of course, the threat of Ono's pressure alone was not sufficient to force Kelly to reduce the government's role in the formation of the Science Council. Kelly must have had other sources of information. Moreover, the Japanese Renewal Committee strongly opposed the "popular front" line of the Association of Democratic Scientists societies. Nevertheless, the appearance of the Ono group at the same time as the Socialist Party cabinet of Katayama Tetsu had a profound influence on Kelly's thinking. The consequence was, ironically, that the Science Council, freedom of thought, and research autonomy were protected from the government while at the same time they were also insulated from leftist influence. In this way Kelly effectively isolated the council and reduced its significance.

NOTES

1 This study is based on documents found in the Science Reorganization and the Scientific Renewal files of the Scientific and Technical Division, Economic and Scientific Section, GHQ. These materials can be found in the National Recording Center, Virginia.

2 Tamiya Hiroshi, "On the Tenth Anniversary of the Founding of the Science Council of Japan" (Japanese) *Gakujutsu Geppo,* no. 5 (May 1959), p. 5.

3 Ibid.

4 This was shortly before Kelly's return to the United States for summer vacation. He hoped to use this trip to get in contact with American research institutions. He had earlier argued that it would have been preferable if the Japanese case were brought to American research groups by Japanese scientists themselves. As a result, to strengthen the appeal, he tried to persuade Tamiya to accompany him. Tamiya, however, proved unwilling.

5 Aochi Shin, "Kaya Seiji" (Japanese) *Chuo Koron* (January 1958).

6 Itakura Kiyotsugu et al., *Nagaoka Hantaro* (Japanese) (Asahi Shimbunsha, 1963), p. 540.

7 Nihon Gakujutsu Kaigi, *Twenty-five Year History of the Science Council of Japan* (Japanese) (Nihon Gakujutsu Kaigi, 1974), p. 265.

8 *An Outline of Scientific and Technical History,* vol. 5 (Daiichi Hoki, 1964), pp. 137–9.

9 Nihon Gakujutsu Kaigi, *Science Council of Japan,* p. 269.

10 *Scientific and Technical History,* vol. 5, p. 128.

19

The worm in the core

Science and general education

PETER S. BUCK
Massachusetts Institute of Technology
BARBARA GUTMANN ROSENKRANTZ
Harvard University

There is no more common belief than that nature holds lessons for us all. As every schoolchild knows, nature is studied in science class. From these common views science teachers invoke a scholar's creed for the young: To understand nature, learn science! The pedagogical maxim fit well with late nineteenth-century wisdom assuring the accessibility of scientific facts. Experiments that allowed students to classify the plants and animals, measure the acceleration of falling bodies along inclined planes, or combine acids and bases to produce salts could be conducted in any classroom. Such demonstrations added to each child's store of knowledge, they provided compelling examples of the scientific method at work, and they gave evidence of the predictable order of nature.

Over the past one hundred years the place of the natural sciences in the common curriculum has become more problematic. Our concern here is with how this happened at the college level. We begin by asking what was taught as science and to what end. Our focus is on the understandings of the natural sciences that emerged with the division of the curriculum into three parts: the humanities, the social sciences, and the natural sciences. We show that throughout the twentieth century, every attempt to include science as an essential component of the common curriculum has foundered on disagreements about the bearing of scientific knowledge on human affairs.

The essential difficulty was prefigured in the remark of a nineteenth-century American college president who, seeking to distinguish his institution from a "technical school," observed that "the students who come here are not trained as chemists or geologists or physicists. They are to be taught the great fundamental truths of all sciences. The object aimed at is culture."[1] In the twentieth century, culture has remained the object, and the fundamental truths of all the sciences the subject, of that portion of the general education curriculum in liberal arts colleges assigned to the natural sciences. An underlying symmetry has been assumed among the humanities, the social sciences, and the natural sciences. As the aims and structure of general education have been formed and

This essay was written while Professor Rosenkrantz was a Rockefeller Humanities Fellow and revised while she was a Scholar-in-Residence at the Rockefeller Foundation Study and Conference Center, Bellagio, Italy.

reformed, counterpart courses in these fields of knowledge have been designed to teach both the rudiments of subject matter and the underlying intellectual tradition in each area.

The term "general education" was widely invoked in the early 1800s, but under conditions that were the reverse of those that have surrounded it since World War I. At the start of the nineteenth century, the American college curriculum was limited in its scope, wholly prescribed, and weighted heavily toward classical studies. Most students were preparing for the professions – law, medicine, or divinity – and their studies were broad but lacked depth in areas other than ancient languages and mathematics.[2] Even with the introduction of the elective system, after the Civil War, general education did not seem to need special attention. But during the next half century, three things changed in American colleges: the students, the faculty, and the state of knowledge.

In 1870 less than 2 percent of the nation's seventeen-year-olds graduated from secondary school and went on to college. As the percentage graduating from high school increased to 3.5 in 1890, 8.8 in 1910, and 16.8 in 1920, college administrators and teachers found it reasonable to assume that the social and ethnic character of their students was also changing.[3] Their sense of educational upheaval was heightened by an increase in their own numbers: In 1870, for example, the Harvard College catalog listed thirty-two professors teaching seventy-three courses; by 1910 the figures were 169 professors and 401 courses. At Columbia thirteen professors taught forty-six courses in 1870; by 1910 the faculty had grown to 182, the number of courses to 673, and the president of the university was expressing serious concern about "the lack of adequate and frequent personal touch between undergraduate and teacher, [and] the failure of the Faculty to inform itself sufficiently as to the content and method of the several courses prescribed by it for the Baccalaureate degrees."[4] Against such a background, it seemed important to compensate for social heterogeneity by emphasizing a common cultural heritage. This meant emphasizing what the major branches of knowledge had in common, at a time when the proliferation of disciplines might have led to a fragmented curriculum.

But elements common to the different fields of learning proved difficult to find. Disciplinary distinctions evolved according to different logics and with different consequences in the humanities, the social sciences, and the natural sciences. The subjects taught in the humanities embodied a tradition that explicitly demonstrated the connection between knowledge and values. In both the humanities and the social sciences, efforts to maintain or create coherence assumed an intrinsically meaningful subject matter. Content and method were not sharply distinguished in order to claim special significance for a particular mode of inquiry. The search for general principles that transcended the specific disciplines was not pressed with conspicuous vigor or marked success. At the level affecting general education, knowledge in the humanities and the social sciences grew horizontally, bringing esoteric topics rather than more specialized information on any specific subject into the common curriculum: Just as Charles William Eliot's classification of " 'elementary knowledge' " into " 'four great subdivisions' " – language, history, mathematics, and natural sciences – evoked images of circles to be divided rather than pyramids to be erected, so too in the twentieth century the advancing curricular frontier moved across new territories, not deeper into old ones.[5]

The physical and biological sciences were expected to fit the same mold and simultaneously provide a comprehensive understanding of nature; an equally comprehensive understanding of human beings and their place in nature; a record of achievements to be studied with profit by everyone interested in progress; and a method of inquiry suitable for general use in interpreting not only the natural world but society and human behavior as well. The effort to realize the first of these premises, however, generated an enormous body of knowledge that significantly altered the meaning of the other three. The proliferation of scientific disciplines required to describe and develop new knowledge undermined the reality of nature as the unified subject matter of generic science. It made it difficult to agree about the essential characteristics of the natural world in which men sought their proper place. It threatened to reduce the history of science either to "rational reconstructions" of scientific inquiry or to jumbles of discoveries with little bearing on each other, much less on the human condition. And it engendered descriptions of scientific method that practicing scientists could not recognize in their work.

These consequences of the growth of scientific knowledge are what made the incorporation of science in the common curriculum so notoriously hard to handle. But they did not shake the universal agreement that in principle the natural sciences should participate fully in shaping educational processes and objectives. Our initial survey indicates that proponents of general education asked surprisingly few questions about the grounds for that consensus. They seldom wondered whether physics could "stand in" for biology in the general scheme, for example; or whether there was an underlying animus shared by all the sciences; or whether, if one existed, it could be grasped apart from the actual doing of science.

American colleges devised various strategies for dealing with general education. Once having acknowledged the obvious, that science explains the world of nature in ways that reflect our culture, each institution accommodated itself uneasily to the implications of that fact. Course content and the organization of the curriculum were determined by local needs and interests more often than by underlying principles. The inclusion of new subject matter in the undergraduate curriculum and the growth of knowledge that occasioned it occurred as the purpose of education changed from preparation for life to training for work. So long as American college students resembled their teachers, the shift in aim raised no problems. But by the turn of the twentieth century, the increased heterogeneity of the undergraduate population had turned the absence of a shared culture into an educational deficit. The closely linked questions of how to overcome that deficit and where to locate adequate guides to principled action were particularly complicated by the expansion of the curriculum to include the rapidly growing and increasingly differentiated natural sciences. Whereas political economy developed comfortably and even flourished within the framework of traditional moral philosophy, chemistry, physics, botany, zoology, embryology, and even mathematics demonstrated that natural philosophy no longer existed.

At a time when it still seemed reasonable to assume that education shaped society through the actions of educated individuals, the contrasting states of moral and natural philosophy posed profound problems. But it was unclear where the source of the difficulty lay: in the proliferation of new disciplines or in the stagnation of older branches of learning. The permanent success of traditional moral philosophy could be called on to lend authority and legitimacy to the new natural sciences, or those new sciences could be used to invest a reformed moral philosophy with previously unobtainable powers. But the two strategies could not be easily pursued together. This was the dilemma that general education programs in the twentieth century faced.

CONTEMPORARY CIVILIZATION

Not surprisingly, the distinction between general education and the general curriculum was first drawn institutionally during World War I, and at an institution – Columbia College – newly conscious of "a social diversity" among its students that made it "not very likely ever again to be a fashionable college *per se.*" Explaining "why we fight" to "the different strains in the membership, particularly the boys of various foreign stocks,"[6] seemed sufficiently urgent to call for a new contemporary civilization. That this was an instant course rather than a Burkean embodiment of cultural traditions, residual, dominant, or emergent, guaranteed its success.

Contemporary Civilization grew out of the "War Issues" curriculum designed for the Students Army Training Corps by a Columbia College committee chaired by the philosopher and graduate dean, F. J. E. Woodbridge. Once the War Issues syllabus was in place, a call for a comparable venture centered on "Peace Issues" led in the fall of 1917 to the first prescribed freshman Contemporary Civilization course.[7] For five days a week throughout the academic year, each freshman was informed "of the more outstanding and influential factors of his physical and social environment." He learned about "the chief features of the intellectual, economic, and political life of today," with particular reference to the ways in which those features depended "on the difference from the past."[8]

Hailed immediately by Columbia's president Nicholas Murray Butler as a notable educational advance, in the immediate aftermath of the war the Contemporary Civilization program was directed toward several obvious student constituencies. Those undergraduates "enamored of the cruder and more stupid forms of radicalism" were expected to benefit from being presented with "the facts relating to the origin and development of modern civilization and the part that time plays in building and perfecting human institutions." Conversely, students "afflicted with the more stubborn forms of conservatism" would profit from gaining an "early appreciation of the fact that movement and development are characteristic of life and that change may be constructive as well as destructive." As for the rest, "the main purpose" of Contemporary Civilization was "to lay a foundation for intelligent citizenship, and to enable undergraduate students to prepare themselves to make decisions concerning public questions with intelligence and with conviction."[9]

Although Contemporary Civilization took a large part of the freshman's time, it made limited staffing demands on the college, taught as it was almost exclusively by faculty from philosophy, history, and economics. In its first decade of operation, it was basically a civics course, offered by men who were scholars enough to "realize how little that is being said and thought in the modern world is in any sense new."[10] That in fact was the central doctrine of Contemporary Civilization, one that was entirely compatible with Butler's insistence that it was "not the purpose of this course to teach or to preach doctrine."[11]

The judgment that there was little new being said or thought was also consistent with a recognition that, nonetheless, the organization of knowledge had changed significantly. The departmental structure of the college and the compartmentalization of intellectual activities that it reflected provided a clear rationale for Contemporary Civilization. The new course was required, because neither the cultural environment from which Columbia students were coming, nor the vocational environment for which college was preparing them, offered sufficient incentives for bringing the traditions of the past to bear collectively and systematically on the tasks of the present.[12] The new curriculum was, accordingly, judged a success after its first ten years when Dean Herbert E. Hawkes concluded that the faculty were "pooling their interests to such an extent that one scarcely realizes that they are representing this or that department."[13]

As Hawkes observed in 1928, the success of Contemporary Civilization made it reasonable to ask whether the same could not be done for the natural sciences. At Columbia College the answer was, at first, not quite, and then, in the end, no. The preceding year President Butler in a section of his annual report titled "Stumbling Science" had forecast the problem.

> Scientific study as an educational instrument has fallen far short of the high expectations that were formed of it and for it when the scientific movement in education began some sixty years ago. Meanwhile the content of the natural and physical sciences has been multiplied manifold . . . The scientific method is everywhere extolled and within certain limits is rigorously applied. Yet the public mind reinforced each year by a veritable army of youth which is marched through scientific laboratories and lecture rooms, museums and observatories, is as untouched by scientific method as if no such thing existed.[14]

Judging by how Columbia's science department proposed to meet their obligations in general education, they too were "as untouched by scientific method as if no such thing existed." Left to deal as they saw fit with the problem of providing instruction for "the man who does not propose to go into medicine or engineering, or to become a zoologist or a physicist," they responded by organizing a sequence of loosely related courses designed to show "what those subjects are about and what place they occupy in the thought of the world today."[15] Insofar as their courses were integrated around a common purpose, it was not to expose students to the scientific method but "to present as systematically as possible those themes of modern science that are of general interest and significance." In principle this was a reasonable way of including the natural sciences in the structure of a general education curriculum that emphasized the historical dimensions of the growth of knowledge. But in practice historical perspectives on the scientific component of contemporary thought proved sufficiently irrelevant to be left to the history department, where the college's only history of science course was taught as an elective. Moreover, in contrast to Contemporary Civilization, which demonstrated the relevance of the past to the present by breaking down barriers between different disciplines and departments, Science A and Science B largely respected disciplinary boundaries. Physics, chemistry, geology, and biology were each taught in separate terms, under titles that only barely concealed their departmental origins: "Matter, Energy, and Radiation"; "Chemical Changes in Matter"; "The Earth, Its Origins and Physical History"; "Living Organisms."[16]

There is no doubt that Science A and Science B showed what physics, chemistry, geology, and biology were about. They may even have conveyed a sense of the place of those disciplines "in the thought of the world today." But they spoke neither of Butler's concern about bringing the scientific method to the attention of the public mind, nor to the demand for historical depth that gave shape to Contemporary Civilization. For instruction in scientific methodology, the college's students had to turn to the social science departments whose curricula, designed "to follow the course in Contemporary Civilization," gave prominence to the "elementary methods," of statistics, that had to be mastered in preparation for "further study of the social sciences."[17] For courses that followed the model of Contemporary Civilization and combined work in the natural sciences with instruction in how their chief features depended "on the difference from the past," undergraduates had nowhere to turn.

In its way, the course on statistical methods in the social sciences was an appropriate response to Butler's call for a curriculum that would display the bearing of science on human affairs. His tacit identification of science with method made it logical for the social rather than the natural sciences to emerge as the disciplines best able to represent generic science in general education. Leaving matters to the social scientists avoided the problems suggested by remarks about the content of the natural sciences having "multiplied manifold." The aim and structure of Contemporary Civilization encouraged the same evasion. For the philosophers and historians who mounted that course, the striking feature of scientific ideas and findings, whether in the past or in the present, was how well they comported with the social and intellectual contexts surrounding them. It followed that the way to understand contemporary science was to subsume it within the historically given and historically transmitted belief systems that comprised the West's inherited cultural capital. But since those belief systems were already accessible in their entirety in texts that related them explicitly and beforehand to the central problems of social life, nothing was to be gained by going through the additional exercise of reassembling them out of the fragments of natural philosophy that could be retrieved from the specific disciplines, before discussing their cultural adequacy in the modern world. Any doubts on that score were effectively removed by the depression, which robbed the question of science in general education of its urgency and made it both necessary and sufficient to focus the resources of Contemporary Civilization on the more pressing issue of "economic security."[18]

THE NATURE OF THE WORLD AND MAN

Although on Morningside Heights the depression of the 1930s ratified the curricular choices of the 1920s, in Chicago identifying economic insecurity as the crucial practical problem of the day was neither a necessary nor a sufficient condition for comprehending how best to organize *The Higher Learning in America.* Although "the professionalization of graduate training and research had moved far enough outside specific educational institutions to place itself on a national basis," the process was far from complete, and its architects remained substantially "dependent upon the individual universities in which they continued to place their professional roots." This was even more true of those who conceived of the university "as a counterbalance to the very professionalism to which it was contributing." They might be "seeking a pattern for the 'general' education of a national elite," but the "localism of American university life successfully obscured the national process under way," forcing repeated consideration of the same basic questions.[19]

The University of Chicago's reorganization was announced by Robert Maynard Hutchins in 1930, more than a decade after the institution of Contemporary Civilization, as if the rejection of arbitrary divisions of subject matter and analysis by discipline were wholly unprecedented. At Chicago, the traditional units in the arts, literature, and science were to be abandoned, because they were "traditional rather than rational or convenient," and the foundations of education were to be laid on "the controlling principles of classification" of knowledge itself.[20] Coming to a college where an elective course on "the nature of the world and man" had been a fixture since 1924,[21] Hutchins studiously avoided making any references to it, choosing instead to proclaim his new curriculum as the embodiment of the lines along which the most advanced scholarly and scientific research in the university was proceeding. "The major task of the University of Chicago is and always will be the advancement of knowledge," he explained; the minor task was therefore to arrange undergraduate education to reflect the patterns of "cooperation in investigation" that the faculty had already developed "to a remarkable degree." This was the particular, compelling claim made for the four-way division of the curriculum that was established in the college: Its parts – biological sciences, physical sciences, humanities, and social sciences – corresponded to the "groupings" into which "the investigations of the faculty were falling naturally."[22]

Fortunately, those natural groupings corresponded to equally natural groupings of subject matter, and in such a way as to indicate unambiguously where the core of general education should lie. Although Chicago came to be popularly identified with "the great books," the principles of organization and selection governing its curriculum derived from a coherence of content and method best exemplified in biology. As taught at Chicago, biological science had the singular virtue that its objectives were built into its subject matter – a point stated unmistakably in the 1932–3 catalog announcing the new college requirements. The introductory general education courses in the humanities, the physical sciences, and the social sciences were all defined by the substantive topics they covered (in the physical sciences, the exercise took twenty-six lines of small print and involved seven categories), whereas biology was succinctly characterized by its "dominating objectives": "(1) To cultivate the scientific attitude of mind through repeated illustrations of the scientific method of attack upon nature's problems: (2) to implant such practical information about biology as is desirable for a citizen in the modern world; (3) to awaken interest in the impressive machinery of the organic world and in the major concepts of biology."[23]

The course was divided into three units, each of which carried a message. The first, "variety and relationships among living organisms," started with a survey of the plant kingdom and proceeded via an account of evolutionary progress to a culminating discussion of the flowering plants. The process was then repeated for the animal kingdom, as the unit moved from the invertebrates to the vertebrates that were relevant to man's own prehuman ancestors, the apparent equivalent of the flowering plants. This led naturally to "a section on anthropology, which reviews our knowledge of prehistoric man and outlines the characteristics of modern human races," a suitable preliminary to the unit's predictable lesson concerning "the details of human reproduction and development in man."[24]

For an innovative course in biology, this was distinctly old fashioned. So was the next unit, "the dynamics of living organisms," in which the body was studied first in health, then in disease. The concluding unit – "the nervous system; consciousness and behavior; evolution, heredity, and eugenics; ecology" – synthesized physiology and psychology, in the interest of evaluating "the possibilities of improving the human race" through biology, again a thoroughly traditional ambition.[25] But for all that these units covered well-trodden ground, and indeed were very much outgrowths of the old elective offering on "the nature of the world and man," they set the model that general education courses in the sciences at Chicago would seek to approximate for several decades. Biology beautifully exhibited the connections between the real world, our knowledge of it, and our ability to act in and on it.

The centrality of the biological sciences in the early Chicago general education curriculum, as both the symbol of unified knowledge and the exemplar of useful knowledge, was confirmed by the prosaic character of the social science courses. Initially organized as surveys of the major disciplines, especially economics, sociology, and political science, they rapidly became "social problems" courses in which the particular problem of matching modes of inquiry to subject matter was not addressed. "The application of theoretical social science to practical issues of public policy" never became a major concern;[26] indeed, little attention was given at all to questions of social theory, as distinct from research methods. For example, where Columbia's course on statistics, following the lead of Contemporary Civilization, centered its attentions on how the student might "apply quantitative analysis in his further study of the social sciences," the comparable Chicago offering in "Statistical Sociology" displayed no such social theoretical range. Its emphases were unreservedly technical, on "methods of collecting, tabulating, and analyzing sociological data, the questionnaire, frequency distribution, graphical presentation, interpolation, interpretation of statistics, the nature of statistical evidence, first considerations of sampling, statistical fallacies, measures of variability."[27]

Having elected to separate social problems from social theory, Chicago social scientists found it increasingly difficult to explain how they were contributing to general education. The state of their field placed them at a double disadvantage with respect to a unified and useful biology. On the other hand, courses built around "an over-all theoretical or conceptual 'synthesis' of the social sciences" could not be taught, because the social sciences were "not yet ready for such a synthesis." Nor, on the other hand, could the point be to offer practical guides to action, it being the feature of a social scientific analysis of an issue that it "never tells the student that such and such a measure is 'the solution' to the problem, which he must accept and carry into action."[28] With these options closed, all the remained for the social sciences was a claim to possession of such "interdisciplinary 'problems' and 'themes' " as seemed "appropriate for general education" by virtue of their relevance to "those things that matter most to the individual." But for social scientists who were unwilling or unable to fashion a distinctive body of systematic theory to bring to bear on "those things," this was not a very promising alternative, since general education in that diffuse sense was "shared with the family, the lower schools, the church, and other social agencies."[29]

In effect, the difficulties facing the social sciences at Chicago paralleled those confronting the natural sciences at Columbia. At both institutions, one field of learning – whether biological science or a combination of history and philosophy – preempted the ground on which objectives, subject matter, and mode of inquiry could be integrated, the possible lines of development for other branches of knowledge were correspondingly restricted: Either they chould become pure problem-solving fields, displaying no special scientific logic in their organization and growth; or they could present themselves as pure sciences, with no particular bearing on human affairs.

The fortunes of the physical sciences at Chicago suggest the depth of this dilemma. They too were not well adapted to the Chicago educational environment, and they were treated quite differently from either the biological or the social sciences until after World War II, when as a consequence of reforms begun in 1943 a new program emerged that, in the view of its architects, for the first time presented each of the major natural sciences "not only as a body of knowledge but also as a field of inquiry." The aim then, as in biology before, was "to illuminate for the student the manner in which knowledge is obtained in each field, as well as the subject matters of the several fields."[30] But initially, the physical sciences were built into the curriculum on other terms, established by Hutchins's celebrated 1936 definition of general education as "a course of study consisting of the greatest books of the western world . . . together with mathematics, the best exemplar of the processes of human reason."[31] Incorporating that view, the inaugural physical science course identified studying mathematics with learning to appreciate the "analytic character and method of physical science." The result – in striking contrast to the biological science offering, where the increased complexities of knowledge and of nature were portrayed as mirroring each other, permitting the intellectual development of the student to proceed along evolutionary lines – was a course in which understanding physical reality and understanding physical science appeared as separate undertakings. "Method" stood apart from content: Near the head of the course, in splendid isolation, unit two focused on "the place, character, and effectiveness of the contribution of mathematics to the development of the physical sciences." Remarkably, but inevitably, the presentation was "descriptive, involving a minimum of mathematical technicalities."[32] At Chicago, the queen of the sciences reigned but did not rule.

It was soon recognized that this was not wholly satisfactory. "Experience with a program characterized by relatively little emphasis on science as inquiry, by a tendency to understand inquiry in the narrow sense of 'method,' and by a reduction of 'method' to the status of the subject matter of a narrative account led to dissatisfaction." In 1943 a new effort was begun "to effect a union of content and method," one that would "avoid narrative discourse on either the conclusions of science or its method, by presenting 'method' as the means by which conclusions are reached, verified, and related in science, and 'conclusions' as the consequences of the application of these means to an appropriate subject matter." The idea, familiar from the biological science course inaugurated a decade earlier, was to assign each student "the task of understanding and following the stages of discovery and verification as the main route available to him for knowledge of the conclusions." The task was to be made manageable by having the "processes of inquiry which characterize science . . . exhibited only concretely and by example" in course materials, "not as a doctrine about method but in the form of reports of problems formulated and data gathered and interpreted which the student would need to grasp."[33]

This meant making extensive use of "primary or original source material," chiefly research papers and monographs. Their employment elicited "much puzzlement," which translated into three broad criticisms of the Chicago physical science curriculum: that it "substituted the history of science for science proper": that it "substituted the sociology of science for science proper"; and that it "turned its back on science proper in order to treat scientific papers as humanistic objects, as creations of the human spirit." To each of these evidently damning objections, Chicago scientists bravely responded, "This is not the case." They were not historians interested in "shifts of research patterns or of principles or of scientific subject matter in succeeding epochs or in different places." They were not sociologists interested in portraying "scientific investigation as largely or most significantly the creation of influences which lie outside science." And they were not humanists interested in scientific works as "works of art," even "in the best sense." They were simply scientists interested in the "elaboration of general truths from the particular facts of nature."[34]

Unfortunately, having distinguished themselves so sharply from historians, sociologists, and humanists, the physical scientists promptly discovered that their field in its pure form had no place in a general education curriculum. Accordingly, they proceeded to build "a dole of humanistic and social scientific disciplines" back into the enterprise. The predictable result was not science but its "liberal equivalent."[35] That entity resisted succinct characterization because the only way to describe it was to start by rejecting as inadequate the criteria – subject matter, method, and purpose – used at Chicago "for discriminating and understanding the differences and similarities of the humanities, the social sciences, and the natural sciences."[36]

At a college that had "chosen to divide its curriculum" into just those fields, this was not a very firm foundation on which to base a "*liberal* program in the sciences."[37] In fact, the sought-after "liberal equivalent" of science represented an uneasy compromise between two contradictory ambitions: to "divide the whole of liberal education into parts and assign to them different duties and responsibilities," and to fashion a science program that would nonetheless express "the principles of liberal education as a whole," albeit "as mediated and modified by the qualities which distinguish science from the humanities and the social sciences."[38] By 1950 at Chicago the identification of those distinguishing qualities had become a major academic exercise in its own right, to be conducted at a level of high intellectual abstraction, rather than by simple observation, as Hutchins had assumed, of the natural groupings into which a research faculty arranged itself in the course of advancing knowledge. The converse problem of defining the principles of all of general education likewise presented itself as a matter for almost pure academic discourse, at least as far as the college's scientists were concerned. Two decades spent developing their "liberal program" had left them with an array of arguments and counterarguments about bodies of knowledge, modes of inquiry, and purposes of education that could be deployed without particular reference to the cultural backgrounds and vocational ambitions of any particular set of students.

HISTORY AND SCIENCE

This was not possible for institutions, like Harvard, that were just beginning general education programs immediately following World War II. In Cambridge the issues that had to be addressed in 1945 were the same as the ones that had called forth Contemporary Civilization at Columbia a quarter of a century earlier, after a previous great war, with the exception that the original Columbia vision of a course of study based on a single, unifying theme no longer seemed viable. The problem Harvard faced was to give structure and substance to "distribution requirements" in a curriculum designed for an increasingly discipline-oriented and profession-bound student body. Noting that "as modern life has come increasingly to rest on specialized knowledge, the various fields of college study have in consequence appeared simply as preparation for one or another position in life, . . . a kind of higher vocational training," *General Education in a Free Society* proceeded to draw the "obvious connection" between this vocationalism and "the state of democracy in which the hereditary moneyed class is less strong and almost all young people have to prepare themselves to make a living." The Harvard committee disclaimed any intention of attacking or defending the careerist behavior it observed, offering instead as its "sole point" the judgment that "the rise of this partly, though not wholly, vocational specialism has tended to take from the college what theoretical unity it had." The result was a clear choice: "If the various fields of study do not represent a common discipline or give anything like a common view of life, then such unity as the college has must come chiefly from imponderable tradition or simple gregariousness."[39]

Simple gregariousness no doubt had its charms, and the weight of imponderable tradition certainly could be felt at a three hundred-year-old college, but neither could compete with the allure of a common, disciplined view of life. As in Chicago, the question was how to introduce unity into a curriculum that was necessarily divided, again into three parts – humanities, social sciences, and natural sciences – and whose units had historically been given decidedly unequal treatment, not only in Cambridge but at most colleges and universities. Harvard's solution was to make a concerted effort to redress the imbalance by devising an instructional rationale that fell symmetrically on the three divisions. As at Columbia and Chicago, general education was to be a preliminary to adult life in much the same way that introductory courses were preparatory to advanced work in the disciplines. Since adults had to live with themselves, with each other, and with physical reality, it was appropriate that the curriculum have three constituent elements. "The study of the natural sciences looks to an understanding of our physical environment, so that we may have a suitable relation to it. The study of the social sciences is intended to produce an understanding of our social environment and of human institutions in general, so that the student may achieve a proper relation to society . . . and, by the aid of history, the society of the past and even of the future. Finally, the purpose of the humanities is to enable man to understand man in relation to himself, that is to say, in his inner aspirations and ideals."[40]

These objectives could only be met through specially designed courses, and all students were expected to take the full measure required in each area, including the one in which their major happened to fall. The humanities lent themselves to courses organized around classic texts, and the social sciences to courses that surveyed "important elements in the Western cultural tradition" or that used historical materials in combination with social theory "to raise and illustrate some of the principal questions of the social and moral sciences."[41] But in the natural sciences it proved difficult to find equivalents to *The Iliad, The Odyssey, The Aeneid, The Divine Comedy,* and *Paradise Lost,* five works the study of which comprised a well-spent term of work in Humanities 2 in 1954–5, the year the full general education program went into effect.[42] No one believed that studying falling bodies for a semester would provide the grounding in natural science required for life in a free society. The alternative to encouraging that kind of work *in* science was to develop courses *on* science – "Principles of Physical Science," "Understanding the Physical World," and "Principles of Biological Science" being the original entries.[43]

One obvious, although not necessarily consciously appropriated model for this arrangement was the conventional introductory survey course in the conventional social science department, "Principles of . . ." Another equally obvious, if again unconscious, model was the nineteenth-century college curriculum that "taught the great fundamental truths of all sciences," rather than the small collateral truths of a single science. In contrast to Columbia, where Science A and Science B embodied disciplinary distinctions, the Harvard natural science courses aimed at principles that were not discipline specific. Charged with representing "reasonably broad syntheses within the areas of science and mathematics," they were designed "to convey some integrative viewpoint, scientific method, or the development of scientific concepts, or the scientific worldview." It was also expected that they would show "the various means by which science progresses," especially "the fructification of one science by another," and the "progression from description to analysis and synthesis and from the qualitative to the quantitative."[44]

Creating the University Committee on the Objectives of General Education in a Free Society in 1943, President Conant directed its members "to venture into the vast field of American educational experience" and to make recommendations not just for Harvard but for the nation as a whole.[45] In discharge of this mandate, the committee turned first to Pericles and Plato, who provided the epigrams with which *General Education in a Free Society* opened. Columbia and Chicago, although somewhat closer than ancient Athens, were cited only incidentally, an omission that if rectified might have discouraged the reaction Conant expected from "some who open the book": " 'General education,' they may exclaim, 'what's that?' "[46]

A more appropriate question would have been, What's new? for just as at Columbia, whose faculty saw Contemporary Civilization breaking down disciplinary barriers through courses of study structured according to chronological sequences, Harvard relied on the progress revealed through history in its search for unifying purposes and ideas. What was novel was the extension of that line of reasoning to the natural sciences. Not so much disregarding as inverting distinctions that were coming to seem so important at Chicago, Harvard announced that "the claim of general education is that the history of science is part of science. So are its philosophy, its great literature, and its social and intellectual context."[47]

This was an elegant, economical solution to the problem posed at Chicago of how to divide the curriculum and assign different duties to different fields of learning without leaving large parts of science out of the picture. The boundaries separating the humanities, social sciences, and the natural sciences were to be preserved by the simple device of making science proper a derivative of its liberal equivalent. Elaborating on the significance of the history, philosophy, literature, and contexts of scientific knowledge, the committee on general education insisted that "the contribution of science instruction to the life of the university and to society should include these elements since science includes them. A science course so constructed as to encompass these elements makes an important contribution to general education. It need not, by that token make a poorer contribution in science. One can defend the view that it is all the better science for being good general education."[48]

It took little more than a decade for all of this to seem like so much wishful thinking, no longer sufficient to maintain even the appearance of success. In 1959 a faculty committee chaired by psychologist Jerome Bruner explicitly rejected the assumptions that had guided the initial organization of the natural science curriculum. In their place two new aims were set out. The first was to "communicate a knowledge of the fundamental principles of a special science," instead of presenting principles common to whole classes of sciences. The second was to "give the student an idea of the methods of science as they are known today," not as they had been viewed in the past. These were minimum requirements. It was the committee's view that unless and until they were met, "little [was] served by historical or philosophical treatments of the development or significance of science. Science itself . . . [was] the central concern of the student of science – not its history, or its philosophy, or its sociology, useful as these may be in explicating the nature of science."[49]

New general education science courses were needed, organized this time around "a few specific but significant topics." The requisite "wise choice" of topics was to be made "by outstanding specialists in the sciences," who were expected to know which topics would provide the desired "sense of depth of scientific inquiry." Although the examples of good topics cited by the Bruner committee – evolution, elementary particles in physics, the nervous system and its function – did not all immediately indicate how or why, it was also expressed that the result would be a raising of the mathematical level of the science done in general education, it being "principally by the use of mathematics that one may go deeply and quickly into a scientific problem."[50]

This general approach had the double advantage of being as good for the faculty as it was for the student. Although the Bruner report opened with the obligatory extended discussion of educational purposes, modes of thought, and bodies of knowledge, the committee acknowledged that questions about "how courses in Natural Science are to be taught" were secondary to "the most pressing problem concerning scientific instruction in Harvard College." That problem was "one of staffing," and it had two parts. First, research scientists were awfully busy. What with the demands made on them by "the ever-increasing tempo of scientific research" and by the "arduous tasks" thrust upon them "as consultants to the Government on problems of national security and scientific policy," they had little time for teaching. Humanists and social scientists were apparently not so excessively burdened. Two years after Sputnik, the Bruner committee reported that "it cannot be said that a broadly representative sample of our scientists has participated in the teaching of Natural Science, as it can be said of the group of scholars who have taken part in the teaching of elementary courses in the Humanities and Social Sciences."[51]

The solution to this part of the staffing problem in general education was obvious, and the Bruner committee stated it succinctly: The size of the science departments had to be increased. The second part of the problem was slightly more complicated. It was that "the content of instruction in Natural Science is often further removed from a scientist's research interests than is the corresponding content of a General Education course offered by a scholar in the Humanities or the Social Sciences." The proposed reorganization of the curriculum solved that problem too, because it allowed scientists to structure their general education offerings around "topics of direct concern to them in their research."[52] The hope was that in this way the "misunderstanding and alienation" that had grown up around courses "carried out with a strong historical emphasis" would be significantly reduced.[53] There was, of course, the question of how or whether the new general education would differ from the standard introductory course of the individual science departments. This issue troubled faculty in the humanities and social sciences, but it was viewed with equanimity by natural scientists who saw no reason why courses on matters of genuine scientific substance should not do double duty, as the Bruner report suggested, and "be suitable both as general education and as an introductory course for prospective concentrators."[54]

In 1959 it seemed reasonable to believe that each and every specific science could function equally well in bringing students to appreciate generic science for "its mode of access to nature, the analytic methods it employs to achieve an economy of description and understanding, the techniques it invents for rendering concrete observation into systematic theory, its powerful logic of verification, the deep philosophical dilemmas that it has recently posed." These were all matters of considerable "cultural and human relevance." To grasp them and to see science as "a creative and rational enterprise" was to "understand one of the forms of man's excellence."[55]

SCIENCE AND HISTORY

The difficulties that overtook President Conant's history-centered natural science curriculum might have been predicted in 1949, had Harvard been as conscious of MIT's efforts to reform general education as MIT was of Harvard's. To economist Rupert Maclaurin, the problems of the two schools seemed exactly "reverse": "Here the necessity is to humanize the scientist; there it is to scientize the humanist."[56] But the solution in each place was initially the same: develop courses in the history of science that would encompass its social and intellectual contexts; or, in the language of MIT's 1950 announcement of its new School of Humanities and Social Studies (later, Social Science), establish a general education program centered on "the relation between science and technology on the one hand, and man and his institutions on the other."[57] This seemed an ideal way to prepare prospective scientists and engineers "to become more nearly totally useful citizens," John E. Burchard explained, a year before becoming the school's first dean, because at an institute of technology where it was "neither possible nor desirable to expose these professional people to every possible experience in [the] enormous field of human conduct," the history of science and technology had the unique merit of being a field that would capture "the interest [of] the non-humanities faculty."[58]

Except for the "non," Burchard's characterization of whose interests had to be aroused applied equally well to his neighbor institution. Although the curricular burdens placed on the history of science and technology at Harvard and MIT were reversed, with the surrogate for science at the former being asked to serve as a vehicle for teaching about human conduct at the latter, the apparent symmetry was misleading. The problems being addressed at the two schools were not so much mirror images of each other as identical. Both were seeking to provide students with "an awareness of the interrelations of the scientific, technical, and literary cultures, and a sensitiveness to the diverse forces that motivate the thoughts and actions of people."[59] Both were also seeking to do that without abandoning their special commitments to their respective institutional cultures and constituencies. James B. Conant would have understood James R. Killian's remark about MIT's new School of Humanities and Social Studies: "We shall remain an institution of limited objectives, offering as we have for many years a program centered around science, engineering, architecture, and management."[60]

By 1950 MIT in fact had had a general education program for some thirty years. Its first courses in the field were inaugurated in 1919, exactly at the time when Columbia was instituting Contemporary Civilization, and to much the same end. In Cambridge as in New York, "liberal and humanizing studies" (or "liberal and humanitarian subjects," as they were also called) had to be provided for excessively career-minded undergraduates who did not share the values of their teachers. To be sure, MIT students were not so prone to either the "stupid" radicalism or the "stubborn" conservatism that plagued their counterparts at Columbia. From World War I to World War II, the failing in Cambridge was always simply a matter of the fact that engineers "often mature quickly on the technical side, but slowly on the social, cultural, and philosophical side."[61] But that only made it all the more important to defend the "principle" that their general education courses "should be non-professional," because only in that way could students "get the greatest possible cultural value from them."[62]

As MIT's conception of a professional engineer changed, so did its understanding of the "general principles and objectives" underlying nonprofessional work in the humanities and social sciences. By the mid-1930s the general studies curriculum had grown to include fifty-four courses; third- and four-year undergraduates were free to meet the "non-vocational" course requirements by selecting whatever mix of subjects appealed to their "particular personal tastes."[63] To a faculty Committee on the Humanities appointed in 1935, this seemed an inadequate state of affairs, because the lack of "interrelationship or continuity" among the courses encouraged more " 'browsing around' on the part of the student" than was consistent with sound educational policy.[64] At a time when the institute was discovering that systematic instruction in the substance of science was a prerequisite for successful engineering practice, it seemed appropriate to induce students to order their nonvocational studies in a similarly coherent way. Accordingly, in 1936 the general studies courses were regrouped into four categories – history of science and thought, history of civilization, literature and the fine arts, and social science – and undergraduates were strongly advised to confine themselves to subjects within one of the four.

The Committee on the Humanities acknowledged that each of these groups contained "a considerable variety of subjects." The history of science and thought included courses on geology and organic evolution, for example, as well as on "principles of sanitary science and public health," but not on the history of engineering, which appeared under the history of civilization. The social sciences extended from contemporary political institutions to the inevitable treatment of "biological reproduction." Literature and fine arts encompassed "Dante in English," "the Bible as literature," and "design in manufactured products."[65] But the apparent heterogeneity did not mean that the humanities at MIT, like the natural sciences elsewhere, embodied a unitary method of thought that transcended a manifoldly multiplying content. Just as engineers in the 1930s needed to know not the principles, methods, themes, philosophy, history, great literature, nor the social and intellectual contexts of science, but science itself, so too the point of general education was to offer "the student an opportunity to elect a general field of interest and then give him the opportunity to pursue that interest to the end that he may eventually obtain a fairly adequate and comprehensive knowledge of *the subject matter*."[66]

The notion that general education was like science, in the sense that its subjects could be organized in such a way that the students would acquire it as a body of knowledge, did not survive World War II. By 1943 a Committee on Engineering Education after the War was already thinking about "the humanistic-social stem" of the curriculum along much the same lines as the Harvard Committee on General Education in a Free Society would soon follow in thinking about natural science. Now, detailed knowledge of the subject matter seemed "less important for engineering students than . . . the acquirement of the ability to understand, to analyze and to express the essential elements of an economic, social, or humanistic situation or problem, and to appreciate the implications and relationships of such problems to the life and work of an engineer."[67] This approach had the double virtue of being good for students and consistent with available faculty resources. Just as at Harvard, where teaching about science required something other than narrow scientific expertise, so at MIT "erudition" had become a "secondary consideration" for instructors in the humanities and social sciences. "Mental maturity, perspective, sympathetic perception of the student's present viewpoint and of his career aims, interpretive ability, and ability to awaken interest and elicit active participation are primary."[68]

MIT's solution to the problem of general education was neither more nor less successful than Harvard's or Chicago's or Columbia's, although a certain pathos surrounded it that was absent from the elite liberal arts colleges. Predictably concluding its discussion of the humanities and social sciences with an expression of concern about the "civic responsibilities of engineers," the Committee on Engineering Education after the War suggested that translating those responsibilities "into concrete terms" only meant that "engineers should be counted in full proportion among groups which read books, discuss ideas, hear music, attend the theatre, appreciate works of art, sustain churches, advance philanthropy, man welfare activities, serve as trustees, and lead civic movements." If the stated objectives of Columbia, Chicago, and Harvard were significantly less modest, that was only because no one had ever suggested that the "growth in numbers and influence" of their graduates "might be a menace to social enlightenment and cultural advancement."[69] Yet this was also one of the dangers that general education in science at the liberal arts colleges was designed to meet. Lacking any special commitment to training scientists or engineers, recognizing that in the twentieth century managing society and understanding nature were separable, and knowing that the former rather than the latter was the proper business of cultured people, Columbia, Chicago, and Harvard assumed as a matter of right and responsibility the burden of teaching generic scientific principles and methods that practicing scientists and engineers were themselves not expected to grasp.

CONCLUSION

The general education curricula that we have examined had much in common, even though they were designed over a span of years that extended from the end of World War I to the end of World War II. Dissatisfaction with courses in the natural sciences generated repeated efforts to tailor their content so that they would make sense to faculty in science departments, yet still resemble the science that social scientists and humanists understood. The solutions that won temporary favor all evaded the question of how scientific subject matters and scientific traditions could be combined into a coherent whole. Except insofar as general education courses offered by science faculty were indistinguishable from introductions to advanced study in the several disciplines, matters of genuine scientific substance were simply not taught. Instead, in the interest of emphasizing social and cultural dimensions of scientific knowledge, the sciences were presented as branches of other subjects. Biology became a social science, physics a mundane part of metaphysics, and the principles of natural science the equivalent to the natural laws that classical economists had turned into the principles of their science.

In attempting to refashion educational goals, Columbia, Chicago, and Harvard described their missions as if the fortunes of "a free society" depended on how one of its colleges imbued its undergraduates with a disciplined view of life. New proposals for the natural science component of general education were presented after each perceived crisis with reference to grand intellectual and cultural schemes transcending parochial accommodations to the professional interests of the faculty and contests over staffing and budgets. In practice each institution proceeded as if it had to invent general education entirely on its own, and the chronology of educational advance was consequently marked by anomalies: Chicago embarking on its biology-centered program just as Columbia was abandoning its attempts to work the natural sciences into the structure of Contemporary Civilization; Harvard introducing its history-centered natural science courses at the same time as Chicago was defending itself against the charge of substituting the history of science for proper science. Yet the internal politics of curricular change in the natural sciences at these three institutions were largely the same. In each case the major initiatives were presidential rather than professorial, and in each case their frustration was the result of science departments reclaiming control over their fields.

This was true even at Chicago, where, from the start, a field of science occupied a privileged position in undergraduate education. Robert Maynard Hutchins's argument, that in reorganizing the college he was merely framing the curricular equivalent of established patterns of faculty cooperation in research, did not long conceal the fact that the consequence of his proposal was to reduce the power of existing departments by creating a separate college teaching facility outside established departmental jurisdiction. At Columbia and Harvard, where no scientific discipline took on the kind of paradigmatic importance accorded biology at Chicago, the conflicts between presidential visions and scientific self-interest were still more decisive. Science A and Science B defended departmental prerogatives challenged by Nicholas Murray Butler's admonition to Columbia's scientists to quit stumbling; at Harvard the report of the Bruner committee on science constituted a predictably negative response to James Bryant Conant's suggestion that "historical study could yield a new sort of understanding of the structure and function of scientific research."[70]

It is tempting to follow the lead of the Bruner report and see its rejection of the old natural science curriculum in 1959 as a sign of the broader triumph of the scientific disciplines in the mid-twentieth century American research university. But the opening paragraphs of the report indicate that with regard to undergraduate education, especially for students who were not concentrating in the natural sciences, the science faculty were not entirely sanguine about the responsibilities that followed when they appeared on the lecture podium as bearers of culture. It was one thing to acknowledge that "science represents one of man's principal avenues to knowledge," but another to shape a syllabus around "its mode of access to nature," or "its powerful logic of verification" and "the deep philosophical dilemmas that it has recently posed." Having whetted the intellectual appetite with the suggestion that "to understand science as a creative and rational enterprise is to understand one of the form's of man's excellence," the Bruner report took the pledge and dedicated itself to more sobering fare. In general education the natural science courses would simply meet a kind of distribution requirement, communicating "a knowledge of the fundamental principles of a special science" and "an idea of the methods of science as they are known today." Once this "minimum requirement" was met the temptation to substitute the history or philosophy or sociology of science would be gone. As with more personal resolutions to resist corrupting influences, what was not to be imbibed was far clearer than the proposed course of action.

Inspirational prose barely covered unease on all sides. In retrospect it is hard to imagine the science faculty finding fulfillment in the creation of new courses that systematically played down "the technological or social implications of scientific work" in favor of "elementary instruction . . . conducted by the study of exemplary topics." The Bruner committee insisted that it was absolutely necessary "to enlist the voluntary cooperation and support of scientific departments in the General Education Program," concluding that otherwise "the requirement in Natural Science [would] have to be replaced by offering some other form of instruction." But the final recommendations wisely did not rely entirely on volunteers, at least not from existing staff. Instead, an expansion of the science departments was urged, involving the addition of teachers who would be attracted to general education as the Bruner committee understood it. "More intense concentration on a small set of exemplary topics and flexible course organizations" – these features that promised to engage students were also supposed to be "the very things" that would make the new curriculum appealing to the right sort of scholar.[71]

In point of fact the Bruner report seems to have had no operational consequences. Although its recommendations were adopted by a voice vote of the faculty in 1959, five years later the Harvard science departments had not expanded, and the natural science part of general education was not noticeably different from what it had been before. Of the six elementary courses offered in 1958–9, four were still in the 1963–4 catalog with their descriptions essentially unchanged. One course, "Historical Introduction to the Physical Sciences," had disappeared; another, "Principles and Problems of the Biological Sciences," had been replaced by two new offerings: "The Nature of Living Things" and "Man's Place in Nature."

Although the survival power of Harvard's history-centered natural science curriculum cannot be denied, neither can the fact that the Bruner committee had effectively undermined the chief rationales for its existence. For that matter, in framing its case for the inclusion of the natural sciences, as distinct from their history, philosophy, or associated technologies, in the common curriculum, the committee had come close to demolishing any and all arguments for general education in any area. If the problem that distinguished the natural sciences from other fields was the distance between course content and faculty research interests, then it surely followed that in the social sciences and the humanities – where courses were by implication not so far "removed from the working life" of the people teaching them – there was no real reason to believe that what was being presented in the general education curriculum could not be equally well taught in introductions to the several disciplines.

To pursue that line of argument would have involved seeing the Bruner report as an effort to patch a sinking ship that should have been sunk long before, the ship being general education as a whole, not just its natural science component. Although such views were expressed, they were not encouraged at Harvard in 1959, and the fact that the Bruner committee had left the common curriculum without a clear educational mission that was uniquely its own passed largely unnoticed. It is difficult to avoid the conclusion that this only shows how disinclined the faculty was to take seriously any argument about general education that turned on the special needs of the natural sciences. With one crucial difference, this meant that Harvard was left with a general education curriculum that resembled MIT's, in the sense that both had evolved as if there were no better way of meeting the cultural needs of undergraduates than by introducing them to the traditional humanities and social sciences. The difference, of course, was that at MIT the requirements devised to that end were paralleled by the equivalent of general education requirements in the natural sciences themselves: a year of calculus, a year of physics, and a year of chemistry had to be taken by all freshmen in 1959, including students planning to major in fields outside science and engineering.

Although MIT catalogs have predictably identified such requirements as prerequisites for success in the institute's various degree programs, the need for extensive preparation in the sciences has also long been explained in the language of general education. Undergraduates in 1959 were told that the science courses required of them were "intended to acquaint the future citizen, whatever his profession, with an important part of our heritage from the past."[72] Twenty years later, neither the requirements nor the rationale had changed appreciably: In 1979 the required science courses formed "an essential part of the background" that the school's graduates had to acquire, if they were to function effectively "as citizens in a world deeply influenced by science and technology."[73]

The Bruner report had opened with a similar observation, that "the increasingly important role that science plays in shaping the lives of men makes it urgent that graduates of Harvard College should have some idea of the discipline which is ultimately responsible for this influence."[74] But in practice the outcome at Harvard did not accord with that principle. Just as Columbia three decades earlier had concluded that it could proceed without the natural sciences in developing its program in Contemporary Civilization, so too at Harvard the absence of "science itself" from the common curriculum in 1959 might well have aroused no comment whatever, had it not been for Sputnik. As it was, the problem of teaching science to all Harvard undergraduates soon gave way, as it had at Columbia, to more urgent questions concerning the future of American democracy.

NOTES

1 Quoted in I. Bernard Cohen, "Science and American Society in the First Century of the Republic," Second Annual Alpheus W. Smith Lecture, Department of Physics and Astronomy and the Graduate School of the University, Ohio State University (Columbus: 1961), 24.

2 Russell Thomas, *The Search for a Common Learning: General Education, 1800–1960* (New York: 1962), 76.

3 Theodore R. Sizer, *Secondary Schools at the Turn of the Century* (New Haven, Conn.: 1964), 199. The assumption would have been more reasonable, had the growth in secondary school enrollments not been accompanied by a significant decline in the ratio of high school graduates to college students. By 1940 only a quarter of the 50 percent of the seventeen-year-olds who finished secondary school were entering college, and they do not seem to have been making excessive demands on the older liberal arts institutions. In that year, for example, Harvard admitted 85 percent of the 1,234 candidates who applied for admission.

4 Columbia University, *Annual Report of the President and Treasurer* (New York: 1908), 26. The figures for Harvard were kindly provided by Phyllis Keller, Associate Dean of the Faculty of Arts and Sciences, Harvard University. Those for Columbia were compiled directly from the university's catalogs and are only approximate, given the vagaries of cross-listed courses and joint appointments.

5 Quoted in Hugh Hawkins, *Between Harvard and America: The Educational Leadership of Charles W. Eliot* (New York: 1972), 240.

6 Frederick Keppel, *Columbia* (London: 1914), quoted in Daniel Bell, *The Reforming of General Education* (New York: 1966), 20.

7 Bell, *Reforming of General Education,* n. 6, 14.

8 Columbia University, *Bulletin of Information,* XX: 11 (1919–20), 35–6.

9 Columbia University, *Annual Report of the President and Treasurer* (New York: 1920), 41–2.

10 Nicholas Murray Butler, quoted in Robert Maynard Hutchins, *The Higher Learning in America* (New Haven, Conn.: 1936), 80.

11 Columbia University, *Annual Report* (1920), 42.

12 Ibid., 87–8.

13 Columbia University, *Annual Report of the President and Treasurer* (New York: 1929), 68.

14 Columbia University, *Annual Report of the President and Treasurer* (New York: 1928), 24–5.

15 Columbia University, *Annual Report* (1929), 69.

16 Columbia University, *Bulletin of Information,* XXXVI: 27 (1936), 104.

17 Columbia Univesity, *Bulletin of Information,* XXIX: 29 (1929), 92.

18 Columbia University, *Bulletin of Information,* XXXVI: 27 (1936), 47.

19 Barry Karl, *Charles E. Merriam and the Study of Politics* (Chicago: 1975), 162–3.

20 Robert Maynard Hutchins, "The Reorganization of the University," *The University Record,* XVII: 1 (1931), 2,5.

21 Thomas, *Search for Common Learning,* 76.

22 Hutchins, "Reorganization of the University," 5.

23 *Announcements of the University of Chicago,* XXXII: 12 (1932), 49.

24 *Announcements of the University of Chicago,* XXXII: 22 (1932), 7.

25 Ibid.

26 Milton B. Singer, "The Social Sciences," in *The Idea and Practice of General Education: An Account of the College of the University of Chicago by Present and Former Members of the Faculty* (Chicago: 1950), 130.

27 *Announcements of the University of Chicago,* XXXII: 12 (1932), 416.

28 Singer, "The Social Sciences," 125, 135.

29 Ibid., 125.

30 Joseph J. Schwab and Merle C. Coulter, "The Natural Sciences," in *The Idea and Practice of General Education,* 154.

31 Hutchins, *Higher Learning in America,* 85.

32 *Announcements of the University of Chicago,* XXXII: 12 (1932), 56.

33 Schwab and Coulter, "The Natural Sciences," 156.

34 Ibid., 161–2, 165.

35 Ibid., 165.

36 Ibid., 163.

37 Ibid., 178.

38 Ibid., 150.

39 *General Education in a Free Society: Report of the Harvard Committee* (Cambridge: 1945), 38–9.

40 Ibid., 59.

41 *Harvard University Catalogue* (1954–5), 182–3.

42 Ibid., 178.

43 *Harvard University Catalogue* (1946–7), 240.

44 *General Education in a Free Society,* 221.

45 Ibid., xiii.

46 Ibid., viii.

47 Ibid., 222.

48 Ibid.

49 "Report of the Committee on Science in General Education" (February 1959), 1.

50 Ibid., 7, 9.

51 Ibid., 4, 5.

52 Ibid., 5, 7.

53 Ibid., 6.

54 Ibid., 4.

55 Ibid., 1.

56 Rupert Maclaurin, "Proposal for an Endowment for a Division of Human Relations at Massachusetts Institute of Technology" (August 1947), 4.

57 Press release from the News Service, Massachusetts Institute of Technology, December 17, 1950, 3.

58 John E. Burchard to James R. Killian, December 27, 1949.

59 News Service, MIT, December 17, 1950, 3.

60 Ibid., 1.

61 "Memorandum from the Committee on the Humanities for Presentation to the

Faculty" (May 20, 1936), 1, 2; "The Humanistic-social Stem," excerpt from the Report of the Committee on Engineering Education after the War (January 1944), 1.

62 "Memorandum," 1.
63 Ibid., appendix, 1.
64 Ibid., 1.
65 Ibid., appendix, 1–3.
66 Ibid., 1. Emphasis added.
67 "The Humanistic-social Stem," 3.
68 Ibid., 1.
69 Ibid., 4, 5.
70 See Thomas Kuhn, *The Copernican Revolution* (Cambridge: 1957), ix.
71 "Report of the Committee on Science," 9.
72 *Massachusetts Institute of Technology Bulletin* (1959–60), 168.
73 *Masschusetts Institute of Technology Bulletin* (1979–80), 2.
74 "Report of the Committee on Science," 1.

20

The pre-history of an academic discipline

The study of the history of science in the United States, 1891–1941

ARNOLD THACKRAY
University of Pennsylvania

The history of science is now an established academic discipline. The subject is nourished in many countries, but it is in the United States that the history of science has found its most capacious home. Departments, institutes, "offices" or "programs" devoted to the discipline – often in combination with cognate subjects like the history of technology or medicine, or the philosophy or sociology of science – flourish in such universities as California at Berkeley and Los Angeles, Chicago, Harvard, Indiana, Johns Hopkins, Pennsylvania, Princeton and Wisconsin. The number of graduate students belonging to the History of Science Society – over 300 at present – continues to grow. A wide variety of posts is advertised nationally; 61 were announced for the year 1980–81, including 25 holding out the prospect of permanent tenure. Governmental agencies as diverse as the National Science Foundation, the Congressional Research Service and the Department of Defense employ historians of science, and the National Aeronautics and Space Administration, and the Department of Energy, have elaborate historical programmes. Significant centres of research, and places of employment, are also to be found in specialised libraries, in archives, and in science museums. Even the oil companies are interested.

Specialised journals exist in profusion. Between 10 and 20 such periodical publications, with titles ranging from the *Archive for History of Exact Sciences* to *Studies in History of Biology*, are now edited in the United States. So many and so various are the books published that a catholic selection of examples would still prove invidious. If few of them can claim the popularity enjoyed by Professor Thomas Kuhn's *Structure of Scientific Revolutions* or the wide use promised for the *Dictionary of Scientific Biography*, many will stand as sound works of scholarship and as evidence of the establishment of the history of science as an academic discipline. That establishment is also indicated in the report of the John Simon Guggenheim Memorial Foundation stating that, in the strenuous competition for fellowship awards in the 1970s, "the newer fields of linguistics and the history of science have consolidated their position".[1]

The intellectual and institutional developments that led to the academic establishment of the history of science – developments that seemed highly unlikely

This essay was originally published in *Minerva*, 18 (1980), pp. 448–73. Reprinted by permission.

Table 1. *Ph.D. (or equivalent) degrees conferred: by five-year periods, 1871–95*

Period	Number of degrees conferred
1871–75	89
1876–80	192
1881–85	276
1886–90	574
1891–95	1150

Source: U.S. Bureau of the Census, *Historical Statistics of the United States. Colonial Times to 1970* (Washington, D.C.: Government Printing Office, 1976), Series H 751–765.

only 40 years ago – have yet to be analysed. By way of preliminary, this essay explores the pre-history of the history of science over the period from 1891 to 1941. In those years the subject became known as a field of learning in North America. It was viewed favourably by many influential scientists. Yet their rhetorical and individual commitments to the history of science did not lead to its academic establishment. Institutional provisions for the subject were very scanty and, on the eve of the Second World War, the history of science seemed destined to wither away.

THE "FIRST GENERATION" OF AMERICAN PH.D.s

The founding of Stanford University and the University of Chicago in 1891 continued the movement begun at Johns Hopkins University and encouraged the spirit that was already stirring faintly in several older, slower moving universities such as Columbia, Harvard and Pennsylvania.[2] While the number of recipients of bachelor's degrees increased less than fourfold in the period 1876–1916, the number of doctorates awarded grew by a factor of more than 20. In the five years between 1891 and 1895 alone, American universities awarded 1,150 doctorates, which was more than they had given in the preceding quarter of a century (Table 1).

By the mid-1890s a "first generation" of American doctors of philosophy had emerged in the world. Some of these scholars received a certain amount of pre- or post-doctoral training in Europe. In addition, an undetermined number of Americans obtained doctorates at European universities. Many of these had also studied in the United States. Together they constituted a group of holders of the doctorate sharing a common sense of identity, widened horizons, and stocks of foreign tradition on which they could draw in their own specialised teaching and research and in promoting the institutional fortunes of their subjects.

The first generation soon became active. The development of national learned societies formed around the major disciplines was accompanied by changes in the organisation of higher education. By the early 1900s, the pattern of departments specialised by disciplines which included within themselves a variety of more differentiated specialities, was well established. University teaching became an attractive career for intellectually ambitious young persons. It offered prospects of advancement and it permitted concentration on special fields of interest. Universities began to recognise achievements in research and advanced by small steps to provide material facilities and financial resources for that research. The greater size of the student bodies, both at the undergraduate and graduate levels, required more teachers and larger departments. At Harvard, for example, the number of graduate students rose from none in 1870 to 429 in 1908, while there was an expansion in "other teachers and research fellows" from 14 at the beginning of the 40-year period, to 416 towards its end. Similar patterns could be observed at the other major universities of the country. The income of universities grew rapidly. The enlarged departments permitted more specialisation in the teaching of particular subjects, and this was a great help to teachers who were committed to research and ambitious to make progress in their work.[3]

Spokesman for scientific research found everywhere at hand what was to them persuasive evidence of the scientific basis of much of the country's new prosperity. The generation of electric light and power and the wide use of the telephone could be made to testify to the point. The provisions for organised research by the General Electric Company in 1900, by Du Pont in 1902, and by the American Telephone and Telegraph Company in 1907 were widely noted, and increasingly imitated. The care of health offered similar opportunities for scientific work. The rapid growth of cities and an increase in immigration together created the awareness and the justification for the public health movement. "Scientific" procedures and scientific knowledge promised to provide the answers: the prestige of science, both as a method of thought and action as well as a body of knowledge, was greatly reinforced. Further arguments for research were drawn from the growing triumphs of the new "scientific medicine" practised in the teaching hospitals associated with universities.[4]

The prestige of scientific medicine, the potentialities of scientific research and the moral power believed to inhere in the progressive virtues of scientific method together constituted a combination appealing to the philanthrophic dispositions of some rich men. The Carnegie Institution of Washington, the Carnegie Corporation, the Rockefeller Institute and the various Rockefeller endowments were the result of this persuasion. In the years immediately before and following the First World War, their directors, their staffs, and their grants added a powerful new ingredient to the forces shaping the direction of the natural sciences.[5] The scale of expanding resources exemplified in this munificence was paralleled in countless lowlier enterprises. The profession of science grew proportionately (Table 2). By the time the United States entered the First World War, a small élite from among the first recipients of the doctorate from American universities had become famous and influential, not only in pure science but also in scientific medicine, in governmental science, in the administration of new and old

Table 2. *Industrial research laboratories and scientific societies*

A. Number of industrial research laboratories by five-year periods

Year	Number
1894–98	18
1899–1903	39
1904–08	89
1909–13	136
1914–18	277
1919–23	441

B. Membership in the American Association for the Advancement of Science

Year	Number
1875	807
1900	1,920
1910	8,010
1920	11,500

C. Membership in the American Chemical Society

Year	Number
1890	238
1900	1,710
1910	5,080
1920	16,600

Sources: Bud, Robert F., Carroll, P. Thomas, Sturchio, Jeffrey L., and Thackray, Arnold. *Chemistry in America, 1876–1976: An Historical Application of Science Indicators* (Dordrecht, Holland: D. Reidel, in press), *Proceedings* of the A.A.A.S. for the years listed: Skolnick, Herman and Reese, Kenneth M. (eds.), *A Century of Chemistry* (Washington, D.C.: American Chemical Society, 1976), p. 456.

universities, in the creation of learned societies, and in the development of philanthropic institutions.

Four illustrations must suffice. David Starr Jordan, who received a master of science degree at Cornell in 1872 and a doctorate at Butler University in 1878, had a distinguished career as an ichthyologist and as president successively of Indiana and Stanford Universities. He declined the secretaryship of the Smithsonian Institution in 1906, but served as president of the American Association for the Advancement of Science in 1909 and of the National Education Association in 1915. J. Playfair McMurrich, who graduated from the University of Toronto in 1879 and was awarded a doctorate at Johns Hopkins in 1885, taught anatomy at Haverford College and Cincinnati, at Michigan, and at Toronto, where he eventually became the first dean of the graduate school. He was president of the American Society of Naturalists in 1907, and of the American Association for the Advancement of Science and the Royal Society of Canada in 1922. Simon Flexner took a medical degree at the University of Louisville in

1889, and went on to a major scientific career. He served as president of the Society of Experimental Biology in 1906, director of the Rockefeller Institute from 1903 to 1935, trustee of the Carnegie Institution from 1910 to 1914 and the Rockefeller Foundation from 1913 to 1928, and as president of the American Association for the Advancement of Science in 1920. John Campbell Merriam received a bachelor of science degree at Lennox College in 1889 and was awarded the doctorate at the University of Munich in 1893; he rose to be professor and chairman of the newly founded department of paleontology at the University of California at Berkeley in 1912. He was subsequently chairman of the National Research Council in 1919 and dean of faculties at the University of California in 1920. He resigned this last position to become president of the Carnegie Institution of Washington. In Washington he also became chairman of the committee on government relations of the National Academy of Sciences for many years, and he concluded his administrative career by membership of the Science Advisory Board of President Franklin D. Roosevelt.[6]

These four men were in a distinct minority in their generation, but they embodied its dominant ideals. Although their primary interest lay in scientific work and in the administration of research, they had a pronounced sense of the scientific tradition of which they were the beneficiaries. They wished to see that tradition celebrated. It was quite natural for all four to serve as members of the organising committee of the History of Science Society, in 1924. In this latter role their gifts, ambitions, and successes were representative of that group of scientific "patrons" who were to leave a mark on the institutional patterns of this fledgling speciality.

THE HISTORY OF SCIENCE AS A VISIBLE ACTIVITY

The history of science first emerged as an identifiable activity in the United States in the 1890s. A number of connected causes may be discerned.

Though the number of doctors of philosophy wholly trained in the United States was growing apace, many of the best educated, most able and most affluent students still journeyed to Germany either for their doctorate itself or for a period of training. American students thus became aware of the idealistic implications of such books as Ernst Mach's *Die Mechanik in ihrer Entwicklung, historisch-kritisch dargestellt,* which was published in Leipzig in 1883, and of the writings of Wilhelm Ostwald. The reflective historical arguments of these works provide materials to which members of the "first generation" referred in order to sustain their self-confidence and to develop arguments in favour of science as pure knowledge.

In addition to this, the major American universities conducted their first systematic courses for graduate students in the 1890s. The obvious danger that these seminars might breed a race of narrow specialists was foreseen by some of the scientists who taught them and they wished to guard against it. The scientific disciplines were still somewhat defensive in the face of the criticisms, explicit and unspoken, coming from the older academic specialties such as classics, philology, and moral philosophy. University scientists welcomed a humanistic rationale that could both meet these criticisms and help to establish the higher justification for science. And this, it appears, helped lead to regularly offered courses on the history of the various scientific disciplines, sometimes for graduate students and sometimes for advanced undergraduates.

The Massachusetts Institute of Technology was the scene of one such initiative. In 1887 W. T. Sedgwick began lecturing on the history of biology and C. R. Cross on the history of the physical sciences. By 1905 their two courses had been combined into a regular series on the history of science, given by Professors Sedgwick and H. W. Tyler. Sedgwick, Cross and Tyler were all members of the "first generation." Sedgwick was a graduate of Yale and of Johns Hopkins and, in 1887, a rising young associate professor. In due course he was to become chairman of the department of biology and public health at the Massachusetts Institute of Technology, where he acquired a very distinguished reputation, and to serve as president of numerous scientific societies including the Society of American Bacteriologists (1900), the American Society of Naturalists (1901), and the American Public Health Association (1914–15). Cross was educated at, and headed the physics department of the Massachusetts Institute of Technology. Their colleague H. W. Tyler was educated at Massachusetts Institute of Technology and at Erlangen, where he was awarded a doctorate in 1889. He became in turn secretary of the Institute, chairman of its mathematics department and, in 1916, general secretary of the American Association of University Professors.[7]

A similar pattern was followed at Harvard, where T. W. Richards had been awarded a doctorate in 1888. After a year abroad, Richards began a lifetime of brilliant service on the teaching staff at Harvard. His standing was confirmed when, in 1914, he became the first American to be awarded a Nobel prize and his considerable influence was enhanced when, in 1921, his only daughter married J. B. Conant. As a young instructor in 1890 Richards introduced what proved to be a regular series of lectures on the history of chemistry – lectures which led one of his many gifted pupils, L. J. Henderson, to launch in 1911 a survey of the history of the physical and biological sciences.[8] Stanford from its very beginning provided teaching in the history of chemistry, while the University of Chicago in 1892 offered courses in the history of astronomy and chemistry and later attempted "a very unique and ambitious plan for the fostering and development of the historical courses in science".[9] At the University of Pennsylvania, Edgar Fahs Smith began in 1896 by lecturing in the history of chemistry;[10] while historical courses in one or other fields of science were also given at the University of California at Berkeley, and at Cornell, Illinois, Johns Hopkins, Michigan, North Carolina, Northwestern, and Yale Universities.

By the early 1900s, it became common to discuss explicitly the aims of such teaching. The purpose most often avowed was "the much desired humanizing of science". That aim was linked with the recruitment into science of undecided students, the advancement of bold claims regarding the place of natural science and "scientific method" in modern culture, and the desire to integrate the increasingly differentiated scientific specialities. There was no discussion of the history of science as an academic discipline or of the problems to be dealt with in the application of historical methods to the natural sciences, in which the past was less important than the future. Instead, the educational needs of students of science and the moral and cultural concerns of scientists together defined the proper criteria which should govern the growth of the subject. Understandably, an undercurrent of conviction was present that "history as now taught is altogether too largely devoted to the political phase of the development of civilization".[11] What their proponents believed to be "more fundamental movements", namely "scientific things of the most widespread influence" were not receiving their due. Not many went as far as C. R. Mann, professor of physics at the University of Chicago, but many were probably in sympathy with his robust belief that "the history of science as I see it is the history of civilization since the fall of Rome".

The developing discussion of the importance of science and scientific method was accompanied by a slow, steady attempt to broaden perspectives drawn from the history of partiular sciences to the history of the sciences or perhaps even the history of science. The outbreak of the First World War in Europe was used to give special urgency to the subject. In November 1914, Walter Libby, one of the keener enthusiasts and a "proto-historian of science", published an article adumbrating the by-then familiar themes, but with a new twist:

> One must recognize that science is international, English, German, French, Italian, Russian, all nations cooperating in the interests of racial progress. Accordingly, a survey of the sciences tends to increase mutual respect and to heighten humanitarian sentiment. The history of the sciences can be taught to people of all creeds and colors, and cannot fail to enhance in the breast of every young man or woman, a faith in human progress and good will to all mankind.[12]

The message was taken up by F. E. Brasch, another of the field's hopeful proponents. Summarising the widely held consensus, he asserted that the history of science "represents a strong reactionary movement from the over-materialistic and specializing tendencies of the age" and that "this reaction finds its development in the German school of science, where historical method in the study of the sciences, theoretical and empirical has been practiced". *Science* was duly persuaded to publish Brasch's plea for the subject, in the form of a 16-page survey of "The Teaching of the History of Science". Brasch argued that:

> The final message of the history of science is to show the high plane of science – that which has given life, stability, truth and wealth – in its universal activities and its established international character as the arbiter of the future of men and of peace.[13]

Brasch urged in 1915 that a section of the American Association for the Advancement of Science be devoted to history of science. The same year saw the arrival in America of George Sarton, a young Belgian scientist who was committed to a secular faith in science as the highest form of human activity and a guiding thread of history. Sarton too may be classed as a proto-historian of science. At one and the same time he was fleeing the German invasion of his native land and looking for ways to carry out his recently chosen commitment to the study of the history of science. Like Brasch and Libby, he lacked any institutional setting for his vocation. Unlike them, he possessed the driving determination of the true prophet. Behind him in Belgium, Sarton had left his home, his library, and his fledgling journal, *Isis: Revue consacrée à l'histoire de la science.* Before him lay hope and uncertainty. The prospect of America's involvement in the war foredoomed any rapid action on his plans for a full-scale institute for the history of science as it did Brasch's more modest ambitions. However, American wealth and American convictions about the value of science were sufficient to keep Sarton in the country; American experience in the war also strengthened Brasch's case.[14]

THE SCIENTIFIC ÉLITE AND NEW LEGITIMATIONS

The First World War was a catalyst of certain tendencies in American society and American thought. Trends previously unnoticed but no less real were brought into focus and sharp relief. For some members of the educated classes, the war led them to see the United States as a prospective beneficiary of the promotion, cultivation and application of science. Washington as the seat of the federal government began to preoccupy their minds. The natural sciences were assigned great responsibilities in this process of social advance. The National Academy of Sciences was stirred to life, the National Research Council was formed, and many individuals became very active in them.[15] The intermingling of the well-being of the country and the progress of science was a common article of faith among American natural scientists in this period.

The ten years following the war were years of exultant self-confidence and callow optimism, coupled with some apprehension about the new type of society which was coming into existence. A small and generally unregarded group of literary publicists, radicals and collectivistic liberals denied the value of the prosperity, the hedonism, and the still-persisting piety of American culture and society. Their criticisms acquired few converts to their outlook but they made some scientists feel the need to justify the pattern of society to which they were contributing. One ready answer lay in the affirmation of the value of science. Science, it was said, sustained the technology which was such a crucial element in America's progress and promised to relieve mankind of the evils of poverty and the pain of illness. The scientific method and the scientific outlook could also create a culture free from irrationality.[16]

The continued prosperity in higher education during this period was of a piece with this condition and mood. Professional education thrived. Four undergraduate years became the prerequisite for admission to medical school, with consequent effects on enrolment in courses in the natural sciences. The number of doctorates in those sciences increased rapidly.

The expanded scale of operation of academic science had its problems. It required the acquisition of additional financial resources by university administrators and senior academics. As spokesmen for their intellectual domains, they had to enlarge public appreciation of science as a distinctive strand of modern culture. They had to seek out wealthy and powerful patrons. They had of necessity to justify their own, their benefactors' and their protégés' commitment to the natural sciences. The conflicting interests of the patrons, the scientists, and the users and beneficiaries of scientific knowledge had to be brought into harmony. Many academics were in agreement with reflective businessmen and politicians, publicists and clergymen in their acceptance of the faith in science. Faith in science was expressed in a more popular version of a philosophy of science and in an active history of science. In the words of one exponent of the latter: "The progress of science and research has become an increasing and dominating force in our civilization. And it is, therefore, very natural that following this progress the spirit of reflection should take some definite form."[17] The history of science could, among other things, serve to unite the timeless laws of nature and the historical idea of progress. Within that idea, science was the source and seedbed of the age to come. The faith was shared by persons in quite disparate fields – George Sarton and Herbert Hoover both come to mind – but it found its guardians and spokesmen among the élite of the first generation of recipients of the doctorate in scientific subjects.

By the 1920s, that minority of the first generation who had risen with science were at the peak of their powers. For them, organising science, speaking for science, making representations for science before governmental officials and businessmen, and exercising judgement about the sciences and scientists had replaced active research. To them, the history of science became valuable in a variety of ways which were probably not so highly appreciated by the less distinguished members of their generation who were also interested in the subject. The new interest displayed by the scientific élite was to prove decisive in facilitating the forms, and shaping the content, of the history of science as an organised activity.

One attractive feature of the history of science was that it was a record of the successes of science. It displayed the heroic achievements of great scientists of the past and was a possible basis for confidence in the continuation of the achievements of scientists in the future. The history of science was an argument on behalf of science; it showed that it was not simply a momentarily useful set of devices but was one of the most grandiose of all the undertakings of the human race, with an illustrious past and limitless promise for the future. In addition to this, knowledge of the classics of science and a discriminating bibliophilia helped to create a universe of discourse which scientists could share with those rich businessmen who were amateurs of learning, with educated politicians, and with the trustees of foundations. The history of science also had other merits for the scientific élite. It could serve as a surrogate for active research for administrators in universities and government who wished to be acquainted with scientific acheivements but who were not specialists. George Sarton said as much in 1922:

The history of science is an encyclopaedic discipline. It will stand or fall as we accept or not the possibility of encyclopaedic knowledge. . . . It would not be unreasonable to expect university presidents to be encyclopaedists. . . . The president to whom the intellectual leadership [will in the future be entrusted will] need possibly an encyclopaedic education, and in any case a scientific one. . . . The encyclopaedic training would be the ideal one for a college president and for any man whose scientific duties are not specialized.[18]

Sarton's words aptly fitted such patrons of the history of science as William H. Welch. Welch's position included the presidencies of the American Association of Physicians in 1900, of the American Association for the Advancement of Science in 1906, of the American Medical Association in 1911, of the National Academy of Sciences from 1913 to 1916, and the acting presidency of the Johns Hopkins University in 1914; he refused to accept a permanent appointment to that post. He also served as trustee of the Carnegie Institution from 1903 to 1934 and as president of the board of scientific directors of the Rockefeller Institute from 1901 to 1933. He was a member of the council of the subsection of the history of science of the American Association for the Advancement of Science (1920), and he was successively a member of the council from 1924 to 1926 and again from 1927 to 1929, vice-president in 1930 and president in 1931 of the History of Science Society. Very fittingly, his final academic role was to be the founder and director of an institute devoted to the history of medicine.[19]

The career of Edgar Fahs Smith illuminates the ramified connections of the history of science. Smith came from an obscure Pennsylvania-Dutch family. He graduated from Gettysburg College in 1874 and received a doctorate from Göttingen in 1876. He went on to become a professor of chemistry at and then provost, *i.e.*, president, of the University of Pennsylvania from 1911 to 1920. He was president of the American Philosophical Society for six years. He was three times president of the American Chemical Society: in 1895, 1921, and 1922. He declined to stand for the governorship of Pennsylvania, but was on intimate terms with the leading circles of the Republican Party. He served as president of the state's electoral college in 1925. He was also a connoisseur of the history of science and a noted bibliophile. He was a founder of the history of chemistry section of the American Chemical Society in 1921 and was successively a member of the organising committee and council (1924–26), vice-president (1927) and fourth president (1928) of the History of Science Society. The values which underlay his concern with history, and his bibliophilia, are well expressed in a passage from his *Old Chemistries:*

The criticism that chemistry is absolutely commercialized is frequently heard and, further, that it is the commercial value of the science alone that claims the thoughts of chemists. Such views are widely prevalent. But other ideas exist . . . discarded "old chemistries" bring many other messages – messages in history, in philosophy, in economics, in social relations, in art, in international relations, in literature, and in a wide and extensive culture.[20]

The history of science benefited from the bibliophilic tastes and the encyclopaedic aspirations of some of the most eminent American scientists in the 1920s. David Eugene Smith received the doctorate from Syracuse University in 1887, and went on to serve as professor of mathematics at Teachers College, Columbia University, and as president of the international commission on the teaching of mathematics from 1908 to 1920. As author or co-author of more than 150 text books, he acquired a considerable fortune. Smith was the most active of the moving spirits in the organisation of the History of Science Society. He served as its secretary in 1924 and 1925, vice-president in 1926, and president in 1927. Smith was a bibliophile – his library of 10,000 volumes in the history of science is now at Columbia University. He linked the worlds of high science, high culture, and wealth. Through him, plans were drafted to persuade George Kunz, vice-president of Tiffany and Company, and Archer M. Huntington, cousin-in-law, stepson, and brother-in-law of Henry E. Huntington, a multi-millionaire railway magnate and philanthropist in his own right, to supply the financial support for a research institute in the history of science. Through him, George Plimpton, a very successful banker, chairman of Ginn and Company, trustee and treasurer of Barnard College, trustee of the World Peace Foundation and president of the trustees of Amherst College, was persuaded to join the History of Science Society and to become a member of its council.[21]

The outlook of the history of science was progressive, idealistic, internationalistic, and patriotic. It appealed to the expert; it prized objectivity, it implied the desirability and inevitability of progress; it was a paean to the role of science in furthering the good of mankind. It seemed to say that science could help in the achievement of the highest ideals while remaining apolitical. These features made it very attractive to many educated, public spirited and wealthy Americans. Despite this powerful support in intellectual curiosity and in ideals, wealth and influence, the history of science had great difficulty in acquiring the cognitive and professional identities which mark an established academic discipline.

THE HISTORY OF SCIENCE SOCIETY: PARTIAL INSTITUTIONALISATION WITHOUT ACADEMIC ESTABLISHMENT

The history of science easily achieved an imperfect kind of institutionalisation between the two world wars. In April 1919 a formal call to organise was published in *Science*. By the end of the following year a "history of science subsection" was established in Section L of the American Association for the Advancement of Science. Sufficient energy was available not only to sustain this subsection but also to engender the creation of the history of chemistry section of the American Chemical Society in 1921 and the American Association for the History of Medicine in 1924, in addition to the History of Science Society itself.

The letter of December 1923 announcing the formation of the History of Science Society was signed by an organising committee of 37 individuals.[22] If to their number we add the 10 officers appointed in the first five years who were not among the original committee, there were in all 47 persons in the organising group (Table 3). That group was made up of three subgroups. The largest,

Table 3. *The organising group of the History of Science Society*

Subgroup	Number	Age range in 1924	Number with foreign birth or educational experience	Number of members of National Academy of Sciences	Presidencies of other national societies pre-1924	Active commitment to internationalism
Patrons						
Administrator	9	55–74	8	7	24	6
Research worker	17	54–73	13	6	10	4
Sub-Total	26	54–74	21	13	34	10
"Proto-Historians of science"	9	40–57	5	0	3	0
"Sympathisers"	12	35–55	6	2	2	0
Total	47	35–74	32	15	39	10

Sources: Information derived from the *Dictionary of American Biography*, the *Biographical Memoirs* of the National Academy of Sciences, *American Men of Science*, the *National Cyclopaedia of American Biography*, obituary notices in *Isis*, etc.

containing extraordinary talent and power, was that of the "patrons". Its 26 members – all scientific administrators and active scientists from the "first generation" – were in their mid-fifties or older. They had extensive European experience as students; 21 out of 25 had studied in Europe; and 10 of them had strong and explicit commitments to peace and internationalism. They also had enormous influence within and outside the academic community. Thirteen of them were in the National Academy of Sciences, one was a Nobel laureate. Together they held 34 presidencies of national scientific societies including the American Association for the Advancement of Science, the National Academy of Sciences, the American Chemical, Mathematical, Philosophical, Physical, and Psychological Societies, the American Societies of Anatomists, Naturalists, and Orientalists, the American Neurological Association, the American Council of Learned Societies, the National Research Council and the Royal Society of Canada.

A clear difference in age, prestige and influence separates this subgroup from th next, that of the proto-historians of science. These nine men were relatively young – five were in their forties – they had less experience of Europe, and they did not enjoy eminence in learned societies. They did however play significant roles as editors and as enthusiastic propagandists for the history of science to which their tenuous academic careers were closely linked. They were in fact mainly "outsiders" *vis-à-vis* the academic profession. Frederick Barry, for example, who was born in Massachusetts in 1876, was trained as a chemist at Harvard under T. W. Richards. He received the degrees of bachelor of arts in 1897 and master of arts in 1909. He was awarded the doctorate in 1911. He spent one year as research assistant in chemistry at the Carnegie Institution, before moving to Columbia University as an instructor. He was not granted permanent tenure as a chemist and "for six years after 1917 he continued researches in Columbia laboratories, but as an independent scientist." In 1923 he had rejoined the Columbia academic staff, teaching the history of chemistry. In 1928 he became "associate professor of the history of science," though he had little else to show.[23]

Frederick E. Brasch was one year older than Barry, and even more marginal to academic life. He attended university at both Stanford and Berkeley, but graduated from neither. His special fields were mathematics and astronomy, and he was employed briefly at the Harvard College Observatory from 1902 to 1904 and at the Lick Observatory on Mount Hamilton in 1905. The next seven years were spent in "private study in the history of science". He was then variously employed in Stanford University Library, the John Crerar Library in Chicago, the James J. Hill Library in St. Paul, Minnesota, and in the National Research Council, the Carnegie Institution and the Library of Congress in Washington, D.C. He was George Sarton's first American disciple, enrolling as a special student in his course in the history of science at Harvard University in 1916–17. He was repeatedly disappointed in his hopes for an academic appointment and for support for his research. He retired to California and to private life while still in his fifties. His fluctuating fortunes did not prevent, though they scarcely helped, his propaganda on behalf of his chosen field. His survey of the teaching of the history of science in the United States in *Science* in 1915 established his position as an organiser. In the period after the First World War he was able to animate the commitment to the history of science of Section L of the A.A.A.S., which he served as secretary from 1921 to 1924 and 1925 to 1928. He was also a faithful factotum of the History of Science Society. He held the unsalaried posts of treasurer and assistant secretary from 1924 to 1928 and corresponding secretary from 1928 to 1938.[24]

The difference in standing between the "patrons" and the "proto-historians" was to some extent blurred by the scattering of "sympathisers". This last subgroup was composed of physicians interested in the history of medicine, and, by extension, science; scientists of lesser accomplishments and prestige than the patrons; and a cluster of young academic historians. The intermediate position of the sympathisers was not, however, sufficient to bring together the patrons and the proto-historians; they remained apart. The separateness of these two subgroups from each other, accounts for the contrast between the lofty rhetoric of the patrons and the difficulties of the proto-historians in establishing their subject on a serious academic basis.

The history of science was generally regarded as a subject of secondary importance. This was impressed on Frederick Brasch when, in 1922, he sought a fellowship from the National Research Council to allow him to work on a history of American astronomy to be completed in two volumes. From Yale, his admirer Frank Schlesinger wrote:

> there is no question concerning your ability to write an acceptable history of astronomy in America. . . . The only question in my mind is whether astronomers ought to ask the Research Council to spend money that otherwise might be free for other astronomical purposes . . . I should like an opportunity of discussing the matter with some of my colleagues. My feeling is that a work of this kind, however valuable it may be, might well be undertaken in your spare time . . .

A similar letter was written by a Princeton astronomer. It read:

> You are doing decidedly desirable and useful work. . . . I am very glad to express my approval of it . . . in my judgment it does not seem desirable to recommend that the Carnegie Institution should appoint a research associate to deal with this problem. . . . I believe that other more important uses could be found for the funds of the Institution. . . . I hope that you will recognize that my decision is based upon what appeared to me as general principles, quite independent of any personal considerations.[25]

The history of science was regarded as a reasonable subject but only to be pursued avocationally and not worthwhile enough to command the financial support which could be used for scientific research.

Another example is provided by George Sarton's determined but never-realised ambition to create an Institute of the History of Science and Civilisation. Sarton's plans for an institute dedicated to pure scholarly research were set out at length in *Science* in 1917. The scheme was perfectly in keeping with the aspirations of American academic life in the period immediately following the First World War. Several of the "patrons" of the history of science took some interest in the plan; in

1924 William H. Welch for one thought that the "institute for the history of science . . . if properly located, conducted and supported" would be an important undertaking. As it turned out, however, it was Welch, a patron, not Sarton, a proto-historian, who was in due course rewarded with an institute. Similarly, another "patron", James Breasted, already one of the most famous Egyptologists in the world and who was the second president of the History of Science Society (1925), obtained his own Oriental Institute at the University of Chicago. Sarton's plan was probably closer to the aims expressed by those members of the scientific élite who supported the History of Science Society but in 1924 he was still an immigrant, he held only a very marginal academic appointment, and he was the youngest of the struggling proto-historians.[26]

Only in the rarest instances could personal considerations offset the reigning principle that available funds should be spent on scientific research while the history of science was properly a matter for amateurs. This was not to say that such history was thought to be unimportant: it might after all claim to be "the history of civilization since the fall of Rome". Rather, the ideals expressed in and through the affirmation of this history were not to be advanced through professional historical scholarship but through the practice of rigorous research in the natural sciences. Continuous, systematic, detailed and specialised historical research was of no great urgency to the most powerful patrons of the history of science. Indeed, the history of science was intended to offset the specialisation of research.

INTERESTS, VALUES AND INTELLECTUAL PROGRAMMES

The History of Science Society was a useful platform from which to express certain beliefs common to the "patrons" of the new subject.

The "apolitical" politics of scientific internationalism was one common value, reflecting as it did the cosmopolitan experiences, European heritage and American pride of the patrons. By the 1920s the forces nourishing that internationalism included not only the traditional belief in a "republic of learning" beyond the confines of nationality, but a desire to extend wartime links and to overcome the disappointments born of the failure of Woodrow Wilson's grand designs.[27] The vastly increased wealth of the United States together with improved transoceanic communications encouraged the patrons in their internationalism. The programmes of the Rockefeller Foundation in medicine and public health sent American men of science to all parts of the globe. A belief in the international character of scientific truth which transcended the boundaries of particular cultural traditions was consistent with experiences of administrators like Welch and Flexner at this time.

By the 1920s, Americans had ceased to go in large numbers to Europe to study for the doctorate although American postdoctoral studies in Europe continued. There was a sense of close affinity with European scientists and above all a conviction that American and European scientists both shared a common tradition. The establishment of international societies devoted to the history of science and medicine was, in part, a consequence of this belief; it also helped to sustain it.[28] The wish of the patrons to locate American science within a wider context meant that in its early years the History of Science Society moved quite naturally from one annual meeting wholly devoted to colonial American science to another dealing exclusively with Isaac Newton.[29] An internationalistic vision also encouraged concern with science outside Western civilisation. This interest was, moreover, perfectly compatible with a definite belief in the superior merits of the "Anglo-Saxon" tradition. George Sarton was deeply convinced of the importance of the achievements of Islamic science, but he readily admitted that:

> My chief reasons for choosing English [as the sole language for *Isis*, after the First World War] . . . are my faith in the Anglo-Saxon conception of life, and also my love of and my hope in the younger civilizations of the world: first of all, this United States, then also Canada, Australia, South Africa, and New Zealand . . . I believe that as far as the diffusion of common sense in politics and the spirit of fair play are concerned these young nations are the hope of the world. They are, on the whole, less trammeled by precedent and by prejudice, and more capable of working out radical ideas in a conservative way. I trust that the ideal of the New Humanism – that is, the reconciliation of science and art, truth and beauty – will find a more appreciative audience among them.[30]

The values at stake were exemplified in the life of George Ellery Hale, who was a member of the organising committee of the History of Science Society. Hale had been educated at the Massachusetts Institute of Technology and at the University of Berlin and, after serving as professor of astronomy and director of the Yerkes Observatory at the University of Chicago, became director of the Carnegie Institution's Observatory at Mount Wilson, California. He combined, to an extraordinary degree, high scientific achievements, administrative skill and energy, political acumen, and an internationalistic outlook. He was the recipient of many honours from European institutions including the Académie des sciences of Paris, and the Royal Society of London. He founded and edited *The Astrophysical Journal.* He played an important if not the most important role in creating the California Institute of Technology, the Henry E. Huntington Library and Art Gallery, and the National Research Council. He raised funds for a permanent building for the National Academy of Sciences. Always an internationalist by conviction, his position as foreign secretary of the National Academy of Sciences from 1910 to 1921 gave him a base for much activity. From it he helped to launch the International Research Council, later to become the International Council of Scientific Unions, and the International Astronomical Union. To Hale, it was self-evident that the history of science was the history of progress, of the accumulation of objective knowledge and of an internationalism superior to the corruption of nationalistic politics. Science, unfolding successively deeper truths about nature, was also bringing about greater economic prosperity and moral progress for mankind. Hale was happy to encourage George Sarton over the years and to support the History of Science Society, but his efforts to raise funds went to more urgent subjects, more closely connected with scientific research.[31] Research on the history of science, while admirable in theory, was in practice hardly an urgent task.

Beliefs in progress and internationalism were thought to be confirmed by the history of science, and provided common ground for otherwise anthithetical elements. The progress of science and the grandeur of its history were compatible with the main strands of political belief, among academics of that time, from a devotion to democratic socialism and collectivistic liberalism to a championing of private enterprise, free of governmental control.

One group in the History of Science Society espoused ideas such as the "democratic optimism" and commitment to Fabian "social engineering" of James Harvey Robinson, J. M. Cattel and John Dewey or the "free thought" characteristic of H. E. Barnes and G. L. Burr. In contrast, L. J. Henderson, J. Playfair McMurrich, D. E. Smith and W. H. Welch were much closer to the "cautious and conventional" beliefs of John C. Merriam. There is no indication that political and religious differences ever caused tension over the proper goals and procedures for the History of Science Society.[32] James McKeen Cattell, who was another of the founding generation of the History of Science Society, for example, was a socialist. He was a founder of the American Psychological Association, of the *Psychological Revew*, of *American Men of Science*, and of *School and Society*. He transformed *Science* and *Popular Science Monthly* and was president of the American Association for the Advancement of Science in the year the History of Science Society was founded. For Cattell, belief in the high value of the history of science and the possibilities of moral progress went hand in hand with his commitment to the merits of democratic socialism. At the same time, these beliefs about the values which were thought to be furthered by the history of science would equally well be linked with a flexible conservatism as in the cases of Merriam and Edgar Fahs Smith. As a trustee of the Carnegie Foundation for the Advancement of Teaching, Smith severely criticised Cattell's views on politics and economic policy. The "thirty-third degree Freemason" and the forthright socialist held markedly different political beliefs. However, the history of science provided a platform in praise of progress on which they could comfortably stand together.

The activity of the History of Science Society was a rhetorical affirmation of the supreme value of modern science and scientific method. Its espousal of the need for a "new humanism" brought it into an uneasy alliance with certain historians who wished to break away from traditional political historiography. The rationale for history of science as a "new humanism" could indeed be interpreted as a denial of the value of the traditional humanistic disciplines which studied classical texts and which implied that the past was at least equal in value to the present if not superior to it. Sarton certainly did not hide his intention:

> To focus the system of education upon scientific method and knowledge is . . . to replace a one-sided vision of the world which no longer corresponds to the available knowledge of our time by one which is more adequate; to replace incomplete humanities by complete humanities; to replace an ignorant and insecure idealism by one which is . . . solidly anchored on all the reality we have yet been able to grasp.

It followed that "the center of gravity of historical studies must be displaced". The traditional historian's "absurd pretense to control historical studies" must necessarily give way before the claims of that "new humanism" which found its centre and meaning in the history of science.[33]

This disparagement of the traditional humanistic disciplines as studied in the modern university did not gain the sympathy of either the political historian nor the classical scholar. It was more acceptable if not entirely welcome to those concerned with such fields as oriental history and Egyptology which were not as internally specialised or differentiated as the humanistic studies of European cultures and who were not committed to belief in the superiority to the present of the past which they studied. On the other hand, it did find sympathy and a warm response among those professional historians who were in revolt against what they thought was an excessively narrow tradition of political history.

James Harvey Robinson's view of history as the history of the progress of the rational mind entailed the inclusion of previously neglected social groups and realms of thought into its subject matter. As early as 1912, one of his students at Columbia University, Martha Ornstein, wrote a pioneering doctoral dissertation on *The Role of Scientific Societies in the Seventeenth Century*. In the foreword of 1928 to the published version of that dissertation, Robinson asserted that "The distinctive fruit of man is his power of accumulating knowledge . . . [I]n the early part of the seventeenth century . . . experimental science began to result in what Bacon calls 'the kingdom of man,' in which knowledge acquired with the most scrupulous precautions against mistakes would continue to increase indefinitely, meanwhile ever bettering man's estate."[34] Robinson's rhetoric was very consistent with the scientism of the organising group of the History of Science Society. His *The Mind in the Making* and especially *The Humanizing of Knowledge,* both written in the early 1920s, might even have been taken as a manifesto for the History of Science Society. Together with his disciple – Harry Elmer Barnes – the two of his pupils at Columbia – Preserved Smith and Lynn Thorndike – Robinson formed an articulate bloc of "sympathisers" within the organising group. Barnes was eloquent about the importance of science and technology to the historian, while Smith wrote a work of European intellectual history since the Renaissance which gave much prominence to science. Thorndike had the greatest accomplishment as an historian of science, and perhaps because his theme of *Magic and the Rise of Experimental Science* required many volumes for its completion and because it dealt with more remote periods it was regarded as legitimate by conventional historians.[35]

Despite these efforts, the history of science found no secure place in the schedule of activities of academic historians. The "new historians" were themselves a very tiny group within the historical profession. And, between the two world wars, the number of students specialising in history was neither especially large nor expanding. Opportunities for innovation were in short supply, and hotly contested. The inability to interest academic historians was accentuated by the excessive claims of the proponents of the history of science in America. To become part of academic history, the subject would necessarily have to renounce its pretence that "the history of science is the only history which can illustrate the progress of mankind".[36] Although this claim was agreeable to scientific statesmen and was pleasing to the proto-historians of science in their many struggles, academic historians did not like it. Forced to choose between science and history as proper patron within the departmentally organised university, the history of science came to depend on the more powerful, more visible and more numerous body of natural scientists, but this dependence cut it off from departments of history and from graduate students trained in historical methods.

Sarton was of course the most active of the propagandists on behalf of the history of science. He was also the most tireless organiser of its necessary institutions. He created *Isis,* the journal maintained by the History of Science Society. He produced the long succession of "critical bibliographies" in the subject. He worked on what was designed to be the definitive *Introduction to the History of Science.* He produced a stream of arguments for his chosen field. His formulations of the importance of the subject, and of the moral growth of science, were harmonious with the conception of the value of science generally accepted by its most eminent scientific spokesmen. This was true of his declarations about encyclopaedic knowledge, or about the merits of the Anglo-Saxon tradition, or when he said that:

> The true internationalism toward which the unity of knowledge and the unity of mankind are steadily driving us, will constitute an immense progress. This progress will be largely due to the development of positive knowledge and scientific methods . . . whatever happens at the surface, the essential unity of mankind is a reality which will be attested in the future as it has been in the past by a small elite of scientists. These will pursue their sacred task, undisturbed – even as bees pursue theirs – in spite of wars and other cataclysms. Thus, thanks to them, the secret advance of mankind may go on indefinitely.

In so far as the study of the history of science could enhance the prestige of science by showing its antiquity and in so far as it demonstrated the progress of mankind leading to its contemporary elevation, it received benign encouragement. But it was not thought that relatively recent work required historical treatment. The value of that science was self-evident. Recent and contemporary things did not have to be studied historically. Thus one of George Sarton's own ambitions was "to carry on simultaneously research on ancient science and on nineteenth century science".[37] The first intention was realised while the second quietly failed. Scientists and administrators spoke of the pervasiveness of science and its historic role in shaping the modern world but, as patrons of the history of science, their interests were rather in history as a cultural adornment and a source of bibliophilic pleasure and of communications on esoteric subjects. Despite the affirmative declarations regarding the virtues of the modern world, the research reported in *Isis* ran rather in a different direction (Table 4).

In the days in which the history of science was being organised and Sarton was seeking to define his own intellectual tasks, he often spoke of "my book on nineteenth century physics". In practice, more remote problems and periods were to occupy his scholarly energies. The history of science avoided the recent

Table 4. *Articles in* Isis: *1920–29*[a]

Period	Number	Area	Number
Ancient and medieval	34	Arabic	9
		Indian	4
Sixteenth and		Chinese	2
seventeenth centuries	36	Egyptian	2
		Italian	2
		American	2
Eighteenth century and later	15	English	1

[a]n = 85 for period and 22 for area; the great majority of articles cannot be classified geographically.

past. It is thus understandable that Sarton and Lynn Thorndike, who were the two most gifted, energetic, and productive of the proto-historians of science, should have done their work so far from the history of the scientific activity which the patrons united in praising as the source of all that is best in modern life. The eminent humanistic scholars in the organising group, such as Charles H. Haskins and James Breasted, seemed to have felt a little tension between their austere conception of scholarship and the more grandiloquent praise of scientific progress by some of the other members in their group.[38]

THE NEW EQUILIBRIUM

A new phase in the formal establishment of the history of science occurred between the two world wars. Already in 1924 text books were being written and many courses were offered in universities. Five historians of science held academic appointments in their capacity as historians of science. Frederick Barry, 38 years old, was associate professor of the history of science at Columbia; Florian Cajori, 55, was professor of the history of mathematics at the University of California at Berkeley; Walter Libby, 57, was professor of the history of science at the University of Pittsburgh; George Sarton, aged 40, was lecturer in the history of science at Harvard, and Lynn Thorndike, then 42, was professor of history at Western Reserve.[39] The subject now had powerful patrons in the academic world. The American Association for the Advancement of Science recognised the legitimacy of the history of science, and the American Chemical Society was willing to establish a division for the history of chemistry. Learned societies were formed both for the history of medicine and for the history of science. The organising group of the History of Science Society included academic deans from Harvard, Columbia, and Toronto Universities; departmental chairmen from the California Institute of Technology and Northwestern University; presidents of foundations and academic administrators. The expanded size of the academic profession and the magnified role of science in American life together gave promise of, and offered context for, the development of academic programmes in the history of science. Yet such programmes did not materialise.

Instead, a new equilibrium was established with the subject enjoying much rhetorical but little financial or institutional support. It was able to maintain, but not to enlarge its learned society; individual membership of the History of Science Society reached 446 in 1925, increased very little in the 1930s, and by 1941 was down to 418.[40] The "proto-historians" had few students in a field so devoid of the necessities of academic success, above all opportunities for employment in the field. The new subject was unable to acquire the institutional provision necessary for its establishment and consolidation as a continuously and intensively cultivated body of learning. It had practically nothing of the institutional arrangements for regular lectures and seminars, or for Ph.D. programmes or fellowships, or for departments with their chairs on permanent appointment. It was taught as an adjunct subject in existing departments which were not very friendly to it. The patrons had not envisaged anything like the academic establishment of the subject and in the 1930s scientific statesmen faced far different problems. By then a new subject could not be inserted into the existing array of disciplines which were concerned to prevent the further attrition of what they already had.

The major universities grew somewhat in size, even during the inhospitable years of the depression, but the history of science did not share in that growth. When retirement or death removed Florian Cajori from Berkeley, Walter Libby from Pittsburgh and Frederick Barry from Columbia, they were not replaced. At Harvard, it took George Sarton until 1940, when he was 56 years of age, to gain an appointment on permanent tenure. The savour of even that victory was lessened by the decision of the Carnegie Institution, the following year, that its financial support for the history of science at Harvard would cease with Sarton's retirement.[41] Equally discouraging was the situation with respect to his young protégé, I. Bernard Cohen. In response to Sarton's pleas, his old friend and supporter L. J. Henderson could only say in a letter of 18 June 1941:

> I am terribly sorry to tell you that I can find no way of doing anything for Cohen. This is the conclusion of long consideration and careful conversations . . . the difficulty is that both the Carnegie Institution and the University have taken the position that they cannot, as a matter of principle, commit themselves . . . I can think of nothing else that you could do except yourself to talk with the President, and I should suppose that would probably be unavailing in spite of his deep interest in the history of science . . .[42]

Thus, the subject was honoured but not supported.

Although 1941 was a dark hour for the subject at Harvard and elsewhere, the history of science was being assured of a more prosperous future. The president of Harvard, like many other scientists, was already being drawn into work on the development of an atomic bomb.[43] That bomb was but one of a range of contributions of scientific research to military technology in the Second World War. Those contributions helped create a new appreciation of the importance of science to the national state. The vast extension of the American commitment to pure science, which was one result, brought with it renewed attention to the various intellectual interests which were formed around its boundaries. At the same time, the years since the Second World War have seen a deepening historical consciousness in the United States, and a huge increase in the size of the student population. These three developments together ensured that the history of science would gain enough support to become established as an academic discipline. But in 1941 the history of science seemed doomed to live in intellectual twilight. The subject was caught between the exigent demands and imperatives of university departments organised according to their disciplinary jurisdiction on the one side, and on the other patrons who were more interested in the celebratory legitimation of science than they were in the scholarly study of its history.

NOTES

1 See Olesko, Kathy, "Employment Trends in the History of Science," *Isis,* LXXII, 258 (September, 1981), pp. 477–479, and Ray, Gordon N., in *Reports of the President and Treasurer for 1979* (New York: John Simon Guggenheim Memorial Foundation, 1980), pp. 26 and 32. The fall 1981 issue of the *Journal of Interdisciplinary History* is devoted to "The New History: in the 1980s and Beyond", and carries a full discussion of the history of science.

2 See Veysey, Lawrence R., *The Emergence of the American University* (Chicago: University of Chicago Press, 1965), and Oleson, Alexandra and Voss, John (eds.), *The Organization of Knowledge in Modern America, 1860–1920* (Baltimore: Johns Hopkins University Press, 1979).

3 Many of the major disciplinary learned societies date from the 1880s, *e.g.,* The American Society of Naturalists (1883), Modern Language Association (1883), American Historical Association (1884), American Economic Association (1885), American Mathematical Society (1888), American Political Science Association (1888), Geological Society of America (1888), American Physical Society (1889), and American Chemical Society (1876, reorganised in 1889). For Harvard, see James, Henry, *Charles W. Eliot,* II (Boston: Houghton, Mifflin, 1930), appendices C and D.

4 A recent account is Noble, David F., *America by Design. Science, Technology and the Rise of Corporate Capitalism* (New York: Knopf, 1977), especially chap. 7. See also Hughes, Thomas P., *Thomas Edison, Professional Inventor* (London: H.M. Stationery Office, 1976). On science and medicine, see Rosenkrantz, Barbara G., *Public Health and the State* (Cambridge, Mass.: Harvard University Press, 1972); Fleming, Donald, *William H. Welch and the Rise of Modern Medicine* (Boston: Little, Brown, 1954) and Stevens, Rosemary, *American Medicine and the Public Interest* (New Haven: Yale University Press, 1971), especially chs. 2 and 3.

5 See Fosdick, Raymond B., *The Story of the Rockefeller Foundation* (New York: Harper and Brothers, 1952) and Kohler, Robert E., "The Management of Science: The Experience of Warren Weaver and the Rockefeller Foundation Programme in Molecular Biology", *Minerva,* XIV, 3 (Autumn 1976), pp. 279–306. See also Reingold, Nathan, "National Science Policy in a Private Foundation: The Carnegie Institute of Washington, 1902–1920", Oleson, A. and Voss, J., *Organization of Knowledge,* pp. 313–341.

6 David Starr Jordan (1851–1931), is listed in the *Dictionary of American Biography* (cited henceforth as *DAB*), X (New York: Charles Scribner's Sons, 1937), p. 211. For John C. Merriam, see *DAB,* XXIII, pp. 519–520. Details on Simon Flexner (1863–1946) are given in the *Dictionary of Scientific Biography* (cited henceforth as *DSB*), V (New York: Charles Scribner's sons, 1972), pp. 36–39. In terms of the themes of this essay there were no sharp distinctions between Canada and the United States, as McMurrich's career indicates: J. Playfair McMurrich (1859–1939), *Encyclopedia Canadiana* VI (Toronto: Grolier, 1972), p. 298.

7 William Thompson Sedgwick (1855–1921), *DAB,* XVI, p. 552; Charles Robert Cross (1848–1921), *Who Was Who in America,* I (Chicago: Marquis, 1973), p. 279. An éloge of Harry Walter Tyler (1863–1938) appeares in *Isis,* XXXI, 83 (November 1939), pp. 60–64. The course of Sedgwick and Tyler led to their *Short History of Science* (New York: Macmillan, 1917), a revised edition of which reached its seventh printing in 1958.

8 T. W. Richards to F. E. Brasch, 10 March, 1915: "I began giving a course on the history of chemistry in 1891 . . . The subject is treated from a philosophic and historical standpoint and was the real beginning of our courses given by Dr. Henderson on the history of science." See also the letter dated 31 March, 1915 in which Richards corrected the date to 1890. (Brasch papers, Stanford University Library. Cited hereafter as BSU.) Theodore William Richards (1868–1928), *DAB,* XV, p. 556; James Bryant Conant (1893–1978) was a pupil of Richards and president of Harvard University from 1933 to 1953; Lawrence Joseph Henderson (1878–1942), *DAB,* XXIII, pp. 349–352, was a biological chemist and an influential personage at Harvard. In 1910 Henderson married a sister-in-law of his mentor, Richards. For Frederick E. Brasch (b. 1875), see below.

9 The quotation is from the best guide to these early developments: Brasch, F. E., "The Teaching of the History of Science", *Science,* XLII, 1091 (26 November, 1915), pp. 740–760.

10 Edgar Fahs Smith (1854–1928), *DAB,* XVII, pp. 255–256.

11 This and the three following quotations are from letters of F. R. Moulton – the pioneer of the subject at the University of Chicago – to F. E. Brasch, 8 February, 1909 (BSU) and C. R. Mann to F. E. Brasch, 15 January, 1910 (BSU). Charles Riborg Mann (1869–1942), *Who Was Who in America,* II, p. 343; Forest Ray Moulton (1872–1952), *Who Was Who in America,* III, p. 622. For a characteristic set of arguments for the utility of the history of science, see Mann, C. R., "The History of Science – An Interpretation", *Popular Science Monthly,* LXXII (April 1908), pp. 313–322.

12 "The History of Science", *Science* XL, 1,036 (6 November, 1914), pp. 670–673, Walter Libby (1867–1955) was at the time professor of the history of education at the Carnegie Institute of Technology in Pittsburgh.

13 Quotations from an unpublished letter by F. E. Brasch to the editor of *Science,* 17 January, 1915 (BSU), and Brasch, F. E., "Teaching History of Science."

14 George Sarton (1884–1956), *DSB,* XII, pp. 107–114. See also Thackray, Arnold, "Five Phases of Prehistory, Depicted From Diverse Documents", *Isis,* LXVI, 234 (December 1975), pp. 445–453.

15 See Kevles, Daniel J., "George Ellery Hale, The First World War, and the Advancement of Science in America", *Isis,* LIX, 199 (Winter 1968), pp. 427–437. See also Cochrane, Rexmond C., *The National Academy of Sciences: The First Hundred Years, 1863–1963* (Washington, D.C.: National Academy of Sciences, 1978) and Reingold, Nathan, "The Case of the Disappearing Laboratory", *American Quarterly,* XXIX, 1 (Spring 1977), pp. 79–101.

16 The writings of E. E. Slosson offer an epitome of this secular faith in science. See

Tobey, Ronald C., *The American Ideology of National Science, 1910–1930* (Pittsburgh: Pittsburgh University Press, 1971). See also Hollinger, David A., "Science and Anarchy: Walter Lippmann's *Drift and Mastery*", *American Quarterly,* XXIX, 5 (Winter 1977), pp. 463–475.

17 Circular letter on the 1921 Meeting of Section L of the A.A.A.S., 15 December, 1922 (BSU).

18 Sarton, G., "The Teaching of the History of Science", *Isis,* IV, 11 (October 1922), pp. 225–249. Quotation from pp. 245–247. Sarton often returned to this theme. See his letter dated 3 November, 1938 to James B. Conant, quoted in Conant, "George Sarton and Harvard University," *Isis,* XLVIII, 153 (September 1957), pp. 301–305.

19 The institute was located in the Johns Hopkins University Medical School. Flexner, Simon and Flexner, James T., *William Henry Welch and the Heroic Age of American Medicine* (New York: Viking Press, 1941), chap. 18.

20 Smith, E. F., *Old Chemistries* (New York: McGraw Hill, 1927), p. 89. On Smith, see *Isis,* XI, 36 (June 1928), pp. 375–384; the Edgar Fahs Smith Memorial Collection in the History of Chemistry, presented to the University of Pennsylvania and endowed by Smith's widow, initially contained almost 1,000 rare alchemical and chemical works. Smith was an officer of the French Légion d'honneur from 1923 and recipient of the Priestley Medal of the American Chemical Society in 1926.

21 For details of these campaigns, see Thackray, A., "Five Phases of Prehistory."

22 The letter is reproduced and the founding group listed in *Isis,* VI, 16 (October 1924), p. 5–7.

23 See Barry, Frederick, "A Short Critique of the History of Science", *Columbia University Quarterly,* XXVI (June 1934), pp. 95–111 (September 1934), pp. 259–278; *American Men of Science,* 5th edn. (New York: The Science Press, 1933).

24 *Curriculum vitae* in BSU.

25 Frank Schlesinger (Yale University Observatory) to F. E. Brasch, 29 September and 13 October, 1922. Henry N. Russell (University Observatory, Princeton) to Brasch, 12 October, 1922 (BSU). Schlesinger was born in 1871 and was awarded the doctorate from Columbia University. He was a well-known astronomer and worked at a number of observatories, including the Yerkes Observatory of the University of Chicago, before moving to Yale in 1920. Russell, born in 1879, took a doctorate at Princeton University and became director of the University Observatory there in 1912; *Who Was Who in America,* III, p. 748.

26 See Sarton, George, "An Institute for the History of Science and Civilization", *Science,* XLV, 1160 (23 March, 1917), pp. 284–286 and 399–402; and *Isis,* XXVIII, 76 (February 1938), pp. 7–17. Welch's institute, at Johns Hopkins, was created with a grant of more than $500,000 from the Rockefeller General Education Board. Rockefeller philanthropies provided almost $10 million for Breasted's Oriental Institute at the University of Chicago; see Fosdick, R. B., *Rockefeller Foundation,* p. 238; Gray, George W., *Education on an International Scale* (New York: Harcourt Brace, 1941), pp. 78–82; and Thackray, A., "Five Phases of Prehistory."

27 See Schröder-Gudehus, Brigitte, *Les Scientifiques et la paix: La Communauté scientifique international au cours des années 20* (Montreal: Montreal University Press, 1978).

28 Gray, G. W., *Education,* and Fosdick, R. B., *Rockefeller Foundation,* especially chap. 9. The Société internationale d'histoire de la medicine was founded in Paris in 1921. The Académie internationale d'histoire des sciences was founded in Oslo in 1928. See Sarton, George, *Horus: A Guide to the History of Science* (Waltham, Mass.: Chronica Botanica, 1952), pp. 253–256.

29 *Isis*, IX, 30 (June 1927), pp. 223–225 and X, 34 (June 1928), pp. 333–337.

30 "War and Civilization", *Isis*, II, 6 (October 1919), p. 321.

31 George Ellery Hale (1868–1938), *DAB*, XXII, pp. 270–271. The extensive correspondence between Hale and Sarton, over the years 1916 to 1935, is preserved at Harvard University and at the California Institute of Technology. For instance, Hale wrote on 24 July, 1933 that "I wish it might be possible to realize a dream I have frequently had: of connecting you and *Isis* in some way with the Huntington Library." Hale papers, California Institute of Technology. See also Wright, Helen, *Explorer of the Universe: A Biography of George Ellery Hale* (New York: Dutton, 1966), especially chs. 14, 15 and 18; Kargon, Robert H., "Temple to Science: Cooperative Research and the Birth of the California Institute of Technology", *Historical Studies in the Physical Sciences*, VIII (Baltimore: Johns Hopkins University Press, 1977), pp. 3–31.

32 James Harvey Robinson (1863–1936), *DAB*, XXII, pp. 562–566; John Dewey (1859–1952), *DAB*, XXV, pp. 169–173; George Lincoln Burr (1857–1938), *DAB*, XXII, pp. 75–76; James McKeen Cattell (1860–1944), *DAB*, XXIII, pp. 148–151.

33 "Knowledge and Charity", *Isis*, V, 13 (October 1923), pp. 5–19; "The Teaching of the History of Science", *Scientific Monthly*, VII (September 1918), pp. 193–211.

34 Ornstein, Martha, *The Role of Scientific Societies in the Seventeenth Century* (Chicago: University of Chicago Press, 1928), p. ix.

35 Thorndike, Lynn, *A History of Magic and Experimental Science* (New York: Columbia University Press, 1923–58). On the "new history", see Higham, John (ed.), *History* (Englewood Cliffs, N.J.: Prentice Hall, 1965) and Goddard, Arthur (ed.), *Harry Elmer Barnes, Learned Crusader. The New History in Action* (Colorado Springs: Ralph Myles, 1968). Robinson was vice-president in 1928, Barnes was secretary in the same year, and Thorndike was vice-president in 1927 and president in 1928 of the History of Science Society. Preserved Smith (1880–1941), *DAB*, XXIII, pp. 725–726, was professor of history at Cornell in the 1920s. His *A History of Modern Culture* (New York: Holt, 1930), in two volumes, devotes its first six chapters to the natural sciences, culminating with a discussion of "The Scientific Revolution". On Lynn Thorndike (1882–1965) see *Isis*, LVII, 187 (Spring 1966), pp. 85–89. See also Barnes, H. E., *The History and Prospect of the Social Sciences* (New York: Knopf, 1925), especially pp. 41–45, and Barnes, H. E., "The Historian and the History of Science", *Scientific Monthly*, XI (August 1919), pp. 112–126. Robinson himself was, to the best of my knowledge, the first historian to use "the scientific revolution" as an organising theme. His *The Mind in the Making*, published in 1921, devotes a chapter to the subject. See also I. Bernard Cohen, "The Eighteenth-Century Origins of the Concept of Scientific Revolution", *Journal of the History of Ideas*, XXXVII, 2 (April-June 1976), pp. 257–288.

36 Sarton, George, *The Study of the History of Science* (Cambridge, Mass.: Harvard University Press, 1936), p. 5.

37 Quotations from "The New Humanism", *Isis*, VI, 16 (October 1924), pp. 26–27, and George Sarton to Robert S. Woodward, 10 April, 1918 (Sarton Archives, Carnegie Institution, Washington, D.C.).

38 Haskins' *Studies in the History of Medieval Science* was published in 1924. Charles Homer Haskins (1870–1937) was professor of medieval history at Harvard: see *Isis*, XXVIII, 76 (February 1938), pp. 53–56. See also George Sarton's obituary of Breasted, *Isis*, XXXIV, 96 (Spring 1943) pp. 289–291. On the defensiveness of the humanities in the 1920s and the way Rockefeller grants were "buttressing scholasticism and antiquarianism in our universities", see Fosdick, R. B., *Rockefeller Foundation*, pp. 239–240.

39 Text books included Sedgwick and Tyler, *Short History of Science,* and Libby, Walter, *An Introduction to the History of Science* (Boston: Houghton Mifflin, 1917).

40 See Multhauf, Robert P., "The Society and Its Concerns", *Isis,* LXVI, 234 (December 1975), pp. 454–467.

41 See James Bryant Conant to Sarton, 4 January, 1939 and Sarton to Conant, 23 June, 1941 (Sarton Archives, Harvard University). Thackray, Arnold and Merton, Robert K., "On Discipline Building: The Paradoxes of George Sarton", *Isis,* LXIII, 219 (December 1972), pp. 473–495.

42 See L. J. Henderson to Sarton, 18 June 1941 (Sarton archives, Harvard University). I. Bernard Cohen (b. 1914) persevered and went on to a lifetime of distinguished scholarship at Harvard, where he is now the Victor S. Thomas Professor of the History of Science.

43 James Conant, who had become president of Harvard in 1933, played a highly influential role in the administration of science during the Second World War. See Conant, James Bryant, *My Several Lives: Memoirs of a Social Inventor* (New York: Harper and Row, 1970).

PART IV

Scientific ideas in their cultural context

21

Aristotle, Plato, and Gemisthos

DUANE ROLLER
University of Oklahoma

The activity that came to be called natural philosophy or physics was invented in Ionia (and probably at Miletos) about 600 B.C. Over the next 200 years it developed uniquely within the Ionian dialect group,[1] and together with its offspring, mathematics, it provided the model for what was to be "science."[2] By the fourth century B.C., science had come under the scrutiny of the philosophers. Plato and Aristotle codified and defined what is meant by scientific knowledge and gave it the structure it has had forever more. Both science and scientific knowledge were defined in terms belonging to the Greek culture and hence are peculiarly Greek. Consequently we need not ask if science was ever invented or reinvented elsewhere. There could no more have been a native Chinese science than there could have been a Chinese Parthenon, and an indigenous science in the Western Hemisphere is as unlikely as a pre-Columbian American Indian poet who wrote in heroic hexameter. Wherever science is found it has been transmitted from classical Greece. The comprehension of the origins of modern science must then rest on an understanding of the components of Greek science and the mode of transmission of those components to the modern world.

The Greek philosophical analysis of scientific knowledge identified two realms of discourse important to science. In one of the realms lie statements about sense experiences, about observations, phenomena. These statements – such as "The sky is blue," "Sokrates died," "Children resemble their parents," – comprise descriptions of the natural world stemming from observation. They offer no explanation of the natural world. They come from and are verifiable by observation.

In the other realm of discourse lie statements about concepts. Since concepts are nonobservables, conceptual statements can neither stem from observation nor be verified by observation. Yet it is these conceptual statements that serve to explain experiences. Thus the event, "Sokrates died," is explainable by the conceptual statement, "Man is mortal." It is the conceptual statements that comprise the laws, the principles, the significant scientific knowledge.

These two realms of discourse, the one containing conceptual statements, the other containing statements about experience, are connected by another Greek invention: logic. Logic enables one to deduce from human mortality the mortality of Sokrates and hence explain the unhappy event of his death. For scientific explanation of experience means precisely being able to deduce from conceptual statements the experiential ones.[3] Consequently science, the attempt to explain experiences with things in the natural world, became a search for concepts and for statements about concepts. Since concepts are nonobservables, they must be sought through the mind; science, then, is an intellectual activity.

Plato and Aristotle split on the nature of reality, a split that was sharpened by their intellectual descendants. With Platonism is associated the belief that *concepts* are the intellectual grasping of real forms or ideas and that *things* are crude imitations of that reality.[4] Hence in the Platonistic world the search for truth, for reality, must search for the real ideas and conceptual statements about those ideas, using the only available research tool, the human mind. When, and only when, the idea is grasped and conceived in the mind, the searcher has knowledge. But the Platonist is faced with a serious problem: How is one to recognize true ideas and hence true conceptual statements when they are encountered? The answer lies in the association of truth, reality, and beauty. A real idea is beautiful and so is a true statement about a concept grasped from a real idea.[5] The intellectual search for scientific knowledge, that is, for conceptual statements, thus becomes an intellectual search for beauty. But by "beauty" is inevitably meant "Greek beauty," the beauty of a play by Euripides, of the Hephaistion, of the *Iliad*, and, most of all, the beauty of mathematics. For Plato saw in Greek geometry a body of knowledge that deals with such beautiful ideas as *line* and *sphere* and hence represents one of the few bodies of truth easily available to us. Much of the Pythagorean view was absorbed into Platonism, and the use of mathematics in optics, astronomy, and musical theory made these subjects also favorite Platonic disciplines.

Having found the beautiful truth, the Platonist has no need to verify it further and hence has little interest in connecting true conceptual statements to observational ones, which, after all, are statements about experiences with things that are only crude imitations of reality. Hence pure Platonism is destructive of science, the attempt to explain experiences, for the pure Platonist is not concerned with the imitation world of observables.

The Aristotelian regards things as real and statements about experience with things as real. Concepts are intellectual creations, convenient and even essential in discussing experience but certainly not directly associated with reality. Pure Aristotelianism is equally destructive of science, for its exponent never gets around to the unreal ideas that provide such a fruitful source of concepts to the Platonists.

Science then demands three elements for its existence: the belief in the reality of Platonic ideas, associated with the Platonic philosophy; sufficient belief in the reality of things, associated with the Aristotelian philosophy, to make the explanation of experiences with things a worthwhile activity; the availability of logic to connect conceptual statements with experiential ones. Logic was exposited in the writings of Aristotle, which served as the chief source of logic to later eras. Since antiquity science has never existed without all three of these elements.

The Romans played two major roles in the history of science. First, they erected the protective umbrella of the *pax romana* over the Greek world, permitting the continuation of the Greek culture, including Greek science, throughout the era of Roman political and military strength. Second, because the Romans had relatively little interest in such intellectual matters as science, most of the Greek literature was either not translated into Latin or the translations were not preserved. With the military and economic decay of Rome and the consequent decay of the educational system, interest in, access to, and even knowledge of the Greek writings were almost totally lost. In the realm of science, only a partial version of Plato's *Timaios* and some of the lesser writings of Aristotle survived. Direct knowledge of the Aristotelian philosophy and logic was to all intents and purposes totally lost to the West.

Aurelius Augustinus (354–430) had passed through Neoplatonism on his way to Christianity, itself already considerably influenced by Neoplatonism. As a consequence the Augustinian Christianity that was to dominate the Western intellectual world until the thirteenth century was highly idealistic, turning away from the importance of things to the importance of ideas. This is not, however, to say that Western intellectuals after Boethius (ca. 480–ca. 524) knew the Platonic philosophy. Of Plato's writings, only the partial version of the *Timaios* was available, together with bits and pieces of Platonism filtered through and diluted by various other authors, chiefly Neoplatonic writers.[6] Western Christendom in the Augustinian Age, despite its Neoplatonic idealism, lacked the Platonic philosophy as well as the Aristotelian, and it lacked a knowledge of logic. Thus all three of the elements essential for science were missing in the early medieval West.

Much of the Greek knowledge, including science, survived in medieval Islam. The reconquest of Spain began in the eleventh century and the boundary between Christendom and Islam was steadily pushed southward. By the early twelfth century the Christian world had acquired Spanish Islamic libraries containing, in the Arabic language, Greek and Islamic knowledge. For reasons that are not entirely clear Christian scholars sought out and translated Greek material into Latin, and by the end of the twelfth century, a large, although select, fraction of the extant Greek works was available in some form in Christendom.

It was the Aristotelian philosophy, particularly logic, that entranced the West: Augustinian Christendom was so ignorant of the fundamental intellectual tools developed by the Greeks that it could not even organize its ignorance. The acquistion of the *Organon* of Aristotle changed all of that. It was perhaps the impact of acquiring the extraordinary tool of logic from Aristotle that led the West to so thoroughly accept the Aristotelian conception of reality as well. Or perhaps the burgeoning Western technology made the West amenable to a philosophy that was object oriented. But whatever the reasons, Aristotelianism had a rapid success. Thomas Aquinas's melding of the Christian theology and the Aristotelian philosophy – the baptizing of Aristotle, it has been called – shifted the philosophical foundations of the Western intellectual world. After a number of rearguard skirmishes, Augustinian Christianity retreated into the minority position it has held ever since, and Aristotelian Thomism became dominant.

Thomism offers a theological basis for science within the Christian world by placing reason alongside revelation as a source of knowledge. It also gave science two of the elements that had been missing in the Augustinian world: logic and a belief in the reality of things and, therefore, an interest in the natural world. This basic, fundamental shift in opinion concerning the nature of reality can be seen reflected in the alterations in late medieval painting, sculpture, and literature. A striking example is the shift in pictorial representation of the Madonna and Child. In early versions, the two face the viewer. By the twelfth century, when the flow of Aristotelian knowledge into the West is in full swing, the figures are turned slightly toward one another, but they do not look like human woman and child. This stage is portrayed by two paintings of Madonna and Child together in the Uffizi gallery, one by Cimabue (ca. 1250–1300 or later), painted some time after 1280 for the high altar of Santa Trinita in Florence, the other, commissioned in 1285, by the Sienese Duccio (ca. 1255–before 1315). A third picture by Cimabue's successor, Giotto (1267?–1336/7), and presently displayed in the same room with the other two, was done early in the 1300s for the Church of the Ognissanti in Florence. It has moved vividly into the world of Aristotelian reality. No longer are the two individuals unworldly; they appear as humans appear to the eye.

Niccolò Pisano (1210/20–1278/84) produced the highly stylized pulpit in the baptistry at Pisa using Roman sarcophagi as models. But he then moved on to produce (with the aid of his son) the naturalistic sculpture of the fountain that stands in Perugia between the cathedral and the Palazzo dei Priori. Changes in representation of this sort have long been regarded as an indication of a nascent humanism, but more properly they reflect a growing interest in the sensible world, the world of Aristotelian reality, of which human beings are but one part. The human traits of the mother and child share the artists' interest equally with the rest of the world of experience. Giotto's frescoes in the upper church at Assisi are noted for their so-called realism of portrayal of individuals, but one also finds in them birds and trees, rooms and tables, and even a city, Arezzo.[7]

The gaudy capitals of Corinthian columns degenerated in late antiquity into the simplest of stylizations, and by Romanesque times the simple ornamental designs had lost all resemblance to acanthus leaves. In the mid-twelfth century, design alterations began to occur in cathedral decoration; leaf shapes are distinctly seen, although they do not match the leaves of any known plant. In the next hundred years, these shapes become more complex, as the sculptors drew upon observation of plants for new decorative forms. And in the last quarter of the thirteenth century, the designs of the gothic Corinthian capitals began to display leaf forms copied from nature, leaves that can be specifically identified today by botanists.[8]

By the fourteenth century the European world had turned philosophically from idealism to the view that reality lies in things, and all aspects of the fine arts reflected that shift. The artist by now was engaged in a search to portray the world as it is to the senses and not as it is to the mind. This has falsely been regarded as laying the foundations for scientific inquiry in the West by those who incorrectly regard observation to be synonymous with science. For although Thomism gave to Western thought and to the Western intellectual world two of the missing elements of science, logic and the Aristotelian belief in the reality of things and of experiences with things, it deprived it almost entirely of its third essential element, the Platonic belief in the beautiful reality of ideas. The flood of twelfth-century translations contained only two works of Plato: the *Phaidon* and the *Menon,* translated by Henricus Aristippus, about 1156.[9]

Science had some spectacular successes in the fourteenth century in places where Platonic elements from Augustinian Christianity persisted briefly alongside the new Aristotelian ones and also in the careful logical analysis of earlier knowledge. But in the absence of the Platonic conception of reality, science lacked its creative element, lacked the ability to produce the new conceptual statements that constitute scientific knowledge.

The Aristotelian philosophy gained a powerful ally in that newly invented institution, the university. In its first century the university could nurture an Albert the Great and a Thomas Aquinas, but it rapidly became a conservative institution devoted to the preservation and passing on of the present state of knowledge. Aristotelian logic provided the framework for organizing knowledge taught in the faculties of law, theology, and medicine, as well as philosophy, and the only Platonistic elements to survive in the educational system were meagre ones in the quadrivium, within the arts curriculum. The success of the university as an institution for perpetuating the present state of knowledge assured Aristotelian philosophy, from the fourteenth century on, of sufficient dominance to prevent almost totally the introduction of Platonism into the intellectual world represented by the university.

Under guidance from the university, the Western intellectual had little access to, knowledge of, or interest in Platonism. Some mathematical work from antiquity did become available in the High Middle Ages and prompted study and analysis in the arts faculty, but the immediate consequences for scientific knowledge were relatively small.

In view of the intellectual climate of the West it is difficult to imagine that the sources available in the West for the Platonic philosophy or for Greek mathematics could have served for an unaided revival of either, and we know historically that such a revival did not occur until the injection of Platonism into the West. But how did Platonism survive elsewhere and how did it reach the West?

In general, as the unorthodox fled from the power of the orthodox state church of Rome in the early centuries of that church, they carried Greek ideas, including Platonism, eastward. The explusion of the Nestorians from Edessa in 489 and the closing of the pagan schools of Athens in 529 strongly contributed to this movement.

The factors that tended to suppress Neoplatonism in Christendom, as a serious rival to Christianity, did not appear in the early centuries of Islam. By the ninth century a glittering Baghdad had risen to prominence and the serious collection and translation of Greek manuscripts was under way, as well as the providing of a climate then amicable to scholars.

The city of Harran had been known to the Christians as the "city of pagans," because of the adherence of the Harranians to their peculiar religion. In 830, the Abbasid caliph al-Mamun threatened them with destruction if they failed to convert to a religion recognized by the Koran, having a sacred book and prophet. Some converted, but others declared themselves "Sabians," taking over the name of a recognized but by then lost sect. There is considerable circumstantial evidence that they adopted a collection of Hermetic writings as a sacred book and Hermes as their prophet, an act that may have preserved the works known by the modern name of the Hermetic corpus.[10] These Harranian Sabians made a significant contribution to Islamic intellectual life: the names of twenty-eight have survived, including Thābit ibn Qurra and al-Battani.[11] The Neoplatonism of the Hermetic writings strengthened Neoplatonism in Baghdad.

The innumerable sects of Islam fell into two major groups: a fundamentalist literal-interpretationist group and a dominant liberal wing that claimed the right to apply reason to interpretation of the Koran. The ascendancy of this liberal wing made possible the growth, in the ninth century, of another sect, "the philosophers," Neoplatonists who were technically Moslems but disregarded the Koran or "when obliged to take notice of it, contrived some sort of compromise between their Neoplatonic doctrines and those of Moslem theology."[12] This occurred at precisely the time Harranian Sabians migrated to Badgad. It is with this group that Platonic idealism survived in Islam, in the dilute forms of Neoplatonism and Hermeticism.

Meanwhile, in Byzantium itself Neoplatonism continued at a stronger level than in the West, the educational system seems to have been more effective than in the West, and certainly the stock of Greek manuscripts to be searched was far in excess of anything in the West. As Bagdad rose to prominence as an intellectual center it may have excited Byzantine admiration and envy. At any rate, when the caliph al-Mamun invited to Bagdad a private teacher in Constantinople, Leon the Mathematician (ca. 799–after 869), the Emperor Theophilos (829–42) acted to keep Leon in Byzantium and thereby opened an era of revived intellectual activity in Byzantium. Without Leon "the revival of mathematical studies in the West based on Greek texts is well-nigh inconceivable."[13] "It was during this time that most of the manuscripts forming the vital link in the line of descent from antiquity were written."[14] This activity continued through the reign of Constantine VII Porphyrogenitos, to 959.

There followed a century of relative quiescence in Byzantium and then another outburst of activity began in the mid-eleventh century, when the school of Constantinople was reorganized by Constantine IX Monomachos in 1045. It contained a faculty of philosophy headed by Michael Psellos (1018–78).

At just the time this new revival of learning was occurring in Constantinople, the orthodox wing of Moslem theology won out in Baghdad with a resultant growth of intolerance, and "the philosophers," including the Sabians, simply vanished as an identifiable group.[15] It has been surmised that there may have been a movement of individuals out of Islam back to Constantinople at this time and perhaps a movement of Greek texts as well. For it is in the eleventh century that the Hermetic Corpus shows up in Byzantium; possibly Michael Psellos had a copy, perhaps the only extant copy.[16]

Psellos lectured at the secular school at the Magnaura palace on the quadrivium as well as the trivium, and, interestingly, on the mathematical parts of Aristotle. Further, "he saw in mathematics the connecting links between material objects and ideas, a means of leading students into the realm of abstract thought."[17] Not only did he preserve some otherwise lost works of Proclos and Iamblichus, but he collected the extant writings of Archimedes, a collection that furnished the foundation of William of Moerbeke's translation.

Michael Psellos's student and successor, John Italos, lectured on the "Platonic theory of ideas, on Aristotle, Proclus, and Iamblichus."[18] John's pupil, Eustration, Metropolitan of Nikaia (ca. 1050–ca. 1120), continued Neoplatonic scholarship. Additional works preserved by activity of this period include Euclid's major work, Ptolemy, Heron's *Metrika*, Eutokios, and Nikomachos.

These scholars, with strong Neoplatonic attitudes, had their difficulties with orthodoxy. But these difficulties were far less than realists were to face in the West and a steady widening of difference in attitudes was occurring between East and West. This was intensified by a collection of events, known retrospectively as the Schism of 1054, that through the eleventh and twelfth centuries increasingly separated the Christian church into two churches and the empire into two empires.[19] That "piece of international brigandage," the Fourth Crusade, and the taking of Constantinople in 1204 brought the schism to full maturity and produced an eternal Byzantine mistrust of the Latin West.[20]

The Byzantine court moved to Nikaia and both education and scholarship continued there for the next half century, returning to Constantinople with the court following the expulsion of the Latin barbarians in 1261. The school at Hagia Sophia was revived and we know the names of many of the teachers there, the names of their teachers and pupils, and something about their teaching and scholarly interests. The understandable hatred of the Latins produced a drive for catharsis of the Greek language and a consequent striving for a return to classical forms that resulted in a Greek much more classical than was the contemporary Latin of the West. This in turn intensified the search for manuscripts of classical works and an interest in scholarship that extended beyond the clergy to the nobility, civic leaders, and the civil service. In the fourteenth century there was a nascent Hellenism in Byzantium. Coupled with Platonism and mathematics and despite its being sometimes labeled pagan by conservative theologians, this Hellenism generally gained ground.

It is little wonder that Petrarch could have heard that there were Byzantines who preferred Plato to Aristotle and that Petrarch marveled at this. For the Thomistic synthesis of Aristotelian philosophy and Christian theology was a Western phenomenon. It is thus not astonishing that Byzantine art failed to make the shift toward portrayal of what the senses see that characterizes Western painting in the era of St. Thomas, Cimabue, and Giotto, remaining highly idealistic. Nor is it unexpected that an educated member of the Byzantine nobility such as Manuel Chrysoloras (ca. 1335–1415) should have been embued with an interest in Plato.

Chrysoloras was both a product of the Byzantine educational system and himself a teacher.[21] He visited Italy in 1393 as an emissary from the Byzantine emperor in the latter's perennial and fruitless search for military aid against the Turks. Chrysoloras apparently made an impression on the Florentines and by 1396 he had been invited by the magistrates of Florence to teach Greek at the *studium* in that city. Particularly instrumental in the invitation and appointment were Coluccio Salutati (1331–1406) and Palla Strozzi (1372–1462). One could speculate, from what is known about the Byzantine intellectuals of this period, that the Florentines got a considerable dose of Platonism, along with their Greek verbs, from Chrysoloras, and their actions confirm that speculation.

There are scattered traces of earlier teachers of Greek in Italy in the fourteenth century, but in no case does the teaching of those before Chrysoloras seem to have been effective. Wistful remarks by earlier Italians about the Greek language and classical Greek culture were now replaced by a solid knowledge of both, during and after Chrysoloras's teaching in Florence. He wrote a grammar of the Greek language for his Florentine students, a text that remained in use long after his departure from Italy, and the names of his close associates, disciples, and pupils read like a roster of the notable Italian intellectuals of the first third of the quattrocento.

Historians in the quattrocento itself regarded Chrysoloras's teaching as a major event in the revival of classical learning, and he has been given a role of varying prominence ever since in that recovery. Among those who fell under his influence were not only the Florentine chancellor Coluccio Salutati but two of Salutati's successors: Leonardo Bruni (1370–1444) and Poggio Bracciolini (1380–1459), and a fourth chancellor, Carlo Marsuppini (1398–1453), learned his Greek from a pupil of Chrysoloras's, Guarino da Verona (1374–1460). Chrysoloras's direct influence on educational thought was no less dramatic. Pier Paolo Vergerio (ca. 1370–1444/5), who left his teaching position at Padua to go to Florence and study with Chrysoloras, became the first writer on educational theory in the Italian Renaissance. In his widely read *De ingenuis moribus et liberalibus studiis adolescentiae,* Vergerio advocated the revival of the study of Greek in the curriculum. Guarino da Verona went to Byzantium in 1403 and studied with Chrysoloras there, after the latter had returned from Italy. Guarino established the first humanist school in Venice, the study of Greek literature being an essential part of the curriculum. Vittorino da Feltre (ca. 1378–1446) was perhaps taught by Chrysoloras and certainly by Vergerio and Guarino. He founded at Mantova what may have been the first school directed at inculcating humanistic ideals. Vergerio and Palla Strozzi were responsible for the introduction of the teaching of Greek at Padua.

The alteration in Florentine education after Chrysoloras was also significant, with the teaching of Greek becoming a continuing matter. Roberto Rossi, student of Chrysoloras, became the center of a Florentine school of Greek and

Latin letters that was frequented by (among others) Nicollò de Niccoli (1363–1437), Bartolo Tebaldi, Cosimo de' Medici (1389–1464), and his brother Lorenzo (1395–1440). Vergerio taught Greek at the studium in Florence early in the fifteenth century. Guarino lectured there on Greco-Roman antiquity in 1408. Giovanni Aurispa (ca. 1369–1459) taught Greek literature in Florence in 1425–7 as did Francesco Filefo (1398–1481) in 1429. Carlo Marsuppini, who had learned Greek from Chrysoloras's pupil, Guarino, was appointed to teach at the Florentine studium in 1431.

A number of these scholars went to Byzantium in the decade following Chrysoloras's teaching in Florence: Guarino in 1403, for five years, studying in Constantinople with Chrysoloras; Aurispa went in 1413 and again in 1421–3; Filefo in 1421. They brought back Greek manuscripts for which there was an avid market. Aurispa alone brought 232 manuscripts and became the outstanding bibliographer of the quattrocento. These manuscripts were copied, sold, and translated.

The well-known impetus to an interest in classical Greek literature provided by Chrysoloras has another facet that is perhaps less widely recognized. The excitement in Italy was not solely for Greek manuscripts but particularly for certain kinds of them, reflecting a burgeoning interest in Plato's works and in literary works of a kind quite different from those that had reached the West via Islam. While Chrysoloras was still teaching in Florence, Coluccio Salutati urged Iacopo Giacomo Angeli da Scarperia, a student of Chrysoloras's, to try to procure a complete set of Plato's writings in Byzantium. In 1401 Salutati wrote to Giovanni Conversino da Ravenna (1343–1408), then in Padua, asking him to try to obtain for him Chalcidius's partial translation of the *Timaios* and the Aristippus translation of the *Phaidon*, works that were extant in the West but apparently not available in Florence. No later than 1404 Salutati's friend and fellow student of Chrysoloras's, Leonardo Bruni, began his distinguished career of translating Plato with a new translation of Plato's *Phaidon*, probably at Salutati's request.

Chrysoloras taught by the didactic method and Pier Candido Decembrio (1392–1477) tells us that when his father, Umberto (ca. 1350–1427) was a student of Chrysoloras's, he was given the task of translating Plato's *Republic*.

Among the 232 manuscripts brought back from Byzantium by Aurispa, in the 1420s, was a collection of the works of Plato (lacking the *Republic* and the *Laws*), as well as the *Iliad* and a number of classical Greek plays. A copy of the Plato manuscripts was acquired by Vittorino da Feltre. This was probably also the source of Palla Strozzi's copy of the Platonic writings. Niccolò de Niccoli, disciple of Chrysoloras's and collector of manuscripts (whose collection ended up in the hands of Cosimo de' Medici), also acquired some of Aurispa's manuscripts.

Thus the revival of Greek following Chrysoloras's famous stay in Florence comprised the teaching *and* learning of the Greek language, a transfer of Byzantine educational and pedagogic ideas to northern Italy, and an interest particularly in the Platonic literature as well as in the Greek culture in general. And although many northern Italian cities had access to this new knowledge, it seems to have held a fascination for Florentines that was peculiar almost to the point of being unique. Political and geographical matters may have played some role in this. The capitulation of Padua to the Visconti and the withdrawal of Venice from the conflict made Florence the lonely and undisputed leader of the fight against the Visconti. The successful conclusion of that fight left Florence still in a position of leadership in northern Italy. It has been suggested that Venice, with close seaborne connections with the Greek world, had already solved its needs for commercial communication in ways other than learning the Greek language and culture, whereas Florence was just attempting to open such communication.

Yet despite these factors and their undoubted impact, it is difficult to avoid associating as well an influence of *trecento* literature, architecture, painting, and sculpture with the special interest in the Platonic philosophy, so closely associated with the Greek conception of beauty. Petrarch's love of Greek was largely limited to clutching Greek books he could not read and Boccaccio's proud bringing of a Greek to teach in Florence many have only netted him a fake Greek, a Calabrian monk named Barlaam, who failed the task. Nonetheless these extraordinary authors left a heritage of desire. Arnolfo di Cambio's Palazzo della Signoria and Giotto's Campanile had long stood in Florence and Andrea Pisano's south door of the cathedral, San Giovanni, was a familiar sight to Florentines. The new cathedral, Santa Maria del Fiore, was nearing completion. Byzantine mosaics existed in Tuscany and the Greek conception of beauty had been transmitted through them and probably through the Greco-Etruscan paintings as well to Cimabue and Giotto. Greek beauty was not an alien concept in Tuscany at the end of the trecento.

Meanwhile, in Byzantium a longer-lived contemporary of Chrysoloras, Georgios Gemisthos (ca. 1360–1452), called Plethon,[22] was pursuing twin ideals: a thorough recovery of the classical Greek literature and a complete revival of the Platonic philosophy, both within a framework of Hellenism.

Gemisthos was born in Constantinople shortly after the middle of the fourteenth century and his father was probably a church official. He was educated in the capital and then continued his studies at the Ottoman court, learning about commentaries on Aristotle by Averroës and others and the teachings of Zoroaster. He was back in Constantinople in 1405, where he apparently taught both Markos Eugenikos and the Emperor Manuel and acquired the disfavor of some church officials there, probably because of his theological views. Before 1407 he had left the capital, perhaps in exile but certainly not in imperial disfavor, and had settled in Mistra, in the Peloponnessos, overlooking Hollow Lakaidemonia.

Mistra was originally a Frankish town, established during the Latin occupation of Greece. Unable to hold Constantinople, the Franks established a French Greece and Guillaume de Villehardouin founded the fortress city of Mistra. When the Franks were driven out of Greece, Mistra became a Greek political capital with considerable intellectual and religious autonomy as well as political. Here, in the fourteenth century, there was a development of Hellenism to a degree impossible in Constantinople, for the church regarded Hellenism as essentially pagan because of the esteem in which it held pre-Christian authors. And here to Mistra, in the first decade of the fifteenth century came Georgios Gemisthos. He received two land grants that gave him effectively sovereign rights and hence a base from which to develop his Hellenic dreams.

Byzantine scholars and the Byzantine educational system had preserved a strong thread of Platonic idealism and Neoplatonism, but despite the occasional objections of the orthodox, those who preserved and taught these ideas did so within a framework of idealistic, Augustinian Christianity. Such was not the desire or aim of Gemisthos, who hoped to replace Christianity with a Neoplatonism that stressed the reality of ideas.[23] He taught with great effectiveness and his students became his disciples. Pythagorean ideas surface with these disciples and there is some indication that they formed a sort of Pythagorean secret society devoted to spreading and implementing their master's ideas.[24]

In the 1430s a major effort was launched to heal the schism between the Eastern and Western churches. This was by no means the first such effort, but it was the first time the pope had agreed to the Greek demand for an ecumenical council to deal with the problem. Pope Eugenius also promised his own attendance at the council. Further he agreed to pay the entire expenses of the Greek delegation to attend a council, in Italy. He had good reasons for pressing for reunion: It offered the possibility of extension of papal power at a time when the counciliar movement was threatening to narrow it and indeed even to take the problem of union into its own hands. The Byzantine emperor and indeed all of the Greeks had a more direct reason for desiring union: Turkish pressure made urgent the need for any alliances that might bring military aid. Momentarily this need overrode the deep and justifiable distrust the Byzantine Greeks held for the Latin world.

A council was convoked for Ferrara and in November 1437, the Byzantine Emperor John VIII Palaiologos (1425–48) and a delegation of some 700 Greeks embarked in ships provided by Pope Eugenius IV (1431–47), reaching Venice three months later. The council opened at Ferrara in January 1438. It got off to a bad start and things steadily got worse. The pope was almost totally unable to keep his financial agreement with the Greeks, there was a plague in Ferrara, and the town was under military threat from the Milanese. Discussions of union had gone badly and many of the Greeks became homesick. Collapse of the council seemed imminent when the city of Florence offered massive financial aid if the council were moved to Florence.

Why this Florentine generosity? There were certainly pragmatic reasons, involving both the emperor and the pope, but again it is difficult not to believe that the burgeoning Florentine interests in beauty and in things Greek, and in Greek manuscripts, the almost uninterrupted teaching of Greek in Florence since Chrysoloras, and some knowledge of Platonism were important factors. In any case, the citizens of Florence got more than they could have anticipated out of the bargain. For the 700-member Greek delegation not only included students of Georgios Gemisthos, it included Gemisthos himself.

Gemisthos promptly undertook his profession, teaching, in a series of lectures, expositions, and discussions. We know what he said, for his own summary was preserved and later published. Up to now his enemy had been Christianity, but now he had found a new opponent, Aristotle, entrenched in Western theology and the Western educational system. Gemisthos set about freeing the Florentines from all three. His invidious comparisons of Aristotelianism to Platonism fell on understanding ears.[25]

The ground had been thoroughly prepared for him; after all, he was speaking to an audience containing Chrysoloras's disciples and pupils, and their pupils. It was an audience of avid Grecophiles. Furthermore the emergent Florentine attitude, so amply displayed in the fine arts, enabled the Florentines to readily comprehend Platonic reality, the reality of beauty and of ideas. By now Masaccio (1402–29) had expressed the horror of the explusion from Eden on the wall of the Brancacci Chapel in the Church of the Carmine, the Brunelleschi dome of the new cathedral was completed, and Ghiberti was at work on the golden doors for San Giovanni. Perspective of the geometrical sort, based on that favorite Platonistic discipline, geometry, was in the Tuscan air. The Florentines were amply prepared to learn more about a philosophy that emphasized Greek beauty and beautiful concepts and the reality of ideas rather than of crude things.

Markos Eugenikos, theological leader of the Greek delegation was not only adamant in his objections to the Latin theology, he was a poor diplomat as well. He was effectively replaced by Joannes Bessarion (ca. 1395–1472), who had studied with Gemisthos. Bessarion's diplomatic – and sometimes strained – minimizing of the major differences between the Greek and Latin views aided the Greeks in a reluctant drift toward accepting the Latin position and hence union of the churches. The Greek delegation was under enormous pressure from its political leadership to accept union, for the emperor hoped that once a united Christendom existed, Western Christian princes would provide military aid against the Turks. He was never to learn whether this hope was justified.

Union was declared in July 1439. Most of the Greeks signed; Markos Eugenikos did not. Instead he returned to Byzantium and led a general revolt against the union. Greek fear of the Latin world overrode fear of the Turks and the union was effectively, if not legally, nullified.

Gemisthos's activities during the council greatly increased the interest in and knowledge of Platonism in Florence. It whetted the already-ravenous Florentine appetite for Greek learning as expressed in Greek manuscripts and for the manuscripts themselves and notable collections were built. Cosimo's collection was deposited in the Monastery of San Marco, founding the library there. Palla Strozzi's collection founded the library of the monastery of Santa Trinita.

The council evoked great bitterness, not only between Latins and Greeks, but within the Greek delegation. The signers of the decree of union defended themselves at home with a variety of excuses, particularly claiming duress. They also pointed to those who had misled them, notably Bessarion. When a grateful pope placed him on a salary, Bessarion's position was not improved in Byzantium. After a year there he returned to Italy to stay, was made a cardinal of the Latin church, and led a vigorous life in the intellectual world, collecting Greek manuscripts – a collection that ultimately founded the San Marco Library in Venice. He provided assistance to refugee Greek scholars. Bessarion also devoted considerable time to carefully written responses to attacks on Platonism. These attacks were almost entirely made by Greeks. This is not astonishing, since the only Latins who could read the literature of the controversy were those Italians whose acquisition of a knowledge of Greek had been accompanied by conversion to Platonism. One famous attack was by Georgios Scholarios (ca. 1405–after 1472), a member of the Greek delegation to the Council of Florence, and a supporter of the union, who was converted to opposing it, after the council, by Markos Eugenikos.[26] Scholarios and other Eastern Christians had good reason to suspect both Gemisthos and his disciples of a profound lack of Christian orthodoxy. Gemisthos strongly and steadfastly opposed the union of the churches because he believed it would strengthen the Christian church, which, in all seriousness, he hoped to destroy and replace with Pythagorean, Neoplatonic paganism. Scattered bits and pieces of information indicate that a deep paganism pervaded the views of the School of Mistra and extended even to Cardinal Bessarion, aide and adviser to popes and himself twice considered for the throne of St. Peter, and to the group of refugee scholars that surrounded him in Rome. Little wonder that orthodox Greek Christians held a suspicion of Gemisthos that extended, through him, to both Neoplatonism and the entire Platonistic philosophy. When the only copy of Plethon's *Laws* fell into Scholarios's hands he burned it, saving only a few leaves to demonstrate the paganism of the work.[27] Another attack on Platonism was by Georgios Trapezuntios (1395–1484), whose motive may have been resentment of criticisms of his translating ability by Bessarion and others. Bessarion's reply, in his *In calumniatorem Platonis,* became his most famous exposition on the virtues of Platonism.[28]

This controversy and the documents it produced in the West served to stimulate interest in and spread knowledge of the Platonic philosophy in the mid-fifteenth century. It is from the fifteenth century injection of Platonism into Florence that this third essential element for science enters into Western Europe and thus makes modern science possible. The controversies surrounding the Council of Florence died out fairly quickly, but other events contributed to the spread of Platonism. There have not yet been adequate historical studies of these channels, so they can only be mentioned briefly here.

First were the missions undertaken by Bessarion for the papacy. In 1450 he went to Bologna, then one of the papal states, for a five-year stay and a successful attempt at pacification of that troubled city. The constant turmoil had destroyed the already venerable university at Bologna, which had numbered 10,000 students in the middle of the thirteenth century. But by 1430 it had been closed for three years and when it reopened in the fall of that year the hope was expressed that there would be 500 students by Christmas, a hope that was not realized. Bessarion rebuilt the university, revised its statutes, and attracted teachers.[29] One can only guess at the details of his influence, but it is difficult to imagine this ardent and vocal advocate of the Platonic philosophy passing up an opportunity to improve the position of Platonism in the faculty. We do know that late in the century a Polish medical student, Nicholas Copernicus, fell into the hands of a Florentine Platonist, Domenico da Navarra (1473–1543), then on the faculty at Bologna; we also know that Copernicus found the Ptolemaic theory "not sufficiently pleasing to the mind" and therefore sought a new arrangement.

Meanwhile, another mission had taken Bessarion to Vienna, where he met an arts professor, Georg Peurbach (1423–61). The teaching of the quadrivium included "astronomy," which was limited to calendrical problems (such as how to determine the time of Easter) and astrology. Recognizing the importance of tables of planetary positions in astrology, Peurbach had attempted to determine how the tables were constructed. That in turn led him to the mighty *Almagest* of Ptolemy, which he had available in the twelfth-century translation from the Arabic by Gerard of Cremona (1114?–1187).

Bessarion arrived in Vienna May 5, 1460. When he learned of Peurbach's interests, he pointed out that the Gerard version of Ptolemy's work had come to the West by a torturous route through Islam and probably through several languages. Bessarion urged Peurbach to prepare an abridged and simplified version of Ptolemy's book. He also urged him to return to Italy with him and promised to locate a copy of the book in the original Greek.

Peurbach died shortly thereafter, but his pupil Johannes Müller (1436–76), also known as Regiomontanus, went to Rome with Bessarion in 1461. There he learned Greek and completed the Ptolemaic *Epitome* (as he called it) begun by his teacher.

Regiomontanus put his newly acquired knowledge of Greek to good advantage. He discovered a partial copy of the work of the Greek mathematician Diophantos; he detected translator's errors, due to an insufficient knowledge of Greek, in Jacopo Angeli's popular Latin translation of Ptolemy's *Geography* and wrote notes on the errors; he compiled astronomical tables and lectured and wrote on mathematics. In the 1470s Regiomontanus moved to Nuremberg and there became "the first publisher of astronomical and mathematical literature . . . he sought to advance the work of scientists by providing them with texts free of scribal and typographical errors."[30]

Regiomontanus's activities in Nuremberg were financed by Bernard Walther (1430–1504), who became his pupil. Walther became the first astronomical observer in the West and for over thirty years recorded planetary positions that were widely used by later astronomers. The work of Regiomontanus and Walther established a German school of astronomy that led, through Peter Apian (1495–1552), Philip Apian (1531–89), and Michael Mästlin (1550–1631), to Johann Kepler (1571–1630), whose a priori method is the ultimate in Platonic astronomy and who spoke of Plato and Pythagoras as "nostros genios magistros."[31]

In Florence Cosimo de' Medici had undertaken his famous implementation of Gemisthos's idea that the Academy of Plato should be reopened in Florence, making Florence attractive for students of Plato and producing Florentine Platonists as well. By 1454 there existed a group that called itself the Academy, involving students of Leonardo Bruni and Carlo Marsuppini.[32] Through their efforts Johannes Argyropoulos (ca. 1410–87) was brought to Florence in 1456, by Cosimo, to teach Greek at the studium there, an act that fully Platonized the new academy. For Argyropoulos, student of Gemisthos's, taught the doctrine of Mistra in Florence.[33] His students included Johann Reuchlin (1455–1522), Angelo Poliziano (1454–94), Donato Acciaiuoli (1428–78), and Marsilio Ficino (1433–99).[34]

Ficino began his career by studying Plato in translation and in 1459 he studied Greek with Argyropoulos. In 1463 he came under the patronage of Cosimo de' Medici.

The arrival of the Hermetic corpus in Florence caught Ficino's attention. The Hermetic tradition by now ascribed these middle-Platonistic works to an author of the time of Moses, and regarded both Platonism and Christianity to be descendants and corrupted forms of Hermeticism. Hence the Hermetic texts were of vast interest to Platonic and Christian thinkers alike. Ficino produced a translation of the Hermetic works available to him and then turned to work on the Platonic corpus itself. In 1482 his edition of Plato's writings was published, thus making for the first time, the Platonic philosophy thoroughly available throughout Western Christendom.

In Italy the acquisition, translation, and dissemination of Greek mathematical works burgeoned in the fifteenth century. Italian mathematicians such as Niccolò Tartaglia (d. 1557) and Giovanni Battista Benedetti (1530–90) wrote on motion from the mathematical point of view. Galileo Galilei (1564–1642), the son of a Florentine musician, instigated a battle with the Aristotelian professors of physics and theology over the ultimate truth of the theory that Copernicus had created because the older Ptolemaic theory was "not sufficiently pleasing to the mind." In an extraordinary display of the Platonistic position, Galileo argued over and over again that the lack of observational evidence for the motion of the Earth was unimportant compared to the mathematical beauty of the Copernican theory. Needless to say, Galileo's major work on this subject was in the form of a Platonic dialogue.

The work of these Platonistic mathematical physicists and astronomers was carried on in the seventeenth century by other Platonic mathematicians, such as René Descartes (1596–1650) and Christiaan Huygens (1629–95), culminating in the synthesis of mechanics and astronomy by Isaac Newton (1642–1727). Newton was apparently under considerable influence of the Hermetic tradition, and his idea that "matter attracts matter" seems a prime example of the Platonic view that all objects are besouled.

In the sixteenth century the Hermetic tradition and its Platonic and Neoplatonic elements heavily influenced studies of the nature and properties of matter itself, by men such as Paracelsus (1493–1541) who have been labeled (incorrectly) "mystics," because of their Hermetic views.

The sixteenth century is usually regarded as the period of origin of the modern study of plants, animals, minerals, and human anatomy, a reputation due largely to a number of sixteenth-century books published in those fields. This subject has not yet been adequately studied, but one common characteristic ties practically all of these books together: They contain remarkable drawings printed from woodblocks. Indeed, their frame rests largely on those drawings, all of which were made by professional artists and all of which portrayed what the artist saw with the eye of the mind. The immediate source of this new type of book was the printers, almost all of whom were in the Rhine Valley, and the immediate source of the illustrations was the German woodblock tradition, involving artists such as Albrecht Dürer (1471–1528) and his follower Hans Weidtz. The contribution of the authors of these books to the illustrative material seems to be little in most cases and is demonstrably none in some. What, if any, relation there is to Renaissance sources needs investigation.

The so-called Scientific Revolution then is a happy blend of a firm interest in understanding, in comprehending, the world of Aristotelian reality, an understanding achieved through logic from conceptual statements obtained from the world of Platonic reality. As the medievalists keep telling us, there is no true rebirth, but the Italian Renaissance produced as much of a foundation of modern science as it did of the arts – and for much the same reasons.

NOTES

1 Every natural philosopher from Thales of Miletos (fl. ca. 600 B.C.) to Straton of Lampsakos (fl. 287–69) came from an Ionian dialect group city-state, either an Ionian city or an Ionian colony, with the single exception of Empedokles of Akrages (fl. ca. 450 B.C.). With the decline of Ionia as an important force in the Greek world, natural philosophy ceased to be an important activity for eight centuries.

2 That is, "the attempt to explain natural phenomena by means of postulatory-deductive structures" that characterizes the biological, earth, and physical sciences today. No attempt is being made here to encompass the vast array of quite different activities to which the term "science" is sometimes applied.

3 There is, however, always an illogical step necessary in the logical sequence, for deductions from conceptual statements yield only conceptual statements. Thus, for example, the deduction in Euclidean geometry that the angles of a triangle add up to two right angles applies only to plane trilateral figures composed of three lines, a line being breadthless length. It does not apply to chalk or pencil drawings. The illogical step from a concept to an observable is present in all of scientific knowledge and is taken as an act of pragmatic faith.

4 Plato's terminology is by no means as clear as his modern commentators and translators have made it. A considerable array of Greek terms have been translated as "ideal" or "idea" or "form." I will use the term "idea" to refer to this multiplicity of Platonic terms for the reality to be sought with the eye of the mind.

5 Indeed, where modern translators have used "idea" or "form" Plato often used "the beautiful" or some similar term. See Anders Wedberg, *Plato's philosophy of mathematics* (Stockholm: Almqvist & Wiksell, 1955), p. 142, n. 4.

6 Raymond Klibansky, *The Continuity of the Platonic Tradition during the Middle Ages: I. Outlines of a Corpus platonicum medii aevi* (London: Warburg Institute, 1950), pp. 21–9, "The Latin Tradition." See also his "Plato's *Parmenides* in the Middle Ages and the Renaissance," *Medieval and Renaissance Studies,* I (1943), pp. 281–330, for an example of the indirect transmission of Platonism.

7 Walter Goetz, "Die Entwicklung des Wirklichkeitsinnes vom 12. zum 14. Jahrhundert," *Archiv für Kulturgeschichte,* 27 (1937), pp. 33–73, cites a large number of specific examples of the new attitude he calls "*Wirklichkeitsinnes*" that developed in this period.

8 Denise Jalabert, "La flore gothique, ses origines, son évolution, du XIIe/au XVe siècle," *Bulletin monumental,* 91 (1932), pp. 181–246. Further: "L'évolution de la flore gothique est parallèle à celle de l'architecture" (p. 244).

9 However, copies of the Aristippus translations were not generally available in the High Middle Ages. Klibansky reports only a single copy of the *Phaidon* (at the Sorbonne) prior to the fifteenth century and we know of a fruitless search in Florence, early in the 1400s, for both works.

10 D. Chwolsohn, *Die Ssabier und der Ssabismus,* 2 vols. (St. Petersburg: Buchdruckerei der Kaiserlichen Akademie der Wissenschaften, 1856). See vol. 1, pp. 140–97.

11 Ibid., vol. 1, p. 577.

12 *Hermetica,* edited with English translation and notes by Walter Scott, vol. 1 (Oxford: Clarendon Press, 1924), pp. 97–107.

13 K. Vogel, "Byzantine Science," in *The Cambridge Medieval History,* vol. 4, pt. 2 (Cambridge: Cambridge University Press, 1967), p. 265.

14 Ibid., p. 270.

15 *Hermetica,* vol. 1, p. 107.

16 Both the existence of this copy and its uniqueness are much more certain than is Psellos's possession of it. See *Hermetica,* vol. 1, pp. 25–7.

17 Vogel, "Byzantine Science," p. 272.

18 Ibid.

19 Francis Dvornik, *Byzance et la primauté romaine* (Paris: Les editions du Cerf, 1964), pp. 111–38.

20 Louis Bréhier, *Le schisme oriental due XIe siècle* (Ernest Leroux, 1899), pp. xviii–xxv; Dvornik, *Byzance,* pp. 139–54.

21 Giuseppe Cammelli, *I dotti bizantini e le origini dell'umanesimo: I. Manuele Crisolora* (Florence: Vallecchi, 1941), pp. 41–51.

22 There is a considerable variance in usage of Plethon as a name for Georgios Gemisthos and one finds his name given almost equally as Gemisthos, Plethon, and Gemisthos-Plethon. It is often said that he coined "Plethon," which has much the same meaning as "Gemisthos" in Greek, and added it to his own name to evoke the name of Platon to his hearer or reader. Although this source and rationale of "Plethon" are quite certainly correct, it is unlikely that Gemisthos added it to his name. Rather he used it as a pseudonym signature to his *Laws,* the original and only copy of which was burned by Gennadios II (Georgios Scholarios), Patriarch of Constantinople, between 1460 and 1465. It was again used to sign Gemisthos's *On the differences between Aristotle and Plato,* composed at Florence in 1439. The psuedonym served both to identify Gemisthos as the author and, to Gennadios, to tie the two pagan works together. The ensuing polemic between the two men frequently used the name "Plethon" and made it famous. But it does not appear elsewhere in Gemisthos's lifetime. See François Masai, *Plethon et le platonisme de Mistra* (Paris: 1956), p. 52 and Appendice I, pp. 384–8.

Masai's book is the most authoritative and complete work on Gemisthos and is the source used herein for details of his life and works.

23 Masai, *Plethon,* p. 367.

24 Ibid., pp. 307–14.

25 Gemisthos seems to have had no opposition to the Aristotelian philosophy prior to his arrival in Florence in 1438. In the West, however, he saw Aristotelianism, entrenched both in the Scholastic universities and in Thomistic Christianity, as a serious threat to his aims. His principal work, written at Florence in 1439 and signed Πλήθωνος, is the Περὶ ὧν Ἀριστοτέλης, πρὸς Πλάτωνα διαφέρεσθαι. See Masai, *Plethon*, p. 329.

26 Joseph Gill, *Personalities of the Council of Florence and other essays* (New York: Barnes & Noble, 1964), chap. 7.

27 Scholarios moved through a succession of extraordinary alterations of positions, successively being a student of Gemisthos's at Mistra, a supporter of union, the most violent opponent of union, anti-Platonist, student of the Aristotelian philosophy and its chief exponent in the East, and finally, Gemisthos's most bitter enemy.

28 Henri Vast, *Le Cardinal Bessarion (1403–1472): étude sur la chretiénté et la renaissance vers le milieu du XVe siècle* (Paris: Hachette, 1878), pp. 346–7.

29 Ibid., p. 187.

30 Edward Rosen, "Johannes Regiomontanus," *Dictionary of Scientific Biography*, vol. 11 (1975), p. 351.

31 In a letter to Galileo, October 13, 1597. See Kepler's *Gesammelte Werke*, vol. 13, (Munich: C. H. Beck, 1945), p. 145.

32 Masai, *Plethon*, p. 340.

33 Ibid., p. 343.

34 Mario Emilio Cosenza, *Biographical and Bibliographical Dictionary of the Italian Humanists and of the World of Classical Scholarship*, vol. 6 (Boston: G. K. Hall, 1962–1967) pp. 140–1, lists many students of Argyropoulos.

22

Aristophanes and the antiscientific tradition

RICHARD OLSON
Harvey Mudd College

The interactions between scientific ideas and activities – the scientific tradition – and other important cultural institutions, especially those associated with religious belief, political ideology, and belles lettres, were often characterized by nineteenth- and early twentieth-century scholars in terms like conflict and warfare.[1] Today we often feel very superior to these earlier scholars because we now recognize a vastly greater complexity in these interactions. Religious and scientific ideas have sometimes been strongly supportive of one another,[2] and it seems that the dominant trends in political thought have usually been closely linked to scientific interpretations of the natural world.[3] Yet in transcending the simplistic biases of nineteenth-century historiography, we seem to have lost track of some important themes in Western intellectual history. The scientific tradition has frequently given rise to pressures to transform important institutions or belief structures whose spokesmen have sought to defend against perceived foreign intrusions. So conflict has been a normal and pervasive, though not an exclusive, characteristic of the relationships between scientific institutions and practitioners and other salient elements of the culture in which science has been practiced.

Because antiscientific sentiments seemed to have reached crisis levels during the past two decades, there has been a substantial recent literature attempting to analyse the current malaise. Among such attempts Reinhard Bendix's approach suggests some particularly interesting historical questions. Bendix isolates several categories of current antiscientific thought, which he illustrates with historical examples, suggesting that there have been coherent traditions within the opposition to science at least since the Enlightenment. Among these traditions he associates one with what he calls *imaginative* and one with *conventional* reason.[4] My purpose here will be (1) to briefly characterize these two modes of antiscientific thought; (2) to show that both were expressed in detail by the Athenian dramatist, Aristophanes, in the fifth century, B.C.; and (3) to argue that the modern tradition of antiscience, whose origins we, along with Bendix, usually place in the eighteenth and nineteenth centuries, is rooted at least indirectly and sometimes even self-consciously in Aristophanes' works, the conservative tradition in his *The Clouds* and the imaginative tradition in his *The Frogs*.

SCIENCE AND THE IMAGINATIVE REASON

Modern attacks of the imaginative reason on science begin, at least, with a focus on the relationship among science, poetry, and music. Spokesmen for the imaginative reason, from the time of Percy Shelley's *Defense of Poetry* (1821), have argued that although science can deal abstractly, analytically, and quantitatively with what *is,* only the imagination, expressed through such media as poetry and music, can envision what *ought to be* and provide a stimulus to action in the world. The imaginative reason decries the apparent separation between knowledge and sensibility, between fact and affect, which seems characteristic of science as we know it.

The logic that associates affects and sensibilities with *values,* and knowledge and facts with objective and *value-free* reality – and that identifies the first with music and poetry and the second with science – is problematic. It is part and parcel of that positivistic philosophy now under attack by scientists and nonscientists alike. But these associations have both preceded and outlived their logical justification. They are at least as old as classical antiquity and as modern as today, and their failure to find an explicit formal justification makes them neither unimportant nor incorrect. Their importance is reflected in the autobiographical statements of two nineteenth-century figures who were deeply committed to the scientific life, but who had some sense of its limitations, Charles Darwin and John Stuart Mill. Darwin, for example, reflected that as his life progressed, his mind "seems to have become a kind of machine for grinding out general laws from large collections of facts." He gradually lost interest in music and poetry, and it made him uneasy.

"The loss of these tastes," he wrote, "is a loss of happiness and may possibly be injurious to the intellect, and more probably to the moral character, by enfeebling the emotional part of our nature." Similarly, Mill lamented the impact of the analytic emphasis in scientific thought:

> Now I saw what I had always before received with incredulity – that the habit of analysis has a tendency to wear away the feelings. I was left stranded at the commencement of my voyage, with a well equipped ship and rudder, but with no sail . . . I frequently asked myself if I could go living.[5]

The sense of conflict between science and sensibility – between scientific means and moral ends – expressed so cogently by Mill and Darwin defines much of the intellectual history of the nineteenth century; but it was not until 1869, with the publication of Friedrich Nietzsche's *The Birth of Tragedy,* that the unease felt even by many scientists was made the explicit grounds for a full-scale attack on the modern scientific mentality – an attack pressed today by such writers as the French philosopher-historian Michael Foucault and the American philosopher-classicist Norman O. Brown.

In describing *The Birth of Tragedy* in 1886, Nietzsche wrote:

> What I then laid hands on, something terrible and dangerous, a problem with horns, not necessarily the bull itself, but at all events a *new* problem: I should say today, it was the problem of science itself – science conceived for the first time as problematic – as questionable.[6]

As usual, Nietzsche is deceiving us, or himself, or both, in this statement. For right at the center of his problem of science lay the destruction of Greek tragedy by scientific naturalism and rationalism at the hands of Euripides; and as we shall see later, Nietzsche inherited both the problem and his fundamental analysis of its character from Aristophanes.

SCIENCE AND THE CONVENTIONAL REASON

The agonized cry of the antiscientific conventional reason, like that of the imaginative reason, is that science is simply incompetent to answer the really important questions of life, questions of morality and religion. But spokesmen for conventional reason differ from those espousing the imagination in seeing the font of true wisdom, not in the visions of poets or in the ecstasy of dance and music, but rather in the content of traditions or morality, law, and religion.[7]

Joseph De Maistre, for example, who viewed the French Revolution as the tragic outcome of scientific ideas loosed on French society during the Enlightenment, wrote: "If we do not return to the old maxims, if the guidance of education is not returned to the priests, and if science is not uniformly relegated to a subordinate rank, incalculable evils await us. We shall become brutalized by Science, and that is the worst sort of Brutality."[8] But De Maistre was only one – and one of the least subtle – among many late eighteenth- and early nineteenth-century figures to focus on the disruption of traditional values by science. Edmund Burke, Jean Jacques Rousseau, and their intellectual descendants also realized that scientific modes of understanding challenge accepted answers to questions of religion and morality because the critical and rational methods of the sciences undermine the foundations of unanalyzed faith. Rousseau, speaking of the scientific philosophes of the eighteenth century, made this point with particular force:

> These vain and futile declaimers go everywhere armed with their deadly paradoxes, undermining the foundations of faith and annihilating virtue. They smile disdainfully at fatherland and religion, and devote their talents and philosophy to destroying and debasing all that is sacred among men.[9]

Rousseau does not argue that scientists are *intentionally* subverting religious and moral beliefs. Rather he argues that their methods are corrosive in spite of the scientists' good will – scientific methods are simply incapable of reaching out positively toward things of the heart and the soul, as opposed to the intellect – and they tend to belittle what they cannot understand.

ANTISCIENCE AND THE ATHENIAN POLIS

I want to present just one example of this almost unintended belittling of the nonintellectualizable side of the world by an eminent scientist, both to demonstrate that it is not just a figment of the antiscientists' imagination and to begin to relate these ideas to classical antiquity.

Just before his death in 1912, Henri Poincaré published an eloquent statement of the scientists' creed in *The Value of Science.* Here Poincaré spoke of the beauty of nature studied by science:

Of course I do not here speak of that beauty which strikes the senses, that beauty of qualities and of appearances; not that I undervalue such beauty, far from it, but *it has nothing to do with science;* I mean that profounder beauty which comes from the harmonious order of the parts, and which a pure intelligence can grasp. This it is which gives body, a structure, so to speak, to the iridescent appearances which flatter our senses, and without this support the beauty of these fugitive dreams would only be imperfect, because it would be vague, and always fleeting. On the contrary, *intellectual beauty is sufficient unto itself,* and it is for its sake, more, perhaps, than for the future good of humanity, that the scientist devotes himself to long and difficult labors.

Like many others, Poincaré saw the birth of these scientific ideals in ancient Greece. And he saw their existence as directly related to the historical greatness of the Greeks and their cultural heirs:

If the Greeks triumphed over the barbarians and if Europe, heir of Greek thoughts, dominates the world, it is because the savages loved loud colors and the clamorous tones of the drum which occupied only their senses, while the Greeks loved the intellectual beauty which hides beneath sensuous beauty, and it is this intellectual beauty which makes intelligence sure and strong.[10]

Now the interest of this set of statements, apart from its demonstration that in at least one case the intellectualism of science leads to a self-conscious lowering of the value of a sensuous response to the world, is the way in which Poincaré's biases lead to a very peculiar interpretation of Greek history. Virtually any ancient historian would have to concede that the Greek scientific intellectualism, which Poincaré so eloquently praises, more closely accompanied the decline of the Greek polis than its rise to power. However biased Rousseau was on this topic, and however much he might have overemphasized or distorted a causal relationship between rising intellectualism and declining military might in Greece, he was not totally wrong in charging his readers to

consider Greece, formerly populated by heroes who twice conquered Asia, once at Troy, and once in their homeland. Nascent learning had not yet brought corruption into the hearts of its inhabitants, but the progress of the arts, the dissolution of morals, and the yoke of the Macedonian followed each other closely; and Greece, always learned, always voluptuous, and always enslaved, no longer experienced anything in her revolutions but a change of masters.[11]

Without wanting to imply anything about the causal relationships between the rise of scientific thinking – with its emphases on rationalistic and naturalistic modes of approaching the world – and Greek political fortunes, I do want to argue that Rousseau's perception of a causal connection would not have been foreign to fifth and early fourth-century Athenians themselves. The expulsion of the philosopher Anaxagoras, whatever other motives might have spurred it on, was clearly rationalized with an argument that his new naturalistic interpretations of the heavenly bodies threatened to undermine essential and traditional religious beliefs. And the trial and condemnation of Socrates, while again it may have been stimulated by other political motives, was grounded in the formal assertion that Socrates was a malefactor who meddled in the matters of the heavens and the earth below, who made the worse (legal and moral) argument appear the better, and who taught others to follow his example. Regardless of whether Socrates was, in fact, guilty of what he was charged, an Athenian jury presumably found the teachings of the new learning, which combined naturalistic interpretations of the universe with new logical modes of argument, sufficiently threatening to warrant the explusion, even the execution, of an Athenian citizen.

ARISTOPHANES: CONVENTIONAL REASON AND *THE CLOUDS*

This finally brings me to Aristophanes, in whose play, *The Clouds,* we find the most direct and explicit classical depiction of the new learning as something that threatens to undermine Athenian religion and morality. Scholars have debated how seriously Aristophanes intended his audiences to take some of the ideas I want to discuss. The majority opinion, represented by such otherwise divergent scholars as Gilbert Murray and William Arrowsmith, is that Aristophanes did not have any real deep-seated bitterness toward Socrates or the ideas he put in his mouth. According to this view, Aristophanes was principally an early-day Don Rickles calling forth laughter by indiscriminately attacking and insulting friend and foe alike. Cedric Whitman, who acknowledges the serious moral tone of *The Clouds* as we know it, even suggests that in its original version (now lost), Aristophanes had treated the new learning favorably, and that this is why the play was so poorly received.[12] I tend to disagree with both attitudes. But my main argument is affected neither by Aristophanes' intentions, nor by the possible existence of a quite different earlier version of *The Clouds.* However Aristophanes meant his words to be taken, and whatever might have appeared in the original version, it is clear that both Aristophanes' fifth-century Athenian audience, and a group of critically important eighteenth- and nineteenth-century readers, read out of *The Clouds* that has come down to us a deadly serious questioning of the religious, moral, and political implications of the new learning.

The apparent conflict between the new learning, which we can characterize as in its most essential parts, scientific, and Greek traditions was of enough topical concern in 429 to make it an appropriate subject for treatment at the Dionysian festival in Aristophanes' *The Clouds.* And the degree of this concern was so intense by 399 that Aristophanes implied charges against the real Socrates. Plato's Socrates speaks to the jury of the charges against himself: "You yourselves have seen these very things in Aristophanes' comedy – a Socrates who is carried around in a basket and asserts that he walks upon the air, and a great many other absurdities, of which I am completely ignorant."[13]

What was the specific content of *The Clouds* that could both represent and inflame the unease of at least portions of the Athenian populace about the dangerous implications of the new learning? Here, for brevity, I must select passages from just one side of a portrayal that satirizes elements of both the new and the traditional in Athenian life and education. Very early in the play Strepsiades, the ignorant and slightly corrupt caricature of the traditional Athenian, enters the school of Socrates, intending to learn the new methods of logic that will allow him to cheat his creditors. He meets Socrates, who has been making astronomical observations while suspended in a basket in the air, and he swears by the gods to pay any price for learning the new logic, to which Socrates replies: "By the gods? The gods my dear simple fellow, are a mere expression coined by vulgar superstition. We frown upon such coinage here."[14]

Strepsiades then asks, "What *do you* swear by?" and Socrates offers to teach him the *real* truth about the gods, after which he introduces the chorus of clouds as the only gods there are. Strepsiades responds in two ways. First, he remarks that he thought that clouds were only fog, dew, and vapor. This is, of course, precisely what the Socrates of the play believes. His claim that the clouds are the only gods is not an attempt to dignifiy the clouds, but rather to naturalize and minimize the gods. Second, Strepsiades objects that Socrates has forgotten the great Zeus, to which Socrates responds: "Zeus, what Zeus? Nonsense, there is no Zeus." Again the straight man, Strepsiades, steps in to ask if there is no Zeus then who makes it rain? The ground has thus been laid for Socrates to bitterly satirize those *naturalistic* explanations of natural phenomena that had been offered by Milesian natural philosophers and that were being taught by Sophists in fifth-century Athens.

Socrates begins by explaining that the rain comes from the clouds and that thunder occurs when the clouds bump into one another and roll over one another, moved, not by Zeus, but by the wind.

Strepsiades professes not to understand and Socrates tries another approach:

Take yourself as an example. When you have heartily gorged on stew at the Panathena, you get a stomach-ache and suddenly your belly resounds with prolonged rumbling.

Strepsiades responds:

Yes, yes, by Apollo! I suffer, I get colic, then the stew sets to rumbling like thunder and finally bursts forth with a terrific noise . . . and when I take my crap, why, it's thunder indeed, *pa pa pa pax! pa pax! papapapapax!!* just like the clouds . . . And this must be why the names are so much alike: crap and clap.[15]

Following this magnificent parody, Strepsiades professes to be convinced, swears allegiance to the clouds, and vows: "If I met another god, I'd cut him dead, so help me. Here and now I swear off sacrifice and prayer forever."[16]

In this passage Aristophanes exposes the key issues in the play, the undermining of *nomos* – law, custom, tradition – by *physis* or nature. Socrates has robbed natural events of their dignity by somehow trivializing them, turning phenomena that had depended on the will of Zeus into a grotesque manifestation of the farting of the clouds. Aristophanes seems truly to feel that such attitudes offend the gods, and that Socrates and his ilk deserve to be punished for their offenses. To this end he finishes his play by having Strepsiades, finally cured of his venality, set fire to Socrates' school. As Socrates escapes, Strepsiades beats him with a stick yelling:

Then why did you blaspheme the gods? What made you spy upon the moon in heaven?
Thrash them, beat them, flog them for their crimes, but most of all because they dared outrage the gods of heaven.[17]

While the undermining of religious belief represents the most fundamental pernicious tendency of the new naturalistic learning there are other important tendencies as well, and Aristophanes focuses on the way in which a scientific approach, applied to human affairs, undermines morality and law as well.

Strepsiades gives up trying to learn himself and sends his son Pheidippides to Socrates. The son returns, gets into an argument with his father, and threatens to beat him. To this Strepsiades answers in horror that for the son to strike the father is contrary to all *tradition, law, and custom.* The son then offers to prove by argument that he has the right to strike his father, and the following passages ensue:

Leader of the Chorus: Come, you, who know how to brandish and hurl the keen shafts of the new science, find a way to convince us, give your language an *appearance* of truth.
Pheidippides: How pleasant it is to know these clever new inventions and to be able to defy the established laws! . . . now that the master has altered and improved me and that I live in this world of subtle thought, of reasoning and meditation, I count on being able to prove satisfactorily my right to thrash my father . . .

Who made the Law? An ordinary man like you and me. A man who lobbied for his bill until he persuaded the people to make it law. By the same token, then, what prevents me now from proposing new legislation granting sons the power to inflict corporal punishment on wayward fathers? . . .

If you're still unconvinced, look to Nature for a sanction. Observe the roosters, for instance, and what do you see? A society whose pecking order envisages a permanent state of open warfare between fathers and sons. And how do roosters differ from men, except for the trifling fact that human society is based upon law and rooster society isn't.[18]

Here we see again the battle between *nomos* and *physis* and how it is that the new rhetorical modes that make the worse argument appear the better are tied directly to the new naturalism. Against the older sense of divine and traditional sanctions for the law, we have Pheidippides focusing on the all-too-human politics of legislating and appealing to the demeaning parallels between the social order of chickens and that of human beings.

Strepsiades rails against this, telling his son: "Look, if you want to emulate the roosters, why don't you eat shit and sleep on a perch at night?" But his anger is to no avail, and he is defeated in this instance by the new learning.

I could multiply these examples severalfold, for *The Clouds* presents the new learning as the source of decline in military vigor and physical culture as well as that of religion and fundamental morality, but I hope enough has been said to show how it is that Aristophanes' work could easily be read as an antiscientific broadside focusing on what Rousseau said of the scientists: "These vain and futile declaimers go everywhere armed with their deadily paradoxes, undermining the foundations of faith, and annihilating virtue."

Finally, let me point out that Rousseau's and Aristophanes' attitudes toward the relationship between science and the undermining of traditional morality and religion are tied together by more than parallel conservative temperaments. Rousseau was deeply interested in the theater. A playwright himself, he certainly knew of and studied the works of Aristophanes. In *La Nouvelle Héloise,* Rousseau even explicitly compared Aristophanes and Molière as the first and last dramatists to expose the corruption of the morals of their respective societies. Thus when Rousseau wrote of the nascent learning that had brought corruption into the hearts of the Greeks he had doubtless read Aristophanes' attacks on the new scientific and sophistic learning of fifth-century Athens and had sympathized with elements of them. Even had he not been independently alerted to Aristophanes' attitudes, he would have learned of their links to antiscientific sentiment through his reading of Jonathan Swift's brilliant *Tale of a Tub,* which introduced the criticism of seventeenth-century scientific theories as "edifaces in the air" with a reference to Aristophanes' Socrates, who was "suspended in a basket to help contemplation."[19]

On one major point Rousseau and Aristophanes were in direct and near-total opposition. Aristophanes' Socrates symbolized all of the corrosive elements of the new scientism, whereas for Rousseau, Socrates symbolized the forces of justice, nobility, and humility, which were under attack by the prideful scientists. It is precisely on this issue that post-Rousseauian conventional reason departed company from the imaginative reason in its attitudes toward science. For conservative or conventional thinkers after Rousseau, Socrates the moralist, whose concerns are with the beautiful and the good, stands against scientific thinkers, whose concerns are with naturalism and quantification. Spokesmen for the imaginative reason, on the other hand, continue to see Socrates as Aristophanes portrayed him, and they follow the Aristophonic critique of Socrates from *The Clouds* into *The Frogs,* where it takes on another disturbing dimension.

The transition from what we now see as Aristophanes' conservative attack on science to one associated with the imaginative reason actually begins in *The Clouds* in connection with the incident that sparks the quarrel between Strepsiades and his son. At a supper celebrating Pheidippides' return from Socrates' school, the son refuses his father's request to recite from Aeschylus, declaring that the first great tragedian was "the most colossal, pretentious, spouting, bombastic bore in poetic history."

Strepsiades holds his temper and responds:

> All right son, if that's how you feel, then sing me a passage from one of those highbrow modern plays you're so crazy about.
> So he recited – you can guess – Euripides. One of those slimy tradegies where, so help me, there's a brother who screws his own sister.[20]

In *The Clouds,* Aristophanes does not further develop the contrast between Aeschylean and Euripidean tragedy, but some twenty-four years later, in 405 B.C., he makes this the central focus of his last major play, *The Frogs.* Here, he emphasizes the uplifting, stirring, and activating power of Aeschylean tragedy and compares it to the disillusionment and paralysis brought on by Euripides' realism, naturalism, and intellectualism – a realism, naturalism, and intellectualism tied directly to Socrates' influence.

ARISTOPHANES: THE IMAGINATIVE REASON AND *THE FROGS*

Although there is no question among scholars that *The Frogs* is Aristophanes' most self-consciously serious and "political" play, there remain serious questions about the significance of the fact that over half the play – the portion we will be most interested in – is in the form of a contest over the literary merits of two poets. Similarly there remains a question of how bitterly Aristophanes opposes the innovations of Euripides, just as there is disagreement about his personal attitude toward Socrates.

Cedric Whitman, one of the most perceptive recent interpreters of Aristophanes, argues convincingly that on its most important level the play is about the very survival of the Athenian polity. For him, the literary arguments are almost incidental: "It is clear that the whole literary level is ancillary to a larger design . . . Poets are, as a rule, representative of their times, and it was logical to let Aeschylus speak for the glorious Athenian past, and Euripides for the contemporary decay. To cast the conflict in the form of a literary debate was a scheme which offered infinite opportunity for parody and caricature."[21] Gilbert Murray, on the other hand, insists that the literary theme is Aristophanes' major concern, not a mask for political discourse. And he insists that the contest between Euripides and Aeschylus is not one-sided, but ambivalent. "William James speaks somewhere of a man who threatened suicide because he was only the second best baseball 'striker' in the world; only by such a standard can the *Frogs* be regarded as a condemnation of Euripides."[22]

Again I seem to be in disagreement with the experts. There is no doubt that one of Aristophanes' goals was to entertain his audience and that a literary debate provided him with great scope for having fun. Nor is there any doubt that Aristophanes had at least a grudging admiration for Euripides' intellect and mastery of rhetorical techniques. But according to classical Greek ideals, the function of the poet was principally to educate the community. Aristophanes makes clear his basic agreement with this notion both implicitly and explicitly in *The Frogs.* The very aim of tragic poetry, established overtly as the criterion against which to judge the contest, and agreed upon by Aeschylus and Euripides alike, is "to bring help to the polis by making men better in some respect."[23] Poets are properly seen rather as shapers of society than as reflections of it, and the most telling arguments against Euripides are that he left men less noble and glorious than he found them. Thus the debate about the quality of poetry cannot be divorced from or seen as merely symbolic or a debate about the character of society. It is directly and self-consciously a debate about the quality of that political education that constitutes the very foundations of polis life.

For this same reason, it seems to me that the ultimate triumph of Aeschylus over Euripides does not leave Euripides a celebrated second-best tragedian. It really does leave him the villain of the piece. Thus it is significant that when Aeschylus leaves the contest to return to Athens, it is Sophocles, and not Euripides, that Aristophanes places in the throne to keep it secure until the master's return. In his parting words, Aeschylus charges the lord of the underworld to "see above all that this devil-may-care / Child of deceit with his montebank air / Shall never on that imperial chair / By the wildest of accidents sit."[24]

Once more, however, the impact of Aristophanes' work on the modern antiscientific tradition is less dependent on his intent than on subsequent interpretations. And Friedrich Nietzsche, through whom *The Frogs* has had its greatest modern impact, argued strongly that the revolution in poetic sensibility stimulated by Euripides' new scientific attitudes played a direct and causal role in the decay of Greek society, drawing directly from Aristophanes to make this point.

Let us look briefly into *The Frogs* to see how this could be so, and how the specific and detailed implications of this notion are developed.

The central plot of *The Frogs* is extremely minimal. After the death of Euripides, the last great tragedian, Dionysus, the god in whose honor the Athenian dramatic festivals were held, travels to the underworld in order to bring back the poet who is characterized as most "fruitful and generative." Initially he believes this to be Euripides. In Hades he is convinced that he should judge a contest between Euripides and Aeschylus to determine which author deserves this honor. In the process of the debate, which satirizes both Aeschylean and Euripidean tragedy, Dionysus himself undergoes a political education – or, as Whitman puts it, his own true identity is revealed to him. By the conclusion, Dionysus decides that because only the stirring works of Aeschylus have the power to rally the populace of a demoralized Athens to action, it is he, rather than Euripides, that should be returned to the city.

At the beginning of the contest Aristophanes recalls the conservative argument of *The Clouds* that ties impiety to the new scientific and naturalistic vision of the world. Aeschylus prays to Demeter to give him strength for the contest. Euripides refuses to approach the altar. When pressed he displays the new "coinage" in gods by tossing off the following lines:

> Ether, whereupon I batten, vocal cords,
> Reason, and nostrils swift to sneer,
> Grant that I may duly probe each word I hear.[25]

As the contest opens, Aristophanes also takes a casual swipe at the tendency of scientific intellectualism to seek quantitative and precise knowledge of everything.

> They'll bring straight edges out, and cubit rules, and folded cube frames, and mitre-squares, and wedges . . . Line by line Euripides will test all tragedies.
> What, is it bricks they want?[26]

These bantering introductory shots, however, are merely the prelude to an attack upon scientific naturalism and reason as antithetical to the very aims of tragedy. At the core of this attack lie critical distinctions between ends and means and between norms and descriptions. Like Mill, Aristophanes felt that though scientific analysis might provide a rudder, it could not provide a sail; that although it could tell what life *is* like, it could neither tell us what it *ought to be* like, nor provide that sense of immediacy, urgency, and participation that lifts us from contemplation into action and that is derived from great music and poetry.

Euripides' first great claim is that he brought a realism and naturalism to the theater:

> I put things on the stage that come from daily life and business. Where men could catch me if I tripped; could listen without dizziness to things they knew, and judge my art.[27]

Aeschylus responds by pointing out that this very characteristic has made Euripides' plays publicly dangerous. In particular, Aeschylus attacks Euripides' treatment of women with its focus on what we would now call abnormal psychology, claiming that his treatment of Stheneboea has encouraged suicide as a response to depression. Euripides defends himself with the scientistic intellectuals' standard response, arguing that he is not responsible for inventing his characters – they are such as walk the street or they appear in traditional histories. Very true, responds Aeschylus,

> But the poet should hold such a truth enveloped
> in mystery,
> And not represent it or make it a play. It's his duty
> to teach, and you know it.
> As a child learns from all who may come his way, so the
> grown world learns from the poet.
> Only words of good counsel should flow from his voice.[28]

What people need is not an awareness of their own limitations, perversities, and corruptions, but rather a model to emulate – a lamp to follow rather than a mirror to contemplate. Thus Aeschylus proudly proclaims, "I taught you for glory to long, and against all odds stand fast."[29] It is precisely this kind of guidance that Euripides' scientific realism fails to provide.

Nietzsche too saw a "slavish love of existence" that "left nothing great to strive for"[30] developing out of Euripides' works, and he called upon Aristophanes to support his claims:

> What Euripides takes credit for in the Aristophanean "Frogs," namely that by his household remedies he freed tragic art from its pompous corpulency, is apparent above all in his tragic heroes. The spectator now virtually saw and heard his double on the Euripidean stage ... It was henceforth no longer a secret, how – and with what saws – the commonplace could represent and express itself on the stage. Civic *mediocrity*, ... was now suffered to speak, while heretofore the demigod in tragedy and the drunken satyr, or demiman in comedy, had determined the nature of the language.[31]

The mention of language brings us back to a second of Euripides' major claims in *The Frogs*. For if he did not provide men goals to seek, Euripides was convinced that he provided an important instrumental education: "I didn't rave at random, or plunge in and make confusions," he says.

I taught the town to talk . . . I gave them canons to apply and squares for making verses; taught them to *think*, to see, to *understand*, to *scheme* for what they wanted . . . This was the kind of lore I brought, to school my town in ways of *thought*, I mingled *reasoning* with my art / and shrewdness, till I fired their heart to brood, to *think* things through and through.[32]

Again, Aristophanes turns these proud claims back upon their author, attacking them on the grounds that *talk* and *intellectualized understanding* lead ultimately to a paralysis of will and failure to act effectively in the world. Aeschylus begins the attack by saying:

> True, you have trained in the speechmaking arts nigh
> every infant that crawls.
> Oh, this is the thing that such havoc had wrought in the
> wrestling school, narrowed the hips
> of the poor pale chattering children, and taught the crews
> of the pick of the ships to answer back to their officer's
> nose,
> How unlike my old sailor of yore with no thought in his
> head but to guzzle his booze and sing as he bent to the
> oar.

And Dionysus continues:

> But our new man just sails where it happens to blow, and
> argues, and rows no more.[33]

The last word on this point, however, belongs to the final chorus, which ties Euripides, with his focus on speech and argumentation, both to the renunciation of *true* music and poetry, and to the Socrates of *The Clouds*.

> Go, cast off music, poetry,
> And sit with Socrates and gas!
> Leave the great art of tragedy
> And be – an ass!
> Go, plunge in solemn argument,
> And spend a worthless afternoon
> In quibble, quiddity, and cant,
> And be – a goon![34]

Once more Nietzsche develops Aristophanes' insights. Euripides, he argues, demonstrates the fundamentally scientific attitude that the value and beauty of a thing are tied directly to its intelligibility – remember Poincaré's statement that "intellectual beauty is sufficient unto itself, and it is for its sake, rather than for the future good of humanity, that the scientist devoted himself to long and difficult labors." In this connection Nietzsche follows Aristophanes in tying together Euripides and Socrates – *The Clouds* and *The Frogs*. It was Socrates who first began to teach the world to value knowledge and understanding above all else. Thus, Nietzsche writes, "Euripides was, in a certain sense, only a mask. The deity which spoke through him was neither Dionysis nor Apollo, but an altogether new-born demon called Socrates. This is the new antithesis, the Dionysian and the *Socratic*, and the art work of Greek tragedy was wrecked on it."[35]

Just as Aristophanes' *Clouds* presented to us one of the first clear portrayals of the conflict between the critical naturalism of science and traditions of religion and morality whose origins and justification lie in shared but frequently unanalyzed beliefs, his *Frogs* presents to us the crucial struggle between the rational, intellectual, realistic scientist, whose primary goal is to *understand* or know what *is*, and the imaginative, visionary, involved poet, whose primary goal is to *bring about* what *might be.* Aristophanes clearly championed the latter over the former, not because he felt that the man of science was intrinsically evil or worthless, but because he felt the need for counterbalancing a growing domination of Athenian life by the scientific spirit of Socrates and Euripides. In this he shared and first articulated the basic value orientation of those who speak today on behalf of the imaginative reason against the *domination* of modern life by the spirit of science; for theirs too is a crying out to maintain some concern for poetic visions of the *ought* and ultimate values in a world dominated by empirical and rationalistic knowledge of the *is* and of instrumental values.

NOTES

1 Beginning with David Brewster's *Martyrs of Science* (1841), and culminating in John W. Draper's *History of the Conflict between Religion and Science* (1875), and Andrew D. White's *A History of the Warfare of Science with Theology in Christendom* (1896), the titles of classic nineteenth-century works on the relationships between science and religion best illustrate this theme.

2 The whole literature growing up around the "Merton thesis" makes a series of sociological linkages between religious, economic, and scientific developments. See J. R. and M. C. Jacob, "Seventeenth Century Science and Religion: the State of the Argument," *History of Science, 14* (1976): pp. 196–207, for a summary of this literature. Charles E. Raven, *Natural Religion and Christian Theology* (Cambridge, 1953) illustrates a very different approach, that of the traditional historian of ideas linking theological attitudes to the works of early modern naturalists like John Ray.

3 Though Floyd W. Matson's *The Broken Image: Man, Science, and Society* (New York, 1964) is undeviatingly hostile to the links between scientific ideas and political ideologies, it provides a good introduction to their pervasiveness and importance.

4 Reinhard Bendix, "Science and the Purposes of Knowledge," *Social Research* (1975).

5 Darwin and Mill both quoted by Donald Fleming in "Charles Darwin, the Anaesthetic Man," *Victorian Studies, IV* (1961): p. 219.

6 Friedrich Nietzsche, *The Complete Works of Friedrich Nietzsche,* ed. Oscar Levy, vol. I, *The Birth of Tragedy* (New York: Russell and Russell, 1964), p. 3.

7 For the Greeks it would have been unthinkable to divorce poetry and music from moral, legal, and religious tradition. This is, in fact, why we find the origins of what now seem to be two different traditions in one classical author. It is to the fact that there were different subsequent responses to issues raised in two of Aristophanes' works that we owe the separation of imaginative and conservative traditions. The circumstances that encouraged such a split are discussed later in this chapter.

8 Joseph De Maistre, *On God and Society: Essays on the Generative Principle of Poltical*

Constitutions and other Human Institutions, ed. Elisha Greifer from the 1809 French original (Chicago: Henry Regnery, 1959), p. 54.

9 Jean Jacques Rousseau, *The First and Second Discourses,* ed. Roger D. Masters (New York: St. Martin's Press, 1964), p. 50.

10 Henri Poincaré, *The Value of Science* (New York: Dover, 1958), pp. 8–9.

11 Rousseau, *The First and Second Discourses,* p. 40.

12 Cedric Whitman, *Aristophanes and the Comic Hero* (Cambridge, Mass.: Harvard University Press, 1964), pp. 133–7.

13 Plato, *Apology,* 19B–19C.

14 Aristophanes, *The Clouds,* trans. Wm. Arrowsmith (New York: New American Library, 1962), p. 34.

15 Aristophanes, *The Clouds,* in Whitney J. Oates and Eugene O'Neill, Jr., eds., *The Complete Greek Drama* (New York: Random House, 1938), vol. II, pp. 556–7.

16 *The Clouds,* Arrowsmith translation, p. 47.

17 Ibid., p. 132.

18 *The Clouds,* in *The Complete Greek Drama,* vol. II, pp. 594–5, and Arrowsmith translation, p. 124.

19 Jonathan Swift, *The Prose Works of Jonathan Swift* (London, 1897), vol. I, p. 48.

20 *The Clouds,* Arrowsmith translation, p. 120.

21 Whitman, *Aristophanes and the Comic Hero,* pp. 250–1.

22 Gilbert Murray, *Aristophanes: A Study* (Oxford: The Clarendon Press, 1933), p. 121.

23 Aristophanes, *The Frogs* in *The Complete Greek Drama,* vol. II, p. 971.

24 Ibid., p. 994.

25 Ibid., p. 966.

26 Ibid., p. 962. The order of these passages has been slightly altered.

27 Ibid., p. 969.

28 Ibid., p. 973.

29 Ibid. p. 972.

30 Friedrich Nietzsche, *Complete Works,* vol. I, pp. 89–90.

31 Ibid., p. 88.

32 *The Frogs* in *The Complete Greek Drama,* vol. II, pp. 968–70.

33 Ibid., pp. 974–5.

34 Translation by Cedric Whitman in his *Aristophanes and the Comic Hero,* p. 246.

35 Nietzsche, *Complete Works,* vol. I, p. 95.

23

Carl Voit and the quantitative tradition in biology

FREDERIC L. HOLMES
Yale University

I

In a foreword written in 1957 for the Dover edition of *An Introduction to the Study of Experimental Medicine,* I. Bernard Cohen drew attention to the skepticism of Claude Bernard concerning the application of mathematics to biological phenomena. Calculations based on numerical data, Claude Bernard asserted in 1864,

> are premature in most vital phenomena, precisely because these phenomena are so complex that we must not only assume, but are in fact certain that, beside the few among their conditions which we know, there are numberless others which are still totally unknown . . . I am convinced that, since a complete equation is impossible for the moment, qualitative must necessarily precede quantitative study of phenomena.[1]

Lest Bernard appear shortsighted from the vantage point of present biology, Professor Cohen wisely cautioned that these views "must be read with an awareness of both the general intellectual climate and the state of knowledge when it was written."[2] Few scientists or historians would disagree, in fact, that quantitative studies are futile before there is adequate qualitative information to enable one to measure meaningful phenomena under controlled conditions. The question that remains is whether Bernard had assessed the situation in his own time realistically when he judged that calculations should not be applied in physiology "under present conditions."[3]

Because "measurement has long been considered a hallmark of science properly practised,"[4] historians have paid special homage to those figures of the past who appear to have heralded the eventual quantification of biology. Thus William Harvey's central argument for the circulation of the blood is often celebrated as the first decisive use of quantitative reasoning to establish a physiological process. Santorio Santorio is honored for "the introduction of quantitative experimentation into biological science."[5] Stephen Hales's hydrostatic pressure measurements on plants and animals are prominently treated because they represent the most conspicuous eighteenth-century experimental application, to biological phenomena, of the injunction that "all things are wisely adjusted in number, weight and measure."[6] Historians of science have generally, however, shown more interest in the introduction of quantitative arguments and observations into biology in principle than they have in the achievement of reliable, accurate, and significant measurements. Some of the reasons for this preference are self-evident. Because we have sought especially to identify innovative approaches and broad conceptual shifts, and because we have given priority to the ideas of science over its techniques, we have found the concept of quantitative measurement more exciting than its practice. Because the idea of applying quantitative measurements had to come first, and because relatively inaccurate initial attempts had to precede later improvements, our search for formative events leads us in the same direction. It is all too easy to slip beyond this preference into the prejudice that such first efforts necessarily required greater creative insight or more original and resourceful experimental strategies than did the later attainment of precision or rigorous control. Those of us who do not have intimate knowledge of the complexities of a particular scientific specialty readily assume that the later developments were essentially the work of technicians. Yet the problems involved in achieving precise, reliable measurements in biology have, in fact, often demanded as much imagination, ingenuity, and even courage as has the introduction of the general objective.

One apparent obstacle to the historical discussion of the development of quantitative precision in any science is that the danger of relying on hindsight judgments is especially acute. As Kuhn and others have stressed, there is no consistently applicable criterion for "reasonable" accuracy.[7] Standards vary from one area of science to another, and in any given area they have ordinarily become tighter with passing time. It is all too tempting to assess the quality of past measurements by the degree to which values obtained then approach the latest corresponding values. A more suitable criterion, I believe, is that the measurements attained were sufficiently reliable and discriminating to sustain a fruitful field of research. It is obvious, for example, that the standard of accuracy which Berzelius achieved by the second decade of the nineteenth century in the determination of combining proportions was adequate to support the flourishing development of quantitative chemical analysis, and to provide a sound basis for reasoning about composition. There is a common impression that a comparable situation did not emerge in biology until much later, perhaps after the beginning of the twentieth century.

If we still view nineteenth-century experimental biology as preeminently qualitative, that may be in part because historians regularly treat Claude Bernard as the representative figure of his age and adopt his perspective. Consequently his criticisms of the extravagances committed by contemporaries who were introducing "mathematical calculations" into physiology have overshadowed the serious work these people were carrying out. Today we still read with amusement in Bernard's classic *Introduction,* that when Friedrich Bidder and Carl Schmidt were attempting to measure the nutritional balance of a cat, they counted as excretions the kittens born on the seventeenth day.[8] Bidder and Schmidt's own important treatise, *Die Verdauungssäfte und der Stoffwechsel,* is seldom read. During the period in which Bernard was making his renowned qualitative discoveries concerning nutritional processes, Bidder and Schmidt were in the forefront of a school that attempted to analyze the same kinds of phenomena by characteristically quantitative means. Bernard's unsympathetic attitude did not quell this movement, and in the long run it influenced biology as deeply as did Bernard's more famous achievements.

After 1860 the acknowledged leader of the field that Bidder and Schmidt helped to form was Carl Voit. To Bernard, Voit's work appeared so thin that he regretted having asked his friend Madame Raffalovich to take the trouble to translate for him one of Voit's treatises.[9] Yet by the end of the century Voit was regarded as the "grand old master" of the field of metabolism. Voit reached this position largely through the skill, thoroughness, and tenacity with which he refined procedures and conceptual approaches pioneered by others. I would like to focus on some aspects of his early investigations as nodal events in the process by which quantitative precision entered experimental biology.

II

Owsei Temkin has pointed out that quantitative approaches to physiological problems were rare in Greek physiology.[10] Yet the medical system of Galen, he has also stressed, rested on an implicitly quantitative conception of nutrition, that the amount of food required must be in proportion to the amount of material lost by "insensible transpiration."[11] More recently Jerome Bylebyl has traced the echoes of this theme down to the seventeenth century. The famous "statical" experiments of Santorio represented an effort "to quantify an overall bodily process whose occurrence had been recognized for thousands of years."[12] Bylebyl's illuminating discussion reveals a thread of continuity that can be extended further, because those physiologists and chemists of the mid-nineteenth century who worked to determine nutritional balances traced their own tradition back to Santorio.

By the nineteenth century Lavoisier's theory of respiration, together with the development of methods of combustion analysis for determining the elementary composition of organic substances, and of means for measuring and analysing gases, had provided new foundations for investigating the chemical factors involved in the nutritive circulation of materials. Those who pursued such studies soon realized that for a complete accounting one ought to measure simultaneously the quantities of respired gases, of nutrient materials, and of excretions; but the difficulties involved in incorporating all of these factors into the same investigation seemed insuperable. Respiration experiments had to be short, because they were carried out in closed chambers where the very exchanges being measured caused the atmosphere the animal breathed to become progressively more abnormal. Measurements of the composition of food and excretions required much longer experiments in order to assure that the urine and feces collected derived from material ingested over the same time period, and to reduce the effects of errors due to analytical uncertainties. Consequently two separate lines of research developed. Inferences drawn from one could only be roughly checked by comparisons with results obtained from the other.

A central objective in both types of investigation was to account for the sources and disposition of the four elements that composed the principal classes of plant and animal substances as well as of the respiratory gases. In the case of carbon, hydrogen, and oxygen, experimenters focused their attention on the relative quantities exchanged, in order to test and further elucidate the claim of Lavoisier's followers that the carbon and hydrogen contained in alimentary materials were oxidized to form carbonic acid and water.[13] The role of nitrogen presented a more enigmatic problem. At the beginning of the century the question formulated was how to explain the higher proportion of that element in the substance of animals than in the plant matter that served directly or indirectly as their food.[14] The possibility that atmospheric nitrogen might be assimilated was apparently ruled out during the 1820s by the respiration experiments carried out independently by Pierre Dulong and César Despretz. Both men inferred from their results that there was a net exhalation of nitrogen, amounting to as much as one-quarter of the oxygen consumed. The question remained unsettled, however. Some critics noted that these results were impossible, for within a few days the animals would have breathed out more nitrogen than their bodies contained.[15]

The germinal experiments based on measurements of the elementary composition of food and excretions were carried out by Jean-Baptiste Boussingault, beginning in 1839. Boussingault too addressed the question of whether herbivorous animals receive nitrogen from the air. He fed a cow and a horse uniform diets, sample portions of which he had previously analyzed. Collecting the total excretions of each animal, he determined their contents of carbon, hydrogen, and nitrogen. Obtaining less nitrogen from the excrements than the food should have provided, he concluded that herbivores do not assimilate nitrogen from the air. In 1844, he performed similar experiments on a turtledove and found one-third less nitrogen excreted than the dove ingested. This striking deficit led him now to conclude that a portion of the dietary nitrogen undergoes "complete combustion" and is exhaled as nitrogen gas. This result, he pointed out, conformed to what Dulong and Despretz had found in their respiration experiments, except that the total quantity of nitrogen involved was much less than they had calculated. Boussingault's painstaking experiments became models for many later investigations. His admiring colleague, Jean-Baptiste Dumas, viewed this work as the extension of the rigor of the chemical balance into the study of the questions of general physiology.[16]

In 1842 Justus Liebig's *Animal Chemistry* provided a comprehensive new vision of the way in which physiological phenomena might be given chemical explanations. Among other theories, Liebig asserted that the organized parts of the animal body are formed exclusively by the nitrogenous "albuminous" matters (*Eiweiss*) contained in their nutrients, and assimilated with little chemical change. The nonnitrogenous nutrients – fat and carbohydrate – do not participate in the formation of tissues, according to Liebig, but are oxidized in the blood to provide animal heat. All mechanical work, he maintained, is produced by the *Stoffwechsel*, or "metamorphosis," of the nitrogenous tissue constituents, which are thereby decomposed into a nonnitrogenous residue consumed in respiration, and a nitrogenous portion that is excreted as urea, uric acid, or ammonia. Although highly speculative, Liebig's theory exerted a broad appeal because it provided the first clearly differentiated physiological roles for the main classes of nutrient substances. For his followers he offered an even more enticing prospect, namely, that they would be able to measure the central process of nutrition. "The amount of tissue metamorphosed in a given time," he contended, "may be measured by the quantity of nitrogen in the urine."[17] Thus Liebig's nutritional framework seemed to establish the necessary preconditions for fruitful quantitative experimentation: compounds that appeared externally and could be submitted to ordinary quantitative chemical analysis were taken to be direct indicators of the quantity of nitrogenous tissue materials undergoing the internal *Stoffwechsel.* To make the external measurement more reliable, Liebig himself contributed, in 1852, a new method for the determination of urea by titrating with mercuric nitrate.[18]

The feasibility of the research program Liebig suggested depended on the question of whether nitrogen may be lost from the body in any way other than through the urinary nitrogen. While continuing efforts to resolve this problem, several investigators improved the methods for measuring the gaseous exchanges in a closed chamber. Victor Regnault and Jules Reiset made a crucial advance between 1845 and 1849. Designating an elaborate apparatus that continuously removed the carbon dioxide exhaled and replenished the oxygen supply, they were able to maintain the animals for several days while the gaseous composition of the chamber remained normal. In over one hundred experiments on different classes of animals, Regnault and Reiset found in most cases a very small amount of nitrogen exhaled, but occasionally a little absorption.[19]

At about the same time, Bidder and Schmidt undertook to attain a complete quantitative description of the material exchanges between a cat and its environment. Following the general approach of Boussingault, they measured the elementary constituents of food and excretions; in order to minimize errors due to difficulties in collecting the excretions, they extended each set of measurements over a five- to ten-day period. They were only partially able to combine these measurements with measurements of the gaseous exchanges. Each day they placed the animal in a respiratory chamber for one hour, then extrapolated to estimate the daily quantities of oxygen and carbon dioxide respired, checking in this way the quantities calculated indirectly from the other balance sheets. Their most important advance was to compare the effects of these exchanges caused by different dietary conditions – starvation, the maximum quantity of meat they could feed the cat, a normal meat diet, and the addition of fat or carbohydrate. Through elaborate calculations based on the elementary composition of urea and of meat, Bidder and Schmidt estimated the amounts of nitrogenous and nonnitrogenous nutrients consumed, and the amounts of each assimilated by or lost from the body.[20]

Liebig's most direct disciple was Theodor Bischoff. Beginning in 1853 Bischoff utilized Liebig's new urea method in order to measure the metamorphosis of muscle tissue in a dog over a broad range of conditions. He tested the effect on urea output of hunger and dietary combinations of meat, fat, carbohydrate, gelatine, and salt. One series consisted of a pure meat diet gradually increased over six days from 1,000 to 3,500 grams per day. Bischoff believed that the outcome of this series confirmed Liebig's concept that the urea nitrogen represented exclusively the product of the *Stoffwechsel* occurring in the tissues and that it refuted a view maintained by Bidder and Schmidt, according to which excess nitrogenous nutrients are directly oxidized in the blood. Bischoff encountered a serious setback, however, for large portions of the dietary nitrogen were unaccounted for in the urea. Since Regnault and Reiset's experiments had ruled out the possibility that such large amounts of nitrogen could be exhaled, Bischoff could not account for this deficit.[21]

III

When Carl Voit began his work in nutrition Liebig's ideas served as his "guiding star."[22] After finishing his medical degree in Munich in 1854, Voit attended Liebig's chemistry course and attracted his interest. At Liebig's suggestion Voit then spent a year with Friedrich Wöhler learning the classical methods of organic chemistry. Enthused also by Bidder and Schmidt's work, Voit would have gone next to study with them, but for the fact that Bischoff came to Munich in 1856 and offered him a position as his assistant in the newly formed Physiological Institute. The first project Bischoff assigned Voit was to apply his methods for measuring nitrogen metabolism to a dog with a bile fistula. Soon, however, Voit encountered the technical and conceptual difficulties Bischoff had not yet overcome and decided to try to resolve them. The task occupied much of his career.[23]

Voit undertook first to find out what became of the missing nitrogen in the experiments of Boussingault, Bischoff, and others. One possibility was that urine contained nitrogen in forms other than urea. Voit devised a special method to determine the total urinary nitrogen, but found that the quantities differed little from those obtained with Liebig's urea titration method. Like Bischoff, he found only insignificant amounts of nitrogen in the feces. He was, therefore, "forced back" to the view that nitrogen must be eliminated through the skin or lungs, but since he did not have a respiration chamber available, he could only check that deduction indirectly, by repeating Bischoff's experiments. He followed Bischoff's methods closely, but made several important technical refinements. Noticing that he could easily differentiate feces originating from a meat diet from those deriving from bread, he fed the dogs bread before and after each experimental series. Then he could tell exactly how much of the excretion to include as a product of a determined quantity of ingested meat. He carefully trimmed the fat and tendon off all the meat used, finding that the nitrogen content of the selected pieces of fresh beef was then constant to within 0.3 percent. Finally, he was able to train the dogs to void their feces and urine once daily, so that he could collect them directly in containers and avoid the inevitable losses involved in retrieving them from the floor of a cage. Between September 1856 and May 1857 Voit carried out five experiments on two dogs, each series lasting four days. In each case the quantity of nitrogen ingested in the meat turned out to be nearly equal to that which he obtained in the urea, or else could be accounted for in the gain or loss of weight of the animal. Thus the problem he had hoped to solve had instead disappeared. He believed that the large deficits in Bischoff's experiments must have resulted from Bischoff's use of an old dog with an exceptionally low *Stoffwechsel,* so that its urea had an unusually long time to decompose. Confidently Voit concluded, "We can now with full justification designate the urea as a measure of the *Stoffwechsel.*"[24]

After this initial success Voit and Bischoff extended their investigations together, with Voit carrying out all the experimental operations. Their joint research began in October 1857. In an attempt to establish the fasting level of the consumption of nitrogenous tissue substance they measured the urea output of a dog given nothing to eat for six days. During the next three days they fed the dog starch, and the rate of urea formation fell below the fasting rate, to a daily low representing 176 grams of tissue substance consumed. Treating this figure as the minimum essential quantity of tissue metamorphosis per day, they gave the dog a corresponding quantity of meat daily, together with varying quantities of starch. They were looking for a combination that would just maintain the dog's weight and that might comprise a minimum adequate diet. They then tested the same quantity of meat with increasing quantities of fat and found that 250 grams of fat was enough barely to maintain the weight of the animal. During a second one-day abstinence period, however, the dog consumed only 136 grams of its flesh, so that in two subsequent series they reduced the daily meat ration to 150 grams, combined respectively with sugar and with fat. Next they increased the meat ration to 500 grams, with 250 grams of fat. On this diet the dog gained weight steadily for thirty-two days. It continued to gain on successive diets of 1,500 grams of meat with 250 grams fat, 1,500 grams of meat with 350 grams of fat, and meat alone in quantities varying from 2,200 to 2,900 grams. At the beginning of February 1858, they decided to reduce the daily amounts of meat gradually, while keeping the fat intake at 150 grams, until they reached a level at which the dog would just maintain its weight. After twenty-four days they had come down from 1,500 to 400 grams of meat, yet during the whole period the animal had slowly gained weight.[25] As Voit later recalled, this baffling sequence forced a decisive change in their outlook.[26]

Until then they had regarded the weight of an animal as the direct reflection of the changes in its body resulting from a given nourishment. If its weight remained the same, they assumed that all of the nitrogen contained in its food must reappear in it urine and feces. If they fed an animal 2,000 grams of meat and it gained 500 grams, they supposed that a net quantity of 500 grams of meat was assimilated and that the nitrogen representing the remaining 1,500 grams would appear in the excretions. Yet over the whole period, in this last series, during which the animal was gaining weight, it was excreting more nitrogen than it was ingesting. Similar situations had appeared during several of the preceding series, but Bischoff and Voit only gradually came to realize the significance of the discrepancies. The inverse case, that is, all the older experiments containing a nitrogen "deficit," had led Bischoff and others to believe that nitrogen must leave the body in other ways, but the present situation could not be similarly explained. (They probably did not even consider an older hypothesis, discarded after Boussingault's first experiments, that nitrogen might be absorbed from the atmosphere.) Embracing now a viewpoint that Bidder and Schmidt had already assumed, but which Bischoff had rejected in 1853 because he thought it only complicated the issue, they inferred that the changes in the weight of an animal do not directly correspond to the gains or losses in its nitrogenous constituents, because opposite changes in the quantities of the non-nitrogenous contents of its body – its fat or water – may be taking place simultaneously. Now they began to see that if they fully accepted the postulate that the nitrogen they found in the urine and feces accounted for all the nitrogen lost from the body, then the difference between this total for a given period and the amount contained in the food eaten during the same period was a direct measure of the quantity of nitrogenous substance assimilated into the structure of the body, or, when negative, of the quantity of body tissue consumed in addition to the food. The difference between that net change and the overall change in the body weight served as the basis for calculating the amount of fat or of water added to or lost from the body. As simple as this "truth" was, they afterward related, it had cost them many unexpected experiences to reach it; at first their results only appeared contradictory and chaotic to them. Once they had grasped this essential insight, they possessed "an entirely different foundation for the continuation of the investigation."[27]

After an interruption lasting several months, Bischoff and Voit resumed their investigation at the end of September 1858. They carried out a sequence using pure meat diets, planned apparently in accordance with their new point of view. Starting with a daily ration of 1,800 grams, which supplied adequate nourishment, they descended in six stages to 176 grams, the earlier established minimum rate of consumption. At the level of 1,500 grams daily slightly more nitrogen was excreted than was ingested. The total quantity of nitrogen excreted per day fell with each decrease in the amount of nutrient nitrogen, yet the net loss of nitrogen increased progressively over the same steps. The amount of fat lost from the body according to their indirect calculations became larger as the meat intake decreased. After reaching the minimum level they jumped the ration back to 1,800 grams, then to 2,500 grams, the maximum the dog could eat. Then they returned to 2,000 grams. This series revealed several significant factors. During the first day after each dietary increase, the dog excreted considerably less nitrogen than it ingested, so that nitrogenous substance accumulated in its body; but on the succeeding days the rate of urea formation increased until the nitrogen loss equaled or exceeded the intake. Following dietary decreases the inverse occurred. They noticed especially that when the diet diminished from 2,500 to 2,000 grams of meat the latter quantity was at first insufficient to balance the nitrogen loss, whereas at the beginning of the series 1,800 grams had been more than sufficient. From these patterns they drew the important generalization that the rate of decomposition of nitrogenous matter in the animal does not depend directly on the quantity of nitrogenous nutrients, but rather on the nutritional condition of the animal. The net loss of nitrogen at 1,500 grams, despite the fact that this diet supplied far more nitrogenous substance than the dog consumed in abstinence; the fact that 1,800 grams was sufficient nourishment following a nitrogen-poor diet, whereas 2,000 grams was insufficient following a high nitrogen diet; the gradual rise in urea formation following an increase in daily intake, and the fall following a decrease – all seemed to Bischoff and Voit to imply that as the nitrogenous mass of the body increased or decreased, so did the rate of metamorphosis of these constituents. Thus the same diet might at one time supply enough nitrogen and at another time not, because the requirement itself varied with the changing state of the animal.[28]

From January until June 1859, Bischoff and Voit compared the rate of urea formation when they fed their dog meat alone with the rate when they added sugar, starch, fat, or gelatine to the same quantity of meat. From the results of five such series they concluded that nonnitrogenous nutrients reduce the rate of consumption of nitrogenous substances at all levels of nitrogenous nourishment, but cannot prevent that consumption from increasing again as the intake of nitrogenous food increases.[29]

In October 1858, Voit began on his own to investigate whether other factors can influence the rate of decomposition of the organized tissue constituents. He started with an examination of the effects of coffee, because some investigators had claimed that it saves food by reducing the consumption of nitrogenous substances. For the first series he allowed the dog a free diet of bread taken with a fixed daily quantity of milk. After a month he added a coffee extract to this diet for a second month, removed it again for three weeks, and added it for a final three weeks. Over the whole series the dog consistently lost more nitrogen than it ingested. The overall rate of nitrogen consumption also declined. Voit ascribed the latter effect to the diminishing mass of the nitrogenous tissue constituents rather than to the action of the coffee. During February 1859, he carried out a shorter series in near-starvation conditions, the dog being fed only a small amount of milk. Again he alternated two periods with and without coffee. During each period with coffee the rate of urea production was slightly lower than during the preceding period without it, but Voit again concluded that the decreasing nitrogen content of the body, not the coffee, caused the decline. Finally, in March and April he carried out a third comparative series, using a daily nutritional base of 1,000 grams of meat, a quantity that would "cover as closely as possible the loss" of nitrogen. This time the coffee slightly increased the rate of consumption of nitrogenous substance, "but the differences are so small, that I shall draw no broader conclusions" from them. In the end Voit asserted that coffee does not significantly affect the *Stoffwechsel,* but exerts its influence through the nervous system.[30]

The significance of this investigation extended beyond the role of coffee alone, for either during or as a consequence of this work Voit gained a new general insight into the "laws" of nutrition. During the first two sets of experiments the rate of urea formation changed continuously as the condition of the animal changed, so that he had to try to distinguish the effect of the added factor against a shifting background effect. In the last series there were small fluctuations attributable to similar causes, but the dog entered a condition of near equilibrium between the intake and consumption of nitrogenous substances. Voit could, therefore, discern more easily the small effect of the coffee.[31] By the time he took up the investigation of a second additive, ordinary salt, he had recognized that one must, as a general principle, place an animal in a state of nutritive equilibrium in order to establish whether the addition increased or decreased the *Stoffwechsel.* He had also learned how best to establish such an equilibrium. The animal reached it most quickly on a pure meat diet. When one fed the dog a large daily quantity of meat, most of it accumulated in the body during the first day or so; but as the mass of nitrogenous constituents increased, the rate of nitrogen consumption rose until it balanced the intake. Thereafter, so long as the diet remained constant, the nutritional condition of the animal was stable. Any alteration caused by the addition of another substance to this diet could then be clearly identified.[32]

After becoming aware of the need to establish this steady state in order to test the effects of additional factors, Voit must have seen that the means to attain it already lay implicit in the results of the investigations he had done with Bischoff. In their report on that work they had noted that after an increase in a meat diet the nitrogen output was at first less than the input, but that the output rapidly increased until it equaled or exceeded the latter. They had apparently not seen, however, as Voit now saw, the importance of the fact that when the same diet was kept up the animal soon reached and maintained a nitrogen equilibrium.

IV

When Bischoff and Voit published the massive results of their investigations in 1859, they included calculations not only of the consumption of the nitrogenous nutrient and tissue substances, but also of the amounts of fat and water absorbed or lost. They were aware that the many assumptions they had to introduce made the latter values highly uncertain. Voit had, in fact, already set out, in collaboration with Max Pettenkofer, to provide a firmer basis for these calculations by including in his measurements direct determinations of the gaseous respiratory exchanges;[33] that is, he sought to merge the two previously established lines of balance investigations. Up until then most experiments had dealt either with the food and excretions alone, or the respiration alone, or at least more fully with one than the other, because of the difficulties involved in setting up experimental conditions permitting both sets of measurements to be carried out rigorously at once.

To achieve this ambitious aim Pettenkofer constructed a respiration apparatus on an unprecedented scale. The chamber was like a small room, in which an animal or person could live comfortably for a day or more. He followed the basic arrangement of Regnault and Reiset's design. Air passed continuously through the chamber, the total volume of entering and outgoing air was measured, and samples of each, as well as of the air remaining in the chamber afterward, were collected in order to determine the content of carbon dioxide and water. Pettenkofer took great pains to insure the reliability of the apparatus, especially by eliminating potential sources of leaks. Unlike Regnault and Reiset, he ascertained the degree of accuracy of the system by means of control experiments, burning a wax candle in it to determine the limits for carbon dioxide, and evaporating water in it to test the measurements of water vapor.[34]

Voit and Pettenkofer began using this apparatus for nutritional experiments early in 1861. It took them two years, however, to work out problems concerning the simultaneous measurement of carbon dioxide and water, and to account for anomalous ratios of oxygen to carbon dioxide that occurred in the early trials.[35] Finally, in February 1863 they carried out with notable success a series of experiments from which they obtained a complete balance of the incoming and outgoing elements for a dog maintained in nitrogen equilibrium on a meat diet. The input included the carbon, hydrogen, nitrogen, and oxygen of the food and the inspired oxygen. The output consisted of these elements as contained in the urine, the feces, and the expired carbon dioxide, water, and methane. The total difference between the daily input and output was less than 1 percent of the total mass of material exchanged.[36] Proudly Pettenkofer and Voit announced that their new experimental procedures set a new standard for nutritional investigations. In their balances, they said,

> every single value is ascertained by experiment. All *Stoffwechsel* balances put forward up to now have suffered from the serious defect that for certain factors, partly of the intake, partly of the output, they assumed hypothetical values instead of values actually determined during 24 hours, and consequently left a rather open field for arbitrary interpretations.
>
> The equation which we now have established . . . for the *Stoffwechsel* over 24 hours rests on values which are all actually determined over this time and is, indeed, the first which has ever been established without recourse to hypotheses.[37]

Voit and Pettenkofer had ample reason to be pleased with their new results, for they had reached a goal that numerous investigators had pursued for nearly twenty-five years. Between them they had greatly advanced the art of conducting both of the two types of measurements of the material exchanges of animals with their surroundings that their predecessors had initiated. Then they had been able to combine both types of measurement into one investigation. To do so they had to eliminate the gap between the respective experimental conditions that had previously seemed essential for success in one or the other kind of investigation. It had been difficult to extend respiration experiments beyond a few hours, whereas nutritional experiments had appeared reliable only if they lasted at least several days. Pettenkofer's apparatus not only permitted respiratory measurements lasting twenty-four hours, as those of Regnault and Reiset had, but also improved the accuracy over the period and was spacious enough to accommodate large animals and humans. The larger capacity was crucial, because the larger diets and more copious excretions of larger animals made it possible to measure these factors with less error, over shorter time periods, than with small animals. Equally important, by 1863 Voit had attained precise enough control over the diets and the collection and analysis of excretion that he could obtain regular, reliable measurements for a single day. For the first time, therefore, the two collaborators were in a position to make both sets of measurements rigorously, on one animal, and over the same time period. They had, in fact, established the experimental foundations upon which a whole field of quantitative and energy metabolism flourished during the following decades.

Throughout the years in which Voit was developing the experimental techniques that established him at the center of this emerging field, his empirical investigations were inextricably interwoven with his theoretical efforts to elucidate the internal processes of nutrition. His starting point, as we have seen, was Liebig's scheme of animal chemistry. When he and Bischoff published the results of their early studies in 1859 they incorporated their data into an ingenious elaboration of Liebig's theories. Voit's subsequent investigations, however, forced him gradually to modify, and finally to abandon Liebig's theories. He proposed others in turn, most prominently his controversial distinction between "organ protein" and "circulating protein." I have summarized these developments elsewhere[38] and plan eventually to present a more thorough analysis of the complex interaction between Voit's theoretical aspirations and the course of his research. Here we are concerned mainly with the technical control over measurable, physiological processes that Voit was able to achieve in pursuit of his conceptual objectives. During the 1860s he himself came to stress more and more that rigorous experimental standards were of paramount importance in nutritional studies. To the methods he had perfected, and the quantitative regularities they had revealed, he attributed an importance transcending the particular theories he and others might hold concerning the internal phenomena. In response to criticism of his results he steadily refined those methods. By 1866 he could show that when a dog is kept on a uniform diet its excretion of urea is so regular that the daily quantities deviate from the mean by less than 3 grams.[39] He used such precise results in part to disarm his critics with a display of the rigor of his experimental standards; they served also, however, in his view, to demonstrate that under controlled conditions physiological processes are as lawful as simple physical and chemical processes.

V

The field of metabolic balance sheet investigations in which Carl Voit played so prominent a role was by no means the only route along which quantitative precision penetrated biological experimentation in the mid-nineteenth century. Measurements of the velocity of nerve impulses, of the changes in electrical potential within a stimulated nerve or muscle, and of the pressures, velocities, and gaseous contents of the blood are equally impressive manifestations of the general spread of the "hallmark of science properly done" from the physical sciences into physiology during that era. The nutritional investigations occupied a focal position, however, because they dealt with those material exchanges between organisms and their environments that had been regarded since antiquity as central to the nature of life and that had always been perceived as a problem involving quantities. Voit, more than any other single person, helped to complete the transformations of these quantitative *ideas* into reliable quantitative *measurements* under well-defined conditions.

Voit's standards of measurement were adequate to support the rapid growth of a field of quantitative metabolism whose enduring importance we acknowledge whenever we count calories. The theoretical inferences that he and some of his contemporaries drew from their observations have proven less durable. Not only were specific concepts, such as Voit's organ protein and circulating protein, or its successor, Otto Folin's "endogenous" and "exogenous" metabolism, eventually discarded, but the whole approach of deducing internal processes from the measurement of surface exchanges appears in retrospect to have been "black box" guesswork that could never succeed.[40] But the mastery that Voit and his colleagues achieved over the study of these surface phenomena provided the essential framework within which their successors were able to supercede their nutritional conceptions. The dramatic developments at the beginning of the twentieth century that revealed the "accessory factors" and essential amino acids unsuspected by the physiologists of the nineteenth century, and the "Baustein" concept of nutrition that replaced the nineteenth-century view that animals directly assimilate the nutrient substances, were made possible largely by extensions of the feeding experiment methods developed by Voit and his school. In general these later advances depended on the tight control over diets and over the collection of excreted materials that Voit's investigations had shown to be feasible. More specifically they relied on the fundamental criterion he had established – that the threshold for adequate nourishment is a diet that can maintain an animal in nitrogen equilibrium.

Voit's success was not due merely to the diligence with which he tracked down the sources of error and uncertainty in the measurements of his predecessors. As the detailed chronicle of his early research suggests, there was a close interaction between technical improvements in measuring and elucidating the sources of irregularities. Efforts to make the former more precise merged with efforts to understand the underlying phenomena that produced the variations. The achievement of quantitative precision therefore demanded imagination as well as skill. Later on Voit had to call upon his considerable store of self-confidence and determination as well, for he was required to defend his conclusions repeatedly from charges that his observations were too uncertain to support valid inferences about nutritional processes.

If we return to our original question – whether Claude Bernard's strictures about the whole quantitative approach that included Voit's work were warranted by the state of knowledge when he expressed them – we can affirm that some of the early measurements, as well as the bolder inferences drawn from them, were vulnerable to his criticisms. He proved to be unduly pessimistic, however, in his judgment that the deficiencies he found in the early attempts could not be overcome under existing conditions. Even as he was writing that a "complete equation is impossible for the moment," Voit and Pettenkofer were attaining a kind of complete equation in their balance sheet of incoming and outgoing elements. Furthermore, Bernard's opinion that qualitative studies must precede quantitative studies did not allow for the subtlety of the interplay between measurement and the recognition of qualitative phenomena. That a man of the stature of Claude Bernard so clearly underestimated the long-range potential inherent in the research program of Boussingault, Bidder and Schmidt, and Carl Voit serves only to emphasize how difficult it is for any one person to be attuned to all of the promising trends within his own scientific field.

NOTES

1 Claude Bernard, *An Introduction to the Study of Experimental Medicine,* trans. H. C. Greene (New York: Dover, 1957), pp. 129–30.

2 I. Bernard Cohen, "Foreword," ibid.

3 Ibid., p. 132.

4 Harry Woolf, "The conference on the history of quantification in the sciences," *Isis, 52* (1961): p. 133.

5 M. D. Grmek, "Santorio, Santorio," in *Dictionary of Scientific Biography,* vol. XII (New York: Scribner's, 1975), p. 103.

6 Stephen Hales, *Statistical Essays containing Haemastaticks* (New York: Hafner, 1964), p. XIX.

7 Thomas S. Kuhn, "The function of measurement in modern physical science," *Isis, 52* (1961): p. 166.

8 Bernard, *Introduction,* p. 132. Professor Cohen agreed with Bernard that such a procedure was "absurd." Cohen, "Foreword."

9 Claude Bernard, *Lettres Beaujolaises,* ed. Justin Godart (Villefranche en Beaujolais, 1950), p. 30.

10 Owsei Temkin, "A Galenic model for quantitative physiological reasoning," *Bull. Hist. Med., 35* (1961): p. 470.

11 Owsei Temkin, "Nutrition from classical antiquity to the baroque," in Iago Galdston, ed., *Human Nutrition, Historic and Scientific* (New York: International Universities Press, 1960), p. 370.

12 Jerome J. Bylebyl, "Nutrition, quantification and circulation," *Bull. Hist. Med., 51* (1977): pp. 369–85.

13 See Everett Mendelsohn, *Heat and Life* (Cambridge, Mass.: Harvard University Press, 1964), pp. 166–83; Frederic L. Holmes, "Introduction," in Justus Liebig, *Animal Chemistry* (New York: Johnson Reprint, 1964), pp. xxxvi–xl.

14 Frederic L. Holmes, "Elementary analysis and the origins of physiological chemistry," *Isis, 54* (1963): pp. 62–3.

15 V. Regnault and J. Reiset, "Recherches chimiques sur la respiration des animaux des diverses classes," *Annales de chimie et de physique,* (1849): pp. 305 ff.

16 J. B. Dumas, "Rapport sur un mémoire de M. Boussingault, intitulé: recherches chimiques sur la végétation," *Comptes rendus de l'Académie des Sciences, 8* (1839): p. 54; J. B. Boussingault, "Analyses compareés des aliments consommés et des produits rendus par un cheval soumis a la ration d'entretien," ibid., *71* (1839): pp. 128–36; J. B. Boussingault, "Analyses compareés de l'aliment consommé et des excrements rendus par une tourterelle," ibid., *11* (1844): pp. 433–56; J. B. Dumas, "Rapport sur un memoire de M. Boussingault," *C.r. Acad., 8* (1839): pp. 54–7.

17 Justus Liebig, *Animal Chemistry,* pp. 38–61, 232–3.

18 Holmes, "Introduction," p. xciii.

19 Regnault and Reiset, "Recherches" pp. 299–519.

20 F. Bidder and C. Schmidt, *Die Verdauungssäfte und der Stoffwechsel* (Mitau, 1852), pp. i–v, 291–399.

21 Th. L. W. Bischoff, *Der Harnstoff als Maass des Stoffwechsels* (Giessen, 1853).

22 Carl Voit, "Ueber die Entwicklung der Lehre von der Quelle der Muskelkraft und einiger Theile der Ernährung seit 25 Jahren," *Zeitschrift für Biologie,* 6 (1870): p. 397.

23 Ibid., p. 312; Graham Lusk, *Nutrition* (New York: Hafner, 1964), pp. 103–104; Otto Frank, *Carl von Voit: Gedächtnistrede* (München, 1910), pp. 1, 5–7.

24 Carl Voit, "Beiträge zum Kreslauf des Stickstoffs im thierischen Organismus," *Physiologisch-Chemische Untersuchungen,* I (Augsburg, 1857), pp. 3–40.

25 Th. L. W. Bischoff and Carl Voit, *Die Gesetze der Ernährung des Fleischfressers durch neue Untersuchungen festgestellt* (Leipzig, 1860), pp. 207, 43, 201, 182, 130, 53, 154, 98, 101, 104, 107, 110, 145, 88, 121, 126, 91, second table facing p. 304. The order of pages represents the chronological order of the experiments.

26 Carl Voit, "Der Eiweissumsatz bei Ernährung mit reinem Fleisch," *Z. Biol., 3* (1867): pp. 68–69.

27 Bischoff and Voit, *Ernährung,* pp. 30–2. The experiments whose results might have led to this conclusion, in addition to the twenty-four-day series, are described on pp. 182–3, 101–2, 88–9. The published recounting of these experiments was, of course, written after Bischoff and Voit had accepted their new point of view.

28 Ibid., pp. 210–13, 56–85.

29 Ibid., pp. 85–8, 165–8, 193–6, 115–18, 161–5, 118–21, 79–82, 112–15, 136–42, 226–41, 135–6, 168–79 (pages listed in chronological order).

30 Carl Voit, *Untersuchungen über den Einfluss des Kochsalzes, des Kaffee's und der Muskelbewegungen auf den Stoffwechsel* (München, 1860), pp. 67–147.

31 Ibid., pp. 109–25.

32 Ibid., p. 31.

33 Ibid., p. 6.

34 Max Pettenkofer, "Über die Respiration . . .," *Annalen d. Chem. und Pharm.,* suppl. 2: (1862).

35 Max Pettenkofer and Carl Voit, "Untersuchungen über die Respiration," ibid., pp. 52–70; Pettenkofer and Voit, "Ueber Bestimmung des in der Respiration ausgeschiedenen Wasserstoff- und Grubengases," ibid., pp. 247–8.

36 Max Pettenkofer and Carl Voit, "Ueber die Produkte der Respiration des Hundes bei Fleischnährung und über die Gleichung der Einnahmen und Ausgaben des Körpers dabei," ibid., pp. 361–371.

37 Ibid., pp. 365–366.

38 F. L. Holmes, "Introduction," in Liebig, *Animal Chemistry*, pp. xcii–cix; F. L. Holmes, "Voit, Carl von," in *Dictionary of Scientific Biography*, vol. XIV (New York: Scribner's, 1976), pp. 63–7.

39 Carl Voit, "Untersuchungen über die Ausscheidungswege der stickstoffhaltigen Zersetzungs-Produkte aus dem Thierischen Organismus," *Zeitschrift für Biologie, 2* (1866): pp. 216–25.

40 See Joseph S. Fruton, *Molecules and Life* (New York: Wiley, 1972), pp. 426–9.

24

Ideological factors in the dissemination of Darwinism in England

1860–1900

MARTIN FICHMAN
York University

More than most theories effecting a significant conceptual revolution in science, Charles Darwin's and Alfred Russel Wallace's evolutionary theory bore directly, if ambiguously, on a host of nonscientific questions. The philosophical, religious, sociopolitical, and semantic factors that together shaped the course of the Darwinian debates in England from 1860 to 1900 are well established in the historiography of Victorian culture. These factors are, however, only retrospectively isolable; for most of the participants in the debates they were never clearly demarcated. And, the manner in which they have become demarcated tells us much about the scientific and ideological revolutions we associate with Darwin's name.

The links between science and ideology pose major methodological problems for the study of scientific change. As Sir Karl Popper has observed, these links have fostered a tendency "to conflate science and ideology, and to muddle the distinction between scientific and ideological revolutions." His effort, however, to provide a viable distinction between the two, while salutory, is predicated upon a definition of ideology – "*any non-scientific* theory or creed or view of the world which proves attractive, and which interests people, including scientists" – that itself is problematical to the historian.[1] Indeed, in the case of Darwinism, the interaction between ideological and scientific developments is sufficiently intricate to render the goal of establishing a rigorous distinction between the two elusive (at the least). Rather, we are faced with the phenomenon of science-as-ideology: the hegemony of the scientific way of knowing.[2] In the nineteenth century, the definition of science – as well as of the scientific community – was undergoing rapid transformation, particularly with respect to biology. The Darwinian debates were as much debates on methodological and sociological issues in science as they were on strictly biological questions. They raised fundamental questions for the philosophy as well as practice of biology and, in so doing, illustrate that complex process by which any age defines (or redefines) the domain of science and fixes, for itself, the (malleable) line of demarcation between science and ideology.

To limit a discussion of the Darwinian Revolution to a period of four decades, and to a single country only, requires a word of justification. The sheer complexity of that dual revolution, spanning as it does a period of nearly 250 years and entailing the rejection of (at least) six deeply entrenched concepts – a short earth's history, catastrophism, automatic upward evolution, creationism, essentialism, and anthropocentrism[3] – renders any focus to some extent arbitrary. Even within that limited period, the style of the debates and the reception accorded Darwinism varied significantly depending on the national, institutional, and ideological affiliations of the participants.[4] The English case is of particular interest, however. For the dissemination of Darwinism in the period following the publication of the *Origin of Species* coincides with a period of significant professionalization of science in that country. The Darwinian debates, I shall argue, contributed to that divorce between natural knowledge and general culture that took place in England during the last decades of the nineteenth century.[5] That it should have been evolutionary biology, literally the most human of the sciences – and one with a great bearing on general culture – that contributed to the triumph of the concept of science as a value-neutral but inherently progressive enterprise (itself an ideological revolution coming under increased critical scrutiny today) is a paradox worth exploring.

Writing in 1863, T. H. Huxley concluded the second part of his controversial *Evidence as to Man's Place in Nature* by declaring:

> Science has fulfilled her function when she has ascertained and enunciated truth; and were these pages addressed to men of science only, I should now close this essay, knowing that my colleagues have learned to respect nothing but evidence, and to believe that their highest duty lies in submitting to it, however it may jar against their inclinations.
>
> But desiring, as I do, to reach the wider circle of the intelligent public, it would be unworthy cowardice were I to ignore the repugnance with which the majority of my readers are likely to meet the conclusions to which the most careful and conscientious study I have been able to give to this matter, has led me.[6]

This passage (which is followed by an explanation that our kinship with the brutes in no way derogates from our dignity) encapsulates the strategy involved in the dissemination of Darwinism in England. Huxley at once establishes a sharp distinction between the (biological) scientific community and the wider intelligent public – a distinction that was only just becoming plausible in England – and defines the conduct of that community as the distinterested and critical pursuit of "truth." It is not my purpose to analyze Huxley's sociology of science (its Mertonian character is obvious),[7] but simply to indicate that by establishing a domain exclusive to science, he is arguing that the conclusions of evolutionary biology, founded on "nothing but evidence," are to be accepted irrespective of any "passion and prejudice." And he makes it clear that among those conclusions is one that specifies that the highest human "faculties of feeling and of intellect begin to germinate in lower forms of life."[8]

No aspect of Darwinism was more sensitive to the play of ideological forces than that which dealt with human evolution, particularly with respect to moral and intellectual attributes. Indeed, the intense interest and controversy engendered by Darwin's and Wallace's theory of natural selection could hardly have arisen if the question of man's descent from the lower animals was not perceived "as a virtually unavoidable consequence" of it.[9] There were, broadly speaking, two avenues open to those who sought to facilitate the acceptance of that theory in a milieu as saturated with religion (however diverse its forms) as was England in the latter decades of the nineteenth century: One could either suggest that evolutionary theory was wholly compatible with some version of providential design or that, conversely, evolutionary theory – as a strictly objective science – was beyond the realm of criticism on extrascientific grounds. In choosing the latter, Huxley (and those who argued as he did) tied Darwinism to an ideology that claimed for science the status of an autonomous and objectively neutral activity. And it is this dual revolution, scientific and ideological, that informs the Darwinian debates throughout our period, particularly as they touched on human beings.

It is now established that the question of human evolution occupied both Darwin and Wallace from the start of their respective careers and was fundamental in the development of the hypothesis of natural selection.[10] However, the wording of the communication to the Linnaean Society in 1858 announcing their joint discovery obscures this fact. And the publication of *On the Origin of Species* the following year continued the silence on man. One may agree with Sandra Herbert that Darwin knew, as early as 1837–8, that by admitting a purely naturalistic origin for human beings he was "breaking a trust" with the established traditional and scientific opinion with respect to man's place in nature.[11] Darwin's continued silence on the subject in the decade following the appearance of the *Origin* is consistent with his public caution, a caution that is usually considered to have paid off insofar as the reception accorded the publication of the *Descent of Man* in 1871 caused no such obvious furor as that which attended the *Origin.*

Yet this explanation is only partly true. By 1871 evolution (though not natural selection as its mechanism) had come to be generally accepted. Much of the initial opposition to Darwinism as being too speculative and insufficiently inductive – an opposition that found eloquent testimony in the "conspiracy of silence" that greeted theoretical discussions of Darwin's work in most of the prestigious learned societies and journals in the 1860s[12] – had eroded sufficiently by 1870 to permit widespread acceptance of the thesis of *Descent* with respect to the evolution of human bodily structure. The situation with respect to the evolution of human mental and moral attributes was far different. To understand why the period 1870–1 is central to the discussion of the dissemination of Darwinism, it is instructive to consider two other works that, together with *Descent,* brought into the open the ideological forces within the Darwinian camp. Alfred Russel Wallace's *Contributions to the Theory of Natural Selection* (1870) and St. George Jackson Mivart's *Genesis of Species* (1871), with critiques of those works by the philosopher Chauncey Wright, reveal much about the complex interaction of science and ideology in the Darwinian Revolution. There is apparent paradox here, too, in that it was Wallace who made the most sustained and brilliant defense of the scientific component of Darwinism (evolution by natural selection acting upon random variations) and who so strongly resisted its ideological component (a strict, value-neutral evolutionary naturalism).

It was Wallace who, in 1864, had been the first to demonstrate that evolution by natural selection could provide a powerful methodological framework for the scientific study of man. Others had argued for a reassessment of man's place in nature based on Darwin's *Origin,* most notably Huxley a year earlier (1863), but no one, including Darwin, had made so explicit the conviction that anthropological issues were the legitimate concern of the evolutionary biologist. Although the evidence for man's great antiquity was widely accepted by the mid-sixties, resistance was still strong toward an explanation of human attributes on the basis of natural selection.[13] In a paper read before the Ethnological Society of London on January 26 – "On the Varieties of Man in the Malay Archipelago"[14] – Wallace used those biogeographical arguments he had employed previously in analyzing zoological distribution to suggest that the present distribution and characteristics of human races in the Malay Archipelago were due to evolutionary forces acting on variable human populations over an extended period of geological time. In particular, he dismissed the special creationist hypothesis – that the chief human races were "created as they now are and where they are now found" – with respect to man, as he had earlier done with respect to the rest of organic nature.[15] Two months later, Wallace boldly extended his argument on human evolution.

"The Origin of Human Races and the Antiquity of Man Deduced from the Theory of 'Natural Selection,' " read before the Anthropological Society on March 1, 1864 (and reprinted in *Contributions*), was, in part, an effort to resolve the controversy between the so-called monogenists and polygenists.[16] Wallace's argument – that though racial differences do, in fact, antedate the historical period, the several races nonetheless derive from a common ancestor – turns upon an ingenious application of the principle of natural selection. According to Wallace's and Darwin's theory, animals and plants are constantly modified by the action of natural selection amid gradually changing environmental conditions. When slow changes of climate or physical geography, for example, make it necessary for an animal to alter its diet or bodily covering, this can only be accomplished by a selection of favorable variations, among the population, in bodily structure and internal organization. But "man," Wallace proposed, "by the mere capacity of clothing himself, and making weapons and tools, has taken away from nature that power of slowly but permanently changing the external form and structure in accordance with changes in the external world, which she exercises over all other animals." Sharper spears and better bows substitute for longer nails and teeth, greater bodily strength or swiftness; warmer clothing and better housing substitute for increased bodily hair during glacial epochs. Because of his unique intellect, man, after a certain point in his evolution, would respond to environmental demands with an unchanged body.[17] This "great leading idea," as Darwin described it in declaring it "quite new" to him,[18] provided Wallace with the solution to the monogenist–polygenist controversy.

Since the "striking and constant [physical] peculiarities" that mark the human races cannot have been produced or rendered permanent *after* the power of natural selection had begun to operate primarily upon mental (and behavioral) variations, they must have existed at a more remote period in human evolutionary history. Wallace argued that it was probable that human beings once existed as a "single homogeneous race without the faculty of speech, and probably inhabiting some tropical region." As a dominant species, but one still subject to natural selection of physical characteristics, early man would have spread throughout the warmer regions of the globe, becoming variously modified according to environmental contingencies. "Thus," Wallace concluded, "arose those striking characteristics and special modifications which still distinguish the chief races of mankind." While these changes had been going on, however, man's mental and moral development would have reached a sufficiently advanced stage to become the principal focus of natural selection. Physical variations would no longer (except in minor instances) be subject to selective action, and the diverse racial characteristics would have become fixed. The subsequent persistence of these racial characteristics throughout recorded history – "the stumbling block of those who advocate the unity of mankind" – is, Wallace concluded, not in conflict with the theory of common descent of all races from a single ancestor.[19]

Wallace's demonstration that man's tools (including language) removed his body from the realm of evolutionary specialization that operates inexorably elsewhere was recognized immediately as a turning point in the scientific study of man.[20] His paper was intended as a vehicle for applying natural selection to a wide range of concerns – the antiquity of man, racial superiority, man's taxonomic rank – with the clear implication that human mental and moral attributes would also be subsumed under a strict evolutionary naturalism. Or so matters stood until 1869, when Wallace's (public) posture as the foremost advocate of the thesis that man's evolutionary history could be reconstructed on the basis of natural selection alone was altered abruptly.

In a review of two new editions of geological treatises by Lyell, Wallace announced that man's intellectual capacities and moral qualities – unique phenomena in the history of life – were not explicable by natural selection.[21] Rather, these, as well as certain physical characteristics, required the intervention at appropriate stages in human evolution of "an Overruling Intelligence" that guided the laws of organic development "in definite directions and for special ends."[22] Ironically, it was Darwin's principle of utility that Wallace invoked to substantiate his claim. In the *Origin,* Darwin had noted that natural selection could produce neither a structure harmful to members of a given species nor a structure that was of greater perfection than was necessary for an organism in any given stage of its evolutionary development.[23] Citing the relative hairlessness of human beings (which would have been a positive disadvantage in prehistory) and the brain, hand, apparatus of speech, and "marvellous beauty and symmetry of his whole external form" (all features that would have been of greater perfection than was necessary on strictly utilitarian grounds for primitive people), Wallace argued that natural selection was insufficient as an explanation for their origin. He suggested, instead, that a "new stand-point [was possible] for those who cannot accept the theory of evolution as expressing the whole truth in regard to the origin of man." Using the analogy of domestic variation – the analogy Darwin had exploited successfully (but for very different ends) in the *Origin* – Wallace argued that just as man had used the laws of variation and selection to produce fruits, vegetables, and livestock, so also "in the development of the human race, a Higher Intelligence has guided the same laws for nobler ends," namely, "the indefinite advancement of our mental and moral nature."[24]

Although his new position was heavily indebted to his growing involvement with the then controversial data of phrenology, psychic phenomena, and spiritualism,[25] the arguments Wallace advanced with respect to the insufficiency of natural selection were presented within the context of what Darwin noted was otherwise an "inimitably good exposition" of natural selection.[26] They were, thus, particularly problematical for the Darwinians. That Wallace's new opinions had become integral elements in his evolutionary theory became clear the following year with the publication of *Contributions to the Theory of Natural Selection* (1870), a collection of several previously published essays and one significant new one.[27] The latter, "The Limits to Natural Selection as Applied to Man," was an elaboration of the views sketched in the 1869 review and was intended to clarify the extent to which his views on human evolution had come to differ radically from Darwin's (whose "Man book" he knew was being readied for publication).[28] His position at this juncture was (and was to remain) anomalous: Wallace was at once the most effective advocate of natural selection as the sole mechanism of evolutionary change (excepting man) as well as a forceful opponent of a complete, ostensibly neutral, evolutionary naturalism. The apparent contradiction inherent in *Contributions* elicited a number of critical reviews from the Darwinians, none of which was as ideologically explicit as that by Chauncey Wright.[29]

Though an American, Wright figures directly in the English debates because of the endorsement of his views by Darwin and his role as one of Darwin's "bulldogs" in the controversy surrounding the *Descent.* Wright considered *Contributions* to be an excellent defense and explication of natural selection, marred, however, by the concluding essay. In setting out his views on man, Wallace held that he was "as strictly within the bounds of scientific investigation" as he had been in any other aspect of his work on evolution.[30] Rather than a camouflage for biological heterodoxy or a pious platitude – Wallace indulged in neither tactic – this assertion is both sincere and serious; rejecting it *tout court* runs the risk of miscomprehending certain aspects of the philosophy of nature that underlay much of Victorian biology. To Wright, however, such a claim was fatuous. Wallace had clearly "laid aside his usual scientific caution and acuteness [in devoting] his powers to the service of that superstitious reverence for human nature which, not content with prizing at their worth the actual qualities and acquisitions of humanity, desires to entrench them with a deep and metaphysical line of demarcation." Though Wallace was hardly alone in his (new-found) evolutionary teleology, his views were a particularly cogent illustration of the difficulties facing even the staunchest Darwinian in accepting fully the naturalistic implications of Darwin's theory for man. Wright did not, nor could not, deny that there were difficulties facing that theory. But the limits posed by Wallace (and others) to the efficacy of natural selection with respect to human beings were not in "favour of any rival hypothesis, least of all that greatest of unknown causes, the supernatural."[31] Rather, they represented challenges to scientific researchers to demonstrate that natural selection would ultimately and completely account for the evolution of human mental faculties. Wright invoked Darwin's terse but pregnant assertion in the *Origin* – that "psychology will be based on a new foundation, that of the necessary acquirement of each mental power and capacity by gradation"[32] – in repudiating Wallace's "principles and analyses of a mystical and metaphysical psychology."[33]

Wright's naturalistic conception of psychology reflects the position of those Victorians, including Huxley, John Tyndall, and W. K. Clifford, who, while dissociating themselves from a crude materialism, nonetheless argued that psychology, insofar as it should become scientific, would have to be limited to the observational methods of the new physiological psychology that lay at hand. Thus, whatever deference was paid to traditional notions of morality, volition, and reason, the (presumed) objective method "would delineate the legitimate subject matter of psychology, not the reverse."[34] And although the distinction between "fact" and "value" had yet to assume a dominant status in the Victorian debates, the style of argument advanced by Wright and others signaled the direction in which the ideological component of the human sciences would develop. That ideology was paramount is confirmed by the curious fact that the decisive role played by psychology in the Darwinian debates was not at the level of the experimental findings of psychobiology (which, though available by the early 1870s, were largely unexploited by the debaters) but, rather, at the level at which psychology entered via the domain of social theory and the philosophies of nature that permeated the scientific ideas at issue.[35]

Wright's critique of Wallace's "Limits" depicts him as a fifth columnist of sorts within the Darwinian camp. In fact, Wallace had intended his essay as an attack on the reductionist biology of Huxley.[36] In a polemical lecture given in Edinburgh in 1868 – "On the Physical Basis of Life" – Huxley had argued that the progress of science, particularly physiology, was dependent on a strict materialist approach and condemned the "alternative, or spiritualistic [approach as] utterly barren, and [leading] to nothing but obscurity and confusion of ideas."[37] Wright's attack on Wallace is similarly polemical and he concluded his review with a denunciation of Wallace's views on the "metaphysical isolation of human nature, [as] based, after all, on barbaric conceptions of dignity, which are restricted in their application by every step forward in the progress of science, [and belonging] to a false conservatism, an irrational respect for the ideas and motives of a philosophy which finds it more and more difficult with every advance of knowledge to reconcile its assumptions with facts of observation."[38] The cogency of Wright's critique and the severity of his language were not lost on Wallace, who conceded to Darwin that Wright "almost converts me from the error of my ways."[39] That "error," which consisted, in part, of denying the mechanistic and nonteleological framework for a biology of the mind that a strict Darwinism seemed to entail, was combated again by Wright the following year in his review of Mivart's *The Genesis of Species.*

Using a providential and saltationist version of evolution (as well as the most powerful scientific criticisms that had been advanced by Darwin's opponents in the previous decade), Mivart had argued that natural selection was neither necessary nor sufficient as the mechanism of evolution.[40] More disturbing was Mivart's attempt to demonstrate the degree to which successive editions of the *Origin* rendered Darwin's own position less than consistent.[41] Darwin considered *Genesis* to be sufficiently damaging to both the thesis of natural selection and his own reputation to have Wright's repudiation of Mivart's claims published, at his own expense, in England.[42] Though Darwin told Wallace that he considered the review defective in certain of its specific biological arguments, he considered Wright's "philosophical" defense of natural selection admirable.[43] Wallace agreed that Mivart required a public rejoinder, but was dubious as to the appropriateness of the Wright review. And while reassuring Darwin that he found most of Mivart's minor arguments as leaving "Natural Selection stronger than ever", he noted that "the two or three main arguments do leave a lingering doubt in my mind of some fundamental organic law of development of which we have as yet no notion."[44] Wallace's further disclosure (to Lyell's secretary) that although he thought Mivart had underrated the power of natural selection, he agreed with Mivart's conclusion in the main – and was inclined to think it "more philosophical" than his own evolutionary theory[45] – renders the Darwinian position as this juncture curious indeed.

What did the debaters signify by "philosophical"? As David Hull has rightly emphasized, Darwin's (and Wallace's) theory was enunciated during the period when the philosophy of science was "coming into its own in England." The Darwinians and their opponents were entrapped not only in a biological controversy but in a fundamental controversy over the nature of science itself. Issues as diverse as the appropriate roles of induction and deduction, the nature of theory formulation and justification, the degree of divine activity in nature, and the distinction between supernatural entities and legitimate (but unproven) theoretical constructs, all entered (either overtly or covertly) into the Darwinian debates, which, because of the "jargon of the period," were often marked by confused, at times contradictory, terminology.[46] It was against such a background that the specific controversy surrounding the *Descent of Man* took place, a controversy that was concerned as much with the attempt to establish a demarcation between science and ideology as it was with the validity or invalidity of Darwin's theory.

Mivart, though (as Darwin phrased it) "stimulated by theological fervour,"[47] pointed clearly to the central significance of the *Descent.* In that work "all possibility of misunderstanding [Darwin's ultimate conclusions on man's origin] or a repetition of former disclaimers on the part of any disciple is at an end, and the entire and naked truth as to the logical consequences of Darwinism is displayed."[48] At issue, of course, was Darwin's and Huxley's contention that there is no difference of kind, but only of degree, between human mental and moral faculties and those of the other animals. For Mivart (as for Wallace) there was an ontological chasm between the two: "Physical science, as such, has nothing to do with man's [moral nature] which is hyperphysical."[49] In contesting not so much Darwin's massive array of data concerning animal and human intelligence and behavior as his "mode of conducting his argument," Mivart held that Darwin had failed utterly in his attempt to demonstrate the evolution of the higher human faculties from those of lower forms. Moreover, Darwin's "errors" were "mainly due to a radically false metaphysical system in which he seems (like so many other physicists) to have become entangled. Without a sound philosophic basis, however, no satisfactory scientific superstructure can ever be reared."[50] Mivart's rejection of Darwin's theory points squarely to the ideology at stake in the Darwinian debates. And that is neither the putative warfare between science and theology[51] nor any particular political variant of the amorphous phenomenon called "Social Darwinism" but, rather, the posited neutrality of science and the authority derived therefrom.

Wright, in countering Mivart's attack on natural selection, countered also his assertion that Darwinism was philosophically unsound. On the contrary, he accused Mivart of forgetting, "like many another writer, . . . the age of the world in which he lives and for which he writes, – the age of 'experimental philosophy.' " Wright's defense of Darwinism, moreover, is made within the context of the broader program of constructing the sciences into "a true philosophy of nature." For Wright, "the constitution of nature is written in its actual manifestations, and needs only to be deciphered by experimental and inductive research." The language of science is, then, the opposite of Mivart's (illusory) attempt to decipher a "latent invisible writing, to be brought out by the magic of mental anticipation or metaphysical meditation." Wright's vindication of Darwin's theory is made, finally, from the "neutral ground of experimental science."[52]

I have emphasized Wright's position because his defense of Darwinism is coupled explicitly with the contention that science must be an autonomous pursuit, free "from all forms of extra-scientific control imposed by a priori metaphysical or theological systems or authorities that arbitrarily extend the domain of ethical sentiments to matters of scientific knowledge"[53] – in short, a science free from ideology. Significantly, Wright praised Bacon not for the methodological canons associated with his name by the nineteenth century (Wright was dubious as to the actual value of Bacon's contribution), but for aiming "at establishing for science a position of neutrality, and at the same time of independent respectability."[54]

Darwinism, precisely because it was susceptible to – indeed invited – ethical and cosmological readings, was central to the broader question of the legitimate domain of scientific inquiry. Wright was perhaps the most extreme of the Darwinians in claiming that Darwin's theory was devoid of any metaphysical connotations, and he subjected evolutionary cosmologies – what he termed "German Darwinism," but which also included the work of English and American evolutionary philosophers, notably Herbert Spencer – to trenchant analysis. He insisted that scientific investigation should avoid "as far as possible the terms which have attached to them *good* and *bad* meanings, in place of scientific distinctness, – terms which have a moral connotation as well as a scientific one . . . Words have 'reputations' as well as other authorities, and there is tyranny in their reputations even more fatal to freedom of thought. True science deals with nothing but questions of facts – and in terms, if possible, which shall not determine beforehand how we ought to feel about the facts; for this is one of the most certain and fatal means of corrupting evidence."[55] Thus, in reviewing volume one of Spencer's *Principles of Biology* (1866) Wright emphasized that evolution meant continuous temporal transmutation of species (by natural selection) but not, as Spencer implied, a "progress which is inherent in the order of things"; the latter's biology is, in actuality, a "cosmological theory – or a theory of the universe in its totality – charged with a mission."[56] In a review of the sixth edition of the *Origin* (1872), Wright argued once again for a discussion of biological questions in that spirit of objective neutrality that he deemed both to "protect an investigation from prejudice" and to guarantee the progress of science.[57]

Wright's positivist defense of Darwinism, and the attendant ideological neutrality of Darwinism as science that such positivism entailed, represents a fundamental aspect of the process of dissemination of Darwinism. And whatever the epistemological status of the postulate of objective neutrality of science – an issue at the epicenter of the contemporary controversies among historians, philosophers, and sociologists of science concerning the nature of scientific revolutions – it is clear that during the latter decades or the nineteenth century the Darwinian debates came to be increasingly conducted in the style depicted by Wright and Huxley.

What conclusions may we draw from the "quiet" controversy over the *Descent of Man?* Viewed against the broader question of the dissemination of Darwinism in England, from 1860 to 1900, certain trends do appear. The most striking is that although evolution continued to gain ground, natural selection as the (major) mechanism for that process fell under mounting criticism. In fact, it was not until 1932, when the classic works of R. A. Fisher, Sewall Wright, and J. B. S. Haldane had been published, that the decisive arguments for natural selection were adduced.[58] The reasons for the persistent resistance to natural selection are complex but turn on the failure to uncover an adequate mechanism of inheritance prior to 1900, and, after the rediscovery of Mendel's laws, on the nature of genetic variation and the process of speciation. With respect to the second aspect of the Darwinian Revolution, what I have termed the postulate of a strict (and ethically neutral) evolutionary naturalism, there appears to have been less resistance. Although many continued to indulge in moralizing nature, Huxley's and Wright's confident positivist defense of Darwinism becomes more commonplace after 1870. Thus, the Darwinian E. B. Tylor could defend the assertion that anthropology would show that "the history of mankind is part and parcel of the history of nature, that our thoughts, wills, and actions accord with laws as definite as those which govern the motion of waves, the combination of acids and bases, and the growth of plants and animals."[59] Commenting later on the highly (politically and religiously) charged atmosphere of the debates on man's place in nature in the 1860s, particularly those between the rival (Darwinian) Ethnological Society of London and the (anti-Darwinian) Anthropological Society of London, Tylor noted that from the 1870s onward those debates were conducted in a more dispassionate tone, and went on to recommend that anthropologists go "right forward, like a horse in blinkers, neither looking to the right hand nor to the left."[60] As George Stocking has remarked, the founding of the Anthropological Institute (an amalgamation of the Ethnological Society of London and the Anthropological Society of London) in the same year that the *Descent of Man* was published, "is surely more than sheer historical coincidence."[61] For at that date, the conviction that the study of man could become a strict natural science was in the ascendant, an ascendancy confirmed by the outcome of the controversy concerning Wallace, Mivart, and Wright. This is not to imply, of course, that the human sciences were not expected to provide practical solutions to social and political questions; the utility of science was a canon few could ignore in late Victorian England. To invoke Tylor once more: "The science of culture is essentially a reformer's science."[62] But because those investigations were coming to be regarded – whether justifiably or not – as epistemologically neutral, their conclusions could carry the authoritative label "scientific."

As I have suggested here, it is significant that the Darwinian debates were conducted in a period when the prestige of science was rising. Indeed, the debates were important agents in effecting the process by which credence came to be attached to views backed by the authority of science, whether those views were or were not in their entirety substantiated by the findings of science. Tylor enunciated the neutralist credo:

> We may hope . . . that the science of anthropology will be worked purely for its own sake; for, the moment that anthropologists take to cultivating their science as a party-weapon in politics and religion, this will vititate their reasonings and arguments, and spoil the scientific character of their work . . . Let us do our own work with a simple intention to find out what the principles and courses of events have been in the world, to collect all the facts, to work out all the inferences, to reduce the whole into a science . . . In this way the science of man, accepted as an arbiter, not by a party only, but by the public judgment, will have soonest and most permanently its due effect on the habits and laws and thoughts of mankind.[63]

The fact that such sciences could be denominated as epistemologically neutral, even as they both reflected and nourished the values of an ascendant middle class in the capital of the British Empire,[64] speaks perforce to the ideological legerdemain that (consciously or not) attended the dissemination of

Darwinism. Earlier, in his outspoken presidential address to the British Association for the Advancement of Science in 1874, John Tyndall announced on behalf of those who were claiming for science a broad area of legitimation, that all "theories, schemes and systems, which . . . reach into the domain of science must, *in so far as they do this,* submit to the control of science, and relinquish all thought of controlling it. Acting otherwise proved disastrous in the past, and it is simply fatuous to-day."[65] Tyndall's views were, to be sure, controversial, as the immediate reaction to his address demonstrates. And the "cult of science" he espoused was to experience a renewal of criticism during the 1880s and 1890s, both from within and without the scientific community.[66] But those views were gradually to prevail and if, by 1900, science had not yet gained the unchallenged authority Tyndall envisioned, its professionalization and growing goverment and public recognition and support as an autonomous entity posed "fundamental questions about the social function of science in modern society."[67]

Huxley provides as fitting a conclusion as he did an introduction to this analysis of the dissemination of Darwinism. Speaking at the opening of Sir Josiah Mason's Science College in Birmingham on October 1, 1880, he argued against those (Matthew Arnold was the specific target) who held that science was "incompetent to confer culture; that it touches none of the higher problems of life; and, what is worse, that the continual devotion to scientific studies tends to generate a narrow and bigoted belief in the applicability of scientific methods to the search after truth of all kinds."[68] Arnold, in reply, disputed not Huxley's claim for the importance of science education in England – Arnold was a vigorous supporter of the achievements of Germany in the field of science education and argued for their introduction into England[69] – but rather his claim for its preeminence. To Arnold, "Those who are for giving to natural knowledge, as they call it, the chief place in the education of the majority of mankind, leave one important thing out of their account: the constitution of human nature."[70] In one obvious sense, Arnold's thrust is misdirected: Far from ignoring the "constitution of human nature," Huxley was actively engaged in ascertaining that constitution. Moreover, it was Huxley's brilliant popularization of science that did much to diffuse the teachings and implications of evolutionary biology (as they bore upon man) among a broad segment of Victorian society. But Arnold was astute in another sense: Huxley (and those who argued with him), by emphasizing the presumed neutral and apolitical character of science, was instrumental in fostering acceptance of the view that one could leave elements of human "nature" aside when practicing science and assessing the authority of its conclusions. The dissemination of Darwinism, in which Huxley took so prominent a part, insofar as it was predicated upon the "fallacy of misplaced neutrality"[71] must be seen, finally, as a major agency in inculcating the tenacious belief that science is an autonomous enterprise – a belief that, by 1900, if it did not completely define, radically altered, our place in nature.

NOTES

1 K. R. Popper, "The Rationality of Scientific Revolutions," in Rom Harré (ed.), *Problems of Scientific Revolution: Progress and Obstacles to Progress in the Sciences* (Oxford: Oxford University Press [Clarendon Press], 1975), pp. 87–8; emphasis in original.

2 The phrase is Stuart S. Blume's; see his *Toward a Political Sociology of Science* (New York: Free Press, 1974), p. 53.

3 Ernst Mayr, *Evolution and the Diversity of Life: Selected Essays* (Cambridge, Mass.: Harvard University Press [Belknap Press], 1976), pp. 292–4.

4 See *The Comparative Reception of Darwinism*, ed. Thomas F. Glick (Austin: University of Texas Press, 1974).

5 Steven Shapin and Arnold Thackray, "Prosopography as a Research Tool in History of Science: The British Scientific Community 1700–1900," *Hist. Sci., 12* (1974), p. 11, "speculate" that the breach can be dated between 1870 and 1900.

6 Thomas H. Huxley, *Evidence as to Man's Place in Nature* (London: Williams & Norgate, 1863), pp. 108–9.

7 See Robert K. Merton, *The Sociology of Science* (Chicago: University of Chicago Press, 1973), pp. 270–8, for the classic statement of the positivist norms of science.

8 Huxley, *Man's Place in Nature*, pp. 109–12.

9 Alvar Ellegard, *Darwin and the General Reader: The Reception of Darwin's Theory of Evolution in the British Periodical Press, 1859–1872* (Göteborg: Elanders Boktryckeri Acktiebolag, 1958), p. 332.

10 H. Lewis McKinney, *Wallace and Natural Selection* (New Haven, Conn.: Yale University Press, 1972), pp. 80–96.

11 Sandra Herbert, "The Place of Man in the Development of Darwin's Theory of Transmutation, Part II," *J. Hist. Biol., 10* (1977), pp. 192–7.

12 Frederick Burkhardt, "England and Scotland: The Learned Societies," in *Comparative Reception of Darwinism*, ed. Glick, pp. 32–74.

13 J. W. Burrow, *Evolution and Society: A Study in Victorian Social Theory* (1966; reprint, Cambridge: Cambridge Univ. Press, 1970), p. 131.

14 Alfred Russel Wallace, "On the Varieties of Man in the Malay Archipelago," *Trans. Ethnol. Soc. Lon.*, n.s. *3* (1864–5), pp. 196–215.

15 Ibid., pp. 210–13.

16 Alfred Russel Wallace, "The Origin of Human Races and the Antiquity of Man Deduced from the Theory of 'Natural Selecton,' " *J. Anthro. Soc. Lond., 2* (1864), pp. clviii–clxx.

17 Ibid., pp. clviii–clxv.

18 James Marchant, *Alfred Russel Wallace: Letters and Reminiscences* (1916; reprint, New York: Arno Press, 1975), p. 127.

19 Wallace, "Origin of Human Races," pp. clxv–clxvi.

20 Loren Eiseley, *Darwin's Century: Evolution and the Men Who Discovered It* (1958; reprint, Garden City, N.Y.: Doubleday [Anchor Books], 1961), p. 313.

21 Alfred Russel Wallace, "Geological Climates and the Origin of Species," *Quart. Rev.* (Amer. ed.), *126* (1869), pp. 187–205.

22 Ibid., p. 205.

23 Charles Darwin, *On the Origin of Species* (1859; facsimile reprint, Cambridge, Mass.: Harvard University Press, 1964), pp. 202–6.

24 Wallace, "Geological Climates," pp. 202–5.

25 Malcolm J. Kottler, "Alfred Russel Wallace, the Origin of Man, and Spiritualism," *Isis, 65* (1974), pp. 145–92.

26 Francis Darwin, ed., *The Life and Letters of Charles Darwin* (New York: D. Appleton, 1887), II, p. 296.

27 Alfred Russel Wallace, *Contributions to the Theory of Natural Selection* (London: Macmillan, 1870).

28 Marchant, *Wallace,* pp. 205, 267.

29 Chauncey Wright, "Limits of Natural Selection," *North Amer. Rev., 111* (1870), pp. 282–311.

30 Alfred Russel Wallace, "The Limits of Natural Selection as Applied to Man," in *Natural Selection and Tropical Nature: Essays on Descriptive and Theoretical Biology* (1891; reprint, Westmead: Gregg International Publishers, 1969), p. 188.

31 Wright, "Limits," pp. 291–3.

32 Darwin, *Origin,* p. 488.

33 Wright, "Limits," p. 300.

34 Lorraine J. Daston, "British Responses to Psycho-Physiology, 1860–1900," *Isis, 69* (1978), pp. 201–2.

35 Robert M. Young, "The Role of Psychology in the Nineteenth-Century Evolution Debate," in Mary Henle, Julian Jaynes, and John J. Sullivan (eds.), *Historical Conceptions of Psychology* (New York: Springer-Verlag, 1973), pp. 180–204.

36 Marchant, *Wallace,* p. 205.

37 Thomas H. Huxley, *Collected Essays* (London: Macmillan, 1893), I, p. 164.

38 Wright, "Limits," pp. 310–11.

39 Marchant, *Wallace,* p. 209.

40 St. George Jackson Mivart, *On the Genesis of Species* (London: Macmillan, 1871), pp. 1–112, 220–42.

41 Peter J. Vorzimmer, *Charles Darwin: The Years of Controversy. The 'Origin of Species' and its Critics 1859–1882* (Philadelphia: Temple University Press, 1970), pp. 225–43.

42 Chauncey Wright, "The Genesis of Species," *North Amer. Rev., 113* (1871), pp. 63–103; reprinted as *Darwinism: Being an Examination of Mr. St. George Mivart's 'Genesis of Species' (1871), with an Appendix on Final Causes* (London: John Murray, 1871); further references will be to the original article, hereafter Wright, "Genesis." On Darwin's role, see Marchant, *Wallace,* pp. 217, 220.

43 F. Darwin, ed., *Life and Letters of Darwin,* II, pp. 323–4.

44 Marchant, *Wallace,* p. 219.

45 Ibid., p. 288.

46 David L. Hull, *Darwin and His Critics: The Reception of Darwin's Theory of Evolution by the Scientific Community* (Cambridge, Mass.: Harvard University Press, 1973), pp. 3–14.

47 F. Darwin, ed., *Life and Letters of Darwin,* II, p. 315.

48 St. George Jackson Mivart, "The Descent of Man and Selection in Relation to sex," *Quart. Rev.* (Amer. ed.), *131* (1871), pp. 25–6. See Charles Darwin, *The Descent of Man and Selecton in Relation to Sex* (London: John Murray, 1871).

49 Mivart, *Genesis of Species,* pp. 285–6.

50 Mivart, "Descent of Man," pp. 45–8.

51 The adversary view of Victorian science and religion should, by now, be moribund. Far more promising from an historiographical perspective are the approaches that seek to recapture the dialectic between science and religion that more accurately describes the position of most of the Victorian debaters. See David B. Wilson, "Victorian Science and Religion," *Hist. Sci., 15* (1977), pp. 52–67.

52 Wright, "Genesis," pp. 65–9.

53 Philip P. Wiener, *Evolution and the Founders of Pragmatism* (1949; reprint, New York: Harper & Row, 1965), pp. 56–7.

54 Chauncey Wright, "German Darwinism," *The Nation, 21* (1875), p. 169.

55 James Bradley Thayer, ed., *Letters of Chauncey Wright, with Some Account of His Life* (1878; reprint, New York: Burt Franklin, 1971), pp. 112–13.

56 Chauncey Wright, "Spencer's 'Biology,'" *The Nation, 2* (1866), pp. 724–5.

57 Chauncey Wright, "Evolution by Natural Selection," *North Amer. Rev., 115* (1872), pp. 2–4.

58 George C. Williams, *Adaptation and Natural Selection: A Critique of Some Current Evolutionary Thought* (Princeton, N.J.: Princeton University Press, 1966), p. 3.

59 Edward B. Tylor, *Primitive Culture,* 3d ed. (London: John Murray, 1891), I, p. 2.

60 Edward B. Tylor, "How the problems of American anthropology look to the English anthropologist," *Trans. Anthrop. Soc. Wash., 3* (1885), pp. 81–95.

61 George W. Stocking, Jr., "What's in a name? The origins of the Royal Anthropological Institute (1837–71)," *Man,* n.s. *6,* no. 3 (1971), p. 386.

62 Edward B. Tylor, *Primitive Culture,* 4th ed. (London: John Murray, 1903), I, p. 453.

63 Tylor, "Problems of American anthropology," p. 94.

64 Gay Weber, "Science and Society in Nineteenth-Century Anthropology," *Hist. Sci., 12* (1974), pp. 281–2.

65 John Tyndall, *Address Delivered before the British Association, assembled at Belfast* (London: Longmans, Green, 1874), p. 61; emphasis on original.

66 George Basalla, William Coleman, and Robert H. Kargon, eds., *Victorian Science: A Self-Portrait from the Presidential Addresses of the British Association for the Advancement of Science* (New York: Doubleday [Anchor Books], 1970), p. 20.

67 Roy M. MacLeod, "Resources of Science in Victorian England, 1868–1900," in Peter Mathias (ed.), *Science and Society: 1600–1900* (Cambridge: Cambridge University Press, 1972), pp. 160–6.

68 Huxley, *Collected Essays,* III, p. 140.

69 *The Complete Prose Works of Matthew Arnold,* ed. R. H. Super IV (Ann Arbor: University of Michigan Press, 1964), pp. 309–13.

70 Matthew Arnold, *Discourses in America* (London: Macmillan, 1885), pp. 100–1.

71 The phrase is Dorothy Nelkin's; see her "Creation vs. Evolution: The Politics of Science Education," in Everett Mendelsohn, Peter Weingart, and Richard Whitley, eds.), *The Social Production of Scientific Knowledge* (Dordrecht: Reidel, 1977), pp. 281–2. Nelkin further notes (p. 282) that when "science becomes a cultural process in which theories develop in close relationship to contemporary values and ideologies," the "concept of neutrality and the distinction between science and its ideological and social content has little significance."

25

Transformations in realist philosophy of science from Victorian Baconianism to the present day

YEHUDA ELKANA
Tel Aviv University

APOLOGY AND THESES

In an area of discourse like the present one, in which several disciplines are involved and the topic is broad, comparative, and cross-disciplinary, one is faced with the choice of either being more precise and less comparative, or, respecting the character of the topic, remaining on the programmatic level. My temper, conviction, and inability to do both made me choose the second option.

The theses of this essay are the following:

Realism means an interaction between nature and the human observer; it relies on a wide spectrum of sources of knowledge (mainly faith, intellect, and the senses – not necessarily in this order of priority); it rejects rigid dichotomies on epistemological grounds. It is Bacon's philosophy. It is also Einstein's.

Twentieth-century realist philosophies of science, even though they are differently formulated, are all rooted in *this* conception of realism where the realism of literature, the theatre, and the arts is cognate to the realism of psychology, psychoanalysis, and philosophy.

The opposites of realism are idealisms of any sort, instrumentalism, and any other form of reductionism. Today's idealistic views, such as positivism, reductionism, or behaviorism, are all based on a nineteenth-century vulgarization of Baconianism.

This broadly inclusive, seemingly complex view of realism is fundamentally more commonsensical and biologically more universal than the reductionist, artificially simplistic behaviorism and sense-data positivism with its more sophisticated and complex intellectualizations. This thesis applies equally to historical research, Quine's philosophy, Premack's research on apes, and analyses of the arts.

Realism fosters progress, while at the present stage the sophisticated positivist-reductionist views became rather sterile, although they, too, in their time, have contributed to progress.

Concepts have undergone paradoxical developments. The meaning of what in the nineteenth century was called realism and idealism has been practically reversed: Nineteenth-century idealism was metaphysical, antiempirical, Hegel centered. Today's idealism is an attempt to reduce artificially all biological, physical, and mental phenomena to empirical data, or to what are thought to be pure facts. On the other hand, nineteenth-century realism was purely empirical, whereas today it is a synthesis of the empirical with the metaphysical.

A last thesis by way of reminder: Knowledge grows through a critical dialogue between competing theories and metaphysics within the body of knowledge, through competition between images of knowledge and between competing ideologies. Therefore, even if I emphasize original Baconianism and its nineteenth-century vulgarization, nevertheless these two competing epistemologies were always present: There were narrow reductionists in Bacon's time, just as there were some genuine realists also in the nineteenth century.

PARASCEVE: PREPARATIVE TOWARD A PHILOSOPHY OF REALISM

The following arguments are intended to put an end to dichotomies that seem to me spurious and misleading, impediments to the progress of knowledge. Before embarking on this methodological strategy, we have to ask, What kind of strategy is this? What is it for? In order to avoid narrowing down or eliminating important issues, if I had to choose between the sin of omission and that of chaotic inclusiveness, I would unhesitatingly choose the second, However, I believe both Scylla and Charybdis can be circumnavigated.

Inclusiveness is so important here, because our *Problemstellung* is inclusive: How does human knowledge change and grow? What is the connection between the growth of human knowledge and its sociocultural context?

As to the more specific issue of realistic philosophy, one is expected to be able to apply to science the same concept of realism as to psychoanalysis, art, and literature. A view that would make an artificial distinction between philosophical reasoning and commonsense attitudes to daily life will be rejected. Thus it becomes very important to abolish spurious dichotomies.

In order to avoid confusion, distinctions are introduced between body of knowledge, images of knowledge, and normative ideologies. Yet it is clear that even these distinctions are not time and context independent: They can be made only in a given framework, at a given time and in a specific culture.

Since I have already developed these ideas in detail elsewhere,[1] I shall here explain myself only in brief: the areas of discourse – what we are talking about – are the body of knowledge. This can comprise nature, society, or the individual. It will include what we call experimental data, theories, and also the relevant scientific metaphysics, that is, statements about our topic that, by their very nature, cannot be verified. The images of knowledge are based on clearly spelled-out views about knowledge. These are *sources of knowledge* (sense data, ratiocination, faith, revelation, authority, tradition, symmetry, harmony, beauty, etc.); *aims of knowledge* (discovering truth, prediction, utility, social status, interest, etc.); the *hierarchy* among the sources of knowledge (whether ideas are regarded as more important than facts, or vice versa; faith as the chief source of knowledge in religious matters, etc.); the *legitimization* of knowledge; *methodologies* of investigation.

Here ideologies refer to normative statements on the moral and political behavior of individuals and societies. It is claimed that ideologies influence the images of knowledge but have no direct influence on the body of knowledge, except, of course, in the trivial case of distortion of truth under dictatorships. The images serve as criteria, however, for selecting among the infinite number of possible problems within the body of knowledge.

Knowledge, then, grows in critical dialogues between competing metaphysics and theories in the body of knowledge, between competing images of knowledge and between competing ideologies. No science, no art, "mature" or otherwise, is ever in a one-paradigmatic stage, that is, it *never* happens that "only a single, acknowledged set of laws, theories and methods of investigation" is prevalent.[2] Competition always exists among theories (body of knowledge) and methods of investigation (images of knowledge).

Now back to our program: to analyze realism in a commonsensical way that holds good for all spheres of human culture. Realism is a "hot" issue not only in philosophy of science, but also in psychology, literary theory, and the world of the theatre. In all these contexts I wish to define realism as an attitude of accepting the world *as is* (as near as possible, to our best ability), which concedes the existence of things unknown, but which also acknowledges the reality of a world of emotions, meanings, symbolic images that in the first place belong to private knowledge but that can be spelled out and shared by others, if enough hard work is invested in communicating them. We will not admit the "reality" of purely private worlds and their meanings, if they cannot be communicated and thus shared by others.

In science and philosophy – areas that aspire to a high degree of objectivity – it is often claimed that, in order to be a realist it is not enough to admit the existence of a world "out there," but that it is also necessary to accept that there is a way of knowing *which* of any two contradictory conceptual frameworks is the true one, *which of the two* different translations is the correct one. If this is accepted, then a philosopher like Quine is a relativist, and – vive la petite différence! – so am I. Now I do not accept this premise, nor do I think that these two beliefs must be held simultaneously. This is not merely a semantic issue. On the contrary, I hold that a belief in a *world out there* and a belief that there is no established way of judging between two opposite theories, translations, or conceptual frameworks go very well together. Moreover, I find that generally the two views are held simultaneously by all those whom I would call realists in the old-fashioned commonsensical and cross-disciplinary sense. Elsewhere I called holding both these views simultaneously *two-tier thinking.*[3] The psychologists wish to treat fantasies as "real," at least in the sense that they are communicable. As Laing explains, the term denotes both real experience of which the subject is unconscious and mental functions that have real effects. These real *effects* are the real experiences. Yet a fantasy is also a "*figment,* since it cannot be touched or handled or seen." Does this mean that electrons are figments, or that genes were so until some years ago? They certainly have *real* effects – at least as real as fantasies. I agree with Laing that the source of confusion is a "dichotomous scheme in which the whole theory is cast." The dichotomy between the "inner world of the mind," the external world of the subject's bodily development and behavior, and hence of "other people's mind and bodies" is dubious. So is the dichotomy between other concepts, such as "inner" and "outer," "mental" and "physical," and so on.

Instead, let us admit that here, as in science, there is both "a world out there" and that our knowing it is *our perception of reality*. Next, I wish to show that this is the realism of Bacon, Einstein, and the realism accepted in literature and theatre.

Francis Bacon's first extant work, a fragment that already contains all the main ideas of his ambitious, enormous great instauration, is his *Temporis Partus Masculus* (the masculine birth of time), written in 1603. Its subtitle is "the great instauration of the dominion of man over the universe" and it is dedicated "to God the Father." The dedication reads:

> God the Word, God the Spirit, we pour out our humble and burning prayers, that mindful of the miseries of the human race and this our mortal pilgrimage in which we wear out evil days and few, they would send down upon us new streams from the foundations of their mercy for the relief of our distress; and this too we would ask, that our human interests may not stand in the way of the divine, nor from the unlocking of the paths of sense and the enkindling of a great light in nature may any unbelief or darkness arise in our minds to shut out the knowledge of the divine mysteries; *but rather that the intellect made clean and pure* from all vain fancies, and subjecting itself in voluntary submission to the divine oracles, may render to faith the things that belong to faith.[4]

The Baconian combination of the roles of faith, intellect, and the senses is all there in this dedication, waiting to be painfully explicated in thousands of pages through a lifetime, only to be misinterpreted and distorted by future generations, and especially to be vulgarized by nineteenth-century Victorianism.

The three sources of knowledge are indeed faith, intellect, and the senses, in a hierarchy that is clear but not simple: Primacy is given to faith, but only for things that belong to faith. Human interests (in the same sense as Habermas uses "interest") are the great misleading factors: Bacon's prayer is that human interests may not stand in the way of the divine; "render to faith the things that belong to faith."[5]

The great source of knowledge about the world is the senses; however, their path is easily blocked. Thus again he invokes the human interests, which work through the intellect, that they allow "the unlocking of the paths of the senses."[6] When this is achieved by cleaning and purifying the intellect from all fancies (later called *idols,* which underwent a complex classification), then the senses interact with nature and *enkindle* "a great light in nature." And again, light and darkness dialectically oppose one another: The very light in nature enkindled by the "unlocked paths of the senses" can easily give rise to such hubris that darkness will "arise in our minds to shut out the knowledge of the divine mysteries."[7] Still, it is the intellect, the source of knowledge, that enables us to check ourselves whether the light is indeed genuine light, preventing the encroaching darkness. Through the intellect we subject ourselves "in voluntary submission to the divine oracles."[8] This is a theory of permanent critical dialogue and mutual control between competing sources of knowledge: faith, intellect, the senses. Healthy attitude is a Galenic balance, a genuine homeostasis between the different sources of knowledge; refusing to fix a rigid hierarchy of images of knowledge and especially the sources of knowledge is genuine, rich, pluralistic: it is *realism.*

Baconian realism is directed not only against innate preconceived ideas, but also against empty formalism. Bacon accuses the philosophers, and especially Plato, of having given us "the falsehood that truth is as it were the native habitat of the human mind and need not come in from outside to take up its abode there."[9] Thus, he says, Plato, Aristotle, and the later Aristotelians, like Aquinas and Scotus "out of their *unrealities* created a varied world."[10] Yet Aristotle's great opponent, Peter Ramus, fares no better: Ramus is a collector of facts who then formalizes and "begins to squeeze in the *rack* of his summary method" the facts he has collected and thus "out of the *real* world made a desert."[11] Plato and Ramus are two extremes: Plato turned our minds "away from observation, away from things"; whereas Ramus, who concentrated on things, attempted to subsume everything "to a summary method" (which is a rack!). It is the nineteenth-century mode of thinking that genuinely and basically moved from words to things; Bacon only opposed the extremes and looked for their areas of interaction. He immediately turned against Galen as symbolizing pure empiricism, pure nature mindedness, so much so that there remained no room for nature being molded by humans. "You [i.e., Galen] would have us believe that only Nature can produce a true compound . . . with the malicious intention of lessening human power."[12]

Mere manipulation of nature, according to Paracelsus, is not acceptable to Bacon either. Such manipulation is a creation of "distinctions, products of your [Paracelsus's] own imagination" by which the unity of nature is torn asunder. Such manipulation confuses and mixes up "the divine and the natural, the profane and the sacred, heresies with mythology."

Experience must not go unheeded, nor should it be betrayed. If evidence drawn from things is subjected to a "pre-ordained scheme of interpretation," experience is betrayed. But experimental evidence is not just available, experimental truth is not manifest. Experimental research involves obscurity and delay. "The evidence drawn from things is like a mask cloaking reality," says Bacon.[13] In order to get at reality, evidence needs to be carefully sifted, not according to predesigned schemes of interpretation, but by an intellect cleansed of sense-blocking "idols." This is the only way to avoid delusions: to think in terms of "sources of true knowledge." *Sources of knowledge* is a Baconian concept par excellence.

Bacon's realism is rooted in his understanding of the complexity of nature and that of the human mind. He proposes to unite human beings with source of knowledge – there is no other way. We cannot rid ourselves of the idols before familiarizing ourselves with nature, because we would not know the idols from reality. Nor can we plunge into the "complexities of experimental science before your [our] mind has been purged of its idols." As if anticipating and attacking Descartes, Bacon says: "On waxen tablets you cannot write anything new until you rub out the old, but with the mind it is not so; there you cannot rub out the old till you have written in the new."[14]

It is Bacon's conviction that "ignorabimus" is neither a statement about nature nor an image of knowledge for genuine realist philosophers, but rather an image of knowledge for the superficial practitioner (especially, in his view, the medical practitioner) who "seeks to transfrom the present limitations of his art into a permanent reproach against nature, and whatever his art cannot achieve he artfully declares to be impossible by nature."[15] It is interesting to note that Bacon is aware that the claim that *nature is such and such* can be a statement about *knowledge* (an image of knowledge). Such a view – held, among others, by Galen and his followers – constitutes a "mischievous limitation of human power," introducing distinctions between nature and man (as between the heat of the sun and man-made fire), which Baconian realism does not tolerate. It is fascinating to contemplate this deeply conceived realism, which rejects all demarcationist extremes. At one pole is the sterile philosophical *certainty* based on the claim that what has so far *not* been discovered *cannot* be discovered, while at the other pole are the alchemists (or the nineteenth-century dogmatic experimentalists) who, unguided by any theory, ascribe every negative result to their own imperfection and continue endlessly repeating their experiments. If we extrapolate from this view, we can substitute for the "philosopher" the contemporary logical positivists and their successors, whereas at the other extreme the alchemist is replaced by the sense-data reductionists and stimulus-response behaviorists. The realist, then and now, stands in between, trying to grapple with nature and rejecting absolutes, demarcations, and the sterile quest for certainty. Yet it was Bacon's destiny that in the nineteenth-century exactly that mindless experimental attitude for which he blamed the alchemists would be ascribed to him.

When Gustav Magnus, the head of the celebrated physical laboratory in Berlin, eliminated all theory from science, or when Poggendorff rejected Helmholtz's epoch-making paper "Über die Erhaltung der Kraft," they invoked Bacon's authority for doing so.

A RELEVANT DIGRESSION ON EINSTEIN

Einstein's realism is the same as Bacon's. It is open-minded, pluralistic, and as close as can be to the commonsense view of reality. It also insists on using all human intellectual capacities for doing science, whether it is called metaphysics or experimental evidence, imaginative or logical. What Einstein refused to do was first to set up some accepted views on realism and idealism, and then to find his own position regarding them. Figuratively speaking, he refused to set up a Scylla and a Charybdis before he had mastered the rules of navigation to avoid them safely. Rather, he chose to ignore obstacles, "to think them away." Einstein was just as great in *thinking up* experiments and situations as in *thinking away* complications. His way of solving paradoxes was to think them away; he simply adopted from any philosophy what suited his purpose.

Gerald Holton[16] has already remarked on this uncanny ability and quoted (albeit in a different context: he was interested in Einstein's views on thinking) the following passage:

[The scientist] must appear to the systematic epistemologist as a type of unscrupulous opportunist: he appears as a realist insofar as he seeks to describe the world independent of the acts of perception; as idealist insofar as he looks upon the concepts and theories as the free invitations of the human spirit (not logically derivable from what is empirically given); as positivist insofar as he considers his concepts and theories justified only to the extent to which they furnish a logical representation of relations among sensory experiences. He may even appear as Platonist or Pythagorean insofar as he considers the viewpoint of logical simplicity as an indispensable and effective tool of his research.[17]

With such an attitude, Einstein could simultaneously hold views such as the following:

[1] The belief in an external world independent of the percipient subject is the foundation of science.[18]

But

[2] All our thinking is of this nature of a free play with concepts.[19]

On the one hand, he admits that what the senses tell us is

[3] the only source of our knowledge.[20]

Yet

[4] In error are those theorists who believe that theory comes inductively from experience.[21]

In order to get our bearings, let us now collect a few quotations that will give these seemingly contradictory views some order and coherence:

Following quotation 1, Einstein says in his paper on Maxwell:

[a] But since our sense-perceptions inform us only indirectly of this external world, our Physical Reality, it is only by speculation that it can become comprehensible to us.[22]
[b] The real is not given to us, but put to us (by way of a riddle).[23]
[c] Sense impressions are conditioned by an 'objective' and by a 'subjective' factor.[24]

In his important, long paper "Physics and Reality" (1936), Einstein has very clear things to say on this issue. As he explained in his autobiography, in multiple sense impressions and memory images, the ordering element is a *concept*, and it is the ordering process that is called *thinking*. Thus,

[d] By means of such concepts and mental relations between them, we are able to orient ourselves in the labyrinth of sense-impressions.[25]

Or again,

[e] It is a fact that the totality of sense experience is so constituted as to permit putting them in order by means of thinking – a fact which can only leave us astonished, but which we shall never comprehend. One can say: The eternally incomprehensible thing about the world is its comprehensibility.[26]

The Cartesian belief in an "unlimited penetrating power of thought" is an "aristocratic illusion," whereas "naive realism, according to which things are as they are perceived by us through our senses," is a "plebeian illusion."[27]

Physical reality represents the process of nature, and our conception of physical reality is constantly undergoing changes. Einstein's criticism of quantum mechanics is that "the quantities that appear in the laws make no claim to describe Physical Reality *itself,* but only the *probabilities* for the appearances of a particular physical reality on which our attention is fixed."[28]

When Bacon says, "The science must be such as to select her followers,"[29] and Einstein says that thinking is more than merely solving problems, but rather a tool for letting our strongest side come to the fore, so that "gradually the major interest disengages itself . . . from the momentary and merely personal,"[30] they mean the same thing – Bacon with regard to society and science, Einstein with regard to the individual and his thoughts. As Holton puts it: "Here Einstein is saying, 'have the courage to take your own thoughts seriously, for they will shape you.' "[31]

The best definition of positivism is given in one of Einstein's later letters to Solovine:

> Herein lies the weakness of the positivists and the professional atheists who feel smug in their conviction that they have been successful in divesting the world not only of the gods but also of miracles. The beauty of it is that we have to content ourselves with the recognition of the "miracle," beyond which there is no legitimate way out. This I have to add explicitly, so that you should not think I have fallen prey – weakened by old age – to the clerics.[32]

BACON'S INFLUENCE AND THE NINETEENTH-CENTURY VULGARIZATION OF HIS VIEWS

Already in the eighteenth century, Bacon's influence was immense. d'Alembert in his "Preliminary Discourse" (1750) and Diderot in the fifth volume of his great encyclopedia, under "Encyclopédie" (1755), refer to him repeatedly. So does Coleridge in his famous essay on method, which he wrote as a preface to the *Encyclopaedia Metropolitana.*[33]

For the great nineteenth-century theorists, the whole new era started with Bacon. Sir John Herschel, in his epoch-making "Preliminary Discourse" often refers to Bacon.[34] Because of his fame, popularity, explicitness, and consistent empiricism, I shall quote a few extracts from this classic of positivism:

[1] Among the Greeks, this point was attained by Archimedes, but attained too late, on the eve of that great eclipse of science which was destined to continue for nearly eighteen centuries, till Galileo in Italy, and Bacon in England, at once dispelled the darkness: the one, by his inventions and discoveries; the other, by the irresistible force of his arguments and eloquence.[35]

[2] It is to our immortal countryman Bacon that we owe the broad announcement of his grand ideal, that the whole of natural philosophy consists *entirely of a series of inductive*

generalizations, commencing with the most circumstantially stated particulars, and carried up to universal laws, or axioms, which comprehend in their statements every subordinate degree of generality, and of a corresponding series of inverted reasoning from generals to particulars, by which these axioms are traced back into their remotest consequences, and all particular propositions deduced from them; as well as those by whose immediate consideration we rose to their discovery, as those of which we had no previous knowledge.[36]

[3] Previous to the publication of the Novum Organum of Bacon, natural philosophy, in any legitimate and extensive sense of the word, could hardly be said to exist. Among the Greek philosophers, of whose attainments in science alone, in the earlier ages of the world, we have any positive knowledge, and that but a very limited one, we are struck with the remarkable contrast between their powers of acute and subtle disputation, their extraordinary success in abstract reasoning, and their intimate familiarity with subjects purely intellectual, on the one hand.[37]

[4] An immense impulse was now given to science, and it seemed as if the genius of mankind, long pent up, had at length rushed eagerly upon Nature, and commenced, with one accord, in the great work of turning up hitherto unbroken soil, and exposing the treasures so long concealed. A general sense now prevailed of the poverty and insufficiency of existing knowledge in *matters of fact;* and, as information flowed fast in, an era of excitement and wonder commenced, to which the annals of mankind had furnished nothing similar. It seemed, too, as if Nature herself seconded the impulse; and, while she supplied new and extraordinary aids to those senses which were henceforth to be exercised in her investigation – while the telescope and the microscope laid open *the infinite* in both directions – as if to call attention to her wonders, and signalise the epoch, she displayed the rarest, the most splendid and mysterious, of all astronomical phenomena, the appearance and subsequent total extinction of a new and brilliant fixed star twice within the lifetime of Galileo himself.

The immediate followers of Bacon and Galileo ransacked all nature for new and surprising facts, with something of that craving for the marvellous, which might be regarded as a remnant of the age of alchemy and natural magic, but which, under proper regulation, is a most powerful and useful stimulus to experimental enquiry."[38]

[5] It can hardly be expected that we should terminate this division of our subject without some mention of the "prerogatives of instances" of Bacon, by which he understands characteristic phenomena, selected from the great miscellaneous mass of facts which occur in nature, and which, by their number, indistinctness, and complication, tend rather to confuse than direct the mind in its search for causes and general heads of induction. Phenomena so selected on acount of some peculiarly forcible way in which they strike the reason, and impress us with a kind of sense of causation, or a particular aptitude for generalization, he considers, and justly, as holding a kind of prerogative dignity, and claiming our first and special attention in physical enquiries.[39]

It was typical for the nineteenth century to see philosophy as a battlefield between realism and idealism.[40]

In Germany, idealism was the dominant philosophy. Kuno Fischer's magisterial multivolume *History of Modern Philosophy* is a history of idealism from Descartes to Hegel. All other philosophies are viewed only as antitheses to the idealistic movement, and only Schopenhauer's thought is fully analyzed.[41]

Fischer paid little attention to other German philosophy. It is more remarkable that he devoted a whole book to Bacon, the Englishman, whom he saw as the founding father and prototype of *realistic philosophy.*

Francis Bacon is still regarded by his countrymen as the greatest philosopher of England; and in this opinion they are perfectly right. He is the founder of that philosophy which is called the realistic, which exercised so powerful an influence upon even Leibniz and Kant, to which Kant especially was indebted for the last impulses to his epoch-making works, and to which France paid homage in the eighteenth century.[42]

The reason is that in Fischer's view (shared by the majority of great German historians of philosophy, e.g., Erdmann and Schwegler, and by Schopenhauer himself), the age of Bacon was the age of *realism.* The idealistic reaction then constituted an "unbroken chain from Descartes to Hegel and Schopenhauer."[43] In their view Bacon's realism was a rather primitive stage of philosophy, typical of and well-attuned to the gross, down-to-earth, rather brutal Elizabethan Renaissance and that later, with the emerging Cartesian idealism, the trends converged. As Kuno Fischer puts it:

If we accurately consider the matter, we shall find that realism and idealism, from the time of their modern origin, have described not parallel but convergent paths, which, at the same time, have met at one common point. This point at which the idealistic and realistic tendencies crossed, as at a common vertex, was the Kantian philosophy, which has taken account of them both and united them in their elements. In this, as indeed in every respect, it has set up a standard, which must serve as a polar star to all subsequent philosophy. If, at the present day, we are asked, how we shall follow the right track in philosophy, we must answer, by a most accurate study of Kant. Since his time there has not been a philosopher of importance, who has not desired to be at once a realist and an idealist.[44]

Thus realism was absorbed by idealism, the much more sophisticated, advanced, scientific philosophy that became dominant. What happened afterward is tragicomic. Whereas Fischer and some others tried to do justice to Bacon's pluralistic, broad-minded, all-encompassing realism, later analysts, who wished to counteract German Hegelian idealism, reacted by looking for a narrower, materialistic, empiricist philosophy that would be consistent with the victorious success-minded natural sciences, do away with nonsensical metaphysics, and rely exclusively on "hard-nosed facts." They seemed to find all they were looking for in Bacon's inductive philosophy. This was the great nineteenth-century vulgarization of Bacon, a joint venture of British, German, and French philosophers. According to them, Bacon became antimetaphysical, empiricist, the evidence of the senses being his sole source of knowledge, to collect facts and to *use* nature his sole aim. He was antimathematical, antitheoretical, materialistic, utilitarian. From here on it was only natural to admit to science only sense data and to attempt to reduce all theories to observation, all mental phenomena to behavior, and all natural phenomena to physics and chemistry. Thus, by the first half of the twentieth century, all philosophical thinkers whom we now call idealistic, like the positivists, reductionists, or behaviorists, considered themselves the heirs of Bacon. It is our claim that they were the intellectual heirs not of Francis Bacon's genuine realism, but of the nineteenth-century vulgarization of his theories, a paradoxical development that centers around the concept of certainty and the quest for certainty.

As repeatedly observed, the late sixteenth-century skepticism (masterfully represented by Montaigne) had to be resolved, and this was attempted by several antithetical doctrines: Charron's blind, unquestioning faith, Descartes's clear and distinct ideas, and the allegedly Baconian reduction of all phenomena to sense data. The latter is wrongly ascribed to Bacon, and even the great historian of ideas, Koyré, fell into this nineteenth-century trap.[45]

Charron's solution is well expressed in his *De la Sagesse* (1601):

> Doubt has the double purpose to keep alive the spirit of research and to lead us to Faith. As reason disposes of no means by which to distinguish truth from falsehood, it follows that we are born to search for truth but not to possess it. Truth abides only in the bosom of Deity, & C. (Falckenberg, "Geschichte der Neueren Philosophie," 1886, p. 34)[46]

Descartes's solution was pure ratiocination and basically mathematical formal. Mathematics and mathematization of nature were the way out of uncertainty.

While in the nineteenth century view Bacon was advocating the certainty of facts against the certainty of mathematical reasoning, the actual realist Bacon was not at all interested in certainty. Here lies the greatest difference between realistic philosophy and its opponents.

> The way out of the *uncertainty of knowledge,* which for Continental thinkers was at that time by far the most important problem, seemed indeed to be solved in a promising manner by exactly that aspect of thought for which the philosophy of Bacon had no appreciation. The latter seemed to be unaware of the important part which the application of mathematics was to play in the extension of natural knowledge as well as in giving it precision and value.[47]

Among the many formulations of Bacon's realism, as expressed in opposition to the quest for certainty and in his demand for open-ended pluralism, the one following his analysis of the aims of knowledge is perhaps most illuminating: According to Bacon, most seekers of knowledge are actually interested in something else. They either "seek mental satisfaction or a lucrative profession or some support and ornament for their own renown."[48] On the other hand, "if among the numerous scientists there be one who seeks knowledge with an honest heart and for its own sake, he will be found to aim at variety rather than verity."[49] Realistic pluralism rather than a quest for certainty, in other words.

Thus *realism* is the awareness of the inherent complexity of the world, of human consciousness, and of the interaction between them; it rejects all sharp demarcations, be they between the physical and the mental, theoretical and observational, human and animal, dead and alive, progressive and degenerative, genetic and environmental, internal and external, between discovery and justification. This list purposely includes many of the most controversial issues in philosophy of science, psychology, and history. I shall not now go into all these spurious distinctions, but I shall attempt to show that all twentieth-century *realist philosophies of science* are rooted in their rejection and are thus truly Baconian. The *opposite of realism* can be any form of idealism, such as positivism, behaviorism, plain instrumentalism, or even a nonpositivistic demarcation mania. All these tend to rely on *one* dominant source of knowledge, either Cartesian clear and distinct ideas or the so-called Baconian empirical evidence. It is the very reliance on *one* source of knowledge only that makes them nonrealistic, and those views that claim to rely on empirical evidence only are the results of the nineteenth-century vulgarization of Baconian philosophy.

When Koyré wrote:

> Thus against the skeptical trend that culminates in Montaigne a threefold reaction takes place: Pierre Charron, Francis Bacon, Descartes. In other words: faith, experience, reason . . . Experience, then, is the remedy that Bacon offers to mankind. The *Novum Organon* has no other goal then to set against the sterile *uncertainty* of reason left to itself the fruitful certitude of well-ordered *experience.* And Bacon's challenging work *On The Advancement of Learning* is a reply, as much by its title as by its contents, to the disillusioned work of Agrippa.[50]

He was actually expressing the typical nineteenth-century view, and he was certainly wrong about Bacon. It was the arts, rather than the natural sciences and historical scholarship, that freed themselves from the nineteenth-century positivistic images of knowledge. Peter Brook, in his magnificent *The Empty Space,* put his finger on this: "On the whole, though, the stifling effects of a nineteenth-century obsessive interest in middle class sentiment, cloud much twentieth-century work in all languages."[51]

Instead of attempting to cope with the complexity of reality, it became the accepted thing to falsely simplify situations in life and in nature, in the wake of the great Victorian equalization, reduced to success and to a scale of values – intellectual and moral – that were easily accessible to the nouveau riche middle classes. This was the price paid for the democratization of knowledge and for introducing egalitarianism into the realm of the mind. In order to achieve this, an extremely difficult, unnatural process of "thinking away" had to be invented: behaviorism and sense-data positivism in psychology and science; elimination of the sense of human individuality in social studies; studies of the individual (whether in the theatre, in psychology, or in the history of ideas), however, have no place in any context.

About the theatre, Brook says: "In New York, and London, play after play presents serious leading characters within a softened, diluted or unexplained context – so that heroism, self-torture or martydrom become romantic agonies, in the void."[52]

In psychology, the great universalist theories of Freud and Piaget dominated the stage, until recently the serious, realistic search for the cultural and social foundations of cognitive development came to the fore. In the works of Bruner, Cole, Vygotsky, and Luria, the complicated "simplicity" of contextless minds has disappeared.[53]

In intellectual history, the sharp division between a history of disembodied ideas (whose principal representatives are Lovejoy and Koyré) and a Marxist deterministic attempt to show an interdependence between ideas and economic forces (Hesse, Bernal, Haldane, Hogben, and many others) is giving way to a new historical sociology of scientific knowledge.

In literature, the same historical development can be discerned. There the concept of "realism" is applicable to both objective external existence and to subjective discoveries of levels of reality, so long as they can be spelled out. As Damian Grant points out, both the correspondence theory of truth and the creative action coherence theory of truth belong to the realm of realism.[54] According to Bertrand Russell, both the semantic and the syntactic conceptions of truth are realistic. This is illustrated in Grant's words:

The inevitably subjective and therefore indeterminate status of reality is powerfully dramatized in Joyce's *A Portrait of the Artist as a Young Man,* in which Joyce follows Stephen Dedalus' developing consciousness of different levels of reality, from the simple sensuous reality of a child's sensations to the liberated reality of the disengaged imagination.

Realism as the *conscience of literature* confesses that it owes a duty, some kind of reparation, to the real world – a real world to which it submits itself unquestioningly. George J. Becker is clearly writing of this conscience when he says that it is not identical with a work of art and is anterior to it. *Realism, then, is a formula of Art which, conceiving of reality in a certain way, undertakes to present a simulacrum of it!*[55]

On the other hand, the nineteenth-century vulgarization and the emergent reduction to a positivist world view are beautifully put in Mr. Bounderby's speech in Dickens's *Hard Times:*

You must discard the word Fancy altogether. You have nothing to do with it. You are not to have, in any object of use or ornament, what would be a contradiction in fact. You don't walk on flowers in fact; you cannot be allowed to walk upon flowers in carpets. You don't find that foreign birds and butterflies come and perch upon your crockery; you cannot be permitted to paint foreign birds and butterflies upon your crockery. You never meet with quadrupeds going up and down your walls; you must not have quadrupeds represented upon walls. You must use, "said the gentleman," for all these purposes, combinations and modifications (in primary colour) of mathematical figures which are susceptible of proof and demonstration. This is the new discovery. This is fact. This is taste.[56]

A DIGRESSION ON FREGE

This principle of keeping philosophical realism rooted in commonsensical realism is not only professed by historians, literary analysts and psychoanalysts; a logician like Frege is also one of its adherents.

From a recent debate on Frege's realism it emerges that he was an anti-idealist (Dummett's view) and that he established a conception of the objectivity of mathematical entities to which he ascribed some ontological status, at least insofar as, according to him, such entities exist independently of our knowledge and exhibit causal efficacy (this is strongly opposed by Sluga).[57] Dummett claims that Frege did not regard abstract objects as real, whereas I agree with Currie's conclusion that Frege is a realist, since for him "thought interacts with the mental world and hence indirectly with the physical world."

In a late (1918) paper Frege says:

[1] What value could there be for us in the eternally unchangeable which we could neither experience nor could have effect on us? Something entirely and in every respect inactive would be unreal and non-existent for us. Even the timeless, if it is to be anything for us must somehow be implicated with the temporal. What would a thought (*Gedanke*) be for me that was never apprehended by me? But by apprehending Thought I come into a relation to it and to me. (p. 37)

[2] For it is absolutely essential that the reality be distinct from the idea. (p. 19)

[3] When one ascribes truth to a picture one does not really want to ascribe a property which belongs to this picture altogether independently of other things, but one always has something quite different in mind and one wants to say that that picture corresponds in some way to this thing. "My idea corresponds to Cologne Cathedral" is a sentence and the question now arises of the truth of this sentence. So what is improperly called the truth of pictures and ideas is reduced to the truth of sentences. (p. 19)

[4] So the result seems to be: thoughts are neither things of the outer world nor ideas . . . A third realm must be recognized. What belongs to this corresponds with ideas, in that it cannot be perceived by the senses, but with things, in that it needs no bearer to the contents of whose consciousness to belong. Thus, the thought, for example, which we expressed in the Pythagorean theorem is timelessly true, true independently of whether anyone takes it to be true. It needs no bearer. It is not true for the first time when it is discovered, but is like a planet which, already before anyone has seen it, has been in interaction with other planets. (p. 29)

[5] In consequence of these last considerations I lay down the following: not everything that can be the object of my understanding is an idea. I, as a bearer of ideas, am not myself an idea. Nothing now stands in the way of recognizing other people to be bearers of ideas as I am myself. And, once given the possibility, the probability is very great, so great that it is in my opinion no longer distinguishable from certainty. Would there be a science of history otherwise? Would not every precept of duty, every law otherwise come to nothing? What would be left of religion? The natural sciences too could only be assessed as fables like astrology and alchemy. Thus the reflections I have carried on, assuming that there are other people besides myself who can take the same thing as the object of their consideration, of their thinking, remain essentially unimpaired in force.

Not everything is an idea. Thus I can also recognize the thought, which other people can grasp just as much as I, as being independent of me. I can recognize a science in which many people can be engaged in research. We are not bearers of thoughts as we are bearers of our ideas. We do not have a thought as we have, say, a sense-impression, but we also do not see a thought as we see, say, a star. So it is advisable to choose a special expression and the word "apprehend" offers itself for the purpose. (pp. 34–5)

[6] A fact is a thought that is true. (p. 35)[58]

Views like Frege's are at variance with the sharp polarization between idealists and primitive realists, predominant in the nineteenth century. His realism is in the Baconian tradition.

Interestingly, Frege the logician is aware of the different aims of knowledge. He begins his paper on thought with the following statement: "The word 'true' indicates the aim of logic as does 'beautiful' that of aesthetics, or 'good' that of ethics."[59]

QUINE AND PREMACK

The Cartesian conception that doubt prompts us to develop a theory of knowledge is widely accepted. The converse – that doubt, too, must be rooted in knowledge – is considered a prioristic and anti-Baconian.

Yet, precisely the view that doubt is prompted by knowledge is in a very profound sense both realistic and much more elementary, commonsensical, and unsophisticated than the presupposition of immediate sense data. Moreover, this is Bacon's view. It is also represented, in widely different areas, by Quine and Premack. Quine has recently pointed out that doubt is rooted in knowledge:

> The basis for skepticism is awareness of illusion . . . Illusions are illusions only relative to a prior acceptance of genuine bodies with which to contrast them. In a world of immediate sense-data with no bodies posited and no questions asked, a distinction between reality and illusion would have no place . . . The positing of bodies is already rudimentary physical science.[60]

Such "rudimentary physical science" is commonsensical realism regarding bodies; it is much more elementary and unsophisticated than an empiricist philosophy accepting only sense data. In other words, the skeptical attitude toward illusions and the wish to eliminate them in order to gain certainty already presuppose the acceptance of an unquestionable reality.

This reality, however, is *not absolute* but *context dependent.* In a conceptual framework in which the world is seen as an agglomeration of physical objects (according to their size – the scaling effect), those physical objects are posited according to the point of view of the observer: To a person entering a classroom, the objects will be other humans, chairs, desks, windows, and so on. To someone who could absorb the world in terms of color patches only, quite different kinds of objects would appear, even before his skepticism about an illusion could be aroused. The same holds for a blind or for an inexperienced (untaught) "observer," whose framework is to "see" the world in terms of sounds: He would posit a different assemblage of objects, reflecting his previous knowledge and the connection of that knowledge with the distribution of sounds. Naturally, all these objects, color patches, sounds, and an infinite number of other aspects of reality are present, but since an infinity can never be conceived, the conceptual framework in relation to which we are behaving realistically serves as our standard (or "rudimentary physical science") for selecting our view of the world.

Quine's assertion that a realistic positing of bodies prior to raising questions about illusions is basically much more commonsensical than the sophisticated elimination of all but sense data finds its parallel thesis in David Premack's attribution of a theory of mind to apes and in his claim that to attribute thoughts and motives to others (i.e., to have a theory of mind) is commonsensical and much more fundamental than the sophisticated behaviorism that eliminates all mentalistic terms and relies on observable behavior only.[61] Both Quine and Premack consider this realist attitude as a primitive form of induction: expectation of repetition as a sequel to an antecedent repetition. Although Quine claims "that the utility of science from a practical point of view lies in fulfilled expectations: true prediction," he admits that this is exactly what happens in simple induction and that the same process holds for animals. This is also Premack's view.

Quine's realism has its roots in a fundamental biological awareness – an almost biological determinism. To him, epistemology is an enterprise within natural science. Therefore, such questions as, How can we elaborate a useful science? or Why does the resulting science work so well? are simply viewed as "scientific questions about a species of primates."[62]

The scientific questions are solved in the following manner: In simple induction, the realization, whether by animal or human, that an occurrence is a repetition of a previous one, depends on their subjective assessment of the similarity between the antecedent and the present occurrence. Such a similarity will be selected from among an infinite number of attributes.[63]

According to Quine, all creatures have an *innate similarity standard* that, in the process of Darwinian natural selection, is adopted if it helps survival. This, indeed, is the answer to Quine's question: "Why should nature, however lawful, match up at all with the dog's subjective similarity ratings?"

Yet I find this answer not exhaustive. As Quine himself admits, these innate similarity standards are subject to change with experience – they are learning dependent; and the capacity for learning is itself a product of natural selection. Thus, we could argue on the same grounds that what is being selected is the ability to produce a great number of original, daring exploratory hypotheses, which are then applied to each situation on a trial-and-error basis until the correct one is found and selected. In other words, does it not follow that a Darwinian explanation leaves underdetermined the choice between a Popperian trial-and-error procedure and a Quinian primitive induction? If so, this does not exclude Darwinian selection as basic for this problem, but it certainly casts some doubt on it.[64]

Instead, we might argue as follows: Let us accept that in any given conceptual framework it is the social consensus about images of knowledge that determines the limits of success, truth, satisfactory explanation, reference, reasonableness, similarity of cases, and so on. The force of the consensus is such that we regard it as the *absolute* indicator in any given framework insofar as we consider true as True, and explanation as Explanation. If this is not granted, then the game is endless: The realist will always try to insinuate another seemingly harmless concept, like Putnam's "reasonable" as a limit to the principle of charity, or like

Quine's innate (though changeable) standard of similarity. When pressed, philosophers will always be ingenious enough to clear yet another concept from the odium of relativism – only to bring in another one, and so on.

But what does it mean that we admit some absolutes into our realistic framework? It is commonly accepted that we live *realistically* insofar as we *act* as if an objective external world existed. However, philosophers generally claim that this is not a philosophical argument. Perhaps it is not, but what does this matter? If we *act*, live, and risk our lives on a daily basis, accepting the reality of the external world and its behavior according to our theories, which we hold to be true and empirically confirmed, and if we nevertheless admit that there is no absolute way to choose with certainty between two contradictory theories as translations, then we are indeed genuine realists and two-tier thinkers. We think, act, and create in the same manner as Bacon or Einstein or Quine or Premack and, to a significant extent, as Premack's chimpanzee Sarah!

REALISTIC PHILOSOPHY OF SCIENCE TODAY

How do the arguments discussed so far affect the philosophical views current today?

An open-minded, fair-thinking, egalitarian, liberal philosopher will generally tend to designate himself a realist or a scientific realist. This is "a good thing" to be. Idealist attitudes like positivism, operationalism, behaviorism are nowadays mostly rejected by philosophers of science and are contraposited to realism. Relativism, though not necessarily an idealist position, is also considered to be the opposite of realism and is generally talked of as "the threat." Relativism is "bad" and should be avoided. But could one not look at it this way: Realism is the golden mean between restrictive idealism and overly permissive relativism. Positivist behaviorist attitudes eliminate too much of what is really going on in the world, and what remains is a sorry sight: an emasculated normative view of the world as it should be in order that we can have certainty in it. At the other pole, genuine relativism sacrifices the notions of truth and certainty, either leaving unexplained fundamental changes in the world (Kuhn) or not even attempting to give an orderly account of the world: Anything goes (Feyerabend).[65] Between these two poles, realism has undertaken to deal with the most interesting and most important questions that have grown in importance during the last two hundred years. In addition to the old question, Is there a world external to knowing subjects? which realism answers with a loud yes, we nowadays wish to account for the development, growth, progress – in one word, success – of Western scientific culture. Why did it (or did it?) succeed? Why is it (or is it?) a unique phenomenon? Could we derive from its history the secret of its success, so that we may transfer it to other cultures, into other societies? Is there a satisfactory epistemological theory of truth for the sciences?

Such courageous questions, relevant to a total world view (the answers to them will influence our political ideology, professional activity, and personal ethos), naturally preoccupy many philosophers as well as anthropologists and psychologists. A theory of truth is inseparably connected with problems of meaning and reference and language. Linguistics and theories on the acquisition of languages present problems pertaining to philosophy as much as to psychology. (Compare Quine's remarks on acquisition of language and its similarity to basic inductive processes in acquisition of knowledge about the world.)

Relativism, a "spectre" to some philosophers, leads us straight to anthropology. The problem of a theory of translation is as much an issue in anthropology and psychology as in philosophy of science. Put simplistically: In order to understand science and the changes it undergoes in different periods and disciplines of our culture (history and philosophy of science), or to understand the natural and human constitution of other societies and cultures (anthropology) and to account for the stages in individual human cognitive and emotional development (Piagetian and Kohlberg-type cognitive and moral psychology), we need a theory of translation that is applicable to all these.

Realistic philosophy of science has to take a stand on all issues that were introduced either into positivism or into relativistic philosophies. Thus it was the logical positivist school that placed great emphasis on a clear distinction between observation language and theoretical language aimed at reducing knowledge to purely observational terms. The relativistic approach emphasized the fact that theories that differed in any conceptual ingredient were logically incommensurable. Again, it was Reichenbach who turned to Mill, borrowing his concepts, and enlarged the gulf between the context of *discovery* and the context of *justification.* Realism had to find an answer to all these problems. The fact that realist philosophers indeed made an honest effort to deal with all these issues resulted in a broad spectrum of views called *realistic* and in repeated attempts to label other philosophers as realists or relativists, or even as cryptopositivists. Moreover, since problems arose from both ends, the theoretical constructs became increasingly complex, cumbersome, and conditioned. Sharp distinctions between observational sentences and theoretical sentences had to be abolished, yet it was found important to maintain some distinction between them. Total incommensurability is considered absurd (we *do* understand each other, and there is *some* continuity in scientific progress), yet it remains to be explained how we compare theories having different conceptual ingredients. The total gap between discovery and justification is no longer admitted, yet we go on looking for a *method* by which at least to justify a result that seems to work well. So this is a genuine issue for realism.

I find the phenomenon of "labeling" less interesting. Suffice it to remember that Quine is sometimes called a realist and at others a relativist, to render the whole labeling exercise ridiculous.

Now I shall try to present a short classification of realist views in an attempt to grasp the essence of realism, relying on the works of Putnam, Hesse, Boyd, and, to a great extent, of Quine.

SO, WHAT IS REALISM?

1. Realists tend to believe in a correspondence theory of truth (Putnam).
2. Most realists oppose idealism like positivism or operationalism (Putnam). For realists, the success of science is not a miracle (which it is for idealists).
3. Realism is an empirical hypothesis based on two principles: (a) Terms in a mature science typically refer; and (b) the laws of a theory belonging to a mature science are typically approximately true (Boyd as quoted by Putnam). According to Boyd, scientists act as they do because they believe in (a) and (b), and their strategy works because (a) and (b) are true.
4. A realist account of the world must satisfy these presuppositions:
   1. Theoretical statements have truth value.
   2. The natural world does not change at the behest of our theories.
   3. The realistic character of our scientific knowledge consists, in some sense, of the permanent and cumulative capture of true propositions corresponding to the world.[66]
5. Criteria for an epistemological theory of truth for theoretical sentences are developed in terms of:
   1. consensus theory of truth for some observation sentences;
   2. a probability theory of degrees of belief for theories;
   3. a principle of charitable translations for alien scientific systems; and
   4. a principle of scientific growth.

   These criteria are sufficient for scientific realism.[67]
6. Will abandoning realism, i.e., abondoning the belief in any describable world of unobservable things, also make us give up our definition of truth? No, says Putnam.
7. Sophisticated realists recognize the existence of equivalent descriptions because it follows from their theory of the world that there are these various descriptions.[68]

Let us look into these points one by one:

1. Yes, certainly most of us tend to agree that there is a correspondence between what we hold true, and what there actually is. This is the oldest and least problematic form of realism, though it was attacked in the nineteenth century on the grounds that it implies a direct comparison between concepts and unconceptualized reality. (On this see Putnam, "19th century revisited," in his John Locke Lectures.)

2. Realist opposition to positivism, operationalism, behaviorism is well founded. Indeed, in order to explain why theories of genes, electrons, stars, and so on work well in predicting and explaining phenomena, the existence of genes, electrons, stars, etc., has to be presupposed.

3. Realism as an empirical hypothesis is more controversial.[69]

Boyd has pointed out that in a positivist philosophy of science it is assumed that later theories are improvements on the theories they succeed, and thus they must incorporate some of the observation sentences of earlier theories. Does it follow that the newer, better theories also imply the at least approximate truth of some of the propositions from the earlier theory? In short, is there a continuity or *convergence* of successive theories? We have seen that realism according to Hesse says just that in 4.3 above.

Yet the historian knows very well that such a continuity or convergence is not always the case – or rather it depends on what aspects of a theory we are looking at. *Some* aspects are always conserved; all of them *never.* Whether we find a continuity or a drastic discontinuity will depend on what aspect we choose to emphasize as important; and importance is a socially determined image of knowledge. Einstein's quantum theory is a continuation of Planck's, if we regard them as successive admissions of quantum behavior of energy in more and more (and finally most) experimental setups. Whereas Planck looked at the quantum nature of the interaction between matter and the electromagnetic field from his theoretical point of view, the Einsteinian quantum hypothesis appeared as a radical break. From the conceptual aspect, Einsteinian mechanics and Newtonian mechanics do not converge. The view of the brain as an enzyme-secreting gland (Schally and Guillemin) does not converge with the theoretical view according to which the brain was anything but a simple gland. Yet physiological and morphological theories of the brain and of other glands do certainly converge. Whether or not there is a convergence between sequential theories is a question of consensus on images of knowledge among workers in the field.

3 and 4. The other problem with these is that of reference. When a new theory replaces an older one, realists will assume that most terms used in the new theory refer to the same objects as the terms in the old theory. How could it be otherwise? Could the referent of the Bohr atom be different from the referent of the Bohr-Rutherford atom? How far back does this identity of terms apply? What distance in time between two theories does warrant a referent in the new theory to be assigned the meaning it had in the old one? Why is this warranted in the case of Dalton's atoms, but not for Democritus's? Why not for aether or phlogiston, as Putnam points out? Putnam's answer is that we have to apply a "Principle of Charity" or of "Benefit of the Doubt," as long as it is *reasonable* charity.

5. Hesse formulates the same problem differently when she speaks of a "charitable translation of alien scientific systems." The difference is that Hesse follows Quine and the anthropologists by introducing the concept of *consensus* (in 5.1). Clearly it is a question of images of knowledge determined by consensus that decides the limit of reasonableness. The consensus could be, for example: The principle of charity applies as long as we do not have to assign a referent to an object that our present-day theory regards as nonexistent. Daltonian atoms have fewer characteristics than the Bohr atom, but no really obnoxious ones (like the little Democritian hooks). A Newtonian aether that could be seen as a force field is much more acceptable than nineteenth-century aether consisting of elementary particles of its own, with special forces acting between them, which are neither gravitational nor electromagnetic. These examples show that the distance in time between theories need not be in linear sequence. They also show that two theories, centuries apart, can look closer to each other and have the same referents to their concepts than two theories separated by only a few years.

6. This question is interesting, for only a realist would worry whether, if he abandons realism, he can still keep to Tarski's definition of truth, just as only a

realist would consider realism as an empirical hypothesis *because* it could possibly be false. Putnam, in his John Locke Lectures, reminds us that "formal logic of *true* and *refers* is captured in Tarski's semantics, but the concepts of *truth* and *reference* are *underdetermined* by their formal logic. The notions of truth and reference can indeed be thought of as defined à la Tarski (for one's own language); but is it only by examining our theory of the world, and specifically examining the connection between truth and various kinds of probability or warranted assentability as they are drawn within that theory itself that one can determine whether the notions of truth and reference we employ are realistic or idealistic, 'classical' or 'intuitionist' " (p. 46).

7. The introduction of the issue of underdetermination of theories brings Putnam to his *sophisticated* realist position. This realist *knows* that there are equivalent descriptions of the world. But his is almost identical with Quine's view on the underdetermination of theories – the same view that often gave cause to his being labeled a relativist. Let us mention a few other characteristics of realism, all of the more sophisticated kind.

In his "Reference and Understanding" Putnam says:

8. "For realists our theories are a map of the world" (p. 100). A map is a *usage.* But both *use* and reference are parts of a total story. "What succeeds or fails is not linguistic behaviour, but total behaviour, and what we are after is the contribution of our linguistic behaviour to the success of our total behaviour."

9. Contemporary realists reject a priori truth.

10. Contemporary realists are skeptical of fixed, unchanging scientific methods.

11. For realists there is no sharp dichotomy between method and content. Moreover,

12. "Reality is not a part of the human mind, rather the human mind is a part and a small part of that reality."[70] This recalls Francis Bacon: "The world is not to be confined (as hitherto) within the straits of the intellect but the intellect is to be enlarged to receive the 'image of the world,' such as it is."[71]

These views are genuinely realistic in the Baconian-Einsteinian sense and are thus no genuine alternative to relativism. It is my *empirical* claim that all thinkers who hold these views are two-tier thinkers (as explained elsewhere); that is, they are realists in a given framework and relativists with respect to the choice of their frameworks. As I remarked earlier, I cannot see the importance of whether this argument is considered a "genuinely" philosophical one.

When realism is juxtaposed to instrumentalism, the problem becomes of central importance for creativity. From the history of science I have learned that no creative act has ever been accomplished in an instrumentalist spirit. Scientists do not toy with ideas (i.e., "God does not play dice" on one of the many levels of meaning of Einstein's saying) – they do not try out first one hypothesis and then another. They are always committed to some kind of scientific metaphysics that to them is a higher authority for judging the eligibility of a theory. Such scientific metaphysics are what Maxwell called "science-making."

The same holds true for creativity in the arts. Once more I quote Peter Brook:

No actor can play a cipher: however stylized or schematic the writing, the actor must always believe to some degree in the stage life of the odd animal he represents.[72]

It is this view of realism that allows us to cultivate the arts and the sciences in the same garden. For it is this nonreductionist, many-sided approach to competing and interacting sources of knowledge that allows us to speak of a realistic approach to nature in the same sense as we speak of reality emerging in a poem or in a play. This *reality* can be brought to life also through rituals or symbols. It is also *reality* in this sense that is evoked by psychoanalysis. It enables us to find *what is there* in a sense that is most meaningful in a given context, and it allows us to vary our source of knowledge according to the specific context. We gain abstract ideas through physical means, but we also gather direct physical experience through abstract means. Yeats's beautiful poem "For Anne Gregory" illustrates this well:

Never shall a young man,
Thrown into despair
By those great honey-coloured
Ramparts at your ear,
Love you for yourself alone
And not your yellow hair.

To this Anne Gregory replies:

But I can get a hair dye
And set such colour there,
Brown, or black, or carrot,
That your man in despair
May love me for myself alone
And not my yellow hair.

And the interrogator then has the last verse and the last word:

I heard an old religious man
But yesternight declare
That he found a text to prove
That only God, my dear,
Could love you for yourself alone
And not your yellow hair.[73]

NOTES

1 "The Distinctiveness and Universality of Science: Reflections on the Work of Professor Robin Horton," *Minerva, XI* (1977), p. 155–73; "Two-Tier Thinking: Philosophical Realism and Historical Relativism," *Social Studies of Science, 8* (1978), pp. 309–26; "Science as a Cultural System: An Anthropological Approach," in Paolo Rossi and Vittorio Mathieu (eds.), *Scientific Culture in the Contemporary World* (to be published in French, English and Italian), in SCIENTIA (International Review of Scientific Synthesis), in press.

2 See Kuhn's definition of a "mature" science in *The Structure of Scientific Revolutions,* (Chicago University Press, 1962).

3 See the works cited in n. 1, this chapter. A beautiful formulation of two-tier thinking can be found in Peter Brook, *The Empty Space* (Penguin Books, 1976), p. 81: "A normal stage action will appear real to us if it is convincing, and so we are apt to take it temporarily as objective truth." Now such a "real" is real all right. Neither can we have more certainty than that nor do we need it.

4 In B. F. Farrington, *The Philosophy of Francis Bacon* (Liverpool University Press, 1970), p. 59.

5 Ibid., p. 59.

6 Ibid.

7 Ibid.

8 Ibid.

9 Ibid., p. 64.

10 Ibid.

11 Ibid.

12 Ibid., p. 65.

13 Ibid., p. 66.

14 Ibid., p. 72.

15 In B. F. Farrington, "Thoughts and Conclusions," *Philosophy of Francis Bacon* p. 73.

16 G. Holton, "What Precisely is Thinking? Einstein's Answer," in A.P. French (ed.), *Einstein: A Centennial Volume,* (Heinemann, 1979), pp. 153–67.

17 P. A. Schilpp (ed.), *Albert Einstein, Philosopher-Scientist* (Evanston, Ill., 1949), vol. II, p. 684.

18 "Maxwell's influence on the development of the conception of physical reality," in Sir J. J. Thomson (ed.), *Maxwell Commemorative Volume,* (Cambridge University Press, 1931). pp. 66–73.

19 Schilpp, *Albert Einstein,* vol. I, p. 7.

20 "Ideas and Opinions by Albert Einstein," based on *Mein Weltbild,* ed. Carl Seelig, and other sources (Candor Books, Souvenir Press, 1954), p. 22.

21 Ibid., p. 301.

22 *Maxwell Commemorative Volume,* pp. 66–73.

23 P. A. Schilpp, *Albert Einstein,* p. 680.

24 Ibid., p. 673.

25 "Ideas and Opinions," p. 29.

26 Albert Einstein, *Out of My Later Years* (Philosophical Library, 1950), p. 61.

27 "Ideas and Opinions," p. 20.

28 *Maxwell Commemorative Volume,* pp. 72–3.

29 B. F. Farrington, *Philosophy of Francis Bacon,* p. 61.

30 P. A. Schlipp, *Albert Einstein,* vol. I, p. 7.

31 G. Holton, "What Precisely is Thinking?" p. 155.

32 Albert Einstein to Maurice Solovine, March 30, 1952.

33 Smedley, Rose, and Rose (eds.), *Encyclopaedia Metropolitana,* vol. I (London, 1845).

34 J. F. W. Herschel, *A Preliminary Discourse on the Study of Natural Philosophy.* The Cabinet of Natural Philosophy, conducted by Rev. Dionysius Lardner, LL.D., F.R.S. L. & E. (Carey and Lea, 1831).

35 Ibid., p. 54.

36 Ibid., pp. 78–9.

37 Ibid., p. 79.

38 Ibid., pp. 86–7.

39 Ibid., pp. 136–7.

40 Thus the first sentence of Kuno Fischer's preface to his *Francis Bacon of Verulam: Realistic Philosophy and Its Age,* English translation by John Oxenford (London, 1857), is, "The theatre of modern philosophy is a field of battle, where two opposite and hostile tendencies – Realism and Idealism – contend with each other in asserting claims to truth" (p. ix).

41 "Hegel's philosophy is looked upon as the dominating philosophy of the century, as its underlying Thought, its main characteristics being that it is speculative and not positive (Comte); that it is metaphysical and not psychological (Beneke); that it is monoistic and not dualistic (Günther and Hermes); that it identifies Thought and Being in contrast to their essential differences (Herbart); that it finds the truly Real in logical thought or reason, not in the unreasoning Will (Schopenhauer) or the 'Unconscious' " (v. Hartmann) (see vol. viii, pt. 2, pp. 1,176ff.). The only promising further development of the Hegelian scheme is seen by Fischer in the philosophy of Lotze, who, as I shall have occasion to explain in the sequel, is historically connected with Hegel through his master, Ch. H. Weisse, and to whom belongs, according to Fischer, a position of unusual importance among German philosophers, his main thesis being defined as the conviction that the world is not a fact, but has also a meaning. Without this latter addition philosophy remains unphilosophical, "standing" in the midst of the darkness and thicket of facts, what Bacon termed the *silva silvarum,* "the forest of forests." See vol. viii, p. 1,176. Prominent in Kuno Fischer's *History* are the intimate relations he established between philosophical idealism and the classical and romantic literature of Germany, of which he has a thorough knowledge and a unique conception, being popularly quite as well known through his writings in literary criticism as through his *History of Philosophy.* John Theodore Merz, *A History of European Thought in the Nineteenth century,* vol. III, *Philosophical Thought,* 1897, part II, fn., pp. 39–40.

42 Ibid., p. xii.

43 Ibid., p. 317. Merz sees a new era, much more objective and pluralistic, coming in the histories of Windelband and Höffding.

44 K. Fischer, *Francis Bacon,* p. X.

45 See his introduction to *Selections from Descartes,* ed. Anscombe and Geach.

46 Merz, III, p. 320, n. 2

47 Merz, *Philosophical Thought,* p. 332.

48 Farrington, *Philosophy of Francis Bacon* p. 76. The same idea was succinctly formulated in one of the essays, "Of Studies," though less critically: "Studies serve for Delight, for Ornament and for Ability."

49 Farrington, *Philosophy of Francis Bacon,* p. 76. The expression "numerous scientists" is a translation of "multitudine scientiam." The expression "*variety rather than verity*" is a translation of "(invenietur, tamen rerus potius) varietatem quam veritatem aucupari." The editor thinks that "varietatem" may be an allusion to Cardan's "De Rerum Varietate," which Bacon used when writing *Sylva Sylvarum. Works of Francis Bacon,* Spedding, Ellis & Heath, vol. III, p. 594.

50 A. Koyré's introduction to E. Anscombe and P. T. Geach (eds.), *Descartes Philosophical Writings, A Selection.* pp. x, xii.

51 P. Brook, *The Empty Space,* (Penguin Books, 1973), p. 94.

52 Ibid., p. 95. As to the plays, Brook is complaining that "whether the emphasis falls on the individual or on the analysis of society has become almost completely a division between Marxists and non-Marxists." Historical writing has begun to emancipate itself from this split.

53 See especially J. S. Bruner's Herbert Spencer Lecture, *Times Literary Supplement;* A. R. Luria, *Cognitive Development* (Harvard University Press, 1976); and L. S. Vygotsky's *Mind in Society* (Harvard University Press, 1978), both edited by M. Cole, and others, and Cole's own numerous works.

54 Damian Grant, *Realism* (Methuen, 1970).

55 Ibid., pp. 7, 14.

56 Ibid., p. 15.

57 M. Dummett, "Frege as a Realist," *Inquiry, 19* (1976), p. 457; Hans Sluga, "Frege's alleged Realism." *Inquiry, 20* (1977), p. 236; Gregory Currie, "Frege's Realsim," *Inquiry, 21* (1978), p. 218.

58 All six quotations are from Frege's *Der Gedanke,* trans. A. M. and M. Quinton as "The Thought: A Logical Enquiry," in P. Strawson (ed.), *Philosophical Logic* (Oxford University Press, 1967), pp. 17–38. Quotation 5 is a genuine precursor of Popper's third world philosophy as Popper himself points out in the references to his "Epistemology Without a Knowing Subject" in his *Objective Knowledge,* (Oxford University Press, 1972).

59 Ibid., p. 17.

60 Quine, "The Nature of Natural Knowledge," in S. Guttenplan (ed.), *Mind and Language,* The Wolfson lecture of 1974, (Oxford University Press, 1975), pp. 67–81.

61 Ibid., p. 68.

62 David Premack, "Does the Ape have a Theory of Mind?" Preprint.

63 "Any two objects share membership in countless classes." Quine, "Nature of Natural Knowledge," p. 69.

64 Incidentally, it is interesting to note here that Darwinian natural selection can be viewed as the basis of the empirical data for the two theories that both fit it and yet are contradictory and underdetermined. What is seen as observational (factual) basis for a theory depends on the context. For a different framework this basis will be a theory to be analyzed and not an empirical basis.

65 Feyerabend claims that he was misunderstood and that he never advocated such a position. This may be correct. There is, however, such a position that, rightly or wrongly, is generally associated with his name. Paul Feyerabend, *Against Method; Outline of an Anarchistic Theory of Knowledge* (New Left Books, 1975).

66 Mary Hesse in *The Structure of Scientific Inference* (University of California Press, 1974).

67 M. Hesse, *Truth and the Growth of Scientific Knowledge* (Philosophy of Science Association, 1976), vol. 2.

68 H. Putnam, John Locke Lectures, p. 51.

69 H. Putnam, in note 1 to Lecture II of the John Locke Lectures says, "Realism is *like* an empirical hypothesis," since (1) it could be false, and (2) facts are relevant to its support or critique. It is not scientific and, as explained in chap. 11 and 17 of vol. 2 of *Mind, Language and Reality,* it is not a hypothesis.

70 H. Putnam, *Mathematics, Matter and Mind,* Philosophical Papers. vol. I, p. vii.

71 Parasceve *iv.*

72 P. Brook, *The Empty Space,* p. 85.

73 Quoted in full at the beginning of J. Bronowski's *The Origins of Knowledge and Imagination* (Yale University Press, 1978, p. 3).

26

Science and the city before the nineteenth century

GEORGE BASALLA
University of Delaware

The claim that the urban setting is the most appropriate one for the prosecution of scientific activity has been made by a number of the best-known thinkers in Western culture. Philosophers, physical and social scientists, and literary figures are among those who have suggested that science and the city have a natural affinity for one another. Some have called attention to this affinity without attempting to account for it; others have advanced elaborate arguments to explain its origins. For an initial acquaintance with this body of literature consider the following examples chosen from seventeenth-century England and fourth century B.C. Greece.

Bishop Thomas Sprat, in a book that recounted the history of the Royal Society (1667) and defended the practitioners of experimental science, offered valuable insight into the possible connections between the newly emerging modern science and the city of London wherein the society had recently been founded. After having characterized medieval philosophy as a way of thinking that resulted from the cloistered life of Scholastics whose observations of nature were circumscribed by the walls of "the Garden of their Monasteries," Sprat turned to praise the much wider interests of Fellows of the Royal Society. Their focus was nationwide and worldwide. Their philosophy was not intended for "the retirements of Schools" but "for the use of *Cities.*" In fact, so close was the identification of modern science with the city that the Royal Society was formed

> to resemble the *Cities* themselves: which are compounded of all sorts of men, of the *Gown,* of the *Sword,* of the *Field,* of the *Court,* of the *Sea;* all mutually assisting each other.[1]

In Sprat's discussion of what might be called the geographic-economic basis of scientific activity he further concluded that because England was an island nation and mistress of the ocean she was "the most proper *Seat* for the advancement of Knowledg." The same vessels that brought merchandise to her shores would also bear scientific data and news of experiments done elsewhere. In the near future this great maritime power could expect to make as important discoveries "in the *Intellectual* Globe" as she had already made in the exploration of the "*Material*" globe. And given its propitious location in England the Royal

Society was destined to become "the general *Banck,* and Free-port of the World" of scientific ideas.[2]

Of all the cities in Europe, indeed of all the cities in world history, London was best situated to serve as the center of modern science. Babylon had clear skies necessary for astronomical observation but it, like Memphis in Egypt, was cut off from sea trade with foreigners carrying goods and ideas; Carthage, open to the sea, was unfortunately better known for its pirates than its philosophers; Rome chose to excel in the arts of statecraft; Constantinople, although no longer barbarous, was confined by the straits of Hellespont; Vienna, a frontier town, had no access to the sea; Amsterdam was a place of trade lacking "in the mixture of men of freer thoughts," Athens and Paris both cultivated the arts of speech and education above all else.

For Sprat, only London remained with all of the advantages and none of the disadvantages of the other great urban centers. London was the capital city of a great maritime empire, it was populated by both gentlemen and traders, it had free and easy access to the rest of the world, and it was a place where all "the noises and business in the World" met. Therefore, London must be the proper home for modern science "which itself is made-up of the Reports, and Intelligences of all Countreys."[3]

Bishop Sprat offers an excellent introduction to our topic: He identified science with the city and then attempted to explain their mutual compatibility. Not all of the sources discussed in this essay will be of this sort. Many times science and the city will be associated without the felt need for further explanation. In other instances the city will be defined as a cultural center with the implication that science is among the many intellectual activities generated by urban living. An entry into this aspect of the subject can be found in the opening scene of Plato's *Phaedrus.*

Phaedrus, who has spent the morning in Athens listening to the orator Lysias, is about to take a walk into the countryside, beyond the city's walls, when he chances to meet Socrates. As the two men converse Phaedrus suggests that they continue his interrupted walk into the country. Socrates agrees and follows the younger man, who leads him through a cool brook into the shade of a tall plane tree. Upon reaching their destination Socrates exclaims over the beauty of the natural setting and commends Phaedrus for having served as an admirable guide. Phaedrus, somewhat surprised, remarks that Socrates is indeed like some stranger to Athens being led by a guide. "I rather think," he continues, "that you never venture outside the gates." To this Socrates replies:

> Very true, my good friend; and I hope you will excuse me when you hear the reason, which is, that I am a lover of knowledge, and men who dwell in the city are my teachers, and not the trees or the country.[4]

For Socrates the city was *the* center of learning and the reason it held that distinction was because it was a place where one could communicate most effectively with other men. Whereas Bishop Sprat offered social, economic, and geographical explanations, Socrates noted rural–urban opposition and emphasized social interaction and communications as primary factors contributing to the intellectual vitality of city life.

After considering these two different, but related, responses to the problem we can now pursue a more ordered and chronological study of science and the city prior to the nineteenth century.

ANTIQUITY AND THE MIDDLE AGES

The city in classical antiquity was generally regarded as the center of intellectual life. Thus it was for Aristotle and for Plato, who took urban cultural supremacy for granted and made little effort to determine precisely what it was about city life that stimulated the intellect. Aristotle in his *Politics* and Plato in the *Republic* were mainly interested in the political nature of the city and the moral responsibilities of its citizens. Their focus on polity and ethics rather than urban cultural life arose from their personal involvement with, and their high regard for, the autonomous city-state. This form of political organization, in which the sovereignty of the state was vested in the free citizens of an independent city, stirred wide philosophical debate in ancient times.[5]

Rome, in its evolution from a small Etruscan community into a city-state and then into an imperial city, was acknowledged to be a great cultural center. Despite their admiration for its culture, Latin authors who commented on Rome were not motivated to analyze the possible connection between its cultural life and the urban situation. As in Greece, politics and citizenship, and not the urban roots of the intellect, were topics for close philosophical scrutiny. The Romans did identify, cultivate, and discuss *urbanitas,* which was the refinement and sophistication one acquired while living in the city. The idea of *urbanitas,* however, was never extended to include the high intellectual accomplishments of city dwellers. It stressed style, fashion, language, and expression and set the urbanite apart from his boorish cousin who inhabited the countryside.[6]

Greek and Roman thinker alike saw little need for the kind of philosophical or sociological analysis that might help to account for the rich intellectual life of their cities. And when Rome declined and St. Augustine offered the prospect of a heavenly city of God to take its place he was even less concerned than his predecessors to explain the cultural achievements of the earthly city. It was not in the Christian West but in the Islamic world of the fourteenth century that we find the first full-scale study of the culture of cities. This study was one part of a broad philosophical and historical investigation undertaken by the great Muslim philosopher and statesman Ibn Khaldūn. It appeared in *The Muqaddimah,* an introduction to his history of Western Islam.[7]

In Ibn Khaldūn's day city and state were separate entities with the latter dominant. As the individual Muslim state gained power it conquered or established a number of cities that it then controlled and maintained within its boundaries.

Because the Islamic city was not the single political-ethical entity found in

Aristotle or Plato it required a different approach for its study. Ibn Khaldūn, well aware of the differences between Greek and Islamic cities, was led to make his inquiry into the essence and structure of human social organizations, including cities, after experiencing personal failures in his political career. His failures, he reasoned, grew out of his ignorance of how best to act in particular, concrete circumstances. History, which was the record and study of specific events, would be his guide in the future.

Before Ibn Khaldūn could hope to understand and write history properly he felt it was necessary to create a science of culture whose subject matter would be man and society. Thus Ibn Khaldūn came to examine the relationship between cultural attainment and urban life as part of his pursuit of the universal principles operating throughout human history.[8]

Ibn Khaldūn's explanation of the salient features of human culture required that a sharp distinction be made between civilized and primitive ways of life. The primary attribute of civilization was the city. It was the goal toward which primitive cultures moved as they searched for a way to satisfy any of the human needs and desires that went beyond the bare necessities of life. According to Ibn Khaldūn there was a natural tendency toward cooperation and communality immanent in all primitive societies. Its inevitable result was urbanization and the creation of political, economic, and scientific institutions that fulfilled the inner longing for power, lust, pleasure, rest, and leisure. Urbanization was not, however, a self-contained process of growth; it was nurtured and protected by a powerful state that acted as a civilizing agent and served a special function as patron of scientific institutions.

In writing on the development of civilization Ibn Khaldūn did not claim that economic and military necessity were key elements in city building. People came together in cities only after they had met their basic needs for survival – food, shelter, and defense. Urbanization was the response to psychological urges that made their appearance in the more tranquil period that followed the acquisition of the essentials of life. Thus civilization, or its equivalent life in the city, was a luxury. It did, however, round off and complete the partially fulfilled culture of the primitives.

Urban economic life, according to Ibn Khaldūn, grew rapidly under state support and authority. The outcome was the transformation of the city into a center of production, trade, and commerce whose main purpose was the satisfaction of human desire for luxury and various kinds of pleasure. The economy of the adjacent countryside continued to be grounded in the production of the necessities of life. The nomads, farmers, and hunters of this rustic economy exchanged their primary goods for products and services offered by the city dwellers. Meanwhile, within the city the growth of the economy and the population called for the creation of the bureaucrat and administrator and the maintenance of a highly diversified and skilled group of artisans who produced specialized goods for the populace.

At this point there was a rapidly growing state and city, a vigorous urban economy, and the possibility for the enlargement of human experience within the city because of the diversity contained within its limits. Then, and only then, did the urbanite's thoughts turn to the cultivation of the crafts and the sciences. As Ibn Khaldūn wrote:

> When civilized people have more labor available than they need for mere subsistence, such (surplus) labor is used for activities over and above making a living. These activities are man's prerogative. They are the sciences and the crafts.

The cultivation of the sciences, a marginal activity dependent on surplus labor, would continue in the city so long as there was leisure to pursue learning, a demand for theoretical knowledge from its citizens, and state and city financial support for schools and patronage of intellectuals. Given these conditions, and enough time to establish a viable intellectual tradition in the society, the sciences could be expected to reach a high degree of sophistication and complexity. Of course, all these developments were confined to the city itself. The sciences could not be expected to take root in the desert or farm lands outside the gates of the city. Those living in the countryside were forced to establish contact with the urban intellectual centers if they sought scientific knowledge. And even then, members of a primitive group with little acquaintance with civilization, would find it difficult to compete with the city folk whose culture had long encouraged the life of the mind.[9]

Before assessing Ibn Khaldūn's views on the city and the intellect it is necessary to introduce some words of caution and explanation. Although this summary has stressed the rise of civilization, Ibn Khaldūn, who thought of the process of urbanization in organic terms, was as deeply interested in the city's maturity, senility, and eventual decay. Catering to desires that went beyond basic needs, the city was a fertile place for economic instability, corruption, disease, and degeneracy of all sorts. And the sciences, which were the last to be added to the city, would be the first to go when social life was disrupted. At a time of decline and disintegration concern would shift to that aspect of life where the sciences have the least to offer: the provision of food, shelter, and defense. Finally, the rural dwellers who participated peripherally in civilized activities were relatively immune from the dangers of decay and chaos that threatened the urbanites.

Another caveat involves the word "sciences," which has been used here without prior comment. Obviously, a fourteenth-century Muslim philosopher's list of the sciences is certain to cover areas of knowledge and scholarship not to be included in the twentieth-century definition of science. For Ibn Khaldūn the sciences included physics, medicine, astronomy, logic, and mathematics as well as alchemy, metaphysics, and studies based on divinely inspired laws handed down from God to the Prophet. Although only a few of the fourteenth-century sciences would be accepted as such by a modern audience the wide discrepancy in the definition of science does not detract from Ibn Khaldūn's original contribution to our understanding of the origins and causes of the variety and vigor of urban culture.

It is not necessary to claim that Ibn Khaldūn was a protomodern social scientist in order to praise his efforts to go beyond the boundaries of his culture and times and seek a set of causal factors determining the appearance of the state and the city, the stimulation of economic growth, the complex process of urbanization, and the cultural attainments of civilized life. In that search science and wider intellectual pursuits were seen as existing on the borders of culture. The sciences marked the highest point of civilization and yet they were lumped together with the excesses and pleasures that would contribute to the collapse of the city and the downfall of civilization. Similarly, the sciences owed their ultimate origin to psychological needs felt by all humans and yet because they were only marginally useful to the urban population, they could be safely jettisoned in the harsher days of urban decay.

In Ibn Khaldūn's thought there is a tension between the intellectual wealth of the city and the ignorance of the countryside, between the rich but corrupt city and the simple virtues of the poor rustics, between the life of contemplation in a sedentary urban society and the life of action in a more primitive setting. This failure to resolve completely rural–urban opposition does not alter the real worth of Ibn Khaldūn's achievement in establishing a rational basis for the study of human culture. City–country dualism is an issue that has not been resolved to this day.

RENAISSANCE AND SCIENTIFIC REVOLUTION

Ibn Khaldūn's theories of urban growth and culture were unknown to thinkers of the Renaissance and Scientific Revolution. One does find, however, an occasional reference in the literature of the period that indicates that some intellectuals were thinking along lines similar to those developed by the Muslim philosopher. Thomas Hobbes, for example, wrote in his *Leviathan:*

> *Leisure* is the mother of *philosophy;* and *Commonwealth,* the mother of *peace* and *leisure.* Where first were great and flourishing *cities,* there was first the study of *philosophy.*[10]

Neither in Hobbes, nor in a work with as promising a title as Giovanni Botero's *Cause della grandezza e magnificenza della città* (1589).[11] does one find a generalized approach to urban intellectual life.

Although both the Renaissance and the Scientific Revolution failed to bring forth a well-elaborated explanation of urban culture they did produce two very important ideas: first, the utopian city of science wherein urban life and science were successfully integrated to bring happiness to all; and second, the mercantile metaphor, which by linking science with commerce called attention to their common basis in exchange.

Utopia is a city, or at least the first utopias conceived during the Renaissance were cities. It was later, in the eighteenth and nineteenth century, that utopian communities were deliberately placed in the countryside far from the troubles and temptations of urban life. During the sixteenth and seventeenth century utopia was a self-sufficient city, or group of cities, of moderate size that was carefully designed to bring the good life to all its inhabitants.

A majority of the Renaissance utopists were convinced that science had a special role to play in the ideal cities they proposed. For some of these thinkers science was one of the many intellectual activities to be found flourishing in utopia. For others science dominated the urban utopian scene and made the good life possible.[12]

Sir Thomas More's *Utopia* (1516), which heralded the arrival of modern utopianism, bore evidence of its debt to Plato's *Republic* by its primary concern with ethics and governance. However, along with Christian humanism that was new to utopian thinking More also brought his conception of the place of science in utopia. Science, conspicuous by its absence in the *Republic,* was not viewed as an agent for social change in *Utopia.* It was seen as a religiously sanctioned way to understand God's creation and as a means to solve some of the practical problems encountered within utopia. The limited role Sir Thomas More assigned to science in his utopia was not perpetuated by later utopists who wrote at a time when the nature and possibilities of modern science were better known.[13]

The three major utopias of the Scientific Revolution, reflecting the spirit of the times, incorporated modern science into their designs for the perfect society. Tommaso Campanella's *Città del Sole* (The City of the Sun) (1602, 1623), Johann V. Andreae's *Christianopolis* (1619), and Sir Francis Bacon's *New Atlantis* (1627) were all urban, scientific utopias. In these works the question of why, or if, science was compatible with urban life was never raised for it was assumed that science and the city had a special attraction for one another and that taken together they formed the foundation of the best possible life for mankind.

Campanella, a heretical Italian philosopher and defender of Galileo, envisioned a utopian city whose physical form, education, and government were all shaped by science. Astronomy and astrology dictated the overall physical plan of the City of the Sun. It was divided into seven gigantic circles – one for each of the seven planets – with a temple at its center containing two large globes depicting the sky and the earth. The inner dome of the temple pictured all of the stars in the heavens from the first to sixth magnitude.

The concentric, circular walls of the city were sturdily built to withstand assault from invaders but they also served a scientific purpose. The surfaces of the walls were covered with scientific illustrations and diagrams transforming them into a huge science textbook readily available for the instruction of passing citizens, especially younger ones. City wall decorations included diagrams and propositions taken from Euclid and Archimedes, paintings of gems, minerals, and geological formations, pictures of all known plants and animals, illustrations of tools and technological processes, and specimens, where practicable, of objects represented graphically. Young boys playing in the vicinity of these educational billboards, and roaming the astronomical streets, were said to absorb a great deal of scientific knowledge effortlessly by age ten! When the youngsters entered school they were well prepared to study the higher branches of mathematics, medicine, and the sciences.

Science education, so central to the daily life of the City of the Sun, was also a requirement for advancement in government. The highest ruler, a metaphysician named *Sol,* governed the city with three princes each holding special responsibilities: *Power* (military arts); *Wisdom* (liberal and mechanical arts and science); *Love* (human reproduction and scientific eugenics). The princes carried out their duties with the help of magistrates, many of whom were trained in scientific specialties. A knowledge of science and technology was considered to be far more important to a political career than was the experience gained from participation in power politics. In this science city of the seventeenth century humans were bred according to scientific principles, they were educated along scientific lines, and they gained political status through the assimilation of scientific knowledge.[14]

Utopian cities of the sixteenth and seventeenth century were located on islands far from well-traveled sea lanes and the malevolent influence of ordinary human societies. Ralph Hythloday visited the lost isle that was More's utopia; a fictitious Genoese sea captain told of his travels to the distant island upon which was situated the City of the Sun. Johann Andreae wrote about his shipwreck upon the shores of a remote island and his subsequent discovery of Christianopolis. Before Andreae was granted entrance to the utopian city he was questioned about his moral and spiritual life and asked to report on the progress he had made observing the heavens and the earth and closely examining other natural phenomena. Having satisfied his examiners on all counts our traveler was admitted to a utopia based on Christian ethics, the pursuit of knowledge, and a respect for manual labor and the rights of workingmen.

Andreae, a German scholar and social reformer who was a student of Michael Mästlin and a friend of Johannes Kepler, created an ideal Christian city that was zoned for different kinds of industry and filled with laboratories for research in the physical and biological sciences. The laboratories were supported by a network of workshops, scientific equipment supply houses, a well-stocked library, and appropriate educational institutions. In a discussion of the architecture of Christianopolis Andreae reprinted the plans of Tycho Brahe's Uraniborg, recalling for his readers and for us a famous Renaissance science city in miniature that was built on the island of Hveen near Copenhagen.[15]

The best known of the three seventeenth-century utopias, New Atlantis, exists as a fragment. Sir Francis Bacon did not complete a full picture of all aspects of life in his utopia. For that reason it has been said that *New Atlantis* is less a utopia than it is a plan for a college or society dedicated to scientific research. Following the utopian tradition Bacon located his ideal city on a remote island, in this case Bensalem, somewhere in the South Seas. And, as might be expected in a scientific utopia, there is a detailed description of the personnel, laboratory facilities, and activities of a great scientific institute, here named Salomon's House.

Although we are told that the Brethren who work in Salomon's House seek "the knowledge of Causes and secret motion of things" the examples of the institute's activities – in agriculture, medicine, and chemical and mechanical technology – prove its strong commitment to applied science. Despite the place

Bacon has reserved for religion in New Atlantis he created the first clear example of an urban technological utopia where the good life ultimately depended on progress in science and technology.[16]

The uncharted seas that protected Bensalem and the other island utopias from the evils of the outside world also served to isolate them from any beneficial influences they might receive from visitors carrying new ideas and goods to their shores. The Renaissance utopists attempted to alleviate the isolation of their utopias by fashioning a social mechanism for discrete, controlled access to information and products of foreign lands. Plato in his *Laws* had suggested that older and wiser citizens, acting as "overseas inspectors," should be sent out from his ideal Cretan city to evaluate and collect new ideas from exotic peoples. Campanella assigned the same task to a group of "explorers and ambassadors." English poet Abraham Cowley, in his utopian *Proposition for the Advancement of Experimental Philosophy,* planned that four "Professors Itinerant" should reside abroad for three years gathering information, specimens, books, and the like.[17] Sir Francis Bacon, however, placed a somewhat different emphasis in his solution to the problem of getting new ideas to utopia.

Coupling commerce with knowledge, Bacon dispatched neither explorers, ambassadors, inspectors, nor professors from New Atlantis but rather "Merchants of Light" who collected knowledge from around the globe. A visitor to New Atlantis was given this explanation of what was considered to be suitable commerce in utopia:

> We maintain a trade, not for gold, silver, or jewels; nor for silks; nor for spices; nor any other commodity of matter; but only for God's first creature, which was *Light.*

The Merchants of Light sailed to foreign countries under assumed names in order to garner "books and abstracts, and patterns of experiments." In identifying light with knowledge Bacon was using an analogy that dated to ancient times.[18]

These well-known lines from *New Atlantis* call attention to the obvious but neglected fact that in early modern Europe a mercantile metaphor was widely applied to what went on in the world of ideas. The metaphor identified intellectual activity with the transactions of merchant traders. In the commercial realm there was the trade of real goods by merchants for profit; metaphorically, that activity was extended into the intellectual realm and ideas became commodities that were exchanged between thinkers. In addition, because Renaissance mercantile capitalism entailed long-distance travel over land and sea the mercantile metaphor covered more than local trading ventures. It took on a world wide perspective with definite cross-cultural overtones.

Merchant-trader, ambassador, explorer, overseas inspector, and itinerant professor all shared one important characteristic: they were *travelers.* If science had become the metaphorical equivalent of commerce, then the traveler was equated metaphorically with the inquisitive philosopher cum scientist who, to paraphrase poet William Wordsworth, voyaged "through strange seas of Thought, alone" seeking new features in the intellectual landscape. The image of the philosopher-traveler is a familiar one in the writings of René Descartes, and Sir Isaac Newton, when humbly assessing his life's work, made oblique reference to it. Newton compared himself to a small boy playing with a pebble or shell on the seashore while "the great ocean of truth lay all undiscovered before" him. The vast ocean that Newton saw in his imagination had been portrayed earlier in the frontispiece to Bacon's *Magna Instauratio* (1620) where it was depicted with ships sailing out in search of knowledge.[19]

An early version of the mercantile metaphor appeared in Alberti's treatise on architecture (c. 1450–70). Writing in praise of the architect-engineer's contribution to human progress and comfort Alberti enumerated his work in clearing swamps, controlling rivers, constructing dikes, building ships, and planning new harbors. These engineering feats facilitated trade, enabling men "to furnish one another with Provisions, Spices, Gems, and to communicate their Knowledge, and whatever else is healthful or pleasureable." A modern commentator on this passage has noted: "In Alberti's mind there is no difference between the circulation of commercial goods and the circulation of ideas."[20]

Some 175 years after Alberti Sir Francis Bacon introduced his version of the mercantile metaphor. He disdained trade in "Provisions, Spices, Gems," elevated the importance of the search for knowledge, and created traveling Merchants of Light to make that search abroad. The Baconian version of the metaphor had a deep influence on Bishop Sprat, who transported the Merchants of Light from the House of Salomon in Bensalem to the Royal Society of London.

In bringing the Baconian merchants into the port of London Bishop Sprat mixed the mercantile metaphor with the commercial and maritime reality of England's busy capital city. He spoke metaphorically when he claimed that the Royal Society was destined to become the "general *Banck*, and Free-port of the World" and that discoveries were soon to be made in the "Intellectual Globe" that would rival those being made in the "Material" one. Conversely, he was thinking about the maritime England of his day, and its ties to the Royal Society, when he predicted that shortly "there will scarce a Ship come up the *Thames*, that does not make some return of *Experiments*, as well as of *Merchandize*." Unlike Bacon, Sprat believed that scientific information *and* silks and spices should be carried in the same vessel. He knew that to a limited extent this was already being done by travelers and seamen who were asked by the Royal Society of London to answer queries about the flora and fauna of the exotic lands they visited.[21]

The significance of the mercantile metaphor goes far beyond its use as an intriguing literary trope in the Renaissance and later. The metaphorical association of commerce and knowledge is directly related to the theme of science and the city. Commerce, since very early times, had been closely identified with cities and towns.[22] The mercantile metaphor carried with it the implication that science was to be found in the city because scientific activity, dependent on the exchange of ideas, was modeled after the urban exchange of goods in commercial transactions.

The mercantile metaphor deserves additional comment lest it be confused with the better-known interpretations of seventeenth-century science, technology, and society proposed by Boris Hessen and Robert K. Merton. They contend that modern science was stimulated by the need to solve certain pressing technical problems growing out of England's maritime and mercantile enterprises. The mercantile metaphor makes no such claim; it draws an analogy between urban commercial and intellectual exchanges without the need of technology as an intermediary. One might say that if Hessen and Merton put forth a socioeconomic interpretation of the development of modern science then the mercantile metaphor suggested a sociological or sociopsychological one.

THE ENLIGHTENMENT

Early seventeenth-century utopists were naive and overly optimistic when they proposed science as the prime mover of urban utopianism. Their views did not prevail in the following century. During the Enlightenment science continued to appear in fictive utopian settings but some of the sense of wonder and hope that filled the cities of Campanella, Andreae, and Bacon had been lost.

Utopist Louis-Sébastien Mercier took many of the Enlightenment's cherished beliefs, thrust them forward in time to the twenty-fifth century, and produced a utopian Paris of the future. His *Memoirs de l'An 2440* (1770), the first *temporally* displaced utopia, paid special attention to science. A visitor to Mercier's utopia was given a tour of the King's Cabinet, a vast national museum dedicated to natural history and invention. Here were displayed all the wonders of nature and artifice in an institution that also sponsored scientific research calculated to advance medicine and technology. The Cabinet was an obvious, and stale, copy of the House of Salomon. Yet there was an important difference. The Baconian establishment dominated New Atlantis, giving it distinction and vitality; Mercier's Paris merely included science as one of its features.[23]

The reader of Jonathan Swift's *Gulliver's Travels* (1726) is in no danger of being subjected to yet another recital of the many useful and amazing discoveries made in a Baconian-inspired research laboratory. During a visit to the Laputan capital city of Lagado Gulliver was introduced to the Royal Academy of Projectors, where he met scientists fanatically engaged in absurd research projects. The miserable city of Lagado, a dystopian version of the Renaissance science city, was founded on scientific principles. Its ragged, starving, and ill-housed population had been promised technological miracles by theoretical scientists who were totally unable to provide the basic necessities of city life. Never had the false hopes and unrealistic promises of utopian science been presented so forcefully to a wide audience as they had been by Swift.[24]

The examples of Mercier and Swift suggest that eighteenth-century utopianism had less to offer in encouraging the study of the connection between science and urban life than did its seventeenth-century counterparts. And should one pursue the primitive utopianism of Diderot's *Supplément au Voyage de Bougainville* or Defoe's *Robinson Crusoe* urban culture would appear to be an even more distant concern for the utopists of the Enlightenment.

On the whole, the French philosophes expressed no special interest in the origins and nature of the culture of cities. They addressed themselves to broad philosophical issues and concepts that had little or no relevance to the social and intellectual forces interacting within the urban environment. Therefore, eighteenth-century France was not a particularly fertile field for speculations about the urban foundations of scientific growth.[25]

A notable exception to this generalization was Voltaire, who wrote enthusiastically about the city, praising it as a place of freedom, commerce, and culture. The pleasure-seeking urban rich, and the industrious, parsimonious, and socially mobile urban poor who produced goods for them, were joined in a commercial relationship that benefited both parties and created a superior urban civilization. It is noteworthy that Voltaire's initial praise was for the city of London, not Paris. London, the home of famous philosophers and scientists as well as great men of commerce, held a strong appeal for the French writer.[26]

The union of urban commerce and culture by eighteenth-century social thinkers was central to their understanding of the city as a center for science. Such a union was widely accepted in Europe and America. The German social philosopher Hermann Ludwig Heeren concluded his treatise on politics with a commentary on commerce and the city that recalled the widest implications of the mercantile metaphor:

> THE FIRST SEATS OF COMMERCE WERE ALSO THE FIRST SEATS OF CIVILIZATION. Exchange of merchandise led to exchange of ideas, and by this mutual friction was first kindled the sacred flame of moral and intellectual culture.[27]

Somewhat similar sentiments were expressed in a letter written by a Baltimore gentleman in favor of the creation of a "commercial capital" in the state of Maryland. Recalling Voltaire's urban trinity the American claimed that "liberty, science, and commerce" were "inseparably connected together" and that they "always took up their chief residences in cities." This was as true for the older metropolises of London, Paris, and Rome as it was for the newer cities of Boston, New York, and Philadelphia. Urban centers, old and new, attracted "the greatest geniuses of the age," who were motivated by ambition and example to spur one another on to do their best work. Therefore, the citizens of Maryland could advance "arts and sciences, commerce and mechanics" if they acted immediately to facilitate the growth of cities.[28]

Beyond the scattered remarks noted here, the most thoughtful and sustained study of commercial civilization in the eighteenth century was made by the Scottish moralists. In the thought of Adam Smith especially, the mercantile metaphor was unconsciously but decisively transformed. What had been an intriguing but loose metaphorical association of ideas and merchandise for Alberti, Bacon, and Sprat became a virtual identity for the Scotsman. At the same time the mercantile metaphor lost its global overtones and took on a domestic air. The philosopher-scientist who was compared to the seagoing traveler and merchant in the Renaissance emerged as a modest city tradesman, perhaps even a cobbler, in Adam Smith's version of the metaphor.

While discussing the advantages of the division of labor in the making of pins, Smith noted that "philosophy or speculation" was like any other employment in that it was the special occupation of a certain group of citizens. In addition, philosophy, like any of the other jobs undertaken by human beings, could be divided into many different branches or specialties: "mechanical, chemical, astronomical, Physical, Metaphysical, moral, political, commercial, and critical." Therefore, the concept of the division of labor was applicable to the intellectual realm.

Smith went on to point out other features common to philosophers and workingmen. The intellectual, the tradesman, and the laborer all shared the natural human propensity "to truck, barter, and exchange one thing for another." To the city marketplace the intellectual brought, for purposes of business, the products of his labor: "thought and reason." However, since it was well known that the intellectual did not produce all of his ideas by himself, the majority of them must have been "purchased, in the same manner as his shoes or stockings, from those whose business it [was] to make up and prepare for the market that particular species of goods." This generalization held true not only for trivial notions but for all the great ideas of "religion, morals, and government." In essence, ideas were commodities manufactured by means of the division of intellectual labor. Like any other goods, pins, for instance, they were available to be bought, sold, or exchanged in the marketplace.[29]

Adam Smith was also influential in establishing the four-stage theory of the evolution of human society. This theory claimed that mankind progressed through four stages of development: hunting, pastoral, agricultural, and commercial. The human society in existence at a given stage was dependent on a single factor, the mode of subsistence, and the fourth stage was considered to be the ultimate one in the developmental process. Two important inferences were derived from this scheme: First, the fourth or commercial stage was the time of the emergence of cities and the rise of the arts and sciences; and second, the intellect could not have been cultivated during any of the earlier stages because food, shelter, and defense were not completely secured until the final phase. As Smith maintained in his history of astronomy, only after order, security, and subsistence were assured did mankind have the leisure to become curious about the motions of the heavenly bodies.[30]

The contention that science can only emerge when mankind has the necessary leisure for it, that intellectual activity of any sort is last on the list of early man's priorities, was not peculiar to the eighteenth century. Ibn Khaldūn evaluated the situation thusly in the fourteenth century and a surprising number of thinkers in the nineteenth and twentieth century embraced this rational, but nevertheless fanciful, reconstruction of the early history of the human race.

All who continue to accept this mythical order of human priorities, an order derived from neither an objective study of primitive peoples nor the earliest human records, might well read the opinions of modern anthropologists. Or, they might consider the remarks made on the subject by one of Adam Smith's contemporaries, Adam Ferguson. Contending that man was a creature of a variety of drives or urges, Ferguson refused to accept the proposition that any one of these urges appeared first in time and took precedence over any other one. He thought it was more likely that an interest in food, shelter, defense, natural phenomena, religion, and art operated simultaneously within humanity.[31] A little reflection will reveal how wise Ferguson was and how absurd it was to suggest that ideas about the universe or its maker, or the urge to decorate an artifact, could only be entertained after a man had a full belly, a roof over his head, and a wall around his city.

Although commerce, and consequently urbanism, defined the highest stage in the evolutionary growth of mankind, Adam Smith had doubts about the overall influence of the city. He thought city life was unnatural, unstable, and dependent; it did not yield the freedom and psychic satisfaction people received from following the natural inclination to cultivate the soil. The city might stimulate the intellect but only the country could give a sense of fulfillment.[32]

Smith's negative view of urban life was reinforced by fellow Scotsman Lord Kames, who criticized the city as an unhealthy, immoral, and politically unstable place. In America Thomas Jefferson made similar complaints and offered the agrarian ideal as the guide for the future of the nation. These eighteenth-century critics were the forerunners of powerful forces in the nineteenth century that castigated the city as a center of physical and moral degeneracy.[33]

The urban critics also called attention to the rural–urban opposition already mentioned in the discussion of Socrates and Ibn Khaldūn. The seventeenth-century utopists had skirted the city versus country issue by focusing solely on the city and requiring all of its inhabitants to work for a short time in the countryside producing agricultural products. However, nineteenth- and twentieth-century commentators on science and the city became increasingly interested in determining exactly what features characterized urban and rural life and why those features did, or did not, contribute to intellectual stimulation. Thus, an understanding of rural culture became as important as the study of urban culture for those attempting to establish a relationship between science and the city.

CONCLUSION

From the fourth century B.C. through the eighteenth century a diverse group of thinkers was intrigued by the possible connection between science and urban life. Most of the theories they proposed to explain the supposed affinity assumed that the commercial life of the city was in some sense responsible for its stimulation of scientific activity. At the center of this speculation stood the mercantile metaphor with its claim that the urban exchange of goods provided a model for the exchange of scientific ideas. Along with the identification of the influence of the commercial realm, pre-nineteenth-century thought on science and the city called repeated attention to rural–urban differences and tensions.

Nineteenth- and twentieth-century intellectuals continued to express great interest in the two major elements that had dominated past thought on urban culture. Their contribution was not the unearthing of additional elements but the perfection and utilization of new analytical techniques for testing the old assumptions. These techniques took the form of sophisticated statistical studies of urban and rural data and the application of sociological and psychological theories to the problem.

Ironically, just at the time when the new techniques and theories were being used new developments in the technology of transportation and communication acted to blur the distinction between city and countryside, making it difficult to determine the proper geographical limits of urban culture. And concurrently, anthropologists began the comparative study of urban culture that revealed that in *some* cases science and the city were not necessarily compatible.

These are problems, however, that were only encountered after the time period of this essay. They have been recounted here to furnish a perspective on later facets of the question of the nature of the relationship between science and urban living.

NOTES

1 Thomas Sprat, *History of the Royal Society,* facsimile reprint of the 1667 London ed. with critical apparatus by J. I. Cope and H. W. Jones (St. Louis, Mo.: Washington University Studies, 1958), pp. 19, 76.

2 Ibid., pp. 64, 86.

3 Ibid., pp. 87–8.

4 Plato, *The Dialogues of Plato,* 3d ed., trans. B. Jowett (London: Oxford University Press, 1892), vol. 1, p. 435.

5 E. Barker, *The Political Thought of Plato and Aristotle* (London: Methuen, 1906), pp. 411–12; Mason Hammond, *The City in the Ancient World* (Cambridge, Mass.: Harvard University Press, 1972), pp. 175–95.

6 Lidia Storoni Mazzolani, *The Idea of the City in Roman Thought,* trans. S. O'Donnell (Bloomington: Indiana University Press, 1967); Edwin S. Ramage, *Urbanitas: Ancient Sophistication and Refinement* (Norman: University of Oklahoma Press, 1973).

7 Ibn Khaldūn, *The Muqaddimah: An Introduction to History,* 3 vols., trans. by Franz Rosenthal (New York: Pantheon Books, 1958).

8 S. M. Stern, "The Constitution of the Islamic City," in A. H. Hourani and S. M. Stern (eds.), *The Islamic City* (Oxford: Bruno Cassirer, 1970), pp. 25–50; Muhsin Mandi, *Ibn Khaldūn's Philosophy of History* (Chicago: University of Chicago Press, 1957), pp. 17–62, 209, 295–6.

9 Mandi, *Ibn Khaldūn,* chaps. I, II, IV; Manzoor Alam, "Ibn Khaldūn's Concept of the Origin, Growth, and Decay of Cities," *Islamic Culture, 34* (1960), pp. 90–106. The Ibn Khaldūn quotation is taken from Ibn Khaldūn, *The Muqaddimah,* vol. 2, p. 434. The connection between intellectual life and leisure is mentioned by Aristotle: 1269a 34, 1329a, 1, 1334a 24.

10 Thomas Hobbes, *Leviathan: or the Matter, Forme and Power of a Commonwealth, Ecclesiastical and Civil,* ed. Michael Oakeshott (Oxford: Basil Blackwell, 1946), p. 436;

Warren E. Gates, "The Spread of Ibn Khaldūn's Ideas on Climate and Culture," *Journal of the History of Ideas, 28* (1967), pp. 415–22.

11 For English translation see Giovanni Botero, *A Treatise Concerning the Causes of the Magnificence and Greatness of Cities,* trans. Robert Peterson (London, 1606).

12 I acknowledge my wide use of an excellent study of science and utopianism: Nell Eurich, *Science in Utopia: A Mighty Design* (Cambridge, Mass.: Harvard University Press, 1967). Also useful is Eugenio Garin, "The Ideal City," in *Science and Civic Life in the Italian Renaissance,* trans. Peter Munz (New York: Doubleday, 1969), pp. 21–48.

13 Sir Thomas More, *Utopia,* trans. Ralphe Robynson (London: J. M. Dent, 1906); Russell Ames, *Citizen Thomas More and His Utopia* (Princeton, N.J.: Princeton University Press, 1949), pp. 86–100; Eurich, *Science in Utopia,* pp. 77–80.

14 Eurich, *Science in Utopia,* pp. 108–20; Marie Louise Berneri, *Journey Through Utopia* (New York: Schocken Books, 1971), pp. 88–102; Tommaso Campanella, *The City of the Sun,* trans. William J. Gilstrap, in Glenn Negley and J. Max Patrick (eds.), *The Quest for Utopia: An Anthology of Imaginary Societies* (New York: Schuman, 1952), pp. 317–47.

15 Eurich, *Science in Utopia,* pp. 120–34; Berneri, *Journey Through Utopia,* pp. 103–26; Johann Valentin Andreae, *Christianopolis,* trans. Felix E. Held (New York: Oxford University Press, 1916); Elisabeth Hansot, *Perfection and Progress: Two Modes of Utopian Thought* (Cambridge: MIT Press, 1974), pp. 80–92; Tycho Brahe, *Tycho Brahe's Description of His Instruments and Scientific Work,* trans. and ed. H. Raeder, E. Stromgren, and B. Stromgren (Copenhagen: E. Munksgaard, 1946), pp. 124–40.

16 Eurich, *Science in Utopia,* pp. 134–44; Sir Francis Bacon, *New Atlantis,* in James Spedding, Robert L. Ellis, and Douglas D. Heath, (eds.), *The Works of Francis Bacon,* vol. 5 (Boston: Brown and Taggard, 1862), pp. 359–413; Robert P. Adams, "The Social Responsibilities of Science in *Utopia, New Atlantis* and After" *Journal of the History of Ideas, 10* (1949), pp. 374–98.

17 Plato, *Laws,* Book 12, sec. 950–3; Abraham Cowley, *A Proposition For the Advancement of Experimental Philosophy,* in A. R. Waller (ed.), Abraham Cowley, *Essays, Plays, and Sundry Verses* (Cambridge: Cambridge University Press, 1906), pp. 251–2.

18 Bacon, *New Atlantis,* pp. 384, 409–10; Dorothy Tarrant, "Greek Metaphors of Light," *The Classical Quarterly, 10* (1960), pp. 181–7.

19 The fragmentary quotation from William Wordsworth is taken from *The Prelude,* Book III, line 63, where he describes a statue of Sir Isaac Newton; the Newton seashore anecdote appears in Sir David Brewster, *Memoirs of the Life, Writings, and Discoveries of Sir Isaac Newton,* vol. 2 (Edinburgh: Thomas Constable, 1855), p. 407. For Descartes's use of the philosopher-traveler metaphor see Nathan Edelman, "The Mixed Metaphor in Descartes," *Romantic Review, 41* (1950), pp. 167–8.

20 Leone Battista Alberti, *Ten Books on Architecture,* trans. Cosimo Bartoli and James Leoni, ed. Joseph Rykwert (New York: Transatlantic Arts, 1966), p. x; Garin, *Science and Civic Life,* p. 42. For ancient references to the metaphorical linking of economic and intellectual activity see Marc Shell, *The Economy of Literature* (Baltimore: Johns Hopkins University Press, 1978), pp. 1–112.

21 Sprat, *History,* p. 86; R. W. Frantz, *The English Traveller and the Movement of Ideas, 1660–1732* (Lincoln: University of Nebraska Press, 1967), pp. 15–71.

22 Robert S. Lopez, "The Crossroads Within the Wall," in Oscar Handlin and John Burchard (eds.), *The Historian and the City* (Cambridge, Mass.: MIT Press and Harvard University Press, 1963), pp.27–39.

23 Louis-Sébastien Mercier, *Memoirs of the Year Two Thousand Five Hundred,* trans.

W. Hooper (Philadelphia: Thomas Dobson, 1795), reprinted with a new intro. by Mary E. Bowen (Boston: Gregg Press, 1977), pp. 214–34. Also see Louis-Sébastien Mercier, *The Waiting City, Paris 1782–88,* trans. Helen Simpson (Philadelphia: Lippincott, 1933), p. 275.

24 Jonathan Swift, *Gulliver's Travels,* ed. Herbert Davis, vol. 11, *The Prose Work of Jonathan Swift* (Oxford: Basil Blackwell, 1941), pp. 137–76.

25 Antoine-Nicolas de Condorcet, *Sketch for a Historical Picture of the Progress of the Human Mind,* trans. June Barraclough (New York: Noonday Press, 1955), pp. 34–8; Judith N. Sklar, *Men and Citizens: A Study of Rousseau's Social Theory* (Cambridge: Cambridge University Press, 1969), p. 110; William Boyd, *The Educational Theories of Jean Jacques Rousseau* (New York: Russell & Russell, Inc., 1963), p. 139; Thomas Cassirer, "Awareness of the City in the Encyclopédie," *Journal of the History of Ideas, 24* (1963), pp. 387–396.

26 Carl E. Shorske, "The Idea of the City in European Thought," in Handlin and Burchard, *The Historian and the City,* pp. 95–8; Voltaire, *Letters Concerning the English Nation,* intro. by Charles Whibley (London: Peter Davies, 1926), pp. 54–6, 65–92, 171–9.

27 Quoted in J. S. Slotkin (ed.), *Readings in Early Anthropology* (London: Methuen, 1965), p. 412. Also see Abbé Raynal, *Philosophical and Political History of the Settlements and Trade of the Europeans in the East and West Indies,* trans. J. O. Justamond (London: J. Mundell, 1798), vol. I, pp. 3–4.

28 Carl Bridenbaugh, *Cities in Revolt: Urban Life in America, 1743–1776* (New York: Capricorn Books, 1964), p. 215.

29 William Robert Scott, *Adam Smith as Student and Professor* (Glasgow: Jackson, 1937), pp. 338–45.

30 Ronald L. Meek, *Social Science and the Ignoble Savage* (Cambridge: Cambridge University Press, 1976), pp. 99–130. Adam Smith, "The Principles which lead and direct Philosophical Enquiries; illustrated by the History of Astronomy," in *Essays on Philosophical Subjects* (Dublin: Wogan, Byrre et al., 1795), pp. 34–5. Also relevant is Yves Goguet, *The Origins of Laws, Arts, and their Progress,* vol. I (Edinburgh: A. Donaldson and J. Reid, 1761), pp. 272–3.

31 Adam Ferguson, *An Essay on the History of Civil Society, 1767,* ed. with intro. by Duncan Forbes (Edinburgh: Edinburgh University Press, 1966), pp. xxi–xxiii; Adam Ferguson, *Principles of Moral and Political Science,* vol. I (Edinburgh: Strahan and Cadell, 1972), pp. 239–40.

32 Shorske, "The Idea of the City in European Thought," pp. 98–100.

33 William C. Lehmann, *Henry Home Lord Kames, and the Scottish Enlightenment* (The Hague: Martinus Nijhoff, 1971), pp. 92–93; Morton White and Lucia White, *The Intellectual Versus the City* (Cambridge: Harvard University Press, 1962), pp. 12–20.

27

Why the Scientific Revolution did not take place in China – or didn't it?

NATHAN SIVIN
University of Pennsylvania

When people learn that there were many scientific traditions in the ancient world, they usually begin wondering why the fateful transition to modern science first happened where it did.[1] Joseph Needham has given the "Scientific Revolution problem" its classic formulation: "Why did modern science, the mathematization of hypotheses about Nature, with all its implications for advanced technology, take its meteoric rise *only* in the West at the time of Galileo?" This affirmation implies that one must investigate the absence of such a revolution elsewhere, and indeed page after page of *Science and Civilisation in China* is given over to "why modern science had not developed in Chinese civilization . . . ?" He adds a second question that bears on this absence, enhancing the interest of the larger inquiry: "why, between the first century B.C. and the fifteenth century A.D., Chinese civilization was much *more* efficient than occidental in applying human natural knowledge to practical human needs."[2]

In two decades of study, teaching, and public lecturing on Chinese science and medicine, I have encountered no question more often than why modern science did not develop independently in China, and none on which more firmly based opinions have been formed on the basis of less critical attention to available evidence. Since those who put forth these opinions are on the whole intelligent and thoughtful, I have gradually been led to suspect that there is more to the Scientific Revolution problem than meets the eye. In this essay I will turn it inside out in order to ask what assumptions about the European tradition of science – assumptions by no means confined to Europeans and Americans – encourage us to take this problem more seriously than its intrinsic merits justify.

ISSUES

In that millennium and a half European civilization was first experiencing a general collapse and then recovering from it. It is obvious that we ought to be

The Edward H. Hume Lecture, presented at Yale University, 1982. This essay was originally published in *Chinese Science*, (1982), *5:* 45–66. Reprinted by permission of the author.

looking at the Western end of Eurasia to account for European inferiority in technology over a span of fourteen hundred years. But there are still other doubts to be expressed in connection with this second question. The natural knowledge that was being applied to human needs was not what we usually call Chinese science.

Early technology did not succeed or fail according to how well it applied the insights of early science. Science was done on the whole by members of the minority of educated people in China, and passed down in books. Technology was a matter of craft and manufacturing skills privately transmitted by artisans to their children and apprentices. Most such artisans could not read the scientists' books. They had to depend on their own practical and esthetic knowledge. What that knowledge was like we can only reconstruct from the artifacts they left and from the scattered written testimony of literate people. Literacy spread considerably outside the elite over the last several centuries, but this did not lead to the substantial use of books to teach craft skills. The classical tradition of medicine was an applied science, but not a mere technology, in the sense that therapeutic decisions were based on a cumulative structure of systematic theory applied to experience. Less exalted kinds of curing were sometimes a matter of techniques, sometimes of rituals, neither of which depended on the abstractions of the classical tradition.

It also seems to me that comparing all of the scientific and engineering activity of one civilization with all that of another conceals more than it reveals, since it is only in modern times that these various kinds of work became closely connected. It is true that between the end of the Roman period and 1400 or so, a Chinese visiting Europe would have found it in many respects technologically backward. At the same time there was probably not a great deal to choose between Chinese and European medical practice before about 1850 (knowledge of anatomy and physiology had little therapeutic application earlier). Mathematical astronomy in China by its last high point about 1300 did not quite reach the general level of predictive accuracy that Ptolemy had mastered eleven hundred years earlier.

I need not dwell on comparisons of this kind. They tell us nothing at all about what we can expect to learn from one culture or the other. After all, no one is proposing that we give up the study of Hellenistic alchemy just because it has become clear that the Chinese alchemical literature is richer in chemical knowledge.[3] What matters is that we are now able to begin comparing several strong traditions of science and technology based on the ideas and social arrangements of different civilizations. All of them must be attentively studied if we want to understand the general relations through history and across the globe between science and culture, science and society, science and individual consciousness. Without that understanding we will remain trapped in our own parochial viewpoints. Historians have more urgent work to do than trying to prove the inferiority of every other culture to the one in which they specialize.

SCIENCE AND SCIENCES

As an example of how studying the Chinese experience can suggest clues about the character of early science in general, I offer the case of Shen Kua 沈括 (1031–1095), one of the most versatile figures in the history of Chinese science and engineering. Just to give a few examples, he is famous for the first discussion of magnetic declination and of printing with movable type, the only application of permutations in traditional Chinese mathematics, a proposal for nightly measurements of the lunar and planetary positions, the first suggestion in East Asia of a purely solar calendar, an explanation of the process of land formation by both deposition of silt and erosion, and an important book on the theory and practice of medicine. In addition to his technical activities, his writing has to be consulted by every student of early Chinese archeology, music, art and literary criticism, economic theory, and diplomacy. He made his early reputation as a land reclamation expert, and was deeply involved as a high official in the 1060's in the most important political reform movement for some centuries.

Shen's combination of unlimited curiosity and involvement in the affairs of his time had a special interest for my own education. For some time, through a series of studies roaming through different historic periods and technical disciplines, I have been trying to piece together bits of answers to a large question that I find boundlessly interesting. How did Chinese scientists in traditional times explain to themselves what they were doing? In other words, what was their understanding of nature and of their relation to it as conscious individuals living in a society? How did the insights of the various sciences hang together to form this understanding? I had gradually formed a general idea of the sciences as defined in early China, but I couldn't see how their insights were combined to form that general understanding. It occurred to me that I might do well to study how the sciences fit together in the mind of a person who was involved in all of them. The obvious person to study was Shen Kua.

The pattern that emerged was familiar in its details but unexpected in its overall shape. One aspect was that there does not seem to have been a systematic connection between all the sciences in the minds of the people who did them. The sciences were not integrated under the dominion of philosophy, as schools and universities integrated them in Europe and Islam. Chinese had sciences but no science, no single conception or word for the overarching sum of all of them. Words for the level of generalization above that of the individual science were much too broad. They referred to everything that people could learn through study, whether of Nature or human affairs (*hsueh* 學), or even more broadly to any pattern that could be apprehended through any form of cognition (*li* 理 and *tao* 道). All of these terms could, for instance, include ethical or religious principles discovered through reflection on authoritative texts. *Li* and *tao* could be grasped through mystical illumination.

Let us consider for a moment the connection between items of knowledge about Nature in Shen Kua's writings. We can trace through a thousand years of comprehensive encyclopedias in China the old division between heaven, earth, and man – the cosmos and its phenomena; the earth and its features, territorial divisions, creatures, and products; man and his institutions, usages, and accomplishments. In this conventional classification astronomy and astrology fell obviously enough under the sky; alchemy and medicine usually fell next to divination under technical skills (a rubric that includes painters and sometimes hired assassins). On the other hand, drugs fell sometimes among products of the earth and sometimes among commodities.

In Shen Kua's memoirs, *Meng ch'i pi t'an* 夢溪筆談 (Brush talks from Dream Brook), there is a rubric called "regularities underlying the phenomena 象數." Under this heading he like many others grouped together physical and numerological aspects of astronomy, astrology, cosmology, and divination, which refract the pattern of physical reality in their various ways. A section called "technical skills 技藝" puts medicine, engineering, and mathematics (including astronomical mathematics), which share purely instrumental value, alongside architecture and games. His chapter on "strange occurrences 異事" sets out his thoughts on the origin of plant fossils, the first recorded description of a tornado in East Asia, an account of his experiment on the formation of rainbows, and similar gems, arrayed cheek by jowl with unlikely hearsay and ephemeral curiosities.

What makes us think of Shen Kua as a scientist was widely scattered through his own scheme of human knowledge (see Appendix). That scheme cohered not on the level of science, but on a much more general level. In his writing, there are no clear boundaries between material that fits the modern conception of science and material that does not. That conception cannot carry us far in the effort to understand what Shen Kua was getting at.

Shen Kua, in the second half of the eleventh century, made his turn on the stage of history at a time when a great upsurge in social mobility was broadening the group that ruled China. Many of these new men were interested in all sorts of practical affairs that well-born people in earlier times would have considered beneath them. It was to some extent a matter of greater versatility in public service. This was after all a time when merit ratings of officials were being based on quantitative measures of efficiency in collecting taxes, reclaiming land, and so on, instead of entirely on virtue, breeding, and orthodoxy, as had been the case earlier. At leisure too, this large group that Shen Kua belonged to was free to indulge curiosity – in an amateur way, of course – about anything in the universe, including technical matters that earlier were fit only for clerks or artisans. Only after Shen's lifetime did this evolving amateur ideal settle on philosophy, the arts, and literature, as the appropriate realms to be universal within, once again leaving the study of the earth and sky largely to the mere technicians. In the eleventh century Shen was only one of a number of polymaths whose scientific and technological interests, however amateur, all emerged in connection with their varied official responsibilities. Su Sung 蘇頌 (1020–1101) and Yen Su 燕肅 (d. 1040) are also well-known examples.[4] The intellectual consistency of Shen's style in scientific thought seems to reflect the consistency of his public career, in which that style was formed.

The astronomer in the court computing calendars to be issued in the emperor's name, the doctor curing sick people in whatever part of society he was born into, the alchemist pursuing archaic secrets in mountain haunts of legendary teachers, had no reason to relate their arts to each other. Philosophers were in no position to define a common discipline for all of them, as Aristotle and his successors had done in Europe, and so philosophers had practically no influence on the development of these special pursuits. I do not mean to deny that people who worked in the sciences embedded their special knowledge in world-views and social philosophies. To the contrary, until very recent times the prospect of being a mere functional unit of scientific manpower, unconcerned with the larger significance of one's work, would have been unimaginable. But these wider perspectives to which each field related were passed down within the sciences themselves. Mathematics, geomancy, and so on, each came equipped with its own cosmic meanings. These were seldom quickly or deeply influenced by contemporary philosophy, but continued to develop in their divergent ways out of the theoretical perspectives with which the sciences began in the Han period (206 B.C.–A.D. 220).

If anyone were going to seek out the common ground of the sciences in China, it was people like Shen Kua, who were mastering them all. But Shen put his own understanding together in ways that did not closely link the various traditions that studied physical nature, and in ways that did intimately associate what today would be considered scientific with what would be called grossly superstitious. The distinction between science, proto-science, pseudo-science, and non-science simply gets in the way of understanding the articulation of Shen Kua's thought. Surely it is necessary to understand thought before one begins to evaluate it.

I would have to say that I failed to find the internal unity of Chinese science that I was looking for in the mind of Shen Kua. By way of compensation, I did learn the importance of an issue that I hadn't paid enough attention to before, that is, the relations of the sciences to other kinds of knowledge.

Let me give one example. Shen devoted about as much space in his memoirs to stories that involve strange happenings, predestination, prognostication, and divination, as to accounts that we would relate to science and technology. He himself does not seem to have viewed these enthusiasms as in conflict with his studies of Nature.

His rational explanation of divination is not very different from the view of its most sensible modern students. It is more sensible, I would say, than Carl Jung's famous attempt to explain the notions behind the Book of Changes by his hazy and un-Chinese notion of synchronicity.[5] Shen seems to have believed that the divination techniques used in this time did not really describe the future or what was happening far away, so much as provide counsel about the unknown and inaccessible. For instance, he explained why the same technique gives different outcomes when used by different people:

> In his discussions of divination to determine locations of houses and tombs and to foretell length of life, Lü Ts'ai 呂才 [d. 665] believed that techniques cannot yield reproducible

results. It is quite true that technique is not reliable in this sense. Still he did not realize that every kind [of divination] is a matter of 'substitutes 寓.' "The ability to respond spiritually and make the truth manifest depends on the person"; thus if two people use a single technique the result of their divinations will be different. The human mind is by nature spiritually responsive, but since it is unavoidably burdened, one must, in order to gain access to it, use as a substitute some thing that does not have a mind. The result of divination can only be explained by what makes one's own spiritual response possible. . . . In fact anything that can be seen, heard, thought about, or speculated upon can be used as a substitute for this purpose. If this seems irrational, is not all good and bad fortune, are not all the mutations of life and death, irrational? Only with someone able to understand the pattern common to all this can one discuss the spiritual response that makes foreknowledge possible.

Let me now rather freely paraphrase this statement in language more familiar to modern readers: Understanding is a matter of introspection. The questions we ask in divination are the ones we have tried to deal with reflectively but can't, because at the time our minds are too burdened for introspection to work. Divination techniques ritually manipulate passive material objects, yarrow stalks and so on, providing an external process or set of images on which to concentrate as a way round that blockage. In other words, prognostication draws indirectly, through ritual, on the power of self-examination.

Access to one's personal future, whether by visionary foresight or by divination, is a perfectly natural phenomenon. It merges into the moral faculties, whose choices condition the future. It merges at the other end into the rational comprehension of the natural order as it is reflected in any authentic experience. Shen did not confuse introspection and observation, nor did he draw a clear line between them. Nor did he need to compare the importance of these two ways of knowing. What finally united the sciences, in other words, was the universal system of knowledge – uniting intellection, imagination, and intuition – of which they constituted only a part. I am arguing that sometimes to look at the sciences alone is to look too closely.

ASSUMPTIONS IN THE SCIENTIFIC REVOLUTION PROBLEM

Now back to the Scientific Revolution problem. It is striking that this question – Why didn't Chinese beat Europeans to the Scientific Revolution? – happens to be one of the few questions that people often ask in public places about why something didn't happen in history. It is analogous to the question of why your name did not appear on page 3 of today's newspaper. It belongs to an infinite set of questions that historians do not make part of their research programs because they have no direct answers. They translate into questions about the rest of the world. The one that concerns us, for instance, translates into "in what circumstances did the Scientific Revolution take place in the seventeenth and eighteenth centuries in Western Europe?"

Why do people keep asking why the Scientific Revolution did not take place in China when they know enough not to explain why their names did not appear on page 3 of today's newspaper? Because the question encourages exploration of a fascinating topic and provides some order for thinking about it. It is, in other words, heuristic. Heuristic questions are useful at the beginning of an inquiry. As we comprehend enough to deal with complicated patterns, heuristic questions tend to grow murky, and finally to lose their interest compared with the emerging clarity of what did happen.

So much for heuristic questions in general. Why do we tend to take this one more seriously than the general run? Somehow the Scientific Revolution problem holds a special urgency.

That urgency is there, I suggest, because this way of putting the problem contains and supports certain Western assumptions, assumptions that ordinarily we do not question. Above all we usually assume that the Scientific Revolution is what everybody ought to have had. But it is not at all clear that scientific theory and practice of a characteristically modern kind were what other societies yearned for before they became, in recent times, an urgent matter of survival amidst violent change. In fact we have made very little progress so far in understanding how Europeans originally came to want modern science and its concomitants, since the attention of historians has been concentrated on how these innovations came about. Nor has much attention been given to what accounts for the gradual and uneven diffusion of modern science within Europe – an issue that does not differ essentially, I submit, from that of its spread to other civilized parts of the world.

There is usually the further assumption that civilizations which had the potential for a scientific revolution ought to have had the kind that transpired in the West and encompassed the sorts of institutional and social changes that are identified with modernization.

These assumptions are usually linked to a belief – or a faith, if you prefer – that European civilization all along was somehow in touch with reality in a way no other civilization could be, and that its great share of the world's wealth and power comes from some intrinsic fitness to inherit the earth that was there all along. Many of those like myself who reject this assumption argue that the privileged position of the West comes instead from a head start in the technological exploitation of nature and the political exploitation of societies not technologically equipped to defend themselves.

Finally there is the assumption that, since modern science has so quickly and thoroughly become international, it transcends European historical and philosophic biases, and is as universal, objective, and value-free as the Nature that it seeks to understand and manipulate.

What seems to be common sense in that last assumption (or in the self-conception that all the assumptions I have mentioned are part of) does not stand up to thoughtful examination. Modern science is still too marked by the special circumstances of its development in Europe to be considered universal.

Chinese science got along without dichotomies between mind and body, objective and subjective, even wave and particle. It is not that, for instance, physical and mental acts were confused, but they were not considered mutually exclusive, to be accounted for by sharply different varieties of discourse. In the West the first two dichotomies were entrenched in scientific thought by the time of Plato. Galileo, Descartes, and others carried them into modern times to mark off the realm of physical science from the province of the soul, which was decidedly off limits to secular innovators like themselves. These distinctions let scientists claim authority over the physical world on the ground that purely natural knowledge could not conflict with the authority of established religion.

Science and religion have long since learned to coexist, but we are still living with these distinctions. If they are European peculiarities, and perpetual sources of trouble at that, why hasn't modern science managed to rid itself of them? It is evidently not a simple matter to root them out. Until we do, there is something to be said for frankly admitting a certain parochialism in the foundations of science. The mathematical equations may be universal, but the allocation of human effort among the possibilities of natural knowledge is not.

Science and technology have spread throughout the world, but that has not made them universal, in the sense of transcending European patterns of thought. In one society after another the encounter between old and new ideas has been abortive, resolved by social change and political fiat. Traditional ideas are simply excluded (on the grounds that they are primitive, superstitious, regressive, fit only for the lower classes, etc.) from the educational systems created to teach a new technical and managerial elite the values of technology alongside its theory and practice.

Modern technology is clearly more powerful than that of traditional societies; but to a larger extent than we generally realize, its strength emerges in application to needs and expectations that do not exist until it generates them.[6] True universality would require modern technology to coexist with and serve cultural diversity rather than consistently serving as a tool to standardize it out of existence.

I am arguing that the notion of a universal and value-free modern science, which has somehow become independent of its social and historical origins, is wishful thinking. It is easy even for an intelligent reader to be led astray on this point. The kernel of certainty from which it arises is seldom defined carefully by those who set out to explain science to non-scientists. It would be foolish to deny that modern science has attained a verifiability, an internal consistency, a taxonomic grasp, a precision in accounting for physical phenomena, and an accuracy in prediction that no other kind of activity shares, and that lay far outside the grasp of early sciences. The rigor that makes these remarkable characteristics possible quickly disappears, however, once the formulation of a law or theory in mathematical equations, matrices of categories, or exactly defined technical concepts and models has been translated into the ordinary language and general discourse of a given culture. That translation into analogies and metaphors steeped in values must precede all public discussion of science, and most philosophic discussion. It even precedes most reflection by scientists on fields outside their own disciplines.

Beyond the narrow, abstract realm in which exactitude is possible, we are no longer insulated from the values and subjective judgments that shape every activity situated within a society. There are, for instance, profound differences between the character of modern scientific activity in the contemporary People's Republic of China and United States which reflect different predominant convictions about the relations between basic and applied science, the relation of both to general culture, the roles of scientists in defining research programs, procedures for planning and supporting individuals' research projects, expectations about the social aims to which scientific work will contribute, the organization and status of professional scientists, the connections of political ideas and scientific knowledge, and the division of national resources between science and other priorities, and between various scientific activities. That certain equations and models are invariant between the two societies is a factor in all these consensuses, but so is the ubiquity of opposable thumbs. Despite the invariance, a given constellation of values will determine that certain laws and hypotheses can be developed further, and that others will be abandoned unless they are among the very few that individuals can explore at their private discretion and their private expense. The great disparity in Chinese and American definitions of psychology is only one particularly obvious example that affects the life and death of particular theories in one society or the other.

So long as there is variation of such magnitude in the balance between the cognitive, practical, normative, and social dimensions of science, such words as "international" and "universal" are out of place. When applied to the narrow, rigorous technical realm of scientific cognition alone, they constitute a modest claim indeed – so modest that the two adjectives can hardly be understood in their customary senses.

Nor can one accept uncritically the idea that modern science is in every essential respect European in its social and historical origins. To those familiar with the science of other cultures, any account of the early history of science is lopsided, and misleading on the most fundamental issues, if it restricts itself substantially to discoveries made and understandings worked out at the Western end of Eurasia; if it loses sight of the constant movement of ideas back and forth between civilizations from the New Stone Age to the present; if it does not adequately consider what Europeans had learned by the seventeenth century about Islamic, Indian, and Chinese science; or if it ignores the impact of exotic technologies and materials on the experiences of Europeans.

FALLACIES OF HISTORICAL REASONING

Growing awareness of the high level of science and technology in ancient China has led to cascades and avalanches of hypotheses from one scholar or another about factors that inhibited the evolution of modern science in China, or characteristics unique to the West that made possible or furthered a major scientific revolution.[7] These often incorporate elementary fallacies of historical reasoning that deserve notice.

For roughly two-thirds of a century, it has been argued that although Ch'ing dynasty thinkers took the world as observable, nominalistic fact, just as Sir Francis Bacon (1561–1626) did, unlike him they did not develop a scientific methodology. Despite the positivist assumptions of such arguments, whether Bacon's scientific method has survived in the practice of contemporary science was not even considered.[8] It was, in fact, largely Scholastic in its origins, concerned with taxonomies rather than theories of natural phenomena, and resolutely unconcerned with mathematical measurement. Of the major early modern attempts to define how physical science might fruitfully proceed it was probably the most sterile, in contrast to Bacon's very influential convictions about the organization and ideology of scientific activity.

This pattern of thought that faults the Chinese for not developing a scientific method that later proved abortive in the West crops up in many other forms. Another example is a well-known sociological study of Han astronomy when it attempts to explain the failure to develop a "unified scientific system." One reason given is that Chinese astronomers "were not interested in applied technical sciences, e.g., in developing theoretical tools which could be used to control the flight of a cannon shell or to direct ships safely across the sea."[9] So much for the first civilization to note the declination of the compass needle. So much for the astronomy of an era more than a millennium before the invention of the cannon. The same lack of interest is prominent in the impetus theoreticians from John Philoponus (fl. ca. 530) to Jean Buridan (ca. 1295–ca. 1358) and others of the School of Paris whose investigations furnished much of the basis for Galilean mechanics. How then did what is presented as a disastrous shortcoming in China fail to prevent in Italy – in fact, according to the conventional wisdom, help directly to bring about – the mathematical study of bodies in motion?

Considered generally, this fallacy amounts to claiming that if an important aspect of the European Scientific Revolution cannot be found in another civilization, the whole ensemble of fundamental changes could not have happened there. The flaw of reasoning that underlies it is the arbitrary assumption, never explicit, never discussed, that a given circumstance amounts to a necessary condition. It is almost invariably arbitrary because if we trace the prehistory of the actual Scientific Revolution backward far enough, in most cases we can find a point when the circumstance is absent in Europe. In that case, on what grounds can it be considered a necessary condition? In most cases one need not go back very far. That is why, despite their currency among Sinologists, in the past generation necessary conditions have practically disappeared from the armamentarium of discriminating historians of science.

The mirror image of this fallacy may be seen in an influential estimate of the

Chou i 周易, the Book of Changes, as a deterrent to science. Here is the way Joseph Needham put it in 1956:

> while the five-element 五行 and two-force 陰陽 theories were favourable rather than inimical to the development of scientific thought in China, the elaborated symbolic system of the Book of Changes was almost from the start a mischievous handicap. It tempted those who were interested in Nature to rest in explanations that were no explanations at all. The Book of Changes was a system for *pigeon-holing* novelty and then doing nothing more about it.

Nearly two decades later Ho Peng Yoke assured us that if Chinese scientists "were fully satisfied with an explanation they could find from the system of the *Book of Changes* they would go no further to look for mathematical formulations and experimental verifications of their scientific studies. Looking at the system of the *Book of Changes* in this light, one may regard it as one of the inhibiting factors in the development of scientific ideas in China."[10]

In these instances one is tempted to counter the arguments with matters of fact. Although Needham's extended discussion treats the Book of Changes predominantly as a static classificatory system of concepts, if we examine its applications to natural philosophy we find that it was most often used to construct dynamic explanations of change. One also looks in vain for a habit among Chinese scientists of constructing mathematical formulations and experimental verifications; if one cannot prove that this tendency was evolving steadily to a certain point, if there is no tangible evidence that without the Book of Changes they would have "gone further," there seems to be no warrant for introducing from modern biology the metaphor of inhibition.[11]

Exactly what does "inhibiting factor" mean in such contexts? Consider one of these often adduced to explain the failure of China to beat Europe to the Scientific Revolution despite an early head start, namely the predominance of a scholar-bureaucrat class immersed in books, faced toward the past, and oriented toward human institutions rather than toward Nature as the matrix of the well-lived life. But in Europe at the onset of the Scientific Revolution we are faced with the predominance of the Schoolmen and dons, immersed in books, faced toward the past, and oriented toward human institutions rather than toward Nature. They did not prevent the great changes that swept over Europe. It would take a more imaginative historian than myself to say whether those changes would have taken place sooner had Scholasticism never existed.

The confusion about "inhibiting factors" is no less a confusion when it has to do with ideas or techniques. One might just as well call Euclidean geometry an inhibiting factor for the development of non-Euclidean geometry, since so long as people were satisfied with it they didn't move on to a new step. But can one argue that non-Euclidean geometry would have developed sooner without it? It is unfortunate to see the remarkably interesting technical language of the Book of Changes, so powerful in systematically relating broader ranges of human experience than modern science attempts to encompass, written off as an obstacle before anyone has taken the trouble to comprehend it thoroughly.

The first fallacy confuses for a cause or necessary condition what is merely a description of an earlier state of a culture, or of a culture's way of doing something. In its complement, as can be seen by the examples just given, the absence of the subsequent state is confused with an inhibitor. One who commits this second fallacy is metaphorically stopping growth that may not have been taking place. Both of the confusions I have described – blaming the earlier state for delaying the later state, and using the absence of something modern at one point to explain the unattainability of modernity later – confound continuity with stasis. They are bad history because they are bad philosophy.

I recur to the assumptions about ourselves that I have discussed earlier, for they are at the root of both these fallacies. They turn the history of world science into a saga of Europe's success and everyone else's failure, or at best inherently flawed and transitory success, until the advent of redemption through modernization.

Joint use of the pair of fallacies makes it easy to prove that the European breakthrough is not simply a fact of history, but was inevitable since history began. Was the horse and buggy a necessary preliminary to the invention of the automobile, or did it delay that invention? Would the automobile have emerged sooner if the buggy had never been invented, so that people would have been dissatisfied with less adequate vehicles? If we find some analogue of the horse and buggy in Europe, by fallacy 1 its absence in China made the invention of some analogue of the automobile impossible. If we find some analogue of the horse and buggy in China, we apply fallacy 2 and make it an inhibiting factor. Thus medieval European impetus theory, abstract and unconcerned with application, was a stage in the evolution of inertial guidance; the presumed unconcern of Chinese thinkers for application proves that inertial guidance could never have originated in East Asia.

This is an infallible formula for reading the strength and power of modern science into the historic past – but only the past of Europe. For the past of other civilizations the test is always anticipation of or approximation to some aspect of early European science, or modern science. Why does the science of early Europe not need to be tested? Because of the *assumption* that its parentage of the Scientific Revolution was unique. Other civilizations shine only as they reflect the light of the European tradition. Or so the prophets of modernization suppose.

I claim, therefore, that the fallacies that so often accompany discussions of the Scientific Revolution problem reflect a set of disastrous assumptions that lie beneath the obvious interest and charm of its surface. They are disastrous because they encourage us to devaluate, without troubling ourselves first to comprehend on their own terms, scientific quests other than the one from which modern science most directly sprang.[12] We now find these assumptions accepted not only in Europe but to some extent in every country in which the history of China is studied.

Why should intellectuals in a non-European country, that owes little of its culture before modern times to Western influence, accept this bias? That is perhaps inevitable, considering that modern education establishes itself (as it did originally in Europe) by teaching the rejection of the traditional past or its demotion to a cultural exhibit that may be of use for nurturing nationalism (and, in the era of cheap package tours, for enticing tourists). Since Japan has had a century's experience with a modern educational system and the self-consciousness it produces, Nishijima Sadao's acute analysis does not come as a surprise:

> The 'static character' hypothesis holds that Chinese society lacked the capacity to form progressively a new era through its own efforts. This hypothesis was afforded particular emphasis by the viewpoint that the modernization of Chinese society was retarded. . . . Originally the 'static character' hypothesis, in company with that of 'Oriental despotism,' was advocated, in contrasts with Western European society, as a notion in polar opposition, for the sake of validating the self-consciousness that came into being with the formation of modern Western European society. That is, it was a postulate to serve as an element in the recognition of the value of modern Western European society. . . . In our country, when we deal with Chinese society from the point of view that makes the formation of the modern ego identical with the equal valuation of individuals in Western European civilization, we are led uncritically to use the 'static character' hypothesis. This has brought about our tendency to be controlled by the inverted logic that makes the goal of understanding Chinese society equivalent to grasping the origins or even the mechanism responsible for the persistence of its 'static character.'[13]

In other words, if one begins with the assumption that the paramount issue in the study of China is accounting for the inevitability of backwardness, one is unlikely to question whether backwards was inevitable, to ask whether there were not in her history prominent patterns of success from which we might learn, or to reexamine the assumptions about the modernized West that organize European history of a crescendo of success (with setbacks, to be sure, adding to the complexity and thus the charm of the crescendo), and that of other civilizations as a tableau of failure. Thus Nishijima states the intellectual convictions that justified and supported not only esteem for European civilization but also Japan's political aspirations in East Asia before and during the Sino-Japanese War.

One more fallacy often appears in connection with the Scientific Revolution problem, when historians select the aspects of the European experience that are appropriate for comparison with other civilizations. I mean the fallacious assumption that one can make sense of the evolution of science by looking at intellectual factors alone, or socio-economic factors alone, according to preference. Some people think of science predominantly as an intellectual quest after truths hidden in nature. They tend to think of China's failure to beat England to modern science as an intellectual failure. Other people, who think of science as primarily a social or economic phenomenon, tend to see the defeat as a matter of Chinese social or economic backwardness. But neither of these exclusive approaches to explanation is adequate. The distinction between intellectual and social factors or between internal and external factors is not out there in the events we study, but in the mental habits and professional associations, in the division of labor, of historians.

DIMENSIONS OF THE SCIENTIFIC REVOLUTION

The Scientific Revolution and its consequences cut across the boundaries of historical specializations. Let me make this clear by defining its important dimensions.

To begin at the intellectual end, the Scientific Revolution was a transformation of our knowledge of the external world. It changed the questions we asked, the means we used to explore them, and the character of the answers. It established for the first time the dominion of number and measure over every physical phenomenon. It not only "combined, with effective logical precision, a theoretical search for common forms of explanation with a practical demand for accurately reproducible results," but envisioned precisely controlled action as the outcome of understanding.[14]

Ernst Gellner pointed out not long ago a particular way in which the European Scientific Revolution is more than a leap to a new form of knowing. It is natural to assume that in science the crucial test has always been "is it true?" But earlier that was only one of several equally important questions: Is it beautiful? Is it conventional? Is it morally improving? Does it lead to perception of the Good? Does it conform to certain esthetic patterns that all truth must, as astronomers up to Kepler believed that celestial orbits must be compounded of perfect circular motions? In science the test of truth has displaced most of these and redefined the others. This demand for truth above all was an appeal to fact – fact that was in principle public, verifiable, morally neutral, invariant with the social circumstances of the observer, immune from interference by magician or god. But the new science did more than appeal to facts. It created facts of that kind for the first time. That is an awesome creation. It took place in Europe between the time of Copernicus and Laplace and has spread across the world since.[15]

The same leap was not taken in seventeenth-century China. Could it in principle have been taken anywhere but in Europe, or any time but then? It would be unbecoming to hide our practically total ignorance.

The Scientific Revolution also meant a continuing redefinition of the connections of natural philosophy (i.e., science) to other kinds of knowledge. It meant a redefinition of man's orientation toward the past and the future. It meant a redefinition of what authority should determine what uses may be made of knowledge. It meant a redefinition of what knowledge of nature is socially desirable, and what socially undesirable. It meant a redefinition of how knowledge ought to be related to human individuality and to the active relations of man and nature.

Galileo and his friends and successors could not have got round the authority of the Church on the strength of ideas alone. That message was conveyed to Galileo by the Congregation of the Index in 1616, and then with drastic finality when he was condemned in 1633. But he and his fellow spirits had begun constructing a new intellectual community outside the old establishment. A hundred years earlier there had been no organized alternative to the Church and its scholastic educational system; then even Galileo himself might have died an archbishop.[16] But in the Counter-Reformation the Church was beginning to be less attractive to the most talented and ambitious (and of course there was less room for those who were attracted). A variety of new careers was emerging. Among them the profession of scientist was being invented. This profession could not provide structures that paid for careers in research for more than an occasional genius until about 1800.[17] Nevertheless its creators assumed from the start, for its amateurs, devotees, and enthusiasts, independent authority to formulate the laws of nature – took it away, in fact, from the Scholastics, for whom science could never be more than a collaborator of faith. Secular learning remade the universities and displaced other ancient institutions as over several centuries of evolution and revolution it formed a technical establishment. As Francis Bacon forecast in his *New Atlantis* at the beginning of the process, the institutions of science ultimately inherited the established church's charisma and its public claims to authority over all valid knowledge. The early modern concession to the Church of authority over matters of cognition, rational decision, emotion, etc., to which I have already referred, was essential but only temporary, as these came no longer to be considered activities of the soul. The soul's province has been narrowed to that of faith, when it appears in scientific textbooks at all.

Freidson and others have shown the formation of a profession depends on a grant of autonomy from society – autonomy not only to set its own standards for admission, competence, and compensation, but also to evolve bodies of knowledge and universes of meaning that only specialists can understand.[18] In Europe natural science could contend for this status because other learned groups already had it. Physics could follow medicine in creating new institutions with the right to regulate their internal affairs. I might remark in passing that no such institutions existed in China. The guilds had no such status. Despite the secure bureacratic status of imperial physicians and astronomers, by 1600 there was no occupational group sufficiently autonomous or coherent to be called a profession. Scientific knowledge, although highly technical, was widely studied by elite amateurs, and the social consensus forbade corporate bodies governed by norms distinct from those to which all subscribed (I will refer below to later changes in this situation).

All of the mental and social changes in Europe that I have summarized above were one process, and none of its parts seem to me separable. I remember how puzzled a classicist colleague once was at why Archimedes didn't set off a scientific revolution. My colleague was convinced that Archimedes had the mathematical tools to invent dynamics; why did he stop short, eighteen hundred years before Galileo? Why didn't someone form a committee and fund an enormous research project so that Archimedes' dynamics and his talent for invention could have saved the Hellenistic world from its enemies? Unfortunately, neither new ideas without human organization to carry them out, nor new associations rearranging clichés, can change the world.

This outline of the Scientific Revolution's many dimensions is meant to suggest how much we are likely to miss if we care only about social factors, or only about intellectual factors, as we survey the situation in China. Until recently, for instance, people concerned with that topic, including myself, have overlooked a significant piece of the Chinese picture, which I will now consider.

SCIENTIFIC REVOLUTION IN SEVENTEENTH-CENTURY CHINA

By conventional intellectual criteria, China had its own scientific revolution in the seventeenth century. This is a point of no small interest if we are meditating about why China could not have had one.

Western mathematics and mathematical astronomy were introduced to China beginning around 1630 – in a form that before long would be obsolete in those parts of Europe where readers were permitted access to current knowledge (in post-Galilean Italy they were not). Several Chinese scholars, among them Mei Wen-ting 梅文鼎 (1633–1721), Hsueh Peng-tso 薛鳳祚 (ca. 1620–1680), and Wang Hsi-shan 王錫闡 (1628–1682), quickly responded and began reshaping the way astronomy was done in China. They radically and permanently reoriented the sense of how one goes about comprehending the celestial motions. They changed the sense of which concepts, tools, and methods are centrally important, so that geometry and trigonometry largely replaced traditional numerical or algebraic procedures. Such issues as the absolute sense of rotation of a planet and its relative distance from the earth became important for the first time. Chinese astronomers came to believe for the first time that mathematical models can explain the phenomena as well as predict them. These changes amount to a conceptual revolution in astronomy.

That revolution did not generate the same pitch of tension as the one going on in Europe at the same time. It did not burst forth in as fundamental a reorientation of thought about Nature. It did not cast doubt on *all* the traditional ideas of what constitutes an astronomical problem, and what significance astronomical prediction can have for the ultimate understanding of Nature and of man's relation to it.

Most important, it did not extend the domain of number and measure in astronomy until it embraced every terrestrial phenomenon (the Jesuits were obliged to conceal from the Chinese that development in Europe). What happened in China bears comparison with the conservative revolution of Copernicus rather than with the radical mathematization of hypotheses Galileo precipitated. In a sense the Galilean breakthrough had nothing to break through in China, where, since no Aristotle had convinced students of nature that mathematical precision could not apply to quotidian terrestrial events, it was freely so applied. Still we do not find anything corresponding to Galileo's conviction that physical realities must be isolated from the flux of sensation by measurement (as in experiments) so that physical truth can be made to manifest itself in straightforward quantitative relations. Shen Kua, in the eleventh century, once argued that measure is an artifact, and possible only when phenomena are isolated from the continuum of nature in observational instruments; but his concern was the design of armillary spheres, and he saw no repercussions for the study of those aspects of nature that man can manipulate.[19]

The most striking long-range outcome of the mid-seventeenth-century Chinese encounter with European science, in fact, was a revival of traditional astronomy, a rediscovery of forgotten methods, that were studied once again in combination with the new ideas and that supported what might be called a new classicism.[20] Rather than replacing traditional values, the new values implicit in the foreign astronomical writings were used to perpetuate traditional values.

Why didn't this conceptual revolution have the social consequences that historians of Western science have insisted we should expect? By the mid-seventeenth century European civilization had had no appreciable political or social impact in China, and astronomy had to make its way on its own merits. The old and new astronomy ceased to be in antagonistic competition, once the predictive superiority of the European techniques was acknowledged.

One is tempted to see the process by which Western astronomy put down its first roots in China as the last major face-to-face encounter of non-Western and European science in world history. By the eighteenth century modern science was crossing national boundaries on the coattails of Empire, and competition between sciences, literatures, religions, etc., on the basis of their abstract merits had become a thing of the past. Even in the seventeenth century, despite the high drama of eclipse prediction contests in the court, the fact remains that the triumph of European computational techniques came about not through a consensus of great minds but by an imperial decision to hand over operational control of the Astronomical Bureau to Jesuit missionaries.

Revolutions in science as well as in politics take place at the margins of society, but the people who made the one in seventeenth-century China were firmly attached to the dominant values of their culture.[21] At the time there could be no students of astronomy motivated to cast off traditional values, and willing to follow ideas where they led even if the society around them fell apart. The foreign techniques, powerful though they were, offered Chinese students no alternative route to security and fame, and the civil service examination system hardly left leeway for one. The only astronomers who could respond to the Jesuits' writings were members of the old intellectual elite. They were bound to evaluate innovations in the light of established ideals that they felt an individual responsibility to strengthen and pass on to the next generation.

The most influential early champions of Western astronomy, such as Mei and Wang, were men of the lower Yangtze region who lived through the Manchu invasion and adopted the traditional role of the loyalist who would not serve a new dynasty 不貳臣. Having refused to strive for conventional careers in a society that in their view *had* fallen apart, they were motivated to spend their lives studying and teaching the new mathematics and astronomy while they used them to master the neglected techniques of their own tradition. They rejected the

Ch'ing present not for a modernist future but to keep alive the lost cause of the Ming for one more generation. Wang Hsi-shan even avoided using the Ch'ing dating system. The ancient model he adopted provided a role for only one generation, the one that was maturing during the transition between two dynasties; he could have no permanently marginal posterity. Despite his superb critical acumen he was the opposite of Descartes, for whom every ancient institution had to justify itself by the new criteria of clear and distinct ideas or be considered a mere vestige.

The principal Chinese proponents of European astronomy in the late seventeenth and early eighteenth centuries – the successors of Wang and Mei – argued that its archaic foundations had originated in China and made their way westward, so that studying it could not be considered a rejection of tradition. They made it one of the foundations of the Evidential Research (*k'ao-cheng* 考證) movement. That movement substituted for the intense moral inquiry and concern with statecraft of the Ming-Ch'ing transition a sustained cumulative application of philology and quantitative research to recover the pristine forms of the classics. An important analysis of this movement by Benjamin Elman demonstrates that, especially in the lower Yangtze region, it did evolve career patterns independent of the civil service, patterns in which success and financial security depended on the regard of peers for one's technical publications. But the great philologists, despite their concern for rigor, were not Cartesians either: the goal of their striving was authenticated, definitive editions of their civilization's most ancient documents.[22]

In summary, over most of the past thousand years Chinese society was sufficiently unified and permeable in its upper reaches that few who had the learning and ambition to formulate alternatives to conventional ideas were relegated to the margins. The classical mode of education ensured that to be an intellectual at all was to take up the burdens of elite culture. The contrast with Japan, beginning in the late eighteenth century, is instructive. There astronomy for some time had been done only by functionaries, who could not even perceive the basic ideas that animated Western science. On the other hand, medicine provided a practically unique avenue of social and economic mobility *within* the rather broad margins of a society that as a matter of principle excluded non-hereditary talent from the top. One finds advocates of Western ideas among medical men from the *rōnin* and townsman classes, whose stake in the Chu Hsi synthesis and other aspects of shogunal orthodoxy was minimal. It was the doctors who championed foreign ideas in other fields of science as well.[23] In the very different social circumstances of China before the twentieth century, physicians were as unimportant in the propagation of European science as astronomers were important.

If then we seek in China those for whom science was not a means to conservative ends, for whom a proven fact outweighed the whole body of millennial values, we do not find them until the late nineteenth century, when they became the first modern scientists in that country. By that time foreigners exempt from Chinese law and backed by gunboats had constructed new institutions and new career lines. These were most attractive to those they had educated, who had no other prospects. We can no longer talk about the encounter of the old and new astronomy. Social and political change had left nothing for the old to do. It became rare as time passed for modern scientists to be aware that their country had had its own scientific tradition. Only in the last generation has that awareness been revived.

CONCLUSION

My frustrations in trying to make sense of science in China arise partly because of the many levels of human activity that have to be encompassed over such a great sweep of time and human experience. They arise partly because the European Scientific Revolution seems to call for an understanding in greater breadth and depth than its historians have insisted upon. Once we keep in mind the many dimensions of scientific change and their complex relations, it becomes less surprising that the Scientific Revolution took place only when and where it did. The process increasingly comes to resemble historic evolution – the sum of human decisions and acts, always to some extent arbitrary and wrongheaded – rather than fate, inexorable determinism, teleology, manifest destiny, an inexorable logic unfolding from ancient Greek rigor, or the hidden operation of some World Spirit.

Looking at these three scientific revolutions – the one we think we know so well in Europe, the one that didn't take place in Archimedes' lifetime, and the one that wasn't what we expect it to be in seventeenth-century China – suggests that we have a great deal to learn about the specific circumstances of each, seen in all its dimensions, before we are ready to tell the world why the first of the three couldn't have happened in other times or places.

I believe that the breakthroughs coming up in the study of Chinese science will be of another kind altogether. They will have to do with understanding in depth and in an integral way the circumstances of people who did science and technology: how their technical ideas related to the rest of their thought; what the scientific communities were – that is, who formed a consensus that certain phenomena were problematic, and that certain kinds of answers were legitimate; how those communities were related to the rest of society; how the responsibility of men of knowledge to their colleagues in science was reconciled with their responsibility to society; what larger ends the sciences served, that kept their laws comformable to the laws of Chinese painting and to the basic principles of moral conduct.[24]

These are issues about which we understand very little with respect to China or to Europe. It will take much further study and reflection on both sides before the comparative history of science is ready to take off. My prognostication is that by that time we will no longer be asking why the transition to modern science did not first take place in China.

APPENDIX
CHAPTER HEADINGS IN SHEN KUA, BRUSH TALKS FROM DREAM BROOK

Number	Title	Begins With Jotting No.
1	Ancient Usages	1
3	Philological Criticism	42
5	Music and Mathematical Harmonics	82
7	Regularities Underlying the Phenomena	116
9	Human Affairs	151
11	Civil Service	189
13	Wisdom in Emergencies	224
14	Literature	245
17	Calligraphy and Painting	277
18	Technical Skills	298
20	The Supernormal	338
21	Strange Occurrences	357
22	Errors	388
23	Wit and Satire	401
24	Miscellaneous	420
26	Materia Medica	480

There are a total of 507 jottings in Brush Talks from Dream Brook, or 609 including its two sequels. For a classification of the book's contents according to field of knowledge treated, see Needham, *Science and Civilisation in China,* I, 136.

NOTES

1 This essay incorporates my current views on a historical issue to which I have returned regularly for some years. No doubt my views on this topic will be different in another decade; all I mean to accomplish with these ephemeral reflections is to transmit the idea that the issue is worth thinking about, to suggest how one might think about it, and to point out that certain ways of thinking about it are so burdened by suspect assumptions that they do not encourage clear explanation. I have addressed one aspect or another in previous writings, to which the reader is referred for documentation: "Copernicus in China," *Studia Copernicana,* 1973, *6:* 63–122; "Shen Kua" and "Wang Hsi-shan," in *Dictionary of Scientific Biography, s.v.,;* "Next Steps in Learning about Science from the Chinese Experience," *Proceedings,* XIVth International Congress of the History of Science (Tokyo and Kyoto, 19–27 August 1974), I, 10–18; Sivin (ed.), *Science and Technology in East Asia* (New York: Science History Publications, 1977), pp. xi–xxi; and "Chinesische Naturwissenschaft: Weber und Needham," in Wolfgang Schluchter (ed.), *Max Webers Studie über Konfuzianismus und Taoismus. Interpretation und Kritik* (München, 1982). Joseph Needham has also provided a summary of our conversations and correspondence on the "Scientific Revolution problem," an interesting attempt to specify differences and similarities in our views, in *Science and Civilisation in China,* vol. 5, part 2, *Spagyrical Discovery and Invention: Magisteries of Gold and Immortality* (Cambridge University Press, 1974), pp. xxii–xxvii. Here I set out my own somewhat different view of the divergences that accompany our very broad areas of agreement. In several points regarding the Scientific Revolution problem I have been anticipated by Wing-tsit Chan, "Neo-Confucianism and Chinese Scientific Thought," *Philosophy East and West,* 1957, *6:* 309–332.

I use "Scientific Revolution" to refer primarily to the transition in the exact sciences between Galileo and Laplace and its wider repercussions by 1800. This is one of several definitions in current use. I adopt it for the purpose of this essay not because it is the best possible definition, but because it is the one most commonly presupposed by Sinologists and laymen who set out to compare developments in China and the West. Needham's usage of the term "Scientific Revolution" is often, but not consistently, broader. No definition can be considered better than a historiographic expedient. Lack of a consensus about the significance of the term has led some historians of science recently to reject its use altogether.

2 Needham, *The Grand Titration: Science and Society in East and West* (Toronto: University of Toronto Press, 1969), pp. 16 and 190.

3 See in particular Needham, *Science and Civilisation in China,* vol 5, parts 2–4 (1974–1980).

4 In addition to frequent references in *Science and Civilisation in China,* see S. Miyasita, "Su Sung," III, 969–970 in Herbert Franke (ed.), *Sung Biographies* (3 vols., Wiesbaden, 1976); Teng Kuang-ming 鄧廣銘 & Wang Chen-to 王振鐸, "Su Sung," pp. 123–134 in Institute for the History of Science, Chinese Academy of Sciences (ed.), *Chung-kuo ku-tai k'o-hsueh-chia* "中国古代科学家" (Ancient Chinese scientists, Beijing, 1959); and Wang Chin-kuang 王锦光, "宋代科学家燕肃," *Hang-chou ta-hsueh hsueh-pao* 杭州大学学报, 1979, *3:* 34–38.

5 Jung, "Foreword," to *The I Ching or Book of Changes* (tr. Richard Wilhelm & Cary F. Baynes; Bollingen Series, XIX; New York: Pantheon Books, 1950), I, iii–vii. The quotation below is from *Meng ch'i pi t'an,* item 145. Lü Ts'ai was a famous diviner and theoretician of divination. The word I have translated "substitute" means more literally "to make something serve as a temporary abode." The sentence in quotation marks is quoted by Shen from the Great Commentary (*Hsi tz'u* 繫辭) to the Book of Changes, A. 12 (compare Wilhelm tr., I, 349).

6 This point has been most persuasively argued in Langdon Winner, *Autonomous Technology: Technics-out-of-control as a Theme in Political Thought* (Cambridge, Mass., 1977).

7 The most significant early contributions to this literature, in order of their appearance, are Jen Hung-chün 任鴻雋, "說中國無科學之原因," (The reason for China's lack of science), *K'o-hsueh* 科學, 1915, *1:* 8–13; Yu-lan Fung, "Why China Has No Science – An Interpretation of the History and Consequences of Chinese Philosophy," *The International Journal of Ethics,* 1922, *32:* 237–263; Homer H. Dubs, "The Failure of the Chinese to Produce Philosophical Systems," *T'oung Pao,* 1929, *26:* 96–109; Derk Bodde, "The Attitude toward Science and Scientific Method in Ancient China," *T'ien Hsia Monthly,* 1936, *2:* 139–160; and Rhoads Murphey, "The Nondevelopment of Science in Traditional China," *Papers on China,* 1947, *1:* 1–30 (for others see the bibliographies of *Science and Civilisation in China,* esp. vol. II). Jen claims that science failed to develop in China after the Han period because of inattention to "the inductive method." Fung claims "it is because of the fact that the Chinese ideal prefers enjoyment to power that China has no need of science . . ." (p. 261). Dubs refutes the silly prejudice that the

character of the Chinese language made systematic thought impossible, but argues that "the result of the absence of mathematical systems was that the Chinese philosophers attacked the world piecemeal . . . by empirical rather than by rational methods" (p. 108). He has nothing whatever to say about Chinese scientists. Bodde considers a number of aspects of attitudes toward science, and is aware of a few isolated scientific accomplishments despite his disregard for the technical literature, but suggests that the most important "retarding effect upon scientific innovation . . . has been the ideographic nature of the Chinese written language" (p. 158). Murphey, dependent upon Western-language sources and influenced by the stereotypes of F. S. C. Northrup, concludes "a naturalistic philosophy which might be called a reliance on the aesthetic continuum . . . clearly had no place for the inductive hypotheses necessary for science" (p. 15). Writings of this sort are full of acute observations, particularly about philosophic attitudes expressed in the early classics, but demonstrate a failure on the parts of their authors to acquaint themselves with the literature of the Chinese scientific traditions. They may well strengthen our conviction that Lao-tzu or Hsun-tzu would have been mediocre biologists or mathematicians, but they do not help us account for the theoretical analyses, mathematical proofs, and programs of empirical discovery so profusely documented in the writings of those actually engaged in studies of nature.

Because of his knowledge of the Chinese sciences and the breadth of his hypotheses, Needham's is the earliest discussion of the Scientific Revolution problem that still commands attention, and is still the best. The most useful critiques of Needham's writings on this subject are, from Sinologists, Bodde, "Evidence for 'Laws of Nature' in Chinese Thought," *Harvard Journal of Asiatic Studies,* 1957 (publ. 1959), *20:* 709–727, and "Chinese 'Laws of Nature': A Reconsideration," ibid., 1979, *39:* 139–155, Chan, and A. C. Graham, "China, Europe, and the Origins of Modern Science: Needham's The Grand Titration," pp. 45–69 in Shigeru Nakayama & Sivin (ed.), *Chinese Science. Explorations of an Ancient Tradition* (Cambridge, Mass., 1973); from a historian of science, Nakayama, "Joseph Needham, Organic Philosopher," ibid., pp. 23–43; from a philosopher, Robert S. Cohen, "The Problem of 19 (k)," *Journal of Chinese Philosophy,* 1973, *1:* 103–117; and from sociologists, Benjamin Nelson, "*Sciences* and Civilizations, 'East' and 'West.' Joseph Needham and Max Weber," *Boston Studies in the Philosophy of Science,* 1974, *11:* 445–493, and Sal Restivo, "Joseph Needham and the Comparative Sociology of Chinese and Modern Science," *Research in Sociology of Knowledge, Sciences and Art,* 1979, *2:* 25–51.

8 Jen, loc. cit.: Joseph R. Levenson, *Confucian China and its Modern Fate. The Problem of Intellectual Continuity* (London, 1958), pp. 3–14; and David E. Mungello, "On the Significance of the Question 'Did China Have Science,' " *Philosophy East and West,* 1972, *22:* 467–478, and my comments on his article in the same journal, 1973, *23:* 413–416. Fung also refers to Bacon in connection with the Scientific Revolution problem, but there the issue is not method but, more pertinently, the relation of science and power (Fung does not specify what kind of power).

9 Wolfram Eberhard, "The Political Function of Astronomy and Astronomers in Han China," pp. 33–70, 345–352 in *Chinese Thought and Institutions* (ed. John K. Fairbank; Chicago, 1957), p. 66.

10 Needham, *Science and Civilisation in China, II,* 336 and 340; Ho, "The System of the Book of Changes and Chinese Science," *Japanese Studies in the History of Science,* 1972, *11:* 23–39. Attempts to explain scientific revolutions by lists of positive and negative factors abstracted from context have been criticized by Robert K. Merton in *Science, Technology and Society in Seventeenth-Century England* (New York, 1970), p. x.

11 Although Needham has given considerable weight to the notion of inhibition, as one would expect of a first-rate biologist he is cautious about using it in relation to processes that he cannot prove were under way. Writers who draw on his work do not tend to be so discriminating. This point is easily demonstrated by examining the perceptive list of 29 "factors inhibiting the emergence of modern science in China and Western Europe" compiled from his writing in Restivo (see note 7), pp. 46–47. In only four of these 29 does Needham actually invoke the concept of inhibition, and all are tautologous or too vague to challenge (e.g., "it is a matter for reflection how far Chinese algebra was inhibited from developments of post-Renaissance type by its failure to produce a sign which would permit the setting up of equations in modern form," *Science and Civilisation in China,* III, 115). In a half-dozen other places Needham uses wording which suggests inhibition, generally in a similarly vague way (e.g., Confucian rationalism and humanism as "fundamental tendencies which paradoxically helped the germs of science on the one hand and injured them on the other," II, 12). In the remaining score of instances inhibition has been read by the sociologist into statements about failure, lack, and inadequacy.

12 Although Needham consciously assumes "that there is only one unitary science of Nature, approached more or less closely, built up more or less successfully and continuously, by various groups of mankind from time to time," he sees this as a reason to study, rather than to ignore, non-European traditions. See his discussion cited in note 1 above.

The sorts of scholars who affirm, without troubling themselves to peruse the Chinese scientific literature, that it could not possibly have any value (see notes 7 and 8) have recently provoked a reaction, equally uninformed, that claims European science was markedly inferior to that of China as recently as three hundred years ago. See John Gribbin, "Did Chinese Cosmology Anticipate Relativity?" *Nature,* 1975, *256:* 619–620, and for a critical discussion, Sivin, "Chinese Cosmology," *ibid.*, 1976, *259:* 249.

13 Nishijima Sadao 西嶋定生, *Chūgoku keizaishi kenkyū* 中國經濟史研究 (Studies in Chinese economic history; Tokyo, 1966), pp. 3–4. Nishijima's remarks are part of an effort to explain the slow development of studies in Chinese agriculture.

14 Alistair C. Crombie, "Science and the Arts in the Renaissance: The Search for Truth and Certainty, Old and New," *History of Science,* 1980, *18:* 233–246, esp. p. 235.

15 Gellner, *Legitimation of Belief* (Cambridge, England: At the University Press, 1974), *passim.* The physicist-philosopher Robert Cohen makes the same point incidentally but a little more broadly when he speaks of "the Galilean turn" as a "rush toward dynamic, functionalized mathematics and abstract quality-stripped epistemology" (see note 7), p. 114.

16 On the earlier position of the Church as a locus of careers open to talent see Alexander Murray, *Reason and Society in the Middle Ages* (Oxford, 1978), pp. 282–314.

17 See, for instance, Arnold Thackray, "Natural Knowledge in Cultural Context: The Manchester Model," *The American Historical Review,* 1974, *79:* 672–709, esp. p. 692.

18 On professional autonomy, Eliot Freidson, *Profession of Medicine. A Study of the Sociology of Applied Knowledge* (New York, 1970), esp. pp. 23–46; on "conceptual machineries of universe-maintenance" and the organizations that support them, Peter L. Berger & Thomas Luckmann, *The Social Construction of Reality. A Treatise in the Sociology of Knowledge* (Harmondsworth, England, 1971, first publ. 1966), pp. 110–146.

19 See "Shen Kua" (note 1), p. 385.

20 Yoshio Mikami, "The Ch'ou-Jen Chuan of Yüan Yüan," *Isis,* 1928, *11:* 125.

21 See "Wang Hsi-shan," pp. 159 and 164, and for further details, Sivin, "Wang Hsi-shan," *Dictionary of Ming Biography, s.v.*

22 Elman, "The Unravelling of Neo-Confucianism: The Lower Yangtze Academic Community in Late Imperial China," unpublished Ph.D. dissertation, Oriental Studies,

University of Pennsylvania, 1980 (UM8107740). Another signal contribution on the influence of astronomy is John Henderson, "The Ordering of the Heavens and Earth in Early Ch'ing Thought," unpublished Ph.D. dissertation, History, University of California at Berkeley, 1977 (DDK-77-31393).

23 Shigeru Nakayama, *Characteristics of Scientific Development in Japan* (New Delhi, 1977), pp. 20–23, which duplicates to some extent his "Japanese Scientific Thought," *Dictionary of Scientific Biography XV*, 728–758, esp. pp. 742–744; Masayoshi Sugimoto & David L. Swain, *Science and Culture in Traditional Japan. A.D. 600–1854* (Cambridge, MA, 1978), ch. 5.

24 The language in which I pose these questions is more or less that of the sociology of knowledge. It is interesting that Crombie (see note 14) has phrased a very similar set of topics, also intended to provide an integrated view of the Scientific Revolution problem in terms familiar to intellectual historians and thus to the majority of historians of science: "conceptions of nature and of science, of scientific inquiry and scientific explanation, of the identity of natural science within an intellectual culture, and the intellectual commitments and expectations that affect attitudes to innovation and change" (p. 234).

Index